# Fungal Populations and Species

# Fungal Populations and Species

John Burnett

OXFORD
UNIVERSITY PRESS

# OXFORD
UNIVERSITY PRESS

Great Clarendon Street, Oxford OX2 6DP

Oxford University Press is a department of the University of Oxford.
It furthers the University's objective of excellence in research, scholarship,
and education by publishing worldwide in

Oxford New York

Auckland Bangkok Buenos Aires Cape Town Chennai
Dar es Salaam Delhi Hong Kong Istanbul Karachi Kolkata
Kuala Lumpur Madrid Melbourne Mexico City Mumbai Nairobi
São Paulo Shanghai Taipei Tokyo Toronto

Oxford is a registered trade mark of Oxford University Press
in the UK and in certain other countries

Published in the United States
by Oxford University Press Inc., New York

A Catalogue record for this title is available from the British Library

Library of Congress Cataloging in Publication Data
Burnett, J. H. (John Harrison), 1922-
Fungal populations and species / John Burnett.
Includes bibliographical references and index.
1. Fungi—Genetics.   2. Fungi—Speciation.   I. Title.
QK602 .B86 2003      579.5'13—dc21      2002030818

ISBN 0 19 851552 9 (Hbk)
ISBN 0 19 851553 7 (Pbk)

10 9 8 7 6 5 4 3 2 1

Typeset by Newgen Imaging Systems (P) Ltd, Chennai, India
Printed in Great Britain
on acid-free paper by The Bath Press, Avon

# Preface

The last 15 years have seen advances in the understanding of fungal population genetics and fungal evolution that were unimaginable even a quarter of a century ago. These results are largely confined to journals and an introductory account which could provide a basis for those coming newly to the topics is not available. I hope that this book will provide such a basis and act as an impetus to both mycologists and evolutionists alike to exploit the numerous techniques now available to increase understanding of these aspects of fungal biology.

Such an understanding is badly needed. For too long, mycologists have isolated themselves from the mainstream of evolutionary thinking and, equally, evolutionists have neglected the fungi. This is the more regrettable because fungi are not only immensely significant as the principal decomposers in the terrestrial ecosystems but are also the largest group of predominantly haploid eukaryotes which probably include a higher proportion of apparently asexual species than any other eukaryotic group. Similarly, fungal ecologists, plant pathologists, those concerned with human and animal mycoses and those concerned with the commercial exploitation of fungi as mycorrhiza or in industrial fermentations of all kinds need an improved understanding of the structure and behaviour of fungal populations if they are to understand their material and develop new methods for the control or exploitation of the fungi.

I hope, therefore, that in addition to providing groundwork for such studies, this book will result in more biologists recognizing the fascination of these organisms and of the many challenges they still hold for investigators. It is time that the fungi took their rightful place in the mainstream of biological thinking!

Few authors can write a book in total isolation and this one is no exception. I am indebted to many people. To my students in the many universities in which I have been privileged to teach and to my colleagues, especially mycological, genetical and taxonomic, for discussions over many years. I am deeply indebted to successive librarians of the Department of Botany (now Plant Sciences) and the Radcliffe Science Library, both of the University of Oxford, for over six decades, throughout which they have responded with unfailing courtesy and helpfulness to all my numerous requests. I am grateful also for reprints and unpublished material recently supplied by Drs Aanen, Gabrioletto, James, Selosse and to Professor Schardl for Table 8.2. I also owe a special debt of gratitude to Dr Richard Ennos of the University of Edinburgh who has read and constructively criticized the complete manuscript to its great benefit, although, of course, I take sole responsibility for what is written here. Lastly, I thank my wife both for her unstinting support over many years and for her contributions direct and indirect to this and my other publications. She has also read the entire manuscript—more than once—improved its grammar and been responsible for several of the illustrations. Without her forbearance and encouragement this book would never have been finished. As ever, I am in her debt.

*John Burnett*
*Oxford*
*March 2002*

# Acknowledgements

I am indebted to the following publishers and other copyright holders for permission to reproduce figures, tables or quotations from their journals or other publications in addition to the individual authors who are acknowledged in the text:

Academic Press, London (*Advances in Plant Pathology*); American Phytopathological Society—APS (*Phytopathology*); Annual Reviews (*Annual Review of Phytopathology*); Blackwell Science (Molecular Ecology); British Mycological Society—BMS (*Transactions of the BMS, Mycological Research*); Colston Research Society, Bristol; CSIRO Publishing, Australia (*Australian Journal of Biological Science*); Ecological Society of America—ESA (*Ecology*); Genetical Society of America (*Genetics*), Kluwer Academic Publishers, Dordrecht (*European Journal of Plant Pathology*); Mycological Society of America—MSA (*Mycologia*); National Academy of Sciences of the United States of America (*Proceedings*); National Research Council of Canada (NRCC) Research Press (*Canadian Journal of Botany*); Nature Publishing Group (*Heredity, Nature*); New York Botanical Garden Press (*Mycologia* to 1998); Oxford University Press; Princeton University Press; Royal Society of London (Proceedings), Society for General Microbiology (*Microbiology*), Springer Verlag, Berlin (Current Genetics); Sinauer Associates Inc; Society for the Study of Evolution (Evolution), Trustees of the *New Phytologist* (New Phytologist), University of Chicago Press (*American Naturalist, Botanical Gazette*).

I am especially indebted to Nick Legon for his photograph of *Pholiota aurivella* (Batsch) Kummer used on the cover.

# Contents

# Introduction

The majority of plant or animal populations studied exist as groups of diploid individuals of limited growth that reproduce sexually. In contrast, the majority of true fungi are haploid, are capable of potentially unlimited growth provided nutrients are available for them to absorb and can reproduce sexually, or propagate clonally, by means of genetically identical mitotically produced spores. Exceptionally, a small group of fungus-like organisms, the Straminipila, included for over a century within the fungi but now known to be independently derived and related to heterokont algae, are diploid. Both groups share a common feature, namely hyphal growth. There are vast numbers of fungi—a conservative estimate is 1.5 million species, or even twice that number—but only about 70 000, i.e. less than 5 per cent, have been described to date.

It is surprising that so little attention has been given by biologists and evolutionists to this major group of haploid Eukaryota. As a consequence, an understanding of their population biology, speciation, phylogeny and evolution has lagged far behind that of other groups of organisms. The single most important reason for such neglect is ignorance. Apart from speculative phylogenies, until recently mycologists have neglected evolutionary aspects and, sadly, for non-mycologists, the unique structure and way of life of the fungi is so very different from that of other eukaryotes that this has been a deterrent to their study. However, increasingly over the last 20 years, extrapolation of studies, such as population biology and genetics, ecological aspects and speciation, from other kinds of organism to fungal situations has been attempted. However, it is still not clear how far fungal behaviour parallels evolutionary behaviour in other eukaryotes.

But rapid change is on the way. As a result of advances in mycology over the last 50 years, the population biology, modes of speciation and even evolution of fungi are now becoming capable of practical and objective examination. These advances include the following.

• The development of the formal genetics of Neurospora, Saccharomyces, Schizosaccharomyces, Aspergillus, Coprinus, Schizophyllum, Phytophthora and several pathogenic species. Although largely concerned with fungal biochemistry, reproduction and morphogenesis, such work resulted in the provision of a range of genetic markers, although only a few have been used in population studies of fungi.
• The advent and application of new genetic markers, both protein markers (isozyme) and DNA markers (hybridization, sequences, RFLPs etc). These markers, together with the increased recognition that control of fungal pathogens can be enhanced by a knowledge of their population biology, has led to a start being made on studies that can be used and developed as paradigms for the population biology of fungi in general.
• The ability to compare and relate DNAs, whether as intact molecules or as sequences, and the application of cladistic methods. Towards the end of the nineteenth century and, sporadically, during the twentieth century, phylogenetic speculations concerning the origins and evolution of the fungi were based almost entirely upon comparative morphology. Moreover, largely due to the continuing dearth of fungal taxonomists, there were and still are serious imperfections in fungal taxonomy, especially at the familial and generic levels, and little attention was given to modern species concepts. Application of these new

1

methods has permitted a more objective approach to be developed to the nature of fungal species and their relationships.

Now that the necessary basic tools for investigating fungal population biology, speciation and evolution have become available, the last 15 years has seen an explosive growth, still in progress, in these areas. In view of the relatively primitive and confused state of fungal taxonomy, much of the present effort is canalized, not unreasonably, into determining phylogenetic relationships. In parallel, provided that sufficient numbers are attracted to work in the area, a greater understanding of the structure of fungal populations and speciation could and should now develop.

This account is intended to provide a contemporary baseline. Few general conclusions can be drawn because current knowledge is patchy and incomplete. Moreover, it needs to be remembered that the fungi cover a taxonomic range at least as great as that of all photosynthetic organisms and appreciably wider than many other well-studied eukaryotic groups. Just as rather different patterns of speciation predominate in green algae from those in ferns or flowering plants, by analogy it is possible that different patterns may be found to occur in different subdivisions of fungi.

# PART I

# Basic mycology

One reason why fungal population genetics has, in general, been neglected by population geneticists and evolutionists is that the fungi differ so greatly in their morphology and general biology from other eukaryotes that they have been treated as a specialist study outside the general run of biological knowledge. Therefore, to assist non-mycological readers, the following chapter provides a basic account of those fungal features relevant to an understanding of their biology as members of populations and as species.

# CHAPTER 1

# Elements of classification, structure, sexuality, biology and genetics of Fungi

The Fungi constitute a eukaryotic phylum comparable in diversity, albeit often on a physically smaller scale, with that of the animals or plants and, numerically, many times larger than the latter. Their structure and biology differ from other eukaryotes in a number of characteristic ways, and some of these differences affect their population biology and genetics. In particular, in the vast majority, the characteristic structural growth unit is the **hypha**, a fine cylindrical cell showing apical growth by means of a wall secreted at its tip. This mode of growth has evolved at least twice.

The three main groups of Eukaryota emerged at least $10^9$ years ago and the Eumycota, the major fungal group, share a common ancestor with the Animalia, distinct from that which gave rise to the Plantae. There are now vast numbers of fungi; a conservative estimate is 1.5 million species, or even twice that number, although only about 70 000+ species have been described to date, i.e. less than 5 per cent. A second and far smaller group, the Straminipila (= Chromista), of perhaps 600 species were originally classified with the fungi. They are now known to be unrelated and, in fact, are derived from heterokont-related algae that developed the hyphal habit, probably acquiring chloroplasts later (Cavalier-Smith 1987, 2001; Barr 1992; van der Auwera *et al.* 1995). They include important and well-known genera such as *Saprolegnia*, the semi-parasitic 'fish fungus', the plant parasitic genera *Phytophthora*, including the species responsible for the great potato blight epidemic in the nineteenth century, and *Pythium*, including the many 'damping-off' fungi responsible for the death of

seedlings. For convenience, both groups are included in this book because of the great similarities in habit, biology and ecology between them and because the Straminipila form an interesting diploid evolutionary contrast to the predominantly haploid Eumycota.

A basic classification of the fungi in this broad sense, i.e. including both groups, is given first. Thereafter, only features specifically characteristic of fungi are described since other aspects are common to all other eukaryotes.

## 1.1 Elements of classification

The classification of the fungi is still in a relatively immature state and undergoes continual change. Within the **Eumycota** four natural groups can be distinguished plus, for convenience, one artificial group.

- **Chytridiomycotina** (or Chytrids), which includes species with a motile phase, possibly polyphyletic.
- **Zygomycotina**, a relatively small and isolated group which includes 'pin moulds' and their allies, almost certainly polyphyletic.
- **Ascomycotina**, the largest group of fungi which includes many common 'moulds', mildews and many of the yeasts; the fungal component of the majority of lichens are members of this group.
- **Basidiomycotina**, a large and heterogeneous group which include mushrooms and toadstools, bracket fungi (conks), jelly fungi, puffballs, earth

stars, truffles, stinkhorns and the obligately parasitic 'rust' and 'smut' pathogenic fungi.
- **Fungi Anamorphici**: many fungi are known for
which sexual reproduction has never been found.
Most produce spore types identical with those classified as Ascomycotina, some are yeasts and a
smaller number possess features characteristic of
the Basidiomycotina. Comparisons of DNA sequences between sexual and similar asexual fungi
increasingly indicate close taxonomic affinities
between them. Therefore it has been suggested that
the latter should be classified with their related
sexual forms. At present this is impractical because
the affinities of most fungi lacking sexuality are not
yet known. Therefore these so-called 'imperfect'
forms (Fungi Imperfecti, Deuteromycotina) are
sometimes grouped together in a fifth artificial
assemblage, which it has become fashionable to
describe as the 'mitosporic fungi' since they only
exhibit the mitotic state. However, the majority of
Eumycota produce spores by mitosis in their
asexual or **anamorphic** phase, with sexual reproduction and meiosis being confined to the **teleomorphic** phase. (Fungi where both phases occur
in the life cycle are described as **holomorphic**.)
Therefore it is preferable to contrast holomorphic
('perfect') fungi with strictly anamorphic ('imperfect') ones. A further practical consideration is that
many anamorphic forms, even when their holomorphic affinities are known, are normally only
encountered in the asexual state in the field.
Therefore these fungi are still best described
in this way as though they were Fungi Anamorphici, the term used here.

The **Straminipila** include two main families.

- **Oomycota** (or Oomycetes), capable of producing
biflagellate zoospores, which include the blights,
'damping-off' fungi and water moulds.
- **Hyphochytridiomycota**, a small group of mostly
marine fungi with biflagellate zoospores.

Currently available information concerning
population biology and speciation comes almost
entirely from a limited number of groups within
the Ascomycotina and Basidiomycotina and from
two genera, *Phytophthora* and *Pythium*, of the
Straminipila.

The distinguishing features of each of the major
groups are briefly set out more fully in Table 1.1
and at the end of this chapter (section 1.7).

## 1.1.1 Convention on the use of names in mycology

Many fungi now known to have a sexual stage
occur most commonly in the field as apparently
asexual fungi. In some cases more than one independent asexual form has been associated with a
single sexual phase, i.e. they are **pleomorphic**.
Initially, such fungi were inevitably described using
anamorphic names and, because their **teleomorphs**
(sexual stages) are hardly ever seen, or recognized,
they are still referred to in this way unless specific
reference needs to be made to the holomorphic
names. For example, the common glume blotch
fungus of wheat has been known since the recognition of its single anamorphic form in 1842 when
it was named *Septoria nodorum*. *Mycosphaerella
graminicola*, an apparently new and distinct teleomorphic form found on litter, was named in 1894. It
was only realized in 1972, 130 years after the first
recognition of these two fungi, that they represented
different phases of the same species (Sanderson
1972). The holomorphic name, based on both the
anamorphic form or forms and the teleomorph, is
usually used only if reference needs to be made to
the fungus in situations when both phases are
detectable or associated. Pathologists still often refer
to this fungus by its anamorphic name since that is
the most evident phase on diseased wheat plants,
although it is sometimes cited as *Mycosphaerella
(Septoria) graminicola*. Similarly, many species of
*Fusarium*, *Aspergillus* and other genera possess
sexual stages but are most commonly met with in
the field or laboratory in the asexual form and
simply referred to by that name. This convention
on the use of names is largely confined to mycologists and needs to be borne in mind by non-
mycologists when reading mycological literature.
In addition, as with all organisms, when the state of
taxonomy is poor, as in many fungal groups, mis-
identification or new taxonomic insights result in
several names being in use to describe both an
anamorphic form and even its teleomorphic form.

**Table 1.1** The distinguishing features of the main groups of Eumycota

| Character | Chytridiomycotina | Zygomycotina | Ascomycotina | Basidiomycotina |
|---|---|---|---|---|
| *Somatic characters* | | | | |
| Hyphae | | | | |
|   Compartmented | − | ± | + | + |
|   Simple pores | − | − | + | − |
|   Plugged pores | − | ± | − | − |
|   Dolipores | − | − | − | + |
| Mycelium | | | | |
|   Rudimentary | + | − | ± | − |
|   Branched | − | + | − | − |
|   Reticulate-branched | − | − | + | + |
| *Asexual reproduction* | | | | |
| Sporangial | | | | |
|   Motile zoospores | + | − | − | − |
|   Non-motile endospores | − | + | − | − |
| Conidial | | | | |
|   Conidia | − | − | + | ± |
|   Microconidia | − | − | ± | ± |
|   Oidia | − | − | − | ± |
| *Sexual reproduction* | | | | |
| Gametangial | | | | |
|   Antheridia/oogonia | + | − | −(+) | − |
|   ±Identical gametangia | − | + | − | − |
|   Proto-apothecia/perithecia | − | − | + | − |
|   Hyphal fusion | − | − | (rare) | + |
| Plasmogamy | | | | |
|   Gametes fuse | + | − | − | − |
|   Gametangia fuse | − | + | − | − |
|   (Micro)conidia/trichogyne | − | − | + | − |
|   Hyphal short dikaryotic phase | − | − | ± | − |
|   ±Long dikaryotic phase | − | − | − | + |
| Site of karyogamy/meiosis | | | | |
|   Motile oospore/2n mycelium | + | − | − | − |
|   Zygospore | − | + | − | − |
|   Ascus | − | − | + | − |
|   Basidium | − | − | − | + |
| Reproductive structures | | | | |
|   Oospore/2n mycelium | + | − | − | − |
|   Zygospore | − | + | − | − |
|   ±Enclosing hyphal growth | − | − | ± | − |
|   Ascoma (Ascocarp) | − | − | + | − |
|   Basidioma (Basidiocarp) | − | − | − | + |

Thus, for example, *Septoria nodorum* was sometimes incorrectly referred to as *Stagnospora nodorum* before the two genera were distinguished. In fact, they are quite unrelated.

To assist in resolving these possible causes of confusion variant names are given in full in the index and some indication of common variants is given when a fungus is first mentioned in the text.

## Abbreviations for the main groups of fungi

**Eumycota**

| | |
|---|---|
| Chytridomycotina | [C] |
| Zygomycotina | [Z] |
| Ascomycotina | |
| Simple | [A] |
|   With perithecium | [AP] |
|   With apothecium | [AD] |
|   Lichenolicous | [AL] |
| Basidiomycotina | |
| General | [B] |
|   Rusts (Urediniomycetes) | [BR] |
|   Smuts (Ustilaginomycetes) | [BS] |
|   Agaricomycetidae (agarics, polypores etc.) | [BA] |
|   Tremellomycetidae (jelly fungi etc.) | [BT] |
| Fungi Anamorphici | [FA] |

**Straminipila**

| | |
|---|---|
| Hyphochytriomycota | [Ho] |
| Oomycota | [O] |

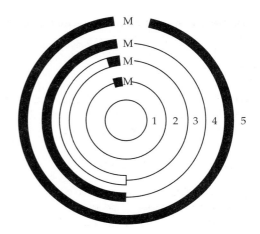

**Figure 1.1** Fungal life cycles. Each circle represents a life cycle and should be read clockwise: M, meiosis; single line, haploid phase; double line, dikaryotic phase; solid line, diploid phase. The three principal life cycles are as follows: 1, asexual; 2, haploid (common in Ascomycotina); 3, haploid dikaryotic (many Ascomycotina have a brief dikaryotic phase but a long dikaryotic phase is typical of Basidiomycotina); 4, haploid–diploid (unusual in fungi but occurs in some Chytridiomycotina); 5, diploid (characteristic of the Straminipila, only very rarely in Eumycota). (From Burnett (1975).)

In addition, as an indicative guide to the taxonomic affinities of each fungus mentioned in the text, their names are followed by abbreviations, in square brackets, for some of the main groups of fungi, as set out in the box.

Examples of possible nomenclatural combinations to be found in the text are as follows.

• *Mycosphaerella* (*Septoria*) *nodorum* [AD], used on first mention in the text, indicating a apothecial fungus (Discomycete) with its anamorphic name in parentheses.

• *Mycosphaerella nodorum* [AD], a holomorphic Discomycete used when attention is being drawn to the sexual phase; subsumes the anamorphic phase.

• *Septoria nodorum* [FA/AD], indicating that the data refer to the fungus in its anamorphic form, or simply as [FA] if the teleomorph is unknown, e.g. *Penicillium chrysogenum* [FA], a penicillin-producing strictly anamorphic fungus.

## 1.2 Life cycles

Three idealized life cycles can be recognized in the Eumycota—the **haplo-diploid**, the **haplo-dikaryotic** and the **asexual**—and two in the Straminipila—the **diplo-haploid** and the **asexual** (Figure 1.1).

The Zygomycotina and Ascomycotina are characterized by the haplo-diploid cycle in which the haploid phase predominates in both time and space and diploidy is confined to the zygotic cell where it is followed almost immediately by meiosis. They are sometimes described as haplontic. The smut fungi (Ustilaginomycetes) amongst the Basidiomycotina are also haplo-diploid, but the majority of the Basidiomycotina, including the rust fungi (Urediniomycetes), are characterized by a haplo-dikaryotic life cycle. In this life cycle, compatible nuclei, which, potentially, will eventually fuse at fertilization, persist after **somatic fusion** (hyphal fusion), dividing synchronously within the same **mycelium**, sometimes as dikaryotic compartments but not necessarily so. Diploidy is but a transient phase in the **basidium** prior to meiosis. Haplo-dikaryotic forms are intermediate between diplo-haploid forms and typical haplo-diploid forms. The dikaryotic phase has been described as a 'detour on the way to diploidy' (Raper and Flexer 1970).

Most asexual species resemble members of the Ascomycotina but a few are known from other groups. They are distinguished here as Fungi Anamorphici.

The diplo-haploid life cycle characterizes the Straminipila. Their mycelium is diploid and bears gametangia in which meiosis occurs; diploidy is restored by the fusion of haploid gametes.

## 1.3 Structural considerations

Fungi range in size from microscopic unicells, such as yeasts, to macroscopic pseudocellular structures such as the massive underground sclerotia (see section 1.3.3) of *Polyporus mylittae* [BA], which are the size of a football and weigh up to 4 kg. Unicells apart, fungal structures are all based on a microscopic cylindrical structural unit, the **hypha**. The ability of a hypha to branch in a regular manner and to grow at its apex results in the development of a hyphal network, the **mycelium**, which is the effective ecological unit.

Most mycelia pass successively through three phases:

- the 'vegetative' phase—growth and exploitation of the substrate
- the anamorphic phase—spore production and dispersal
- the teleomorphic phase—sexual reproduction followed by further spore dispersal.

If the vegetative phase is perennial, one or both of the two subsequent phases can recur periodically, often seasonally, throughout the life of the mycelium.

### 1.3.1 Hyphae and mycelia

The hyphal apex plays an essential role in growth, foraging, nutrient acquisition, sensory recognition and response, and in both somatic and reproductive differentiation. Some fungi exhibit senescence as the result of infection or morbid conditions resulting in death of apices. However, in most fungi, the hyphal tips are potentially immortal.

The hyphae of Eumycota are **coenocytic** (multinucleate and non-cellular) although, apart from most Zygomycotina, they become compartmentalized and strengthened internally by the centripetal

development of centrally incomplete cross-walls or **septa** of various degrees of complexity (Figure 1.2). The frequency, type and complexity of the septa are characteristic of different taxonomic groups, notably the dolipore septum which occurs only in some Basidiomycotina (Table 1.1). Cytoplasmic streaming occurs in all fungi. Many kinds of cell organelle, notably nuclei and mitochondria, can traverse septal pores in appropriate environmental or genetically determined conditions. How far a hyphal compartment is like a cell and acts as a functionally autonomous unit is uncertain (Todd and Aylmore 1985). The hyphae of Straminipila lack well-developed septa. In all fungi, however, complete septa usually delimit the various kinds of spore-bearing structures, or reproductive bodies, from the rest of the mycelium at some stage of their development.

The distribution of nuclei within coenocytic hyphae, or hyphal compartments, is characteristic for each species. Sometimes the apical

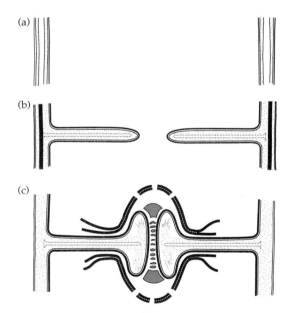

**Figure 1.2** Fungal septa. Fungal hyphae are (a) coenocytic, rarely with cross-walls, in Chytridiomycotina, Zygomycotina and Straminipila, (b) septate with a simple central pore in Ascomycotina, Urediniomycetes and Ustilaginomycetes, or (c) septate with complex septa and dolipores in Basidiomycetes where the endoplasmic reticulum forms a perforated cap (parenthosome) on either side of the pore. (From Burnett (1975).)

compartment contains many nuclei which tend to divide synchronously, with the number of nuclei becoming fewer in more distal compartments, i.e. away from the apex, e.g. *Aspergillus* [A] (Clutterbuck and Roper 1966), and even being reduced to one immediately distal to the apical segment, e.g. *Fusarium* [FA] spp. All such mycelia normally behave as haploid monokaryons (except in polyploid fungi), even when, as in many fungi, the apical compartment, or even all compartments, are multinucleate. In other fungi each compartment, including the apical one, is regularly uninucleate or, after hyphal fusion, regularly binucleate—the **monokaryotic** and **dikaryotic** conditions respectively. The unique condition, the **dikaryon**, occurs in some Basidiomycotina where the two nuclei in each compartment, whether genetically identical or different, show synchronized conjugate divisions. In addition, one nucleus may enter a compartment at nuclear division via a special kind of intercompartmental fusion process or **clamp connection** (Figure 1.3).

Genetically heterogeneous dikaryons occur regularly in certain fungi, e.g. *Coprinus* [BA], and the behaviour of mycelia of such genotypes resembles that of diploid organisms in several ways (Raper and Flexer 1970); monokaryons behave as haploids. In most Ascomycotina and other Basidiomycotina each compartment is multinucleate but their nuclear divisions, especially near the apex, are often synchronized. The nuclei of the coenocytic Straminipila are diploid and distributed irregularly along the length of the usually non-septate hyphae.

Hyphae branch laterally and subapically, and the primary laterals may themselves branch, and so on,

to form a radially expanding mycelium growing at the margin of a homogenous environment on a flat surface, or a spherical pellet growing at the surface of a liquid medium. This simple mycelium is the characteristic growth pattern of coenocytic fungi, i.e. Chytridiomycotina and Zygomycotina. It contrasts with the septate–reticulate mycelium of the compartmented Ascomycotina and Basidiomycotina where, early in the development of the mycelium, intra-mycelial hyphal **anastomoses**, i.e. hyphal fusions, convert it into a marginally expanding reticulated structure. Hyphal anastomoses occur most frequently between terminal and lateral apices and are often preceded by tropistically determined growth responses. In some cases an apex can induce the formation of a new apex up to 10 μm distant from the lateral wall of another hypha with which it will then fuse, but fusion without such induction is at least as common (Aylmore and Todd 1984). Anastomoses of a similar nature can also occur between hyphal tips and spores or similar structures, e.g. the oidia of *Coprinus* [BA] spp. (Kemp 1975). Successful hyphal fusions are associated with extremely rapid and effective mutual adhesion on contact, followed by the dissolution of the walls in the juxtaposed region, potentially permitting passage of cell contents across the junction.

In **intraspecific** fusions, the success or failure, both of the fusion process and the subsequent passage of cell organelles such as nuclei and mitochondria, is normally under precise genetic control (Garnjobst and Wilson 1956; Casselton and Condit 1972; Aylmore and Todd 1984; Todd and Aylmore 1985), although environmental conditions can affect the frequency of fusions (Bourchier 1957).

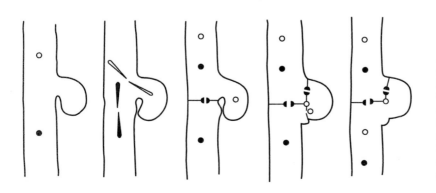

**Figure 1.3** Anastomoses and clamp connections. Anastomoses are mostly, but not uniquely, through tip-to-tip fusions. Many Basidiomycetes possess clamp connections, a special form of hyphal fusion associated with conjugate nuclear division which arises in the typical manner of a hyphal anastomosis. The developmental sequence goes from left to right.

In contrast, control of **interspecific**, or even **intergeneric**, fusions is not well understood. Unsuccessful fusions, whether intra- or interspecific, are affected by a range of different environmental conditions and genetic defects. The conjugants may simply fail to adhere and fall apart, or they may remain in close contact although the walls remain intact. Some apparently normal fusions are followed, once cytoplasmic contact has occurred, by the death of the regions in contact, or even of adjacent compartments as well.

When a mycelium expands radially for a prolonged time, the contents may either be lost from its oldest central senescent region or translocated to actively growing regions of hyphae. The central vacuolated region then ceases to grow, is sealed off and autolyses. Growth patterns reflecting this behaviour occur in nature, e.g. the perennial 'fairy ring' fungi of grassland where the central region of the mycelium has been lost over the years, or in submerged industrial cultures, where fungal pellets become hollow in the centre, in part because of anaerobiosis.

Radially symmetric mycelia are often the exception in nature, since the form of any mycelium reflects the adaptive growth response of its apices to their immediate environmental, especially nutritional, conditions. Therefore growth can differ depending on the location. In nutrient-rich regions, the mycelium is a compact many-branched reticulum, but it becomes reduced to a narrow strand of hyphae, even a single hypha, in nutrient-poor regions, e.g. mycelium of the dry rot fungus *Serpula lacrymans* [BA]. This fungus extends through nutrient-deficient plaster as slender multiple interwoven hyphal strands, but develops a fan of radially expanding hyphae when a more nutritious substrate, such as wood, is reached. In some fungi the hyphal strands become differentiated as **rhizomorphs** within which translocation is highly effective.

## 1.3.2 Dimorphism

A great many Eumycota are capable of growing either as hyphae or as individual yeast-like cells, i.e. they are **dimorphic** (e.g. *Candida* [A] spp.), but in several species the yeast form is the norm (e.g.

baker's and wine yeasts *Saccharomyces* [A] spp., fission yeasts *Schizosaccharomyces* [A] and red yeasts *Rhodotorula* [BT]). Dimorphic change is triggered by alteration in one or more environmental conditions that are characteristic for each species.

The yeast phase is propagated by budding, as in baker's yeast, with the buds often adhering to the parent or daughter cells for some time before becoming detached, or by cell division and separation, as in the fission yeasts. Detached buds or the products of division are normally uninucleate, can be passively dispersed and are analogous to the spores of hyphal forms. Dikaryotic yeast cells are unusual, but they are known and can be induced in some species; they normally propagate by dikaryotic buds only. Diploid yeast cells formed in connection with sexual reproduction are capable of persisting and propagating in this state but normally undergo meiosis immediately to form haploid ascospores.

Dimorphism, natural or induced, is not a universal phenomenon. Many yeasts are incapable of reverting or being induced to revert to a mycelial form, just as some hyphal forms apparently cannot be induced to grow in a yeast-like manner.

## 1.3.3 Perennating structures

In the vast majority of fungi, detached lengths of hyphae, even as small as a single compartment, can regenerate to form a new mycelium identical with that from which the hyphal fragment was derived. This occurs in nature as a consequence of hyphal damage resulting from the feeding of slime moulds, mites and other organisms. Such hyphal fragments do not persist unless they regenerate.

However, somatic differentiation can result in the production of perennating structures such as **chlamydospores** or **sclerotia** (Figure 1.4). Their development is often associated with the depletion of available nutrients, but fungi show much inherent variation in this respect. Chlamydospores develop through the sealing off at, or near, the septal region of at least one, and sometimes several, hyphal compartments. The wall thickens and the chlamydospores may become detached as a result of the drying out and death of immediately adjacent hyphae, or the whole mycelium may degenerate

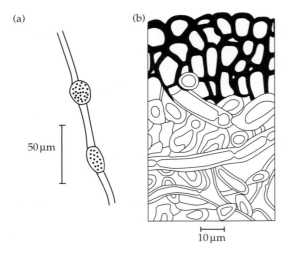

(a)

50 μm

(b)

10 μm

**Figure 1.4** Fungal perennating structures: (a) chlamydospores—modified single compartments, often containing lipid droplets. (b) Aggregations of closely compacted hyphae forming sclerotia, in which the inner cells often contain much glycogen and lipid material, and the outer layers are often melanized and mechanically resistant as shown in the section.

apart from the chlamydospores. These may then be dispersed locally in a variety of ways—through the mechanical displacement of soil particles, downwash of soil water or transfer on the surface, or through the guts, of soil animals. They can remain viable for many years, particularly under dry conditions, although eventually they germinate directly to form a new mycelium.

Sclerotia are multihyphal structures showing some degree of further differentiation. They can arise on soil or aerial mycelia in various ways, usually involving short branching, anastomoses and, generally, the thickening of the cell walls plus adhesion of the hyphae involved. The innermost compartments increase greatly in size or form a mass of distorted tightly packed cells filled with reserve carbohydrates, proteins and oil globules. Often, the cell walls of one or more outer hyphal layers become thickened and melanized, and the cells lose their contents and become compressed to form a hard impermeable outer 'rind'. The dormant resistant sclerotia so formed may be spherical, subspherical or irregular in shape, and, depending on the species and conditions, may range in size from a pinhead to several centimetres in diameter. Once dispersed, they can remain viable for decades and

eventually germinate in response to an appropriate stimulus, usually environmental. In some pathogenic species they germinate in response to material leaked from adjacent host plants. Germinating sclerotia either develop a new mycelium directly or, in many cases, develop a spore- or conidia-bearing structure resulting in localized dispersal.

Since their differentiation involves mitotic divisions only, both chlamydospores and sclerotia, whatever their nuclear content, retain the genetic constitution and ploidy of the hypha(e) from which they originated. Although their spatial dispersal may be limited, their persistence ensures the clonal dispersal of individual genotypes over time.

### 1.3.4 Spores and conidia

Most spores of all types, regardless of their origin, are actively or passively dispersed in space (see section 1.5). Aerial hyphae develop, usually a short distance behind a growing apex, and, a little further back still, specialized aerial hyphae differentiate in a variety of ways to form spore-bearing structures, either as sporangiophores or conidiophores (Figure 1.5). **Sporangiophores** (sporangia-bearing hyphae) of Zygomycotina bear terminally, sometimes also laterally, one or more **sporangia** in which **sporangiospores** are produced endogenously, their nuclei being derived by mitosis. Motile sporangiospores, **zoospores**, are produced by some Straminipila and Chytridiomycotina. **Conidiophores**, or **conidiomata** if they are multihyphal structures, develop conidiogenous cells from which specialized non-motile cells, the **conidia**, are developed externally in a great variety of ways. The same mycelium may develop more than one kind of conidium depending on the environmental conditions or stage of the life cycle. In addition, some produce small uninucleate cells that act as fertilizing agents (**spermatia** or **microconidia**). The conidium is the characteristic spore of the Ascomycotina and related Fungi Anamorphici. Many Basidiomycotina either do not produce spores or else develop spore-like cells (**oidia**) which can only act as fertilizing agents. However, exceptionally, the rust fungi produce a sequence of different spore types at different stages of their life cycle which ensure their effective asexual propagation; some

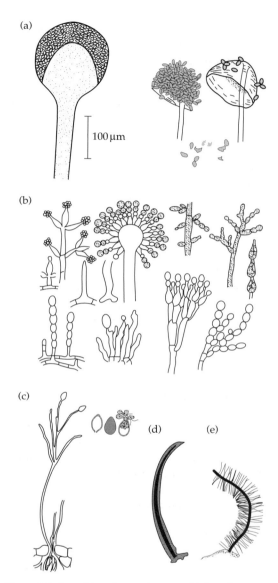

**Figure 1.5** Asexual fungal spores (not to scale). Fungal spores develop (a) internally in a sporangium, as in *Rhizopus* [Z], or (b) externally as conidia on variety of different spore-bearing structures or conidiophores, as in Ascomycotina and related Fungi Anamorphici. (c) Spores of the Straminipila are motile, possessing both (d) a lateral whiplash and (e) a tinsel flagellum. Zoospores of Chytridiomycotina possess one, or rarely several, posteriorly placed whiplash flagellae.

are monokaryotic and derived after meiosis (**basidiospores**), whereas others are mitotically formed and are dikaryotic (**aecidiospores, uredospores** and **teleutospores**).

Spores of all kinds can be uni- or multinucleate. If the latter, the nuclei may be derived either from a single parent nucleus through successive mitoses in the spore, or by the migration of several nuclei into the spore during its formation. Therefore multinucleate spores may be **homokaryotic** or **heterokaryotic**, but this cannot be determined by inspection alone.

Chytridiomycotous fungi produce motile zoospores with a single posterior flagellum that gives them limited powers of movement which may be enhanced by water movements. Straminipilous zoospores bear two different flagella and frequently show tropistic responses to nutrient sources or, in the case of parasites, to their host's roots. All zoospores have a limited lifespan.

Sporangiospores and conidia are liberated actively or passively and are then dispersed passively in various ways. Depending on the species they may remain viable for a relatively short period before germinating. Some species require exogenous nutrients to germinate; others do not.

## 1.4 Sexual reproduction and sexuality

Apart from the Fungi Anamorphici, where it has not been detected, sexual reproduction can occur in fungi provided that environmental conditions are favourable. Genealogical trees based on DNA analysis of various kinds have now shown that many Fungi Anamorphici are apparently closely related to sexually reproducing fungi but have lost all sexuality (see Chapters 7, 11 and 12). The essential features of fungal sexuality are summarized in Figure 1.6.

Regardless of the nature of the structures involved in sexual reproduction, there is a common underlying pattern of physiological co-ordinating processes, namely telemorphosis, zygotropism and thigmotropism prior to plasmogamy and karyogamy. **Telemorphosis**, the induction of potentially conjugant regions at a distance, initiates sexual reproduction. **Zygotropism**, the directed growth of one or both regions to each other, follows and their contact is ensured through **thigmotrophic** interactions, i.e. contact stimuli. These mediating processes are most evident in aquatic fungi and the Zygomycotina. Indeed, they were first described in

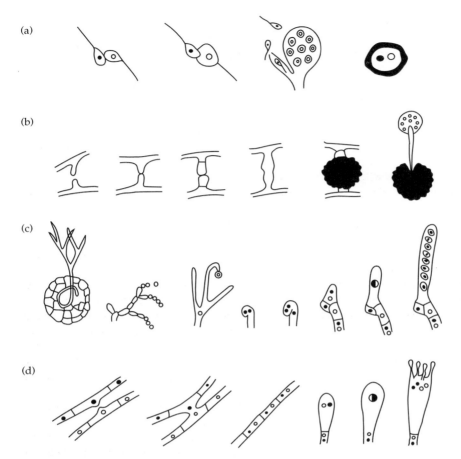

**Figure 1.6** Sexual reproduction in the Eumycota. (a) The gametes of the Chytridiomycotina are mostly motile and may be of similar or unequal size, although in some the 'female' gametes are oogamous. The zygote, however formed, usually germinates by releasing zoospores. All motile cells have a single posterior whiplash flagellum. (b) In Zygomycotina, specialized hyphae (zygophores) are mutually attracted, fuse, develop progametangia and ultimately a zygote where meiosis probably occurs followed by germination via a germ sporangium. (c) In many Ascomycotina, the sites of the apothecia or perithecia are predetermined on the mycelium by small convoluted hyphae, often terminating in a trichogyne which grows towards compatible conidia or microconidia and fuses with them. The 'male' nucleus migrates down the trichogyne and establishes a brief lineage of ascogenous dikaryotic hyphae, the penultimate terminal cells of which develop as asci, where ascospores form after meiosis. Meanwhile 'maternal' hyphae develop around the ascogenous hyphae to form a hyphal weft, a cup-shaped apothecium or a flask-shaped perithecium. (d) In Basidiomycotina, compatible monokaryotic hyphae fuse and the dikaryotic cells thus formed persist for a greater or shorter time as a somatic mycelium. Ultimately hyphae aggregate to form a basidiocarp within which terminal dikaryotic cells, often arranged in a parallel hymenium, develop into basidia on which basidiospores develop after meiosis.

a classic paper on *Phycomyces* [Z] by Burgeff (1924) and later by Raper (1939) for the aquatic straminipilous fungus *Achlya* [O]. **Pheromones** ('sexual' hormones) are implicated in both telemorphosis and zygotropism, and a number of these have now been identified. The basis of the thigmotropic reaction is not fully understood in any fungus but, clearly, some kind of surface molecules are involved that induce an immensely firm binding together of the conjugant regions. Thereafter, some form of wall dissolution occurs, enabling **plasmogamy** (cytoplasmic fusion) to occur and potentially conjugant nuclei to migrate towards each other, eventually fusing at fertilization (**karyogamy**).

Sexual dimorphism is uncommon except in aquatic Straminipila and Chytridiomycotina where gametes or gametangia showing a range of structural differentiation are known.

In the Zygomycotina, sexual reproduction is always gametangial, although most commonly the multinucleated gametangia lack regular sexual dimorphism, e.g. *Mucor mucedo* [Z], *Phycomyces nitens* [Z]. At least some nuclei in the single multinucleated cell derived from gametangial fusion, the **zygospore**, fuse in pairs and undergo meiosis. Zygotes do not germinate readily, but when they do it is by means of a sporangium containing, most commonly, uninucleate spores. Experimental germinations indicate that these can include either or both recombinant and, sometimes, parental nuclei which have undergone neither karyogamy nor meiosis.

In many Ascomycotina, equivalent 'female' structures such as **protoapothecia** or **protoperithecia** develop on the mycelium prior to fertilization. These determine the sites where **perithecia** and **apothecia** (the structures—**ascomata** (sing. **ascoma**) or **ascocarps**—in which meiosis occurs, giving rise to ascospores) eventually differentiate. In some cases they develop sexually receptive hyphae—**trichogynes**. In other species dense masses of highly compacted hyphae, termed **stroma** (pl. **stromata**), develop on the surface of a substrate. They may remain sterile or develop superficial conidia (as in Fungi Anamorphici) or, as in Ascomycotina, perithecia may develop on, or embedded in, the stroma. In all cases the equivalent 'male' differentiation is less specialized. Conidia, microconidia or even hyphae can act as fertilizing agents and fuse with the trichogyne, if present. A 'male' nucleus enters the trichogyne and becomes associated with the 'female' nucleus in its basal region. These sexually compatible nuclei usually undergo a number of conjugate divisions in the dikaryotic **ascogonial hyphae**, the penultimate cells of which form numerous **asci**, i.e. genetically identical elongated cells, the equivalent of zygotes in which nuclear fusion (fertilization) and meiosis occurs. This is followed rapidly by one or more mitoses resulting in the production of eight, or some multiple of four, ascospores. In most Ascomycotina, hyphae of the original 'female' mycelium grow up while this is taking place and surround the ascogenous hyphae to form the ascomata, but in some no ascoma is developed. Two main types of ascoma occur. **Apothecia** developed by Discomycetes are concave and disk-shaped, and the asci form a layer (the **hymenium**) lining the upper surface from which they are eventually able to liberate their ascospores to the air. Pyrenomycetes develop **perithecia**— hollow flask-shaped structures that enclose the asci which develop from basal hyphae. Perithecia may develop on a mycelium or embedded in a solid mass of hypha or stroma. Either whole asci or individual ascospores are exuded or actively discharged from perithecia, thus liberating the meiotic products for dispersal.

Mutant, so-called 'female sterile', forms are not uncommon in several species but they can, of course, act as functional 'males'. 'Male sterile' forms, which can function as 'females', also occur (e.g. in *Pythium* [O]) but, because almost any spore or hypha can function as a fertilizing agent, they are rare.

In some Ascomycotina and most Basidiomycotina, morphologically differentiated specialized reproductive structures are totally lacking and sexual fusion occurs through hyphal fusions between sexually compatible mycelia. (The consequential absence of obviously sexual fusions caused earlier workers to claim that sexuality had been lost in these fungi!)

In Basidomycotina, other than the Urediniomycetes and Ustilaginomycetes, **dikaryotization**, the pairing of sexually compatible nuclei as a consequence of nuclear migration through anastomoses from one or both interacting hyphae, results in previously monokaryotic compartments becoming dikaryotic. Thereafter, the paired nuclei divide synchronously and growth occurs, and it is from this dikaryotic, so-called secondary, mycelium that **basidiomata** ('fruiting bodies', sing. **basidioma**), or **basidiocarps**, develop. These are the toadstools, brackets etc. of common parlance. It is not clear how their sites on the mycelium are determined. Multihyphal structures of varying degrees of complexity develop from the dikaryotic hyphae at these sites but sexual fusion is delayed and restricted to their, usually enclosed, sub-terminal cells. In these zygotic cells, the future **basidia**, nuclear fusion occurs. The diploid nucleus, once formed, undergoes meiosis almost immediately. The four nuclei so formed migrate into exogenously developed

**basidiospores** within which a post-meiotic mitosis can occur giving rise to binucleate basidiospores. Sometimes this division occurs before the nuclei enter the spores when either only four different nuclei migrate, or two migrate, into each basidiospore. They may be genetically the same or different and the genetic consequences vary (see Chapter 7, section 7.2.5).

The details of sexual reproduction in the Urediniomycetes and Ustilaginomycetes are somewhat different although in the former sexual reproduction is preceded by a prolonged dikaryotic phase.

Ascospores or basidiospores representing the four products of individual meioses—a **tetrad**—can usually be isolated. In many Ascomycotina, **ordered tetrads** can be isolated because the meiotic products are arranged linearly in the ascus, reflecting the spindle orientations of successive meiotic and mitotic divisions. These properties greatly facilitate formal genetic analyses. Although a basically similar sequence of processes occurs in the rust and smut fungi, the details are more complex.

In many Basidiomycotina, and far fewer Ascomycotina, the perennial mycelium can persist as a sexually compatible dikaryon for many years. Examples include the following: morels, *Morchella* [AP] spp., a few decades (Burnett, unpublished); a 'fairy ring' toadstool, *Marasmius oreades* [BA], probably hundreds of years (Burnett and Evans 1966); one of the honey fungi, *Armillaria gallica* [BA] (Syn. *A. bulbosa*), perhaps 1500+ years (Smith *et al.* 1992). Such perennial mycelia are capable of developing crops of ascomata or basidiomata annually or even more frequently from the persistent dikaryon. Hence over their lifetime, many thousands of millions of meiotic ascospores or basidiospores may be produced, *all* derived from the original two compatible haploid nuclei.

Straminipila, being basically diploid, produce haploid conjugant nuclei in morphologically differentiated **antheridia** and **oogonia** where meiosis takes place. Both structures are originally multinucleate, but in many species the number of haploid nuclei is reduced in the oogonium, and in the case of *Phytophthora* [O] and *Pythium* [O] to a uninucleate **oosphere**. Fertilization takes place after one or more 'male' nuclei have passed through a fertilization tube developed from the antheridium, which penetrates the oosphere fusing with one or more resident 'female' nuclei. The diploid **oospore** thus formed remains dormant. Germination is usually preceded, or accompanied, by numerous mitoses and is usually by a germ tube which develops into a mycelium. In moist conditions, the terminal region of the germ tube becomes differentiated as a sporangium in which numerous mitoses occur, the products eventually being released as uninucleate diploid motile zoospores. These may be chemotactically attracted to host roots or other materials where they germinate to re-form the diploid mycelial phase.

### 1.4.1 Mating systems

Sexual compatibility is determined by a range of mating systems in fungi (Table 1.2) and many of the variant patterns found in nature are described in Chapter 8.

Many fungi are capable of self-fertilization—**homothallic** or **homomictic** forms—but in many of these cross-fertilization is also possible and certainly occurs in nature. In others—**heterothallic** or **heteromictic** forms—cross-fertilization is obligatory and compatible matings are determined by specific mating-type factors (MTFs). Many normally heteromictic fungi can develop self-fertilizing structures if cross-fertilization fails.

It is convenient to recognize **dimictic** forms, e.g. in Zygomycotina and Ascomycotina, where a *pair* of MTFs control mating and segregate at meiosis as if they were a pair of alleles (variously designated $+/-$, $a/a$, $a/\alpha$, mat $a$/mat $\alpha$ or mat1-1/mat1-2). It is now known that in some Ascomycotina the size and most of the DNA sequences of the mating-type factors are very dissimilar and therefore they have been termed **idiomorphs** rather than alleles (Glass *et al.* 1988). Comparable idiomorphs occur in genera from the same family but not elsewhere, e.g. *Neurospora* [AP], *Gelasinospora* [AP] and other Sordariaceae. In some related homothallic (or homomictic) species only a single idiomorph occurs (Glass *et al.* 1990; Cisar *et al.* 1994). A further complication arises when dimictic forms, as a result of structural or genetical features, possess mycelia that can carry *both* functional MTFs and thus become **secondarily homothallic** (or **homo-heteromictic**).

**Table 1.2** Mating systems of the Eumycota and Straminipila

| System (alternative names) | Number of loci involved | Total number of MTFs identified | Occurrence |
|---|---|---|---|
| *Heterothallism (heteromixis)* | | | |
| 2-Allele morphological | 2? 'male', 'female' | ? | Chytridiomycotina |
| | | | Laboulbeniales |
| 2-Allele physiological | 2 *Mat A1* and *Mat A2* | 2 ?*Aa* and *aa* | Straminipila |
| (di- or bi-allelic, bipolar, dimixis) | 2 *Mat A* and *Mat a* (or *a*/α, $+/-$) | 2 | Zygomycotina |
| | | | Ascomycotina |
| | | | Ustaliginomycetes |
| Multiple-allele (diaphoromixis) | | | |
| Unifactorial (bipolar) | 1 *Mat A1, A2, A3* etc. | Tens | Basidiomycetes |
| Bifactorial (tetrapolar) | 2 *Mat A1–An* and *Mat B1–Bn* | Hundreds | Basidiomycetes |
| | | | ?Urediniomycetes |
| *Anomalous* | | | |
| ?Unifactorial | 1 *Mat A1* etc. | ?Tens | *Glomerella* [AP] |
| ?Bifactorial | 2 *Mat A1* and *Mat A2*; *B1–Bn* | Tens | *Ustilago maydis* [BS] |
| ?Trifactorial | 3 *Mat A, Mat B, Mat C* | 6 | *Psathyrella* [BA] |
| *Homo-heterothallism (secondary homothallism, homo-heteromixis)* | | | |
| Self-fertility regularly imposed on 2- or multiple heterothallism | ? | ? | Ascomycotina and Basidiomycotina |
| *Homothallism (homomixis)* | | | |
| Normally self-fertile | ? | ? | All groups |

The structure of the MTFs of the Zygomycotina is unknown.

Amongst the Basidiomycotina there is a greater diversity of mating systems than elsewhere in the fungi. In many Agarico- and Trmellomycetidae sexual compatibility is determined by *more than two MTFs*, a condition termed, for convenience, **diaphoromixis**. This can be the case even within a single genus, e.g. *Marasmius* [BA] includes homomictic, *unifactorial* (or bipolar) and *bifactorial* (or tetrapolar) *diaphoromictic* species, where the numerous MTFs occur at *one* or *two* loci respectively. The alternative terms refer to the genetically different kinds of basidiospore in respect of their MTF constitution which can be borne on a single basidium—two or four. In diaphoromictic Basidiomycotina, hyphal fusions are sexually compatible provided that *the MTFs at the same locus are different*, whether the fungus is unifactorial or bifactorial, i.e.

• *A1* is compatible with *A2*, or *Ax* with *Ay* in a unifactorial species, e.g. *Marasmius oreades* [BA] basidia bear two spore types *Ax* and *Ay*

• *A1B2* is compatible with *A2B1* or *AxBym* etc. in a bifactorial species, e.g. *Coprinus cinereus* [BA] basidia can bear four spore types *AxBx*, *AyBy*, *AxBy* and *AyBx*.

One species, *Psathyrella coprobia* [BA], even appears to have compatibility determined at three separable loci (Jurand and Kemp 1973). The loci are designated *A, B* or *C* and the different allelic factors are simply numbered, i.e. *A1, A2 . . . Ax*, or *A1B1C1, A2B2C2 . . . AxBxCx*, etc.

Each MTF in a diaphoromictic fungus has a complex genetic structure. They were originally thought to comprise at least two closely linked sub-loci which functioned in concert, i.e. $Ax = [A + A]$ (Raper 1966). In addition, multiple alleles were found at each sub-locus in the three or four fungi that have been investigated. For example, in a world sample of *Schizophyllum commune* [BA], nine and 32 alleles were identified at the *A* factor and nine and nine alleles at the *B* factor (Raper 1966). Allowing for rare crossing over between alleles these specify, potentially, 288 *A* and 81 *B* MTFs, far more than have actually been identified. In fact, the molecular complexity of the MTFs of the two Basidiomycotina that have undergone DNA sequence analysis, namely *S. commune* and *Coprinus cinereus*, is appreciably greater. In *C. cinereus* [BA] there are four sites each of two closely linked loci, each locus with several multiple alleles. However, not all the sites need be represented and the nucleotide distances between the sites are variable (May *et al.* 1991; Kües and Casselton 1992) (see also Chapter X). However, for most purposes in population genetics these complex loci can usually be treated as if they were simply multiple alleles at the *A*, *B* or *C* loci, i.e. *A1*, *A2*, *A3 . . . An* etc., depending on the species under consideration.

Homomictic forms, whose genetics have not yet been investigated, are common in the Basidiomycotina.

Although most smut fungi are dimictic, in one species, *Ustilago maydis* [BS], there is one di-allelic *A* locus and a second locus, *B*, that is multi-allelic. The mating system of the rust fungi has not yet been adequately analysed with certainty. They may be diaphoromictic.

Lastly, although only a few cases are known, some Eumycota appear to undergo sexual reproduction but in fact do not. In some Mucoráceae, sexual reproduction appears to follow a normal course but no meiotic divisions occur in the zygote, e.g. *Sporodinia grandis* [Z]. Nuclear divisions mimicking meiosis and subsequent mitoses occur in some Asco- and Basidiomycotina, but in fact *all* the divisions are *mitotic*. By analogy with the apomictic behaviour of green plants this can be described as **amixis**, e.g. *Podospora arizonensis* [AP] (Mainwaring and Wilson 1968).

The Straminipila show morphological sexual differentiation and may be self- or cross-fertilizing, or even be capable of both. In cross-fertilizing species an incompatibility system is superimposed on the morphological sexuality. It appears to involve *two* allelic MTFs, compatible isolates being determined by the heterozygous and homozygous recessive conditions, i.e. *Aa* and *aa*. However, in many species, e.g. *Pythium* [O], numerous other genes appear to affect the expression of sexual activity, both morphological and physiological, so that 'strong' and 'weak' males or females are often

recognizable and the situation is described as one of **relative sexuality**, e.g. *Pythium sylvaticum* [O] (Pratt and Green 1973). In other species, e.g. *Pythium ultimum* [O], both hermaphrodite and male forms are known. There is apparently no incompatibility system and both self and crossed progeny can arise in nature in various ways (Francis and St Clair 1993). The situation is only partially understood in many Straminipila, and genetic studies tend to be confined to those species where the situation is more clear cut, as in *Phytophthora infestans* [O] (Goodwin 1997), *Bremia lactuca* [O] (Michelmore and Sansome 1982) and *Pythium ultimum* [O] (Francis and St Clair 1993).

## 1.5  Dispersal

The significance for the temporal dispersal of fungal genotypes by perennating structures has already been mentioned (see section 1.3.3), but their spatial dispersal is limited. Spore dispersal, whether the spores are derived mitotically or meiotically, is the principal means whereby fungal genotypes are perpetuated and widely dispersed, and it also increases the potential for outcrossing and gene flow. Numerous mechanisms exist to release sporangiospores, conidia and meiospores, actively or passively, ideally so as to rise above the relatively immobile laminar boundary layer (Ingold 1971) (Figure 1.7). Thereafter, transport may be mediated through atmospheric turbulence, by wind, by water or through carriage by animals.

Even so, the immediate range of dispersal can be very limited (less than a metre to a few metres at most), particularly when the source of inoculum is in effect a point source, when aerial turbulence is low, as at night, or when dispersal is via an animal agency. The great exception to limited dispersal nowadays is dispersal, usually inadvertently, by humans or their machines! However, airborne spores can undergo long-range dispersal in two ways. First, the rapid development of successive mitotic spore generations, interspersed with dispersal phases with a prevailing wind direction, can spread the original mycelium through a series of successive short aerial flights, as in the seasonal

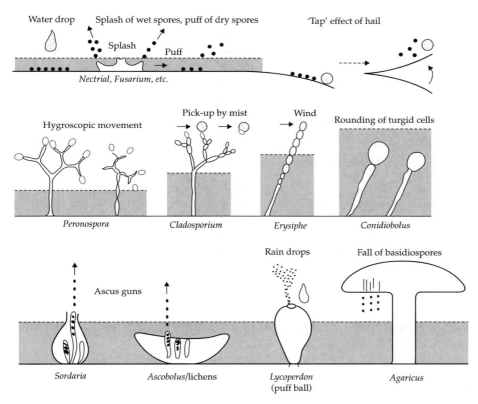

**Figure 1.7** Spore release by Eumycota. For effective dispersal, spores must be released, ejected or carried in some way above the relatively immobile laminar boundary layer close to the ground, indicated by the shaded region in the figure. (Reproduced from Deacon (1997), with permission from Blackwell Science.)

rust fungus *Puccinia* [BR]-path where spores of various *Puccinia* [BR] species are transported from Mexico northwards to the USA and Canadian wheat fields in the spring and later in the season back to Mexico, or the spread of conidia of barley mildew *Blumeria* (*Erysiphe*) *graminis* f.sp. *hordei* [A] across Europe. A second mode of dispersal occurs through the high upward transport of liberated spores so that they are carried over long distances by high-altitude winds and even trans-oceanically by jet streams (Aylor 1990; Nagarajan and Singh 1990). In this way, spore clouds are regularly blown across the North Sea by prevailing westerly winds (Hirst *et al.* 1967). An exceptional and probably very rare example is that rust fungi parasitic on wheat in East Africa have almost certainly been transported to Australia in a single flight, thus introducing novel pathogenic races to that continent (Burdon *et al.* 1982) (see also Chapter 6).

## 1.6 Aspects of fungal biology of genetic significance

Three characteristic biological features of the majority of fungal mycelia are of genetic significance:

• the **clonal** habit
• a high potential for exhibiting **phenotypic plasticity**
• **pleomorphism**—the separation in space and time of the sexual and asexual phases often associated with considerable morphological differences between them.

The importance of other genetic aspects for their effects on fungal populations are still uncertain.

• **Heterokaryosis** and the **parasexual cycle**, phenomena well known from experimental studies, have only been detected rarely in nature but may be important (see Chapter 8).

• Fungal **chromosomes** and chromosome behaviour are still less well understood than in most other eukaryotes. Knowledge of chromosomal abnormalities is limited but increasing, so that their significance for populations is not yet clear.

• Although **cytoplasmic organelles** and other cytoplasmic elements have proved more accessible and amenable to investigation than in other eukaryotes and their genetic role in cellular behaviour has been shown to be important, their significance for populations is still uncertain.

### 1.6.1 Clones

Fungal mycelia are essentially clonal in nature. Although one mycelium may persist for a brief period, occupying only a small territory in an ecological niche, or even the whole niche, at any one time, it usually produces large numbers of daughter clones from mitospores, hyphal fragments, chlamydospores or sclerotia which perpetuate it in both time and space, e.g. *Penicillium* [A/FA] spp. or *Mucor rammanianus* [Z] in the soil. Even if mutation occurs in some nuclei within the mycelium or mitotic recombination occurs at the time of conidium formation, as in *Verticillium albo-atrum* [FA] under laboratory conditions (Hastie 1967), these are relatively rare events, and most conidia perpetuate the original genetic clone or **genet**. Exceptionally, if dikaryons produce uninucleate conidia (de-dikaryotization), as some do (e.g. *Vararia granulata* [BA]), the two components of the clone can become somewhat separated at dispersal, although either the original dikaryon can re-form through hyphal anastomoses or a novel dikaryon can be formed. In general, therefore, dispersed clones from a single ephemeral microscopic mycelium may, as a consequence of spore dispersal, come to occupy the equivalent of a very large and widespread area. If the clonal mycelia are close enough to develop anastomoses, this may result in a single continuous mycelium covering a very large area.

Alternatively, a single perennial mycelium may grow vegetatively as a clone to cover a large area virtually continuously for hundreds or thousands of years. For example, a genetically uniform mycelial clone of *Armillaria gallica* [BA], a fungus which infects and spreads from root to root, has been found to cover an area of 15 ha and, based on its estimated average growth rate, is estimated to be of the order of 15000 years old (Smith *et al.* 1992). How exactly it was formed is not known.

Because of these two factors—the clonal habit and mycelial size—the concept of a 'fungal individual' is difficult to define. For instance, all the toadstool-like fructifications of *A. gallica* [BA] occurring over a wide area are identical genets and may be derived from a single clone. However, if a number of clones intermingle, the fructifications may either all be derived from the same genet or derived from different genets. In the latter case each clone represents a *different individual*. This is not usually obvious. For example, the numerous close-packed adjacent brackets of *Coriolus versicolor* [BA] are usually derived from different genets within the woody substrate, but they develop more or less simultaneously over the same limited site (Burnett and Partington 1957).

Therefore inspection alone does not enable an individual genet to be recognized on sight any more than the affinities of offshoots from it, the **ramets**, can be recognized as identical or different. Of course, similar situations are known in vegetatively propagating higher plants, e.g. the grasses *Festuca ovina* and *Festuca rubra* where widespread intermingled clones, 400–1000 years old, have been described from the Pentland grasslands in Scotland (Harberd 1961, 1962). In such cases, as with clonal fungi, the sampling procedure and subsequent identification procedures to recognize different genets are critical and often laborious (see Chapter 4).

### 1.6.2 Phenotypic plasticity

Phenotypic plasticity, defined as variance in genotypic expression in response to environmental influences, is shown by the mycelia of most fungi by changes in both form and physiology. Mycelial growth in a heterogeneous environment exposes growing hyphal tips in different regions to differing environmental conditions and there may also be appreciable fluctuations in both space and time. Because hyphal tips respond to their immediate environments, their growth and physiological adaptation may differ in different parts of the same mycelium. Examples of such behaviour are diffuse versus directed growth in response to high or low local nutrient availability, or a switch from synthesizing one extracellular carbohydrase to synthesizing another in relation to the availability of the local external carbohydrates.

Phenotypic plasticity of the mycelium resulting in differential adaptation in different regions provides an alternative to evolving a range of variants, each adapted to grow effectively in differing micro-environments. In effect, phenotypic plasticity by canalizing adaptive response into phenotypic change—one genotype but many phenotypes (Stearns 1982)—dampens the effect of selection on the genotype in the mycelial phase.

Reproductive structures, in contrast with mycelium, show much less phenotypic plasticity except in features like overall size. Many of their features, such as morphology and spore dimensions, are remarkably invariant in view of the range of environmental conditions under which they are developed and have long been recognized by taxonomists as more reliable conservative characteristics.

## 1.6.3 Pleomorphism

Another aspect of the adaptive behaviour of fungi is the occurrence of pleomorphism, i.e. the occurrence of morphologically distinct apparently unconnected asexual and sexual phases (**anamorph** and **teleomorph** respectively) of the same species at different times or in different situations. In some species more than one free-living anamorphic phase may be associated with the same teleomorphic phase. The origin of this condition lies in the differing environmental requirements for the development of the different phases. As a consequence, either mycelial growth and conidial production die down to be replaced by ascomata or basidiomata which differ entirely in their appearance, or the teleomorphic phase may be relatively inconspicuous, developing at a different location and often at a different time from the anamorphic phase. Therefore they appear unrelated. For example, in *Septoria nodorum* [FA], the conidial anamorphic

phase causes leaf lesions on cereals and grasses during spring and summer, but the inconspicuous teleomorphic *Mycosphaerella graminicola* [AP] phase develops in the litter, most abundantly in the autumn (Scott *et al.* 1988), although it releases ascospores for a considerable period, infecting winter wheat throughout its growing season (Hunter *et al.* 1999).

One consequence of pleomorphism is that the anamorphic states of many pathogenic species are often the only phase recognized in the field. An important practical consequence for fungal population biology of *not* recognizing the connection between the two phases is that because the life cycle is only partially known it can be wrongly interpreted. Therefore the population biology of a species may be seriously misinterpreted, its taxonomy confused, and genetic change and adaptation incapable of adequate explanation. Consequently, observed variation may not be recognized as being due to meiotic recombination. Pleomorphism combined with appreciable phenotypic plasticity can make it even more difficult both to recognize a fungal taxon and to interpret its behaviour.

## 1.6.4 Heterokaryosis and the parasexual cycle

Mutation in a multinucleate mycelium or hyphal anastomoses between two mycelia carrying genetically different nuclei, including those carrying complementary MTFs but different cytoplasms, can establish:

- a single common **heterokaryotic** mycelium carrying different kinds of nuclei
- a **heteroplasmic** mycelium containing different kinds of cytoplasm and organelles
- a mycelium which is both heterokaryotic and heteroplasmic.

Dikaryotic hyphae are a special case of heterokaryosis and also posses some of the properties of diploid cells, e.g. genetic or physiological complementation between different nuclei having different non-allelic genetically caused nutrient requirements. The control of migration of nuclei and cytoplasmic elements between hyphae at the time they fuse, when a heterokaryon is established, is not uniform and varies from fungus to fungus.

Unilateral, bilateral and limited nuclear migrations have been recorded. Even if no nuclear migration occurs, viruses and mitochondria may also be distributed through hyphal fusions by the same means under specific conditions, mitochondria less commonly than viruses.

Heterokaryosis provides a situation where a parasexual cycle has been demonstrated in some fungi. This is when somatic nuclear fusion occurs, i.e. **diploidization**, in a heterokaryon or dikaryon followed by mitotic recombination in the diploid nucleus and its subsequent **haploidization** through progressive loss of chromosomes to restore the haploid condition. Because most fungi are normally haploid, their nuclear products can be detected if, for example, they segregate into different uninucleate conidia. The consequences of mitotic recombination are different from those of meiotic recombination (Figure 1.8) but, although readily demonstrated in the laboratory, it is not known whether mitotic recombination occurs in nature or, if it does, how widespread it is (see also Chapter 8).

### 1.6.5 Chromosomal DNA and cytology

Fungi, like other Eukaryota, possess linear chromosomes and their DNA is associated with histones. Information concerning the total DNA content and the proportion which is repetitive is only available for a few fungi. Therefore it is difficult to make any reliable generalizations, but in most filamentous Eumycota the DNA content is of the order of about 30–40 Mb whereas that of yeasts is about half that amount. Repetitive DNA does occur but nearly all is accounted for by multiple copies of genes for ribosomal RNA, e.g. 185 copies in *N. crassa* [A] (Krumlauf and Marzluf 1980). Straminipila, in contrast, have appreciably more DNA, more of which is repetitive DNA.

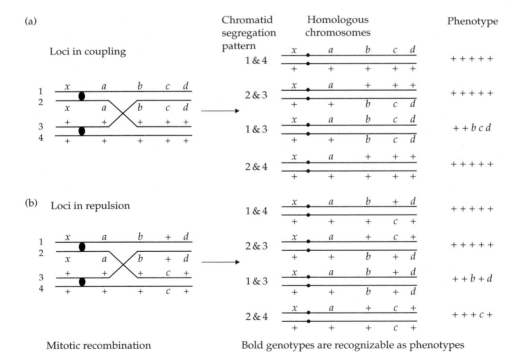

**Figure 1.8** The results of mitotic recombination in a diploid nucleus formed as consequence of diploidization during the parasexual sequence. The locus *x*/ + is to the left of the centromere, and the other locus is to its right. The consequence of mitotic recombination resulting from a single cross-over (the norm) are shown when the loci are in either coupling or repulsion. As the diploid nuclei undergo haploidization the chromatids become separated as chromosomes with the phenotypes shown.

The majority of Eumycota behave as if they are haploid in the somatic phase, but diploid exceptions are scattered throughout the fungi, e.g. *Allomyces arbuscula* [C] (Wilson 1952), *Saccharomyces cerevisiae* [A] (Winge 1935), *Verticillium longisporum* [FA] (Ingram 1968) and some forms of *Armillaria mellea* agg [BA]. (Korhonen and Hintikka 1974).

Polyploidy also occurs sporadically throughout the fungi, although this may simply reflect present ignorance, so that its general significance for the fungi is not known (Rogers 1973) (see also Chapter 8). It has been detected in meiotic chromosome preparations, e.g. *Puccinia kraussiana* [BR] (Sansome 1959), or more frequently by comparisons of depth of staining of whole non-dividing nuclei under constant controlled conditions with presumed haploid material. Recently DAPI (4′-6-diamidino-2-phenyindole), which binds preferentially to regions rich in adenosine and thymine, has been the preferred stain for comparisons between Eumycota, e.g. in the agaric *Pleurotus* [BA] spp. (Bresinsky *et al.* 1987), but Feulgen is effective in Straminipila, e.g. *Phytophthora infestans* [O] (Tooley and Therrien 1991). An essential technical condition for obtaining comparable results is that the amounts of repetitive DNA in different isolates are comparable. However, such techniques are, at best, indicative and, unless there is a known standard haploid, provide little information concerning the degree of polyploidy.

The consequences of mitosis and meiosis, although not their mechanical details, are the same as in other eukaryotes, and fungi exhibit normal Mendelian behaviour. Published fungal karyotypes are often misleading. A distinct metaphase plate has rarely been observed because that phase is either exceedingly rapid or, as in most fungi, anaphasic separation of chromatids of each chromosome occurs independently over a brief period (Aist and Morris 1999). Therefore mitotic chromosome counts are not easy to make, although they can now be obtained from burst germ tube preparations (Shirane *et al.* 1989; Taga and Murata 1994). Most chromosome numbers have been obtained from meiotic preparations, but since many fungal chromosomes are near to, or at, the limit of resolution by light microscopy they may be missed. Also, because many fungi have some additional minute 'B' chromosomes, which can only be detected by pulsed field gel electrophoresis of their DNA, chromosome number is usually underestimated by conventional microscopy. Even gel electrophoresis is not infallible since it fails to separate chromosomes larger than about 6 Mb unless special techniques are adopted. Therefore karyotype variation is difficult to study in fungi. Ideally, both light microscopy and gel electrophoresis methods should be used.

In contrast, the chromosomes of the Straminipila are, on the whole, somewhat larger than those of the Eumycota and have proved amenable to classical staining techniques, e.g. Feulgen staining is effective with *Phytophthora infestans* [O] (Tooley and Therrien 1991). Diploidy is, of course, the rule, but both auto- and allopolyploidy are quite common (Sansome 1987; Tooley and Therrien 1991).

### 1.6.6 Cytoplasmic organelles and other inherited cytoplasmic components

The fungal mitochondrial genome is a circular DNA molecule which is probably present in some tens of copies in each mitochondrion in every cell or fungal compartment. There is much variation in mitochondrial genomes in both size, from 115 kbp in *Cochliobolus heterostrophus* [AP] (Garber and Yoder 1984) to 17.6 kbp in the yeast *Schizosaccharomyces pombe* [A] (Zimmer *et al.* 1984), and gene arrangement. In some cases there is also variation in size and gene order within species. Mitochondrial inheritance, which is predominantly maternal, can be followed more readily in fungi than in most other Eukaryota. Mitochondria can be transferred via hyphal fusions, but often their movement is either restricted to the recipient heterokaryon or no exchange occurs. For instance, with heterokaryosis in *Coprinus cinereus* [BA], two dikaryons develop, i.e. bilateral nuclear exchange occurs but each is characterized by its own resident mitochondria which have not migrated after hyphal fusion (Casselton and Condit 1972). In *Neurospora*, however, although the organelle itself is not transferred, mitochondrial recombination can apparently occur in the fusion zone.

The extranuclear genetic repertoire of fungi is comparable with that of other eukaryotes. Several free and chromosomally bound mobile elements occur:

- retrotransposons such as the *Ty* elements of yeast (Roeder and Fink 1983)
- both nuclear and mitochondrial plasmids, e.g. *N. crassa* [AP] and *Podospora anserina* [AP] (Tudzynski and Esser 1985)
- free double-stranded RNA (ds-RNA) viruses, e.g. *Penicillium chrysogenum* [FA] (Lemke and Nash 1970).

Most of these elements can be transferred either from nucleus to nucleus or from mycelium to mycelium through heterokaryosis, e.g. plasmids which promote senescence in *Neurospora* [AP]. Indeed, even during transient heterokaryosis between different species of *Neurospora* [AP], when no true heterokaryon was formed, the *kalilo* plasmid was transferred from *N. intermedia* [AP] to *N. crassa* [AP] (Griffiths *et al*. 1990). The action of these various elements alone and in concert with the nuclear DNA (nDNA) and mitochondrial DNA (mtDNA) genomes is not always predictable and can mimic mutational processes (see Chapter 7, section 7.1).

## 1.7 A simplified classification of the Fungi

Various classifications of the Fungi exist concurrently. The major ones are summarized by Kirk *et al*. (2001) and a more detailed treatment is given by McLaughlin *et al*. (2001a,b).

### 1.7.1 Mycota

Predominantly basically haploid (although a small group normally diploid) eukaryotic osmotrophic organisms (i.e. capable of absorbing organic materials across their surfaces), generally constructed of microscopic cylindrical cells which grow at their tips, or sometimes as non-motile yeast-like cells.

*Eumycota*
'True' fungi, typically hyphal or, less commonly, non-motile unicells; walls with chitin (rarely, chitosan) and $\beta$-(1,3)-glucans and/or mannans;

spores typically non-motile, rarely motile; sexual reproduction widespread.

**Chytridiomycotina** Forms ranging from simple cells with rhizoids to septate or aseptate hyphal mycelia; chitin in walls; posteriorly uniflagellate, rarely multiflagellate, flagella of whiplash type; sexual reproduction occurs but unknown in many simple forms.

**Zygomycotina** Non-reticulate mycelia of predominantly aseptate hyphae; chitin or chitosan in walls; typically producing non-motile sporangiospores formed endogenously, rarely conidial; sexual reproduction gametangial.

**Ascomycotina** Reticulate mycelia, hyphae septate with simple pores; conidial; mostly with sexual reproduction; products borne in an ascoma (mostly apothecial, or perithecial) where nuclear fusion is most commonly followed by meiosis and mitosis within an ascus to produce typically eight, or multiples of eight, endogenous ascospores, often discharged violently. The classification of the Ascomycotina is only beginning to be resolved but it is still convenient to recognize three morphological groups:
- Archimycetes—yeasts, and supposedly simple forms,
- Pyrenomycetes—'flask' fungi, possessing a perithecium,
- Discomycetes—'cup' fungi, possessing an apothecium.

**Basidiomycotina** Reticulate mycelia, hyphae septate with simple or complex (dolipore) pores; clamp connections, dikaryons and monokaryons occur in some groups; conidia in some groups only; sexual reproduction on, rarely in, basidiomata which range from simple to massive complex structures; sexually compatible nuclei capable of persisting for long periods in association, often as binucleate (dikaryotic) hyphae, prior to fusion and meiosis in a basidium, resulting typically in four apical exogenously produced basidiospores which are most commonly violently discharged. They include three major groups:
- the Urediniomycetes (Teliomycetes, Uredinales)—rust fungi

- the Ustilaginomycetes (Ustomycetes, Ustilaginales)—smut fungi and related forms
- the Basidiomycetes (in two main divisions)

    Agaricomycetidae (Homobasidiomycetes)—toadstools, bracket fungi, puffballs etc. with undivided basidia

    Tremellomycetidae (Heterobasidiomycetes), a smaller group of often gelatinous forms with septate basidia.

**Fungi Anamorphici** An artificial group having as sole feature in common the total absence (or absence of detection!) of sexual reproduction; the majority resemble morphologically and in their spore forms the Ascomycotina, and a few resemble the Basidiomycotina, rarely lacking spores altogether.

*Straminipila*

Derived from a heterokont algal group but lacking chloroplasts; diploid, with a typical hyphal growth habit, or, if unicellular, often producing rhizoids; structural wall glucans include a cellulose-like $\beta$-(1,4)-glucan; many capable of producing motile biflagellate zoospores. Sexual reproduction, where known, oogamous (i.e. involving gametes of unequal size, one being an 'egg').

**Oomycota** Hyphal growth form lacking well-developed septa, with infrequent branching; asexual reproduction by sporangia capable of producing biflagellate zoospores (with one tinsel and one whiplash flagellum) or germinating directly; sexual reproduction by antheridia and oogonia.

**Hyphochytriomycota** Single-celled with rhizoids; asexual reproduction by sporangia producing zoospores with single anterior flagellum; sexual reproduction not known.

A more detailed classification is given in the Species Index, where every fungus referred to in this book is included in its appropriate taxonomic position.

## Further reading

Further detailed information about fungi can be found in the following texts. Elementary broadly based introductions, such as:

Carlile MJ, Watkinson SC and Gooday GW (2000). *The fungi* (2nd edn). Academic Press, London.

Deacon JW (1997). *Modern mycology* (3rd edn). Blackwell Scientific, Oxford.

*General taxonomically based texts such as:*

Alexopoulos CJ, Mims CW and Blackwell M (1996). *Introductory mycology* (4th edn). Wiley, New York.

Webster J (1980). *Introduction to fungi* (2nd edn). Cambridge University Press.

*General biologically based texts such as:*

Burnett JH (1976). *Fundamentals of mycology* (2nd edn). Edward Arnold, London.

Gow NAR and Gadd GM (1995). *The growing fungus*. Chapman & Hall, London.

*Fungal physiology, in a broad sense, is superbly summarized in:*

Griffin DH (1994). *Fungal physiology* (2nd edn). Wiley–Liss, New York.

*Fungal genetics are still best dealt with in two excellent, but now dated, texts:*

Esser K and Kuenen R (1967). *Genetics of fungi*. Springer Verlag, New York. (This is the English edition of *Genetik der Pilze*, Springer Verlag, Berlin, 1965.)

Fincham JRS, Day PR and Radford A (1979). *Fungal genetics* (4th edn). Blackwell Scientific, Oxford.

*Invaluable reference works on virtually all topics, however briefly described, one totally comprehensive and one elementary are:*

Kirk PM, Cannon PF, David JC and Stalpers JA (2001). *Ainsworth & Bisby's Dictionary of the fungi* (9th edn). CAB International, Wallingford.

Ulloa M and Hanlin RT (2000). *Illustrated dictionary of the fungi*. APS Press, St Paul, MN.

# PART II

# Methodology

The methods used for identifying, describing or identifying individuals and populations are no different, in principle, from those used with any other organisms but in practice they need to be specifically adapted to the characteristics and biology of the Fungi.

Apart from some Basidiomycotina, Discomycetes and lichens (Ascomycotina), the majority of fungi are small and not readily identifiable in the field as are larger eukaryotes. Although morphologically very variable, many of their characteristics are microscopic and do not lend themselves readily to short diagnostic descriptions. Satisfactory identification is often a laboratory activity. The growth habit and biology of the fungi also makes the identification of individuals more difficult, as these features can rarely be determined by inspection in the field. A further source of difficulty is that in Ascomycotina and Fungi Anamorphici, an anamorphic state is often common to different genera and species, e.g. the anamorphic conidial form genus *Verticillium* [FA] is common to a number of different teleomorphic genera including *Nectria*, *Nectriopsis*, *Cordyceps*, *Torrubiella*, *Ephemeroascus*, *Calonectria*, and *Hypocrea* [all A]. Hence, physiological, biochemical and molecular features assume a much greater significance as markers with the fungi than with higher eukaryotes.

In fungi, as in other organisms, most variation in natural populations is continuous, i.e. quantitative, rather than discrete, but quantitative inheritance has only been studied in a limited number of cases in fungi. Such traits have rarely been used in population studies. Most of the available genetic information concerning fungi relates to discrete characteristics, but not many of these are detectable or usable in the field. Natural markers are now increasingly being superseded by molecular markers of various kinds, and it is evident that in future population studies they will be used predominantly as the genetic markers.

Since both markers and individuals often have to be determined by laboratory tests, it is essential to adopt an effective sampling technique when collecting wild isolates in order to ensure an adequate representation of the kinds of individuals present in a field population and their range of variation. This, in turn, requires care to be taken in the choice of markers to be used for any particular investigation. In this period of active development, the most desirable and indicative markers and features have not always been used, and indeed some aspects, notably sampling, have received little investigation and less attention than they deserve. Therefore there is still much to be done to improve the existing methodologies described in the following chapters.

Descriptive techniques for defining populations, described later in this section, are those applied to other organisms. Many are available in standard biometrical texts; accordingly, in many cases they are only briefly outlined here.

# CHAPTER 2

# Genetic markers for population studies—I Morphological, physiological, and cytological markers

Traditionally, quantitative and morphological traits have been employed most frequently in the description of species, variants and variation. For population studies, however, there is a need for discrete characters which vary within the species that can be used as markers to recognize or differentiate populations. It is especially useful if the markers used have a known genetic basis, and, indeed, the use of **genetic markers**—discrete differences in genetic information possessed by different individuals within a species—is the preferred option. This is, in essence, an extension of classical genetic analysis where genes and their mutants are employed as genetic markers in studies of inheritance. Fungal geneticists have usually used induced rather than spontaneous mutants in inheritance studies. Such markers have only been introduced infrequently into natural populations so that they are not often employed in population studies. However, spontaneous semi-quantitative physiological traits such as pigment production, virulence factors, fungicide resistance, mating-type factors and somatic incompatibility reactions are usually treated in an all-or-none manner, i.e. as if they were discrete characters, and are employed.

A more precise approach, to be described in Chapter 3 is to employ **molecular markers**—genetic markers detected using molecular methods—such as specific proteins, especially enzymes, or discrete recognizable sequences of DNA whether representing whole genes or not.

Phenotypes of mutant genes or chromosomal mutants, both spontaneous and induced, such as

spore or mycelial colour mutants, provide a limited range of readily detectable natural markers which can be used in the study of fungal populations. Biochemical markers, mostly induced discrete mutants, also occur spontaneously in very small numbers, e.g. in *Ophiostoma multiannulatum* [AP] (Fries 1948), but more elaborate techniques are required to identify them and they have rarely been used in population studies. Nevertheless, a convenient range of naturally occurring morphological, behavioural and physiological markers exist which can be used as markers in population studies. To date, apart from spore or hyphal colour mutants, not much use has been made of morphological markers. However, several useful behavioural or physiological markers exist. They include the different kinds of interactions which can occur between mycelia determined either by mating-type factors (MTFs) or by somatic incompatibility (SI) factors. In pathogenic fungi, attributes such as pathogenicity or resistance to fungicides are additional useful physiological markers.

Classical cytological techniques for chromosomes have only limited applicability in fungi because of both the small size of their chromosomes and their patterns of mitosis which rarely provide visually clear chromosomal separation at metaphase. Spontaneous gross chromosomal changes are detectable at meiosis in favourable circumstances in some Ascomycotina, Basidiomycotina and Straminipila. Although genetic analysis enables rearrangements, translocations and the like to be detected genetically, their identification would be excessively laborious if applied to the numerous individuals of a population. However, the small size of many fungal chromosomes enables them to be treated as molecular markers. They can conveniently be separated electrophoretically, their sizes determined and some structural changes detected.

## 2.1 Morphological markers

The dimensions of structures such as sporangiospores, conidia or meiotic spores, together with overall dimensions of other structures—hyphae, sclerotia, conidiophores, zygotes, ascocarps and basidiocarps—provide directly quantitative data which can be compared and contrasted statistically. The actual numbers and sizes of such structures can also be used in some cases, e.g. sclerotia formed by different species of *Sclerotinia* [A] (Tariq *et al.* 1985).

Growth rates and 'fruiting times', provided that they are measured under standardized conditions, are often more characteristic of particular isolates or strains. The range of variation in growth rates and fruiting times of both dikaryons and their monokaryotic progeny within natural populations of the agaric *Collybia velutipes* [BA](Croft and Simchen 1965) and *Schizophyllum commune* [BA] (Simchen and Jinks 1964; Simchen 1967; Brasier 1970) has been described and provides models of how such a character can be used in population studies, e.g. growth rates of *S. commune* [BA] at three different locations. The shapes of structures and colony morphology when mycelia are grown in culture are often distinctive qualitative characteristics, but if they can be quantified in some way, such as by the fractal transformations of colony shapes, their use is greatly enhanced, e.g. *Macrophomina phaseolina* [FA] (Mihail *et al.* 1994).

Colour mutants, usually spore colour but also hyphal pigmentation, occur spontaneously in most pigmented fungi, the most common being white or colourless variants, e.g. the creamy-white spore mutants of the normally black conidia of *Rhynchosporium secalis* [FA] (McDermott *et al.* 1989). Many induced colour mutants are also **auxotrophic** (i.e. they require a specific additional nutrient in order to grow on minimal medium), but this is not usually the case with spontaneous variants since in most natural situations auxotrophy is probably at a selective disadvantage. Nevertheless, while colour mutants may be prototrophic and mutant colour genes selectively neutral, their mycelia may differ selectively, depending on the rest of the genotype. For example, three naturally occurring prototrophic colour forms of *Fusarium moniliforme* [FA] from China—purple, red and white—differ in pathogenicity and in gibberellin production in the ratio $100 : 50 : 1$ respectively, but all are equally capable of forming heterokaryons with each other (Ming *et al.* 1966).

## 2.2 Mating-type factors (MTFs)

The simplest example of physiological/behavioural markers are the MTFs which determine sexual compatibility (see section 1.4.1). Although only two occur in dimictic fungi, their frequency may differ in different populations, even though in the species overall their numbers are statistically equal. However, in *Ustilago maydis* [BS] and other diaphoromictic Basidiomycotina, multiple alleles occur for one or more mating-type loci and there can be much greater variation in the numbers and frequency of MTFs within populations and species.

The determination of mating type cannot normally be made in the field but involves sampling and testing for compatible gametangial or hyphal fusions between haploid monokaryons in culture against mycelia or monokaryons of known mating-type constitution. Mycelial field isolates of diaphoromictic fungi are usually dikaryotic, or their equivalent, so that it is necessary either to collect basidiocarps and isolate basidiospores, or to de-dikaryotize the mycelium to obtain monokaryons for mating tests. Various methods have been described, but one of the most convenient is to add glycine to the growth medium and isolate the de-dikaryotized hyphae (Leal-Lara and Eger-Hummel 1982). In some groups, e.g. many of the crust-like basidiomycotous Corticeaceae, the dikaryotic hyphae bear uninucleate conidia which, of course, enable large numbers of monokaryons of each mating type to be obtained directly. Alternatively, single basidiospore isolations from basidiomata can be used.

**Table 2.1** The reactions of monokaryons carring different MTFs in a bifactorial diaphoromictic basidiomycete such as the common bracket fungus *Coriolus versicolor* [BA]

| Constitution of monokaryon | a1b1 | a1b2 | a1b3 | a2b1 | a2b2 | a2b3 | a3b1 | a3b2 | a3b3 |
|---|---|---|---|---|---|---|---|---|---|
| a1b1 | − | − | − | − | + | + | − | + | + |
| a1b2 | − | − | − | + | − | + | + | − | + |
| a1b3 | − | − | − | + | + | − | + | + | − |
| a2b1 | | | | − | − | − | − | + | + |
| a2b2 | | | | − | − | − | + | − | + |
| a2b3 | | | | − | − | − | + | − | − |
| a3b1 | | | | | | | − | − | − |
| a3b2 | | | | | | | − | − | − |
| a3b3 | | | | | | | − | − | − |

+ indicates the formation of a dikaryotic mycelium with clamp connections. Note that the MTFs have to be different at *both* loci for compatibility.

MTF identification is usually determined in plate culture and observations are generally restricted to the assessment of effective fusions or, in many Basidiomycotina, to the development of a dikaryon (Table 2.1). This is because there are often unknown chemical, or environmental, requirements for the development in culture of ascocarps or basidiocarps, or an appreciable period of some months before their development. Interfertility or sexual compatibility is assumed thereafter, but, of course, neither hyphal fusion nor dikaryon formation alone is conclusive. However, this is not necessarily a limitation on the study of populations of the same species, since the mycelia of many of these fungi are perennial and all that is needed is to identify either their MTF constitution or whether there are genetically distinct mycelia.

However, when this kind of MTF identification is employed with different or putatively different species it may be misleading, because some hyphal fusions can occur between different, although usually related, species even if they are sexually incompatible, i.e. no further development occurs, as in the polyporaceous *Hirschioporus abietinus* [BA] species complex (Macrae 1967). In some cases, hyphal fusions fail or are greatly reduced between strains which belong to the same species and which are sexually compatible, e.g. *Aspergillus nidulans* [A] (Grindle 1963a,b). This is because hyphal fusions can be affected by genetic factors other than MTFs, notably by the SI factors (see section 2.3.1)

which determine the success or failure of strictly somatic hyphal fusions but not sexual competence.

## 2.3 Somatic incompatibility (SI)

The prevention of hyphal fusions between mycelia as a consequence of their genetic constitution, other than their MTF constitution, is called **somatic incompatibility**. It is also termed 'mycelial incompatibility', 'heterokaryon incompatibility' or, more frequently, 'vegetative incompatibility'. SI is not known to occur in Chytridiomycotina, Zygomycotina or the Straminipila. Its expression differs in Ascomycotina from that in Basidiomycotina.

### 2.3.1 SI in Ascomycotina and Fungi Anamorphici

SI is widespread in filamentous Ascomycotina and related anamorphic forms, where it affects the ability to form heterokaryons between *haploid monokaryotic* mycelia (Leslie 1993). Genetically, the most widespread pattern underlying SI is *homogenic* incompatibility with a multi-locus basis. The alleles involved are designated by either *het*, for heterokaryon incompatibility, or *vic*, for vegetative incompatibility.

• If the *het* or *vic* alleles at *all* SI loci in two isolates are *identical*, then *compatible hyphal fusions can occur*, but

• However, if the *het* or *vic* alleles are *different* at *any one* of the loci then *hyphal fusions are either greatly reduced or may fail completely.* Thus there is no exchange of cytoplasmic content or nuclei and a stable heterokaryon does not form.

It is important to recognize that an incompatible reaction indicates unambiguously that *only* the SI genotypes of the interacting mycelia differ. Conversely, it is *not* necessarily the case that compatible mycelia are genetically identical *save at the loci that determine the SI reaction*, although they may be similar in some cases.

As a consequence of SI, a species becomes divided into a number of somatically incompatible groups, each termed a **vegetative incompatibility group** (VCG) or, less frequently, a **heterokaryon incompatibility group** (HET group). Hyphal fusions within a VCG are normally frequent.

In most cases, but not all, an incompatible SI reaction does *not* prevent normal fertile sexual reproduction between members of different VCGs, e.g. as in *Aspergillus nidulans* [A] (Caten 1972), but in some, e.g. *A.amstelodami* [A], it is impeded. An unusual situation occurs in *Neurospora crassa* [AP] where the MTFs also act as SI factors (Garnjobst 1953, 1955).

### 2.3.1.1 Recognizing SI in Ascomycotina

Specific morphological features are often associated with incompatible interactions. A 'barrage' may develop which can take a variety of forms in different species. Sometimes the hyphae interact to form a zone of vacuolated, dying and often pigmented hyphae or hyphal compartments which may or may not be associated, on each side of the barrage, with a mass of aerial hyphae, e.g. *Podospora anserina* [AP] (Rizet and Esser 1953). These latter hyphae may be normal, distorted or have abnormal cytoplasmic contents. Such a hyphal mass may be the sole expression of the 'barrage', e.g. *Nectria fuckeliana* [AP] (Vasiliauskas and Stenlid 1997). In other cases a line of sclerotia may form along the interaction zone, e.g. *Verticillium dahliae* [FA] (Puhalla 1979). When the mycelia under investigation are known to belong to the same species, it is often sufficient simply to identify the different VCGs on the basis of hyphal interaction patterns without actually determining the numbers of loci and alleles involved.

In *V. dahliae* [FA], where in the USA sixteen VCGs were originally recognized amongst 86 isolates on the basis of their pairing reactions in all combinations (Puhalla and Hummel 1983), a more precise test was developed, namely their inability to form a forced heterokaryon. Two ultraviolet-induced mutants, possessing white and brown microsclerotia respectively, were used but, on successful heterokaryons, black microsclerotia (the normal wild-type colour) developed. A more generally useful comparable technique to identify VCGs rapidly is the use of nitrate reductase (*nit*) mutant strains. Cove (1976a,b) showed that when *A. nidulans* [A] is grown on chlorate medium, chlorate-resistant mutants, which are very often also mutant for nitrate reductase, are readily recovered. If such chlorate-resistant strains are paired on plates with nitrate as the sole nitrogen source, auxotrophs of the *same* VCG form prototrophic heterokaryons and growth is comparable with that of wild-type strains. Mutants from *different* VCGs are incapable of forming prototrophic heterokaryons and no wild-type growth results. When this technique is used, many hundreds of strains can be screened quite rapidly. This technique has been found to be effective with other species of fungi. It was first adopted to determine VCGs in *Fusarium oxysporum* [FA] (Puhalla 1985). Unusually, in this taxon the genotypes of each VCG, as determined by molecular methods, are identical or very similar and each VCG appears to be a distinct clone (Kistler 1997).

However, it should be noted that the number of VCGs detected depends on the method adopted to detect the reaction. For instance, when the nitrate reductase test was applied to Puhalla's original 86 isolates of *V. dahliae* [FA], only three VCGs were recognized, although some could be further subdivided on other features such as 'strength' of reaction, host specificity, virulence etc. (Rowe 1995). Nevertheless, three, or possibly four, VCGs are now generally accepted and these are all that have been detected in over 1000 isolates tested. It follows that a more complete understanding of SI depends upon the identification of the *vic* genes themselves (see section 2.3.1.3).

Puhalla (1985) was also the first to develop systematic numbering for VCGs—a very necessary requirement when large numbers of VCGs are isolated from different hosts over wide geographical regions. He used a four- or five-digit system, with the first three digits indicating the *forma specialis* (f.sp.), i.e. an isolate based on the host species to which it appears to be specialized, e.g. f.sp. *hordei* specialized to *Hordeum* spp. (Barley). The remaining numbers are allocated to the individual VCGs as they are isolated. A worldwide co-ordinating system has been established for wide-ranging fungi with many VCGs (e.g. *Fusarium oxysporum* [FA] with more than 125 known VCGs) to ensure that new VCGs are recognized and duplication avoided (Kistler *et al*. 1998). Thus, for instance, isolates of *F. oxysporum*, f.sp *albedinis* [FA] found on the date palm *Phoenix dactylifera* are prefaced by 017-, and since all are identical for VCG they are designated 017**0**. By contrast, the six VCGs known from *Dianthus*, *F. oxysporum* f.sp. *dianthi* [FA], prefixed by 002-, are numbered 002**0**, 002**1**, 002**2**, 002**5**, 002**7** and 002**8**. Hence, for example, it follows that both the host and the behaviour of two isolates of *F. oxysporum* [FA], numbered 0170 or 0025, can be recognized immediately from the coded description. In many cases, VCGs are associated with particular areas, mitochondrial genotypes or other features, which enables them to be used to analyse the history and population biology of the fungus.

### 2.3.1.2 Heterokaryon self-incompatibility

A possible source of confusion in the determination of VCGs is the occurrence of **heterokaryon self-incompatible** (HSI) strains. In these, the ability to form any hyphal fusions at all is greatly reduced. As a consequence, the numbers of fusions between HSI strains are negligible and very low even within a single mycelium. For example, within HSI strains of *Fusarium moniliforme* [FA] not only were the numbers of laterals reduced but the frequency of fusions was only 0.2–1.1 fusions $mm^{-2}$, whereas within a normal strain the frequency was 6.9–8.1 fusions $mm^{-2}$. This HSI strain is also female sterile. Although fusions within or between HSI strains are low or non-existent respectively, rare fusions can occur between a normal and an HSI strain in *F. moniliforme* [FA]. Nuclei can be exchanged and if the strains happen to be sexually compatible, perithecia can develop on the normal

mycelium. The normal to HSI characteristics segregate 1:1 in such perithecia. Similar *hsi* mutants are known in other species, e.g. *A. flavus* [A], *F. oxysporum* [FA], *F. subglutinans* [FA], *Venturia inaequalis* [FA] and *Verticillium dahliae* [FA]. About 1–2 per cent of field isolations of any species possessing a vegetative compatibility system are likely to be single gene *hsi* mutants (Correll *et al*. 1989; Leslie 1993).

### 2.3.1.3 SI in similar or related taxa

SI behaviour often differs between similar or apparently related taxa. In *Verticillium*, for example, there appear to be more VCGs in the saprophytic than in the parasitic species. Thus there are two and three VCGs in *V. albo-atrum* and *V. dahliae* [both FA] respectively, but 11, 5 and 5–13 in *V. tricorpus*, *V. nigrescens* and *V. lecanii* [all FA] respectively. No VCGs were detected in the diploid *V. longisporum* (Katan 2000). Whether or not this is a general situation is unknown.

Apparently similar species in identical ecological situations may also differ profoundly in their behaviour. For example, superficially similar isolates of *Phomopsis* [FA] from twigs and bark of species of elm (*Ulmus*) fall into two morphologically distinct groups when cultured on malt agar. In culture, one group of mycelia were quite variable in appearance but produced greyish mycelium and abundant conidiomata, whereas mycelia of the other group were remarkably uniform, creamy coloured and rarely produced any conidiomata at all. The former, designated **group 1**, are dimictic, and the latter, designated **group 2** are self-fertile; the two groups are not inter-fertile and when confronted developed a strong darkly pigmented aversion zone (Webber and Gibbs 1984; Brayford 1990a). Single conidial isolates from group 1 frequently developed a distinct zone of antagonism involving pigmented hyphae when confronted; this also occurred between mycelia derived from sibling ascospores. This antagonism was interpreted as an SI reaction. Thus numerous VCGs, as many as 29 in one sample, occurred in the outbreeding group 1 and at least five *vic* genes appeared to be involved to judge by segregation data. Nevertheless, cross-fertilization could occur between different or identical sexually compatible VCGs. In contrast, mycelia developed from different single ascospore isolates of group 2 usually simply intermingled and a clear antagonistic zone rarely developed. Paired progeny from single perithecia showed no interactions. If VCGs occurred at all in the inbreeding group 2, they were exceedingly uncommon (Brayford 1990b). Thus two similar taxa growing in the same habitat differed completely in their sexual reproduction, SI reactions, and, hence, population structure.

### 2.3.1.4 SI genetics

Recording the compatibility reaction is often sufficient for the purposes of population analysis. However, if information concerning their genotypic structure and diversity is known, as well as the numbers of loci involved, their chromosomal locations and the numbers of alleles at each locus, further analysis can be made. For example, such information has been used to investigate problems of the origin of, selection for, and structure of VCGs and can also be used to describe population structure.

The genetic basis for VCGs has only been investigated in a few fungi. For example, *Neurospora crassa* [AP] has at least 11 *het* loci distributed over five linkage groups. Eight loci have two alleles and two have multiple alleles (Leslie 1993). *Het* genes were detected by the use of duplication, producing strains used as testers on natural isolates from the wild. Duplications homozygous for *het* genes are normal in growth but, if heterozygous, they are initially inhibited in growth and produce a characteristic dark pigment on media containing phenyl alanine and tyrosine, although they eventually recover. They also give an indication of the site of the *het* loci which can be confirmed and refined by appropriate crosses including known loci putatively linked to the *het* loci (Mylyk 1975, 1976). Once the position of the *het* loci was known, it later proved possible to identify and isolate the equivalent nucleotide sequences. In *A. nidulans* and other fungi the *het*/*vic* genes were identified by an appropriate series of crosses and back-crosses between different isolates. Eight *het* loci are known in *A. nidulans* [A]; one of these, *het-C*, possesses three alleles and there are four at *het-B* (Croft and Dales 1984). Similar situations have been found in other species.

Since these loci, if unlinked, can recombine freely at sexual reproduction, many more VCGs can be generated than are usually detected. For instance 10 loci, each with two alleles, could generate over 1000 VCGs, and if some loci have multiple alleles the potential numbers would be increased further. In practice, such high numbers have not been isolated. For example, the numbers of VCGs detected in the chestnut blight fungus *Cryphonectria parasitica* [AP] at an isolated site over four successive years (1982–1985) were 6, 11, 38 and finally 48. The most plausible explanation of this increase in VCGs is their generation through successive annual rounds of recombination (Anagnostakis and Kranz 1987).

In fact, the numbers of VCGs in a population will depend on three factors:

- the numbers of *vic* or *het* loci involved
- the allelic variability at these loci
- the extent of sexual recombination within the population.

Thus if the level of allelic variability is comparable in two populations, the numbers of VCGs can give an indirect measure of sexual recombination within them, provided that migration can be excluded. A much finer analysis can be made of the population structure using *vic* genes rather than VCGs, as has been done for *Cryphonectria parasitica* [AP] (Milgroom and Cortesi 1999). These authors were able to demonstrate that there were far fewer *vic* genes in European samples compared with those from North America, confirming the nature of the founder populations in Europe as a result of the introduction of the fungus from North America.

The molecular analysis of the *het-C* locus of *Neurospora* [AP] and some taxonomically related species has provided further insight into the maintenance of the *het* gene in populations. Three allelomorphs, *het-C*$^{OR}$, *het-C*$^{M}$ and *het-C*$^{GR}$, were identified and cloned (Saupe *et al.* 1996). They differed in three characteristic sequences determining a 34–48 long amino acid sequence. The three alleles were found to occur with approximately equal frequency in both a world sample and in a single population of *N. crassa* [AP]. This suggests that they were maintained in a balanced polymorphism

(Wu *et al.* 1998). Investigations of other heteromictic species of *Neurospora* [AP] and the taxonomically related genera *Sordaria* [AP] and *Gelasinospora* [AP] revealed that they too carried the same *het-C* alleles with similar characteristic sequences. This suggests that these alleles are extremely ancient, probably predating the origin of these three genera, and that the polymorphism is probably also ancient (see Chapter 9, section 9.4.3.2).

### 2.3.1.5 *Complex SI behaviour*

VCG expression can be much more complex. A unique range of SI interactions is known from the Dutch elm disease fungus *Ophiostoma ulmi* [A], where somatic incompatibility appears to be under multigenic control. Expression ranges from complete compatibility **c** through four grades of incompatible barrage reaction: **w**, *wide* barrage; **n**, *narrow* barrage; **l**, *line* (a thin line of barrage mycelium); **lg**, *line-gap* (a narrow gap).

The **c** reaction, as usual, indicates that all *vic* genes are identical in both partners, the **w** reaction, which is most common, reflects few or no *vic* genes in common, and the **n**, **l** and **lg** reactions are due to increasing numbers of *vic* genes in common. Multigenic control was demonstrated by crossing two isolates showing a **w** reaction, selecting the two isolates with the weakest reactions from 20 F1

isolates, and successively back-crossing the weakest progeny to the weakest parent of the preceding cross from samples in each case of 20 progeny. Within four back-crosses all the progeny were compatible. These together with other crosses suggest that, in addition, multiple alleles exist at the **w** locus and also that epistasis occurs between the **w** and **n** loci (Brasier 1984).

A further complication is that cytoplasmic transfer of **d-factor** (a dsRNA mycovirus) is differentially prevented in matings of identical VCGs, namely 4 per cent transfer in **w** confrontations, 50 per cent transfer in **n** confrontations, 66 per cent transfer in **l** or **lg** confrontations and 100 per cent transfer in the compatible **c** reaction (Rogers *et al.* 1988).

The main features of the SI reaction in *O. ulmi* are set out in Table 2.2.

## 2.3.2 SI in Basidiomycotina

The commonly detected SI reaction is restricted to different paired secondary heterokaryotic mycelia (which are often but not always dikaryons) in many Agaricomycetidae, i.e. toadstools, brackets and the like. The phenomenon has also been termed **intraspecific antagonism** (Rayner and

**Table 2.2** SI reactions in *Ophiostoma novo-ulmi* ssp. a*mericana* [AP]

| Reaction type | Physiological features | Probable genetic similarities/differences between paired isolates |
|---|---|---|
| **w** | Full vegetative compatibility<br>Strongly different<br>Restricted cytoplasmic transfer of d-factors<br>Strong mycelial penetration effects | All or most *vic* genes<br>Different **w** alleles |
| **n** | Strong vegetative incompatibility<br>Some mycelial penetration effects | Most *vic* genes different<br>Common **w** allele |
| **l** | Slightly restricted cytoplasmic transfer<br>No nuclear migration<br>Little or no mycelial penetration | A few (>1) *vic* genes different |
| **lg** | Probably unrestricted cytoplasmic transfer<br>No nuclear migration or mycelial transfer | Difference at only one *vic* locus<br>Different **lg** alleles? |
| **c** | Full vegetative compatibility<br>Free cytoplasmic transfer of d-factors<br>No nuclear migration or mycelial penetration | All *vic* genes the same |

Modified from Brasier (1984), with permission from the British Mycological Society.

Todd 1978; Worrall 1997). Thus, unlike the Ascomycotina and Fungi Anamorphici, SI is shown *only* between *secondary (dikaryotic) mycelia* in Basidiomycotina and is *not* shown between primary homokaryotic, or monokaryotic, mycelia.

As in Ascomycotina, however, non-self-anastomosing isolates have been detected in a very few Basidiomycotina, e.g. *Rhizoctonia solani* [BA] (Hyakumachi and Ui 1987). As in the Ascomycotina, the most informative reaction, genetically, is that for incompatibility when, of course, the genotypes are different. Although compatibility is a reflection of an unspecified specific closeness of the interacting mycelia, it *cannot* be equated with exact genotypic identity.

Adverse interactions are usually detected by pairing isolations of secondary mycelia in plate culture but they can sometimes be detected in the field as well. For example, in wood-inhabiting fungi such as the polypore *Coriolus versicolor* [BA], the occurrence of dark pigmented interaction lines between adjacent mycelial-infected wood provides a rapid means of detecting different, individual and somatically incompatible mycelia (Rayner and Todd 1978) (Figure 2.1).

The pairing technique has proved particularly useful when large numbers of basidiocarps or mycelial structures such as rhizomorphs and mycelial fans are detected within the same area, e.g. *Armillaria* [BA] spp. (Worrall 1994). At some sites in this example different species of *Armillaria* [BA], as well as different dikaryons within each species, occurred in the same area. It proved possible to identify both the species and the probable areal limits for each genet by distinguishing the genetically identical adjacent mycelia of the same species.

In general, a useful 'working rule' is that:

Genetically *distinct individuals of the same species* are distinguished by the secondary mycelium pairing test for SI, whereas *different species* are separated by pairing representative monokaryotic tester isolates from each anticipated species with unknown secondary mycelia isolated from a site.

When the latter are conspecific with any of the tester monokaryons, those monokaryons carrying compatible MTFs become converted into secondary mycelia through invasive nuclear migration. This is indicated by observing either the development of clamp connections on the tester monokaryons or,

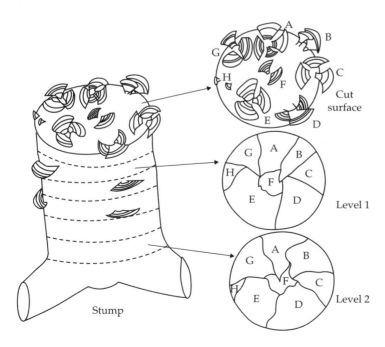

**Figure 2.1** Diagram to illustrate the natural SI interactions of *Coriolus versicolor* [BA] in an infected tree stump. The dikaryotic basidiocarps A–H on the surface of the stump arise from mycelia infecting the wood below. Darkly pigmented boundaries arise where the opposed margins of incompatible dikaryons meet. (Reproduced from Rayner and Todd (1979), with permission from Academic Press.)

when there is a visually detectable difference between the mono- and dikaryotic mycelia, by observing the change from one morphology to the other.

### 2.3.2.1 Recognizing SI in Basidiomycotina

SI pairings do not always distinguish between closely related genotypes (Kile 1983; Korhonen 1978b), and in experimental studies SI reactions have not always been an infallible guide to fungal individualism (May 1988; Rizzo *et al.* 1995). For instance, in *Coprinus cinereus* [BA], clear-cut results were not always obtained. In some cases, only weakly developed zones of antagonism occurred between different dikaryons and, sometimes, primordia genetically identical with one of the two paired dikaryons developed in the zone. The geographical origin of dikaryons showing these effects appeared to be implicated but the sample was not large enough for a firm correlation to be established.

An unambiguous demonstration that isolates which behave with the SI test as if belonging to the same SI group are, in fact, genotypically distinguishable has been provided by Jacobson *et al.* (1993) using the boletoid fungus *Suillus granulatus* [BA]. They demonstrated that comparisons using randomly selected molecular sequences (random amplified polymorphic DNA (RAPD); see Chapter 3, section 3.3.3) provided better genotypic discrimination than did SI reactions (Figure 2.2).

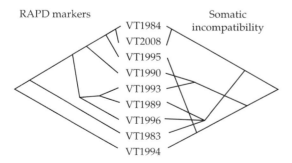

**Figure 2.2** SI groups are not genetically identical in *Suillus granulatus* [BA]. Dendrogram comparing identity of groups recognized by their SI reactions with those recognized by RAPD markers. (Reproduced from Jacobson *et al.* (1993), with permission from the National Academy of Sciences of the USA.)

Further support was provided by Guillaumin *et al.* (1996) after a careful comparison of different methods for distinguishing genets of the three species of honey fungus, *A. gallica*, *A. ostoyae* and *A. cepistipes* [all BA]. *A. mellea* agg. [BA] provides an example of morphological differences which can be used for identifying secondary mycelia with species of *Armillaria* [BA]. Secondary mycelia produce relatively little aerial mycelium, often develop a reddish-brown crust and appear 'flat' or crustose, whereas monokaryotic isolates produce abundant white aerial mycelia, appearing 'fluffy' (Ullrich and Anderson 1978). Therefore the conversion of a 'fluffy' tester monokaryon into a 'flat' or crustose mycelium can be used as a diagnostic test of specific identity between a tester and an unknown secondary mycelium, provided that their MTFs differed, i.e. are compatible. If the unknown is a *different* taxon, the tester is *not* converted to the 'flat' crustose condition.

Using this technique, 36 genets were identified using SI confrontations amongst 71 isolates from beech forests in central France. Guillaumin *et al.* (1996) compared this degree of genet discrimination with that achieved using RAPDs, MTFs, restriction fragment-length polymorphism (RFLP) patterns of mitochondrial DNA (mtDNA) and isoenzyme patterns (see Chapter 3 for molecular techniques).

Isoenzyme differences distinguished clearly between the species. In just over 8 per cent (three cases) of all isolates, two subgroups could be recognized within an SI-determined genet using RAPDs, although two of the genetically different SI genets were identical in their MTF composition. However, this is not unexpected since the MTFs do not affect the SI reaction. Although only a single mtDNA pattern occurred in each genet, some mtDNA patterns occurred in more than one genet so that 18 types could be distinguished by this means.

One interpretation of this finding is that it could indicate a common ancestry for those SI genets whose mtDNA patterns are identical. Thus, using different techniques, individual mycelia in any population of several species can be recognized at different levels, both specific and intra-specific. Amongst the latter, individual dikaryotic genotypes provide the finest degree of discrimination closely followed by that determined by SI reactions.

Thus, in general, the SI test does distinguish the vast majority of genotypically distinct individuals and has the advantage that it is a relatively rapid and valuable analytical tool. Therefore it is reasonable to suppose that the vast majority of homobasidiomycotous individuals that occur in nature as dikaryons or their equivalents behave as separate genets.

### 2.3.2.2 Complex SI

Just as in the Ascomycotina, more complex kinds of SI interactions occur, e.g. in *Rhizoctonia* [BA] spp. In many species of this genus, the hyphal compartments are multinucleated and lack clamp connections, and so recognition of the genetic and cytological status of both heterokaryons (dikaryons) and homokaryons (monokaryons) is difficult. Distinct 'anastomosis groups' (AGs), whose taxonomic status is debatable, occur. At least 12 AGs have been recognized to date in the taxonomically poorly defined *R. solani* [BA] complex. However, detailed examination of AG8 of *R. solani* [BA] in Australia (MacNish *et al.* 1993a) has led to the recognition of four reasonably well-defined states.

- C0 Isolates belong to different AGs — Hyphae show no reaction and simply grow past each other
- C1 Isolates have only a distant genetic relationship — Hyphae may make contact and this may be followed by death of hyphal compartments
- C2 Isolates are genetically different belong to the *same* AG — Wall fusion occurs accompanied by death of fused and adjacent hyphal compartment; there is often a visible zone of lysis
- C3 Genetically identical or very closely related — Complete compartmental fusion without lysis or any compartmental death.

Within AG8 it has also proved possible to recognize a number of patterns of extracted electrophoretically distinct pectic enzymes—so-called 'zymogram groups' (ZGs) (see Chapter 3, section 3.2)

which apparently represent different clones. Using both these and the anastomosis reactions it was possible to demonstrate eight distinct ZGs: AG8 (ZG1–1) to AG8(ZG1–8). In nature, a single ZG normally dominated each characteristic bare patch in pasture caused by this fungus (MacNish and Sweetingham 1993). In the rare cases where two ZGs occurred, they *never* coalesced and there was a distinct apparently fungus-free zone between them. Moreover, when one mycelial type died out, the area it had occupied was *not* colonized by the adjacent type, e.g. AG8(ZG1–1) and AG8(ZG1–5) (MacNish *et al.* 1993b). This phenomenon in soil resembles the dark demarcation lines recorded in wood at the boundaries of mycelia of *C. versicolor* [BA] carrying different SI genes (see Figure 2.1), i.e. akin to a barrage reaction. However, the potential diversity of reactions shown by *Rhizoctonia* [BA] spp. and the fact that different AGs may show different anastomosis patterns clearly indicates that such reactions should only be used with caution for population work with these species. The SI situation in similar taxa that have not received careful prior investigation of the system need to be investigated before SI can be used for genet discrimination.

### 2.3.2.3 Monokaryotic homogenic incompatibility in Basidiomycotina

In addition to the SI reactions commonly recorded, Kemp (1995) has proposed a *homogenic* incompatibility system that affects *monokaryons* only. By making a complex series of intercrosses between progeny of a small number of species of *Coprinus* [BA], notably the bispored bipolar diaphoromictic *C. bisporus* [BA], in the laboratory, he has shown that it can be detected. Kemp has also investigated the theoretical consequences of such a system in a tetrapolar diaphoromictic species and demonstrated how it could be detected. The consequences of this system for monokaryons within a population are relatively slight, with only a few of the possible monokaryotic fusions being prevented. It is not yet clear how significant this system is under natural conditions and, indeed, because of the complexity of the crosses involved to detect it, it is not very amenable to investigation. Under certain conditions it could lead to abrupt reproductive isolation in *C. bisporus* [BA] (see Chapter 12, section 12.3.4).

### 2.3.2.4 *Spatial extent of SI genets*

The numbers of genetically distinct genets within a given area is variable, being affected by the biology of the species concerned and the prevailing ecological conditions which, not surprisingly, may also affect the areal extent of each genet, as does its age. For example, the frequency of genets of *Armillaria* [BA] species, which are facultative pathogens of tree roots and stems, can range from as little as $1\,\text{ha}^{-2}$ to $50\,\text{ha}^{-2}$ (Worrall 1994), and their sizes tend to fall into a Poisson distribution. This range of sizes may perhaps reflect the age and fitness of different genets, the larger being older and perhaps fitter, although in some fungi, e.g. *Hebeloma cylindrosporum* [BA] (Guidot *et al.* 2001), size is apparently determined genetically. On the other hand, ecological conditions may also be significant since genet size differs in this fungus at inland and coastal sites (Gryta *et al.* 1995; Debaud *et al.* 1997).

In contrast the numbers of different genets of the oyster mushroom *Pleurotus ostreatus* [BA], a common decomposer of hardwoods, in an area of mixed deciduous forest totalled 56 from a sample of 60 dikaryons and the areal limit of each genet was more uniform and quite small, less than 1m in diameter in some cases (Kay and Vilgalys 1992). This is a not uncommon pattern of size distribution and has been interpreted as arising from the limited dispersal of basidiospores that give rise to comparatively short-lived mycelia.

### 2.3.2.5 *SI genetics*

The genetic basis for SI in Basidiomycotina is not well understood and has only been incompletely determined for a few basidiomycotous fungi. From the few results available, the numbers of loci controlling SI in Basidiomycotina seem to be fewer than in the Ascomycotina. In *Phellinus gilvus* [BA] only a single locus appears to be involved (Rizzo *et al.* 1995) and this may also be the case in *P. weirii* [BA] also (Hansen *et al.* 1994). From a preliminary analysis only two closely linked genes appear to control SI in *Armillaria ostoyae* (Guillaumin 1998). In *Heterobasidion annosum* [BA], pairing reactions of various kinds suggest that in this species SI is controlled by three or four loci acting independently and heterogenetically, i.e. on the basis of differences between multiple alleles at one locus at least (Hansen *et al.* 1993). Two kinds of SI interaction are shown by *Collybia fusipes* [BA]: strong, with a clear antagonistic zone bordered by heavily pigmented hyphae, and weak, with only lightly pigmented hyphae and minimal antagonism. The strong reaction appears to be determined by a single locus, the weak reaction by one to three loci. However, many quantitative discrepancies suggest that both reactions may be affected by modifying genes (Marçais *et al.* 2000). The situation is not unlike that found in *P. gilvus* [BA] which also shows a strong reaction controlled by one locus but with some quantitative variation as well. It will be evident that the genetic basis of SI in Basidiomycotina is not yet fully understood.

## 2.4 Pathogenic markers

The majority of fungi are not pathogenic; an appreciable number are capable of both a saprotrophic and a parasitic lifestyle, and only a few specialized groups are obligately parasitic. The interaction between potentially pathogenic fungi and their hosts is variable. Fungal isolations of the same species may differ in their **virulence**, i.e. their ability to infect particular host cultivars because of an interaction between specific genes which they carry and specific genes carried by the host—the so-called 'gene-for-gene' interaction. This situation is common in but not unique to obligate pathogens. Pathogens may also, or alternatively, differ in the degree of infection of the host as a result of a quantitative manifestation of the pathogen's fitness, i.e. its **aggressiveness**.

### 2.4.1 Virulence characters

Virulence characteristics of pathogenic fungi were utilized quite early to distinguish what were described as *formae speciales* (f.spp.), i.e. intraspecific groups of individuals restricted to a characteristic host, or small group of hosts. For example, several f.spp. were recognized amongst the black (or stem) rust of cereals and grasses, *Puccinia graminis* [BR]

(Eriksson 1894), namely *Puccinia graminis* f.sp. *tritici* [BR], individuals characteristically adapted to wheats (*Triticum*) but capable of weakly infecting barley and many wild grasses, *P. graminis* f.sp. *avenae* [BR], adapted to oats (*Avena*), *Dactylis glomerata* and a few wild grasses differing from those infected by the f.sp. *tritici* and *not* occurring on wheat, and *P. graminis* f.sp. *secalis* [BR], characteristically adapted to infecting rye.

Further subdivisions followed when **physiological races** or **pathotypes** were recognized within each *forma specialis* (Stakman 1914; Stakman and Levine 1922). A pathotype (often called a 'race' in the pathology literature) includes all those individuals that show the same reaction pattern when inoculated onto a selected group of host cultivars described as 'differentials'. Therefore pathotypes can be distinguished by their differing reaction patterns. In essence, this is a means of classifying fungal variation in terms of pathogenic phenotypes, but it is important to recognize that for each pathotype the *only* character in common, albeit a phenotypically complex one, is its reaction pattern on the selected cultivars. In other features—effectiveness of infection, growth rate, time to sporulation, intensity of sporulation, spore germinability etc.—there can be, and is, variation between individual isolates. Thus a group of isolates by virtue of their specific virulence genes may be classified as belonging to the same pathotype or race but may differ amongst themselves in their aggressiveness. Pathogenic success is, of course, determined by both these attributes. Variants of an original pathotype are formally distinguished when they are biologically or practically significant, e.g. recognizable by a field pathologist, or persisting for an appreciable period. For example, two variants of *P. graminis* f.sp. *tritici Race 15* [BR], namely *15A* and *15B* were recognized. *Race 15* was the original race, defined by its virulence pattern and recognized in the USA, *15A* was identified in Japan, but differed in the size of its infection pustules, while *15B* was recognized later, in 1937 in the USA, because it differed in its virulence to some tester wheats from the original race. In fact, *Race 15* eventually virtually disappeared from the American wheat fields but from 1950 was replaced, especially in Canada, by *15B* (Stakman and Harrar 1957). The sampling

and identification of races of cereal rusts over successive seasons in the USA and Canadian wheat belts has demonstrated population fluctuations in respect of pathogenicity which are often correlated with the predominant cultivars grown (Figure 2.3) (Stakman and Harrar 1957).

From the outset, sampling wheat fields throughout a season or over several seasons enabled changes in the pathogen population to be described in terms of its pathotype composition. The sensitivity of the procedure to detect new pathotypes is, of course, determined by the range of isolates sampled and the differential cultivars chosen for testing. At first, differentials were selected in the light of local or regional conditions, but, as the widespread distribution of some major pathogens was recognized, the selection of differentials became a matter of continental or international agreement. Moreover, as specific host resistance genes were identified and located chromosomally, differentials were classified and selected on the basis of the specific resistance genes they carried. Thus it is now possible to describe pathotypes of *Puccinia graminis* f.sp. *tritici* [BR], for example, in terms of their complementary specific virulence attributes on a worldwide comparative basis. This has enabled matters such as migration of existing pathotypes, their mutation or the origin of new pathotypes to be investigated with greater precision.

### 2.4.1.1 Virulence genetics

Sometimes the genes involved in the gene-for-gene relationship need to be identified. Flor (1942) first analysed the relationship between different cultivars of flax (*Linum usitatissimum*) and different isolations of flax rust (*Melampsora lini* [BR]). Single loci were involved and avirulence was shown to be dominant to virulence in this fungus, a situation found in many crop pathogens. Further work showed that multiple alleles are sometimes involved. Host resistance genes are usually dominant. Unfortunately, several exceptions to this simple situation are known (Christ *et al*. 1987). For instance, avirulence can be masked as a consequence of epistasis, more than one locus may be involved and, in dikaryotic fungi, heterokaryons cannot always be distinguished from homokaryons. Nevertheless, a number of loci and alleles have

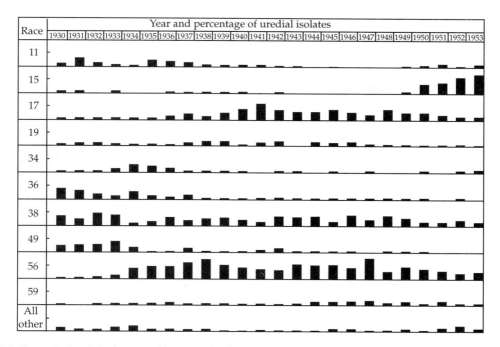

**Figure 2.3** Changes in the relative frequency of important physiological races (pathotypes) of *Puccini graminis* f.sp. *tritici* [BR] in the USA from 1930 to 1953. The height of each row is equivalent to 100 per cent presence. (Based on Stakman and Harrar (1957).)

been identified and located which provide powerful tools for describing pathogen populations and their changes in response both to cultivars and over time, e.g. in Barley mildew *Blumeria graminis* f.sp. *hordei* [A] (Limpert *et al.* 1990).

On the whole this technique has been confined to horticultural or agricultural crops and their pathogens, but a gene-for-gene situation has been shown to exist in a few natural populations of wild plants, e.g. *Erysiphe fischeri* [A] on *Senecio vulgaris* in the UK (Clarke 1997) and natural flax rust *Melampsora lini* [BR] on an arboreal native wild Australian flax (*Linum marginale*) (Burdon and Jarosz 1991, 1992; Jarosz and Burdon 1991, 1992). It may well be that gene-for-gene relationships are widespread in natural populations of wild plants (see Chapter 9, section 9.4.2).

### 2.4.2 Aggressiveness

Differences in aggressiveness are widespread amongst pathogenic fungi and often account for new epidemics; for example the cause of the current European epidemic of Dutch elm disease,

*Ophiostoma novo-ulmi* [AP], is appreciably more aggressive than *O. ulmi* [AP], the cause of the earlier epidemics of the 1920s and 1930s. Differences in aggressiveness are usually measured by means of a metric character such as, in this case, percentage defoliation of infected elm saplings over a set period. Therefore 'aggressiveness' can be treated like any other quantitative character. Table 2.3 illustrates how different measures of pathogenicity of *F. oxysporum* f.sp. *lycopersici* [FA] to tomato (*Lycopersicum esculentum*) can be treated and how such information can be used to describe isolates or even populations.

## 2.5 Fungicide resistance as a marker

Many pathogenic fungi, especially those infecting crop plants can, or have, developed resistance to various fungicides, often on a single-gene basis. Where resistance is not spread throughout the species, or where resistance has developed at different loci, such genes can potentially be used as markers. Obviously, they are of particular value

**Table 2.3** Aggressiveness shown by *Fusarium oxysporum* f.sp. *lycopersici* [FA] to different tomato cultivars

| Morphology of strains[a] | Average disease severity[b] | |
| --- | --- | --- |
| | cv. Bonny Best | cv. Marglobe |
| Fluffy | 10.4 | 7.3 |
| Fluffy–sclerotial | 8.7 | 6.3 |
| Intermediate fluffy | 8.3 | 4.5 |
| Intermediate appressed | 6.3 | 4.1 |
| Appressed/slimy | 4.7 | 3.0 |

[a] In culture, single spore cultures fell into the five broadly defined morphological types.
[b] Disease severity was meaured on an arbitrary scale from 0 (no disease) to 15 (severe disease).
Adapted from Burnett (1975).

for studies of the response of pathogen populations to fungicide treatments. For example, major gene resistance to benzimidazoles is widespread amongst Eumycota and the relatively recently developed resistance to metalaxyl by *Phytophthora infestans* [O] is associated with increased fitness (Kadish and Cohen 1988). Some fungicide resistance is more complex genetically; for instance, multigenic control is believed by some to be involved in resistance to ethirimol by *Erysiphe graminis* [A] (Bent 1982) but, because of the genetic complexity and range of phenotypic responses, this is less useful in analysing natural populations.

## 2.6 Cytological markers

Cytological markers have not been used very often in fungi and, indeed, fungal cytology has been neglected when compared with the attention given to other eukaryotes (Sansome 1987). There are two reasons for this: first, the small size of fungal chromosomes and their general inability to be resolved clearly at mitosis; second, their frequent inability to stain readily with standard cytological stains employed with eukaryotes, a disadvantage that has now been overcome in some species. However, in the last decade, the small size of fungal chromosomes has been put to good use since it enables them to be separated electrophoretically as DNA molecules. Therefore classical chromosome cytology has only limited uses with fungi and, increasingly,

electrophoretic chromosome profiles are being employed. Both methods have their drawbacks.

### 2.6.1 Chromosomal techniques

Chromosome numbers of fungi have been recorded from the late nineteenth century, often on sectioned ascal or basidial material. However, they could not be relied on until effective squash methods were introduced for the ascomycotous *Neurospora* [A] in the 1940s (McClintock 1945), when reciprocal translocations were first observed in a fungus. Polyploidy was first demonstrated in *Allomyces* [C] (Wilson 1952), which was shortly thereafter shown to include two autopolyploid series (Emerson and Wilson 1954). A decade later, diploidy and polyploidy were demonstrated in the Straminipila in *Pythium* and *Phytophthora* [both O] (Sansome 1961, 1965).

The methods developed for meiotic chromosomes of *Neurospora* [A] spp. have been applied, with considerable trial and error, to other filamentous Asco- and Basidiomycotina. The best results have been obtained using aceto-orcein or a modification of iron–haematoxylin, often after prior fixation in ethanol–acetic acid/lactic acid, e.g. *Neurospora* [A] (Barry 1966; Raju and Newmeyer 1977; Raju 1978). However, to date, yeasts and Zygomycotina have proved to be quite intractable and few Chytridiomycotina have been examined. In some cases, chromosome numbers are underestimated since very small chromosomes, which are not uncommon in fungi, cannot be resolved or detected (see section 2.6.2).

Ultrastructural observations of meiotic pachytene have also been used for chromosome counts. The first reliable cytological count of the chromosomes of baker's yeast *Saccharomyces cerevisiae* [A] was obtained by the laborious method of examining ultra-thin sections of meiotic cells electron microscopically and counting the numbers of synaptonemal complexes (Byers and Goetsch 1975). Another method applied to some rusts has been first to detect the pachytene stage of meiosis after treatment with an epifluorescent dye such as mithromycin or 4–6-diamidino-2-phenylindole (DAPI) and then to section it ultrastructurally. This has given much higher chromosome numbers than had been obtained from the classical squash technique, e.g. $n = 18$ for *Puccinia graminis* f.sp. *tritici* [BR] (Boehm *et al.* 1992) compared with $n = 3$ using a squash technique (McGinnis 1956). Mitotic chromosomes at division and those of the Fungi Anamorphici, which lack a meiotic stage, are virtually unstudied. However, by taking advantage of the large numbers of mitotically dividing nuclei in the germ tubes of germinating spores or conidia it has proved possible to obtain counts from burst germ tubes using the DAPI technique. Fungal chromosomes are rarely aligned clearly on a metaphase plate, but by using this method they are separated and spread over the surface of a slide so that counts are possible, e.g. *Botrytis* [FA] (Shirane *et al.* 1989), *Alternaria* [FA] (Taga and Murata 1994) and *Fusarium solani* [FA] (Taga *et al.* 1998). However, it seems probable that electrophoretic methods are more likely to be applied in future despite their shortcomings (see section 2.6.2).

Translocations, inversions (mostly pericentric), insertions and duplications have been detected cytologically but more frequently identified through abnormal segregation of genetic markers, often in stocks exposed to mutagenic treatments such as X-irradiation (Käfer 1977; Perkins and Barry 1977). Spontaneous chromosomal aberrations have been recorded sporadically in fungi but have only rarely been investigated, e.g. *Sordaria fimicola* [A] (Cox and Parry 1967), apart from those known in *Neurospora* [A] (Perkins 1997). Cytological aberrations appeared to be infrequent in extensive collections of *Neurospora* [A] species from the wild, with only four reciprocal translocations having been recorded from over 4000 isolations, i.e. only 0.1 per cent (Perkins and Turner 1988). This kind of frequency seems to be true generally throughout the filamentous fungi, at least (Perkins 1997).

Polyploidy appears to show a sporadic distribution amongst both the Eumycota and the Straminipila, although this may simply reflect present ignorance. The general significance of polyploidy for the fungi is not known (Rogers 1973) (see also Chapter 8).

Only a few cases of polyploidy, such as those already mentioned, have been detected using classical cytological methods. Most examples of putative polyploidy in fungi are less well documented than any of these. They depend largely on comparisons of the depth of staining of whole non-dividing nuclei under constant controlled conditions using, most recently, DAPI which binds preferentially to regions rich in adenosine and thymine, e.g. in *Pleurotus* [BA] (Bresinsky *et al.* 1987). An essential condition for this technique is that the amount of repetitive DNA in different isolates is comparable. At its best, this technique is indicative and provides little information concerning the degree of polyploidy.

At present, less than 200 reliably determined chromosome numbers are known from the fungi, although that number could be increased by a systematic study of meiosis in Ascomycotina and Basidiomycotina using proven squash techniques. Nevertheless, the inherent and technical difficulties in detecting chromosomal features in fungi by classical cytological methods renders them of little use as genetic markers in most population studies. This accounts for much of the growing use of electrophoretic methods, which provide limited but more readily accessible data more easily.

### 2.6.2 Electrophoretic chromosomal techniques

The cytologically intractable chromosome number of Baker's yeast *Saccharomyces cerevisiae* [A], $n = 16$, was one of the first determined using pulse-field gel electrophoresis (PGFE) (Carle and Olson 1984, 1985). This technique separates DNA molecules in agarose gels by subjecting them to electric currents that alternate in direction. Chromosome resolution depends on the fact that the time taken for a DNA molecule, i.e. a whole chromosome, to change direction is proportional to its size; large DNA molecules take longer to reorientate than small ones. Hence the duration of the pulse and the size of each chromosome is critical in separating them. Chromosomes which can be separated by this means range in size from 170 to 4000 kb. Large chromosomes (2–3 Mb or more) require quite low voltages and prolonged electrophoresis times. For instance, the pin mould *Absidia glauca* [Z] which has some large chromosomes, one of size 7 Mb, requires $1.7\,\mathrm{V\,cm^{-1}}$ for periods as long as 10–14 days, and higher voltages result in the degradation of such large chromosomes (Kayser and Wöstenmeyer 1991). Frequently, only the smaller chromosomes are detected by this method; for example, PGFE will not resolve the two largest chromosomes of *Neurospora* [AP] (Orbach *et al.* 1988) (see 2.6.3).

Originally, chromosomal material was separated enzymatically from protoplasts but this has proved unnecessary. Now, fungal cells, or macerated hyphae, are first suspended in isotonic medium to give small unclumped fragments and are then resuspended and spun to form a pellet before embedding in molten agarose. Plugs from this are digested with proteinases in sodium dodecyl sulphate and EDTA at 50 °C for 24 hs to degrade DNA-binding and other proteins, inhibit nucleases, and destroy cell membranes prior to electrophoresis (McCluskey *et al.* 1990).

Commercial systems most often use a modification of transverse alternating field electrophoresis (TAFE) which has some drawbacks; in particular, the linear separation is not uniformly proportional to the size of the DNA molecule and the bands are often curved.. The preferred modification, contour-clamped homogenous electric field (CHEF), has 24 electrodes arranged hexagonally with orientation angles of 120° or 60°. This ensures that the bands of separated molecules are straight, not curved, and are at right angles to the long axis of the gel, and that their mobility is truly proportional to their size (Chu 1989). Recent summaries of the methodology include those of Mills and McCluskey (1990) and Skinner *et al.* (1991), and a list of those fungi for which such chromosome data are available to date is given by Zolan (1995).

Band number is related to the chromosome number. Variation in band size or density can be due to two chromosomes of very similar size occupying the same site, aneuploidy or polyploidy (both diploidy or higher ploidy), and variations in length can be due to gross chromosomal rearrangements, e.g. aneuploidy in yeast (Bakalinsky and Snow 1990), polyploidy in *Phytophthora megasperma* [O] (Howlett 1990) or translocations in maize smut *Ustilago maydis* [BS] (Kinscherf and Leong 1988). Because chromosomes of the same size, regardless of their genotype, will co-migrate to the same site, it is best to test for this and to identify specific loci on each chromosome by using known probes derived from fragments obtained by either RFLP or RAPD techniques (see Chapter 3). If the bands have

been stained with ethidium bromide, as is usual, their density can be determined, and this often assists in detecting two coincident chromosomes. Another useful technique is to construct plasmids, each carrying a DNA fragment which can be used as probes with the DNA gels in the same way as with RFLPs (see Chapter 3, section 3.3.1), as has been done with *Septoria tritici* [FA] (McDonald and Martinez 1990b).

Alternatively, chromosomal gel bands can be eluted individually and used for preparing RFLPs or RAPDs which can be probed as just described and thus the gene loci can be identified. Since the procedure can be carried out on parental hyphal material together with that from individual basidiospore tetrads or ascospores, both chromosomal behaviour and genetic segregation can be followed with precision, e.g. *Leptosphaeria maculans* [AP] (Plummer and Howlett 1995).

Population studies are also possible. Chromosome length mutations have already been applied to analysing populations of *Septoria tritici* [FA] in the field (McDonald and Martinez 1991a) and to the somewhat puzzling chromosomal changes which occur during human infections by *Candida albicans* [A] (Zolan 1995).

Electrophoretically identified karyotypes usually remain remarkably stable through successive mitotic stages, but in some yeasts changes have been detected after prolonged rounds of mitoses. For example, some morphological changes became detectable after 55 successive divisions of a near-wild diploid oenological strain of *S. cerevisiae* [A]. After 275 divisions, 15 out of a sample of 30 cells examined had major chromosome length changes although this had no obvious phenotypic effect except an inability to form viable ascospores. A haploid yeast treated in the same way showed no detectable chromosomal changes after 412 divisions (Largo and Vezinhet 1993). Therefore it was suggested that the changes in the diploid arose from rare mitotic recombination. Dramatic changes in lengths and numbers in the meiotic products are far more frequent, e.g. the basidiomycotous yeast *Filobasidiella* (*Cryptococcus*) *neoformans* [BT], industrial strains of *S. cerevisieae* [A] (Boekhout and van Belkum 1997; Codón *et al.* 1997) and *Leptosphaeria maculans* [AP] (Plummer and Howlett 1993). These are often the inevitable cytological consequences of pairing of unequal chromosomes, or of those carrying translocations, deletions or insertions. In some cases (e.g. yeast), they are due to mutational events caused by transposable elements, such as *Ty1* or *Ty2*, which undergo inter-chromosomal translocations, or multiple *Y'* subtelomeric regions which give rise to asymmetrical homologous recombination or both effects (see Chapter 6, section 6.2.2).

A further novel discovery has been that of very small supernumerary B chromosomes, or 'dispensable' chromosomes, which almost certainly could not be resolved by light microscopy. An example is the small chromosome (1600 kb) in *Nectria haematococca* [AP] which appears to carry the gene for pisatin demethylation activity, rendering those strains which carry it highly pathogenic to peas (Van Etten *et al.* 1994). This B chromosome is somewhat exceptional in having a known function associated with it; the majority appear to be functionless. Putative B chromosomes are now known to occur in a number of fungi, based mainly on size and irregular numbers in different strains, but a possible source of confusion with them is the presence of linear plasmids which occur in many fungi and are of comparable size (see Chapter 6, section 6.2.5).

Electrophoretic chromosome data can clearly provide a further kind of useful genetic marker for fungi, particularly if they can be combined with data derived from RFLPs when specific loci can be located on particular chromosomes.

## 2.6.3 Karyotyping procedures

There is no single procedure that provides complete information on the size, numbers or morphology of fungal karyotypes. This is clearly demonstrated by a comparison of classical cytological methods with the germ-tube burst method and PGFE, using *Nectria haematococca* [AP] ( = *Fusarium solani* [FA]) (Taga *et al.* 1998). Despite excellent resolution of the numbers and morphology of stained meiotic chromosomes, their number was underestimated when compared with counts on PGFE material. On the other hand, the latter method failed to resolve chromosomes larger than about 6 Mb and uncertainty arose when chromosomes of similar size

co-migrated. However, PGFE did enable chromosomal polymorphisms to be detected with greater precision. The germ-tube burst method provided better counts than either meiotic preparations or PGFE for this fungus and reasonable information on chromosome morphology. Whether it would be possible to resolve B chromosomes by this latter method is not clear. In future a combination of the germ-tube and electrophoretic methods seem most likely to provide the maximum amount of information most economically for Ascomycotina and Fungi Anamorphici; these are the only procedures available for the latter. It still remains to be demonstrated whether the germ-tube method is effective with Chytridiomycotina, Zygomycotina or Basidiomycotina.

## Further reading

The literature on MTFs is developing rapidly as a consequence of the application of molecular approaches. The following are useful summaries.

Glass NL and Kuldau GA (1992). Mating type and vegetative compatibility in filamentous Ascomycetes. *Annual Review of Phytopathology*, **30**, 201–4.
Kües U and Casselton LA (1992). Fungal mating type genes—regulators of sexual development. *Mycological Research*, **96**, 933–1006.
Raper JR (1966). *Genetics of sexuality in higher fungi.* Ronald Press, New York.

*Vegetative incompatibility in Ascomycotina is reviewed by:*
Glass NL and Kuldau GA (1992). Mating type and vegetative compatibility in filamentous Ascomycetes. *Annual Review of Phytopathology*, **30**, 201–14.
Leslie JF (1993). Fungal vegetative compatibility. *Annual Review of Phytopathology*, **31**, 127–50.

*Vegetative incompatibility in Basidiomycotina is reviewed by*:
Rayner ADM and Todd NK (1979). Population and community structure and dynamics of fungi in decaying wood. *Advances in Botanical Research*, **7**, 333–420.
Worrall JJ (1997). Somatic incompatibility in Basidiomycetes. *Mycologia*, **89**, 24–36.

*Pathogenicity and the gene-for-gene relationship are admirably reviewed by*:
Crute IR, Holub EB and Burdon JJ (1997) *The gene-for-gene relationship in plant–parasite interactions.* CAB International, Wallingford.

Day PR (1974). *Genetics of host–parasite interaction.* W.H. Freeman, San Francisco, CA.

*Fungicide resistance is reviewed by*:

Heaney S, Slawson D, Hollomon DW, Smith M, Russell PE and Parry DW (1994). *Fungicide resistance.* British Crop Protection Council, Farnham.

*Meiotic chromosome cytological methods are reviewed by*:

Mills D and McCluskey K (1990). Electrophoretic karyotypes of fungi: the new cytology. *Molecular and Plant–Microbe Interactions*, **3**, 351–7.

Raju NB (1978). Meiotic nuclear behaviour and ascospore formation in five homothallic species of *Neurospora. Canadian Journal of Botany*, **56**, 754–63.

# CHAPTER 3

# Genetic markers for population studies—II Molecular markers

A wide range of natural molecules exist such as particular secondary metabolites like the lichen acids which are produced by the fungal component, the whole protein content or specific polymorphic enzyme proteins. Unfortunately, characteristic species-specific molecules only occur sporadically amongst the fungi so that their use as markers is limited, although they are often characteristic of higher groups such as genera, sections or families.

A molecular marker, as now generally understood and as used here, is confined to a discrete difference in the genetic information possessed by two individuals within a species that is *detected by molecular methods*. Such markers can be detected in various ways—indirectly by assaying a primary or secondary product of a gene, or directly by analysing DNA or some specific part of it. Amongst the former, polymorphic enzyme proteins are the most common and most useful naturally occurring specific markers and are still widely used. However, since 1980, with the rapid technical advances in semi-automated separation and isolation of DNA sequences, whether from nuclear DNA (nDNA) or mitochondrial DNA (mtDNA), nucleic acid markers have become available. A wide range of possibilities exists: whole chromosomes, extracted nDNA, particular sequences representing whole functional regions such as the ribosomal DNA (rDNA) gene, shorter lengths associated with a marker gene or arbitrarily selected lengths. Double-stranded RNA (dsRNA) has also had some limited uses, as have plasmids. No single type of DNA marker is ideal, but many possess valuable attributes such as being of frequent occurrence, highly polymorphic, evenly or randomly distributed within the genome, selectively neutral or lacking epistatic effects. Many are co-dominant in expression, enabling the homozygous state to be distinguished from the heterozygous state in diploids or dikaryons. Technically, nucleic acid markers are often amenable to fast and easy assay with high reproducibility. Many are capable of permanent storage, thus permitting access long after they were originally sampled. DNA techniques have the added advantage that nucleic acids can often be extracted from herbarium or preserved specimens, enabling comparisons to be made between contemporary and much older material (Bruns *et al.* 1990). A further benefit of DNA technology, already described in Chapter 2, is the ability to treat chromosomes as large DNA molecules and separate them electrophoretically. In addition, chromosome electrophoresis, combined with other nDNA molecular markers, enables the chromosomal location of markers to be determined as well as the structure and location of chromosomal aberrations.

Enzyme and DNA markers are increasingly the markers of choice because of the following drawbacks shown by other markers for the majority of fungi:

- a lack of large numbers of identified and well mapped genes
- a lack of reliable chromosome cytology
- a lack of knowledge of the sexual phase
- difficulty in determining the genetic constitution quickly and reproducibly in culture.

## 3.1 Specific molecules

Various molecules have been employed as adjuncts to identification in different groups, such as secondary metabolites in the Xylariaceae (Whalley and Edwards 1995), but the best known and most frequently used are the 600 or more lichen substances (Culberson and Elix 1989), including

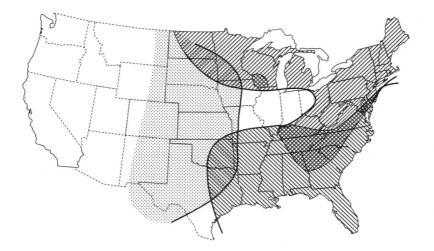

**Figure 3.1** The distribution of the three chemotypes of the lichen *Parmelia plitii* [AL] in the USA: diagonal lines sloping left, stictic acid form; stippled area, salacinic acid form; diagonal lines sloping right, fumar-protocetraric acid form. (Based on Hale (1974).)

the remarkable lichenic acids, notably the depsides and depsidones. They are produced by the fungal lichen partner, most abundantly in the presence of the specific algal partner but also, in some cases, by the mycobiont in pure culture (Hamada 1989). Many are characteristic either of a particular species or of populations within a species, e.g. stictic, salacinic and fumaroprotocetraric acids in *Parmelia plitii* [AL] in the USA (Hale 1974) (Figure 3.1).

Other molecules which have been used as general or specific markers for different fungi are comprehensively reviewed by Frisvad *et al.* (1998).

## 3.2 Proteins and polymorphic enzymes

Extracts of total soluble whole-cell proteins from different isolates have been compared using starch gels or polyacrylamide gel electrophoresis (PAGE) either as linear gels, e.g. *Verticillium* spp. [FA] (Milton *et al.* 1971), in the presence of sodium dodecyl sulphate as a denaturant or, less commonly, by two-way electrophoresis. The gels are often stained with Coomassie blue. While two-way electrophoresis is of some use in distinguishing different species, e.g. *Phytophthora* [O] spp. (Brasier 1991), *Endothia gyrosa* [AP] and *E. (Cryphonectria) parasitica* [AP] (Stipes *et al.* 1982), or subspecific entities, e.g. ecotypes or races of *Gremmeniella abietina* [AP] (Petrini *et al.* 1990), the most effective

method for making comparisons between isolations is the linear electrophoresis of a range of specific polymorphic enzyme proteins.

Two classes of enzymes are available, originally defined as **isoenzymes (isozymes)** and **alloenzymes (allozymes)**. The former are different molecular forms of the same enzyme, usually encoded by different loci; the latter are different forms of an enzyme determined by different alleles of a gene. Regrettably, this distinction is often neglected and the terms are now used synonymously to describe different forms of an enzyme determined by different alleles of a single gene. The differences in the enzyme proteins are basically in their amino acid composition and conformation. Their quarternary structure is determined by the cohesive forces binding their polypeptide chains, and these are affected by forces such as hydrogen bonding, disulphide bridges and hydrophobic interactions. The net effects of these various attributes is to affect their size, shape and net charge, which affects their mobility in a porous gel under an electric field. Positively charged proteins migrate to the cathode and negatively charged ones to the anode. Thus they can be separated by the direction and speed with which they migrate through starch (zymograms) or acrylamide gels. After suitable specific staining they appear as individual bands or, if the enzymes have previously been radioactively labelled, they can be separated by autoradiography.

A number of general methods are available (Tanksley and Orton 1983; Hames and Rickwood 1990) and several of those used with fungi have been listed (Micales *et al.* 1986; Stasz *et al.* 1988). In practice, to obtain the best discrimination, modifications such as the buffer chosen, its pH, the value of the current, the duration of electrophoresis and the protein stain employed often need to be determined for each species. A considerable number of enzymes have been investigated in this way, e.g. *Trichoderma* [FA] spp. (Stasz *et al.* 1988) where 63 enzymes were assayed. A far smaller number (such as five or eight, or even fewer) is more usual, either because only a minority show clear and reproducible separations or because the number of enzyme systems which can be used in any one species is often limited.

One advantage claimed for isozyme electrophoresis is that since most of the enzymes investigated are regulatory, they are presumably highly conserved. Therefore their variation is indicative of major evolutionary events, such as speciation, so that the technique is most useful for investigating taxonomic relationships. It has certainly been used frequently for this purpose.

Enzyme electrophoresis suffers from some disadvantages in genetic studies.

• Enzyme expression can be affected by the conditions under which the fungus is growing and the age or region of the mycelium sampled (Franke 1973), but these factors can be controlled.
• Enzymes represent *transcribed* nDNA *only*, and so they give no information about non-transcribed regions.

It is often claimed that because of the limitations of the genetic code only *one-third* of amino acid substitutions can be detected electrophoretically and so the total variation may be *underestimated*. However, this is not a serious restriction, since the principal concern is to normally obtain a *relative* measure of the variability under investigation.

The usefulness of enzyme electrophoresis depends on the number of loci detected and the complexity of the banding pattern which has to be interpreted in terms of the genetic constitution of each isolate and the structure of the protein. Incidentally, it needs to be borne in mind that the

subunit structure of fungal enzymes is not necessarily the same as that of similar enzymes in other organisms (Darnall and Klotz 1972). For example, hexokinase is usually considered to be monomeric but appears to behave as if it is dimeric in *Uromyces appendiculatus* [BR], although the presence of a duplicated locus cannot be excluded entirely (McCain and Groth 1989). In a haploid homokaryotic fungus, *single* bands are detectable for *each* isozyme regardless of the subunit structure of the protein—monomeric, dimeric, or tetrameric—provided that the allele produces a functional protein. Of course, a non-functional null allele will not be represented at all. In contrast, the patterns can be more complex in a haploid, heterokaryotic, or dikaryotic fungus and in diploids, depending on both the subunit structure of the enzyme and whether the fungus is homozygotic or heterozygotic at the locus involved (Figure 3.2).

A monomeric protein is represented by one of two possible bands in a homozygotic isolate, or by

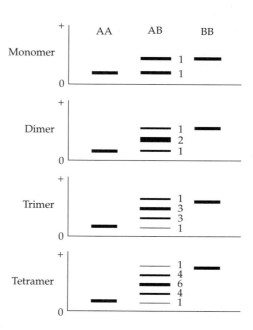

**Figure 3.2** Isozyme banding patterns. The diagrams illustrate the banding patterns expected from a single locus for homozygous and the heterozygous isolates whose enzyme proteins differ in the number of subunits involved. The homozygous patterns are also those expected from haploid monokaryons. (Reproduced from Murphy *et al.* 1996), with permission from Sinauer Associates Inc.)

both bands if, for example, the isolate is hetero-zygous. However, a dimeric protein can produce three different dimers AA, BB and AB, with the last being a heterodimer which will be represented by a distinct intermediately positioned band in a hetero-zygote. Similarly, for a tetrameric protein of two polypeptides, five tetramers are possible—the homodimers $A_4$ and $B_4$ and the three heterodimers $A_3B_1$, $A_2B_2$ and $A_1B_3$. It can be seen that if more than two loci are involved, the banding pattern could become even more complex.

Null alleles also alter the pattern. For example, if a null allele is involved in a dimeric protein, only two bands will be identifiable, those of the active and the hybrid dimer; the latter occurs at the same site as the former but is usually somewhat thinner or less dense. Null alleles normally occur only when loci are duplicated in haploids or in poly-ploids, e.g. esterases and phosphoglucomutases of *Uromyces appendiculatus* [BR] (McCain *et al*. 1992).

The inferred genetic bases just described have not often been confirmed by controlled crosses but this does not prevent the patterns being utilized solely as patterns, which can then be compared statistically for their similarities or differences. This is not a serious drawback, however, for if the mode of inheritance of the variation is understood it can often be confirmed from naturally produced pro-genies and population genetic analysis can be applied (see Chapter 5).

## 3.3 Nucleic acids

The use of DNA markers has developed, and is still developing, at an immensely rapid pace. The earliest methods depended upon isolating DNA in appreciable quantities and comparing intact DNA molecules in various ways (section 3.3.1). Of course, whole DNA molecules are not of particular value as markers in population studies, apart from individual chromosomes (see Chapter 2, section 2.6.2), but they are of some value in distinguishing or assessing relationships of higher taxa. On the other hand, particular regions may be of value for comparisons, for sequencing and for use as markers in populations. The introduction of restriction endonucleases to produce nucleotide

fragments of different lengths—restriction frag-ment-length polymorphism (RFLP)—pioneered this development (section 3.3.5.2). The fragments were isolated and separated by electrophoresis on appropriate gels.

These methods are both laborious and time consuming and only possess limited analytical properties, and they have been almost entirely replaced by more rapid methods requiring far smaller amounts of the initial DNA. These methods depend on the amplification of selected sequences through the use of the polymerase chain reaction (PCR) and its successive variants. This technique also enables the nature of the sequences selected for amplification to be determined to some extent by the PCR procedure adopted. The amplified products are separated on their size or mobility by electro-phoresis on a variety of gels designed to react most effectively with the kind of amplified product produced (section 3.3.2). A further development from the PCR technique, the random amplified polymorphic DNA (RAPD) technique (section 3.3.3) in which synthetic primers of arbitrary base sequences are used to amplify products from DNA, is now one of the most widely used contemporary techniques (Williams *et al*. 1990). It has a number of variants designed for specific purposes. A notable development depends on the detection of micro-satellites—short-sequence repeats (SSRs)—that are both frequent and ubiquitous in animal, plant and fungal genomes. This technique is being used increasingly. In addition, the amplified fragment-length polymorphism (AFLP) technique (section 3.3.4), in effect a combination of the RFLP and RAPD techniques, has been shown to have addi-tional advantages over both. There are a variety of ways of analysing the products obtained by either fragmenting whole DNA or amplifying shorter sequences of it (section 3.3.5). Each method has its advantages and disadvantages and differs in its suitability for different purposes, so that selection needs to be exercised depending on the use being made of the genetic analysis of DNA (section 3.4).

In general, the effectiveness of all these methods depends upon the extraction of pure uncontamin-ated DNA with the minimum of damage and pre-ferably in high yield. Several reliable methods now exist for isolating total DNA, nDNA, mtDNA or even partially degraded DNA in larger or smaller quantities and for separating them, when necessary.

Isolation of DNA from fungi is usually most effective from hyphae/mycelium in the logarithmic growth phase in liquid culture or from spores. After breakage of the cell wall the contents are usually lysed in an appropriate buffer according to various protocols, e.g. general (Garber and Yoder 1983), for nDNA (Lee *et al.* 1988; Weising *et al.* 1994), or for mtDNA (Hudspeth *et al.* 1980; Bruns *et al.* 1988). It is highly desirable to ensure that the DNA preparation is as uncontaminated as possible, especially with mtDNA. A common cause of contamination of DNA samples in fungi is the presence of polysaccharides and polyphenols, but this and other causes of poor yields, such as fragmentation and degradation, can be avoided by using a CsCl technique (Yoon *et al.* 1991). To avoid contamination from cell wall material, some workers start the procedure from protoplasts. DNA isolation from lichens or the fungal component of mycorrhizal roots needs great care since contamination from the root or algal-partner DNA needs to be avoided, and metabolites, such as polysaccharides or polyphenols, can both contaminate and interfere seriously with DNA purification.

Since total DNA is most usually extracted, the mtDNA, which only forms a very small fraction of the total, can be separated after extraction, although with a chance of contamination from total cell DNA. Pure mtDNA is best obtained by separating mitochondria from cell contents by ultracentrifugation prior to DNA extraction, and then isolating the mtDNA from them. Mitochondrial DNA exists in several copies in each mitochondrion as small circular DNA molecules ranging from 27 to 115 kb in Ascomycotina and from 43.3 to 174 kb in Basidiomycotina. Those of the Straminipila are intermediate, ranging from 36.2 to 57 kb. In the few fungi investigated, the structural arrangement of the DNA is variable and rearrangements of gene order are common and frequent. Of course, it is important to remember that normally, although not exclusively, mitochondria are inherited maternally and that they do not always migrate from one mycelium to another even though a heterokaryotic mycelium may develop.

Whole DNA is not easy to manipulate so that, for example, comparison of different DNA molecules can only be achieved, ideally, by sequencing the whole DNA complement. Although this provides a reliable measure of DNA similarities and differences, it is a very long and daunting task even for the much smaller mtDNA molecule. To date, the only complete fungal sequence that has been obtained is for *Saccharomyces cerevisieae* [A] (Feldman 1999). Therefore for rapid practical purposes the application of sequencing, or size measurements, to the comparison of shorter lengths is a much more practicable approach (see section 3.3.2.1). The first method to be developed was to partition the DNA into shorter sequences using restriction endonucleases producing fragments of different sizes—restriction fragment-length polymorphisms (RFLPs)—but this has been largely replaced by the use of specific amplified fragments obtained by PCR and its variants. With their introduction, attention can and has been directed to analysing particular regions of the genome. Mitochondrial DNA has been targeted in fungi but usually together with more comprehensive analysis of nDNA and has not been exploited to the same extent as in some diploid eukaryotes. In fungi, particular attention has been paid to various nDNA regions, notably the nuclear ribosomal region (rDNA) and specific regions within it such as the internal transcribed spacers (the ITS1 and ITS2 regions), the intergenic spacer (the IGS region), the 17 S, 5.8 S, 25 S and 5 S genes, and the region including the ITS1–5.8 S–ITS2 regions, often simply referred to as the 'ITS region'. Attention is increasingly being given to particular genes, notably the $\beta$-tubulin locus and a number of loci determining specific enzymes, e.g. chitin synthases and laccase. Microsatellites, which have become increasingly popular for genetic analysis in many diploid eukaryotes, have only just begun to be investigated in a few fungi but no doubt their use will increase.

Finally, it should never be forgotten that all these procedures involve the possibility of contamination by extraneous DNA during processing, whatever the material. Plastic gloves should be worn routinely and, indeed, this also protects an investigator against reagents such as polyacrylamides which are cumulative neurotoxic agents.

### 3.3.1 The use of whole DNA molecules

Once obtained, whole DNA, whether mtDNA, nDNA or total DNA, can be subjected to a variety of analytical techniques, the choice of which reflects the nature of the information required. Basically, however, one DNA molecule is compared with another. Base composition as measured by thermal denaturation or caesium chloride ($CsCl_2$) buoyant density determinations together with DNA hybridization has been used for comparisons at the species level or higher (Kurtzman 1985). The $CsCl_2$ method is said to be the more accurate technique (Flossdorf 1983). Thermal denaturation studies require extremely pure samples and exceedingly precise control over the temperature range, usually 65–85 °C. The melting curve is followed spectrophotometrically and the midpoint of denaturation ($T_m$) and the temperature range ($\Delta T$) over which 17–83 per cent of the derived melting profile occurs is measured (Mandel and Marmur 1968). From these values the equivalent percentage G + C can be derived from tables (Marmur and Doty 1962), e.g. as in the thermal denaturation of strains of *Penicillium* [A/FA] (Paterson *et al.* 1990) (Figure 3.3).

Two methods are available for measuring the degree of molecular homology between two or more DNAs. One involves the pairing of single-stranded DNAs in free solution. In the other method, one of the samples is immobilized on a solid support, such as a membrane of nitrocellulose fibres or nylon, whereas the other sample is radioactively labelled and in free solution. Renaturation at an optimum temperature is followed spectrophotometrically in the former method (Jahnke 1984) and by determining the amount of free labelled renatured homologous DNA as a measure of DNA homology in the latter method (Jahnke and Bahnweg 1986). The spectroscopic method is probably quicker but requires more material than the solid-support method. Neither gives very reliable measures of homology at low levels but at high levels the data tend to become comparable and probably more reliable (Jahnke and Bahnweg 1986).

The real problem with these methods is the interpretation of the results, since the degree of similarity, or DNA homology, which is considered meaningful is arbitrary. For instance, with DNA hybridization data, in some cases (e.g. *Neurospora* [AP] spp.) very high complementarity values (80–98 per cent) were found between those conidial species showing some degree of interfertility (Dutta 1976), and with lower values there was complete sterility. By contrast, even though different species of the yeast genus *Kluveromyces* [A] showed only 20 per cent homology, there was considerable interfertility between them (van der Walt and Johannsen 1979). Because of these kinds of discrepancy, comparisons of whole DNA homology need to be taken with caution.

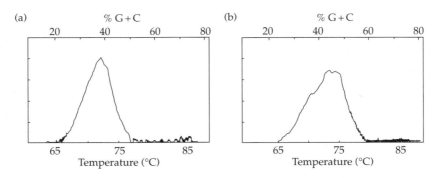

**Figure 3.3** Thermal denaturation curves of DNA across the temperature range 65–85 °C of (a) *Penicillium roquefortii* [A] and (b) *P. chrysogenum* [A]. Curve (a) is approximately symmetrical, indicating a random distribution of single copy sequences, but the shoulder in curve (b) results from a non-random distribution of amplified sequences. The $T_m$ values are 73.2 °C and 72.6 °C, and the percentage G + C values are 41.5 per cent and 39.9 per cent respectively. (Reproduced in part from Paterson *et al.* (1990), with permission from the British Mycological Society).

A more effective way of analysing whole DNA samples is to fragment the material enzymatically and either separate the fragments by electrophoresis or isolate and sequence them, but both processes are laborious (see section 3.3.5).

### 3.3.2 The polymerase chain reaction

Rather than fragment the DNA, the original PCR technique enabled a specific sequence of DNA of known length to be replicated (amplified) at will. The amplified sequence was then compared with comparable amplified sequences from other fungal isolates. The technique and its variants have become one of the most powerful tools available for genetic analysis and, combined with methods capable of providing high-quality uncontaminated DNA from small initial samples, is the preferred approach. It can be used for a variety of purposes: to detect and identify fungal mycelium, for taxonomic and phylogenetic analysis, for gene cloning and expression, and for assessing genetic diversity. A useful summary of its applications has been provided by Bridge *et al.* (1998).

The underlying principle is to expose denatured 'template' DNA to complementary single-stranded oligonucleotide primers, 10–20 bases long, in the presence of a thermostable DNA polymerase. The sequences of the primers complement those on opposite strands flanking the 'template' region which it is desired to amplify. Therefore prior knowledge is required of the 'template' sequence which it is intended to amplify (Saiki *et al.* 1988).

In a typical protocol, the 'template' DNA is first denatured at 94–95 °C and the primers are annealed to the single-stranded 'template' DNA by reducing the temperature to about 36–50 °C. Finally, the temperature is raised to the optimum ($T_\mathrm{m}$) polymerase activity (about 72 °C), when the enzyme extends the 3'-end of the DNA-primer hybrid towards the far end of the primer binding site, thus replicating the chosen sequence (Figure 3.4). The precise temperatures employed depend on the particular polymerase used and need to be determined empirically.

The thermal cycle is then repeated, starting with the double-stranded DNA that has just been formed. Consequently, *both the original and the newly replicated strands* are again replicated. Further amplification of the 'template' sequence is achieved by 25–30 repetitions of this cycle, which results in an exponential increase in the numbers of replicated strands of the 'template' region between the 5'-ends of the primer site.

Great care is required to optimize and keep constant all conditions. Various polymerases, e.g. *Taq*, AmpliTaq (Weising *et al.* 1994), can be used but all should have the property of being thermostable at the different temperatures of the cycle, which is usually automated using a thermocycler. When the amplified DNA is transferred to agarose it can be separated as bands which can be stained with ethidium bromide or, if the DNA has been isotopically labelled, autoradiography can be used. Several detailed protocols are available (e.g. Innis *et al.* 1990).

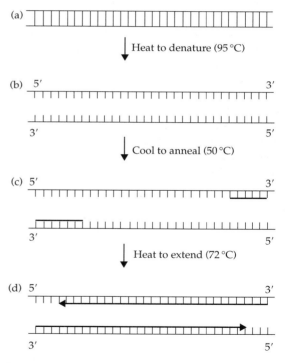

**Figure 3.4** Diagram to demonstrate one cycle of the PCR procedure which is repeated 20–30 times: (a) intact double DNA helix; (b) DNA strands separate with heat treatment; (c) primers anneal to each 'template strand; (d) new DNA strands synthesize in alignment and will act as further template strands when the process is repeated. (Reproduced from Birt and Baker (2000), with permission from Blackwell Science.)

The PCR technique has four great advantages.

- It requires very little DNA—as little as that contained in one nucleus of *Neurospora* [AP] (Lee and Taylor 1990).
- Particular kinds of sequences can be isolated by choice, although later modifications require no *a priori* knowledge of the sequence to be amplified;
- Particular regions of DNA can be amplified with relative ease for a variety of purposes—a single primer can provide numerous fragments.
- Since DNA can be readily stored for long periods, the amplified DNA sample, if retained, is available for reference back or for further study and, if the quantity is reduced by such studies, more can be regenerated using the same technique on the residual material.

The technique does have some drawbacks and very careful control has to be exercised at all stages of the procedure to provide consistent and comparable results.

- DNA extraction: a number of substances which may be co-extracted at the same time are known to inhibit the PCR procedure. Polysaccharides and polyphenols derived from cyanobacteria, green algae or higher green plants associated with a fungus as in lichens, infected plants or mycorrhiza are particularly troublesome (Ekman 1999).
- Optimal annealing temperature $T_m$: this depends on the constitution of the primers used and has to be determined empirically.

It is well worth ensuring that the optimum temperature is achieved. Published protocols are available (Rychlik *et al.* 1990, 1991) and can also be obtained at the website http://alces.med.umn.edu/rawtm.html. An approximation can be obtained for any primer from the relationship

$$2(A + T) + 4(G + C) \text{ in the primer sequence} = T_m \,^\circ C.$$

A useful guide to resolving a range of problems has been provided by Roux (1995).

### 3.3.2.1 Primers for PCR

If primers are selected from a conserved region of DNA, they can be used with a wide variety of organisms for purposes of comparison between them. Primers from variable sites, being more likely to be unique, will tend to restrict the amplification to specific organisms or even to specific alleles. Primers have been designed, for example, for different regions of fungal rDNA, such as

ITS2  5′-TCCGTAGGTGAACCTGCGG-3′
  for the ITS between 17 S and 25 S

5SA  5′-CAGAGTCCTATGGCCGTGGAT-3′
  for the ITS between 25 S and 5.8 S

and are effective with a wide range of fungi (White *et al.* 1990; Henrion *et al.* 1992). Specific primers for the ITS regions of particular groups of fungi have also been developed, e.g. for Basidiomycotina (Gardes and Bruns 1992).

An increasing number of protein-encoding sequences related to specific genes in several fungi, principally Ascomycotina and Basidiomycotina, are available in the databases GenBank and EMBL. They provide a means whereby primers specific to a particular region of a gene's nucleotide sequence can be selected and constructed, e.g. *Neurospora crassa* [AP] (Carbone and Kohn 1999). For example, the database GenBank at http: //www.ncbi.nlm.nih.gov/ was searched, using the tools developed by the National Centre for Biotechnology Information, for encoding sequences in *N. crassa* [AP] with distinct exon–intron boundaries and the intron-containing nucleotide sequences were recovered. All other non-redundant sequence matches in the GenBank database were recovered and presented visually in a graphical form aligned to the initial sequence. Lengths 300–500 bp long which spanned or more introns and showed the maximum number of hits to the target sequence were then selected and appropriate primers were synthesized. In this way primers were developed from *N. crassa* [AP] for the rDNA repeat ITS region and parts of the translation elongation factor 1$\alpha$, calmodulin and chitin synthase 1 genes as well as genes encoding the actin and *ras* proteins. Using these primers, successful amplification was achieved in *Aspergillus nidulans* [A], *Fusarium solani* [FA], *Ophiostoma ulmi* [AP], *Podospora anserina* [AP],

*Sclerotinia sclerotiorum* [AP] and *Trichophyton rubrum* [FA] (Carbone and Kohn 1999). These primers are evidently effective with a range of Ascomycotina and ascomycotous Fungi Anamorphici, and this is a common experience. In general, the closer taxonomically two or more species are, the more likely a primer selected for one will be effective with the others. The procedure can be used with any other fungal group for which data exists in these databases.

### 3.3.3 Random amplified polymorphic DNA (RAPD)

The drawback to the original PCR technique, namely that the sequence of the 'template' DNA had to be known, has been overcome with the RAPD (sometimes abbreviated as RAPD–PCR) technique. This employs *synthetic primers* of (9)–10–(13) arbitrarily chosen base sequences, usually with at least a 50 per cent $G + C$ content. The effectiveness of such arbitrary primers depends on the statistical chance that, somewhere in the 'template' DNA which is being probed, complementary sequences will occur as inverted repeats enclosing a short stretch of DNA (Bachmann 1994). The fragments obtained are separated as usual on agarose gels and stained with ethidium bromide. Sometimes the sequence can be related to a particular phenotype but this is not common. An example using the arbitrary primer, 5'-GGCATCGGCC-3' is the identification of both aggressive and non-aggressive phenotypes and other genetic variants of *Leptosphaeria maculans* [AP] (the anamorphic stage of *Phoma ligam* [FA]) (Goodwin and Annis 1991).

With diploids or in crosses, most RAPDs behave as *dominants*; the alternative allele fails to produce a fragment so that no band is developed, i.e. the recessive's allele is recorded as a 'null' allele. Their usefulness is therefore limited with diploid Straminipila such as *Phytophthora* [O] spp., dikaryotic Basidiomycotina or polyploids because the homozygous dominant condition cannot be distinguished from the heterozygous condition. By combining the RAPD technique with single-strand conformational polymorphisms or denaturing gradient gel electrophoresis (see sections 3.3.3.3

and 3.3.3.4) the problem can be overcome even in dikaryons, since the markers are co-dominant in this condition and both alleles can be recognized (Hayashi, 1991). These are the two most sensitive and convenient techniques for separating sequences differing by as little as a single nucleotide. Direct methods are also being developed for diploid and, presumably, dikaryotic mycelia to increase the number of co-dominant RAPDs, which could also obviate this problem (Davis *et al.* 1995) (see section 3.3.3.1). The RAPD technique causes no problems with haploid Eumycota, although all independently isolated null alleles are *not* necessarily identical. This is because different changes such as chromosome rearrangements or nucleotide substitutions could lead to different kinds of loss, any of which could include the region of interest in common. On the other hand, large numbers of polymorphic markers greatly facilitate the analysis of genetic differentiation in haploid populations and it then becomes possible to identify asexually produced clones unambiguously (Huff *et al.* 1994; Peever and Milgroom 1994).

One practical drawback to the RAPD technique is that it is susceptible to slight variations in the reaction mixtures, such as the concentration of the primers or template material, resulting in a loss of reproducibility. While this can be overcome within a single laboratory through rigid standardization, reproducibility between different laboratories has sometimes proved difficult (Penner *et al.* 1993). Standardization is absolutely essential when comparisons are involved.

A number of variants of the RAPD techniques have been developed for specific purposes and are now briefly described.

#### 3.3.3.1 *Template-mixed RAPDs*

One of the problems with RAPD is posed by the fact that it only recognizes dominants. The template-mixed technique (Davis *et al.* 1995), which has not yet been applied to fungi, increases the yield of codominant RAPDs when dealing with diploids or dikaryons. In essence, the method depends on using a heterozygote with selected primers to produce an additional band which runs more slowly than the normal dominant band. This is a heteroduplex resulting from an association of single-stranded material during the processing of a RAPD. The best way to select the most effective markers is to start with a mixture of equal amounts of DNA from each parent and to process this with different primers until a combination is found which gives a heteroduplex band reliably. Of course, this is not always possible with field material where the parental types are likely to be unknown.

### 3.3.3.2 Sequence-characterized amplified regions (SCARs)

It is often convenient to identify a specific sequence that may, for example be widespread throughout the distribution of the species. SCARs (Paran and Michelmore 1991) permit specific sequences to be identified and have been developed for *Blumeria graminis* f.sp. *hordei* [A] (McDermott *et al.* 1994). A large number of synthetic arbitrary primers, each 19–21 bases long, were first screened for good and highly consistent amplification and polymorphisms, in some cases through 11 successive conidial generations. From these, specific primer sequences were selected that would detect and amplify either specific loci (SCARs) or a locus where seven alleles had been detected which was associated with a differing number of variable tandem repeats (a VNTR locus).

### 3.3.3.3 RAPD with single-strand conformational polymorphism (RAPD–SSCP or SSCP)

It is often found useful to be able to discriminate between sequences differing by as little as a single nucleotide, such as when comparing the same locus in different isolates. The RAPD-SSCP technique (Hayashi 1991; Burt *et al.* 1996) depends upon the fact that changes in sequence alter the conformation of single-stranded DNA and this affects its electrophoretic mobility on native polyacrylamide gels. Sequences differing by a single nucleotide can be detected with this technique. Therefore, if a single locus has been isolated as a strand, allelic differences can be readily detected. Monomorphic fragments about 600 bp long, obtained by the RAPD or PCR techniques, are re-amplified using a single primer and the single-strand amplified fragments are separated on a gel. Since single strands are employed, the markers are effectively co-dominant and therefore both homozygous and heterozygous

individuals can be recognized. This technique has been applied successfully to several fungi, e.g. *Coccidioides immitis* [FA] (Burt *et al.* 1996, 1997) and *Suillus pungens* [BA] (Bonello *et al.* 1998). Other techniques (e.g. denaturing gradient gel electrophoresis) are equally sensitive.

### 3.3.3.4 Denaturing gradient gel electrophoresis (DGGE or CDGE)

This sensitive technique (Myers *et al.* 1986) discriminates between sequences differing in size and also, of even greater value, between those differing by as little as a single sequence. Thus fragments of similar length but differing sequence can be distinguished and this is of particular value in separating allelic sequences. An acrylamide gel is employed which can be progressively denatured by chemical denaturents, such as urea and formamide, whose concentration can be varied to obtain the optimum for separation. For example, to test for separation of ITS amplified fragments from *Peniophora gigantea* [BA], a range of 20–80 per cent was tested, with 40–55 per cent denaturation giving optimum separation or 35–50 per cent for analysing heteroduplex bands derived from heterozygotes (Vainio and Hantula 2000). When only a single optimal concentration is employed, e.g. 30 per cent used to discriminate between ITS region fragments from two similar *Melampsoridium* [BR] taxa growing on birch and alder (Kurkela *et al.* 1999), the technique has been described as constant denaturing gel electrophoresis (CDGE).

### 3.3.3.5 Variable numbers of tandem repeats (VNTRs)

Highly variable loci with many tandem repeated alleles at each locus are widespread in fungi, including *B. graminis* [A], and this technique assists in detecting and resolving them. Very simple oligonucleotides such as $(GATA)_4$ have been employed to detect these highly variable loci which give a many-banded 'fingerprint' (Paran and Michelmore 1991; Weising *et al.* 1994). Since VNTRs have many alleles at each locus, they are useful for identifying cloned or very closely related organisms which exhibit identity or great similarity between their banding patterns. Such patterns can then be measured and compared using some form of similarity index (e.g. Lynch 1990). This technique is of immense value and can be used and developed for a variety of circumstances. For instance, the locus in *B. graminis* f.sp. *hordei* [A] referred to in the previous section is probably the variable length spacer of rDNA. This method enables all the specific loci to be precisely identified so that it should be possible to build up a standard reference set of RAPD markers for this fungus which could be used to analyse barley mildew populations worldwide (McDermott *et al.* 1994).

### 3.3.3.6 *Simple sequence repeats or microsatellites (SSRs)*

Microsatellites—simple sequence repeats (SSRs)—are short tandem repeats of di-, tri- or tetranucleotides differing in both composition and numbers of repeats, which can range from 10 to 50. Some are compound but do not usually include more than two nucleotides, e.g. three mononucleotide $(A)_n$ and one dinucleotide $(CA)_n$ motifs in *Microbotryum violaceum* [BS] (Bucheli *et al.* 1998). The length of the fragments reflects the number of motif repeats, with different lengths behaving as alleles. For example, in 50 human-derived isolates of *Aspergillus fumigatus* [FA], the microsatellites $[CA(CA)_{25}]$, $[(CA)_2 \ C (CA)_{23}]$, $(CA)_8$ and $(CA)_{21}$ were detected as 8, 10, 10 and 17 alleles respectively (Bart-Delabase and Bretagne (1997) quoted by Latgé *et al.* (1998)). They can be isolated using a PCR technique with primers including the repeat motif and the flanking sequences and those available can be found in MICRO-SAT at microsat@sfu.ca. SSRs can be separated on polyacrylamide or denaturing polyacrylimide gels to give greater resolution. Dinucleotide SSRs can suffer from a 'ladder' effect of shadowing bands, although the primary allele band is usually darker. This effect is thought to be a result from slipped strand mispairing during *Taq*-mediated amplification.

Using known primers, SSRs in fungi were first detected in several human pathogenic fungi, e.g. *Cryptococcus neoformans* [BT], *Candida albicans* [A] and *Saccharomyces* [A] species. As a result of a sequence search in the two main databases they are now known to be widespread in Eumycota. Thirty-one eumycotous species revealed 626 SSRs; the most common motif repeats were $(AT)_n$, $(A)_n$ and $(AAT)_n$ (Groppe *et al.* 1995). These authors used a known primer for detecting SSRs to identify a pair

of alleles in a PCR-produced fragment from *Epichloë typhina* [AP] on *Bromus erectus*; one included an $(AAG)_{18}$ motif repeat, and the other included an $(AAG)_8$ motif. Subsequently a range of primers have been developed for *Microbotryum violaceum* [BS] that have enabled variation to be detected in isolates from different hosts (Bucheli *et al.* 1998, Shykoff *et al.* 1999) (Table 3.1).

### 3.3.3.7 *Random amplified microsatellites (RAMs)*

A rather different technique has been developed by Hantula and coworkers (Hantula and Müller 1997; Hantula *et al.* 1997) to detect and amplify sequences of *two* microsatellites and the region between them close enough for them to be amplified as a single entity and polymorphic for different lengths of DNA between their distal ends. Defined primers can be readily identified (Zietkiewicz *et al.* 1994). For instance, two fairly common repeated nucleotide sequences in fungal microsatellites are the tandem repeats AT/TA and the triple repeats AAC/GTT. Primers of length 10 bp based on these have been synthesized, e.g. formulated as

$$5'\text{-}\alpha\beta\alpha(ACA)_5 \text{ or } 5'\text{-}\gamma\delta\gamma(GT)_5G$$

where $\alpha$ = G, T or C, $\beta$ = G, A or T, $\gamma$ = G, A or C, and $\delta$ = A, T, or C.

**Table 3.1** Primers for various microsatellites effective with different fungi

| Species | Primer (5' to 3') | Repeat motif |
|---|---|---|
| *Epichloë typhina* [AP] | CGCACAATACGTCAGCTACGAATG | $(AAG)_{18}$ and $(AAG)_8$ |
| | CCTGAATCAACTTTGCTATCAGGC | |
| *Microbotryum violaceum* [BS] | GTAGCCACCTCCCATCCC | $(AG)_{15}$ |
| | CGGTGTCGAGTTCCTGAC | |
| *M. violaceum* [BS] | AAAACCCAAGACGACTGACGC | $(A)_{11}$ |
| | TTCCTTCGATGCAGCCTC | |
| *M. violaceum* [BS] | GTCGTTCTCGCTTCTCTC | $(AG)_{15}$ |
| | GGGGCTCGTGAAGCCG | |
| *M. violaceum* [BS] | GACCCAGTGTCCACCAATCC | $(AG)_{20}$ |
| | GATCCACCTTATCTTCTTC | |
| *M. violaceum* [BS] | CCCCACAGACGGTATGCTGC | $(AG)_{15}$ |
| | CGTGACACCCTTCCTGCCGC | |

Data from Groppe *et al.* (1995) and Bucheli *et al.* (1998); see also Giraud *et al.* (2002)

The method has been employed to observe different patterns between geographical races of *Gremmeniella abietina* [AD] as well as intra-racial variants. The length differences behave as distinct alleles and the technique enables co-dominant markers to be separated (Hantula *et al.* 1996). This primer was also shown to be effective on a number of unrelated fungi such as *Neurospora crassa* [AP], *S. cerevisieae* [A] and *Ustilago maydis* [BS].

### 3.3.4 Amplified fragment-length polymorphism (AFLP)

Amplified fragment-length polymorphisms (AFLPs) (Vos *et al.* 1995; Majer *et al.* 1996) are produced by a method which effectively combines the RFLP and PCR–RAPD techniques. First, the DNA is exposed to *two* restriction endonucleases which produce a very large number of DNA fragments. Next these are ligated through their 'sticky' ends to oligonucleotide adapter molecules which will act as primer binding sites. Next, the primers, with some homology to the adapters, are used to amplify the attached adapters/fragments. Appropriate physical conditions prevent any fragment to fragment re-fusions and the structure of the primers prevents the restriction site being reconstituted. In order to restrict the numbers of fragments to be scored, the primers have, arbitrarily, one, two or three additional nucleotides, designated $+1$, $+2$ or $+3$ respectively, attached beyond the $3'$ end extension. The type of primer used is adapted to the size of the genome targeted—the larger the genome, the larger the number of added nucleotides. Therefore the sequences amplified are those fragments with one sticky end produced by each endonuclease and the number amplified depends upon the length of the $3'$ end. The fragments can be separated on simple agarose gels or preferably on a denaturing polyacrylamide gel which better ensures adequate separation of the numerous bands. For a fungal genome of about $4 \times 10^4$ kb, the technique is said to produce more than 150 000 fragments, mostly over 500 bp long. Thus AFLPs generate many more fragments more rapidly than other methods and from many different DNA regions.

The following advantages are claimed for the AFLP technique.

- It is applicable to small samples of DNA of the order of 50 ng or less, e.g. 0.1–0.5 ng DNA from a single Glomalean spore (Rosendahl and Taylor 1997).
- It has all the useful features of the RFLP technique—large numbers of independent loci, neutral and some co-dominant expression (bands of different lengths)—plus an appreciably larger number of distinguishable fragments. Typically, some 50–70 bands can be separated on each gel, so that many more fragments can be screened simultaneously than with RFLP.
- It is highly selective and has a high level of reproducibility.
- It is extremely rapid.

A drawback to the method compared with RAPD or RFLP is that it requires higher-quality DNA and is technically more complex. However, ready-to-use kits are now available and the procedure can usually be completed in 2 days.

A typical example is the application of the technique to both *C. fulvum* [FA] and *Pyrenopeziza brassicae* [AD] (Majer *et al.* 1996, 1998). The endonucleases employed on extracted cDNA from the fungi were *Eco*RI followed by *Mse*I when the first digestion was complete. The mixture was then treated with biotinylated *Eco*RI and *Mse*I adapters in the presence of T4 DNA ligase and ATP. The following adapter sequences were used.

*Eco*RI adapter:   Biotin-3′ CTCGTAGACTGCGTACCC
                   CATCTGACGCATGGTTAA 5′
*Mse*I adapter:    3′ GACGATGAGTCCTGAG
                   TACTCAGGACTCAT 5′

Although the *Mse*I fragments were more numerous, *only* the biotinylated fragments were utilized since these could be selected easily by adding streptavidin beads to which they bound. The fragments were resuspended and exposed to PCR cycling, including either $^{32}$P- or $^{33}$P-end-labelled $+2$ *Eco*RI or $+2$ *Mse*I primers, in the usual way, and the amplified fragments were run on a 5 per cent denaturing polyacrylamide gel to separate them. Various combinations of primer (Vos *et al.* 1995) were used and these detected about 160 polymorphisms in *P. brassicae* [A] and 20–30 polymorphisms in *C. fulvum* [FA] in the sample tested. Clear separations were obtained and each segregated independently.

It has been claimed that AFLP fragments of the same size from different isolates are likely to represent the same genomic region, and this was borne out in this investigation. If this is generally true, then it gives an added advantage to the method. Evidence comes from another study using the bean anthracnose fungus *Colletotrichum lindemuthianum* [FA]. Gonzalez *et al.* (1998) compared the AFLP-derived data with that from RAPD using the same DNA samples. Ten RAPD primers gave 62 bands, whereas four AFLP combinations gave 230 bands. The latter method also proved to be more consistent. The correlation coefficients between genetic similarity data derived from the samples, allowing for differences in sample size, showed a wide range.

AFLP compared with RAPD data: $r = 0.15$
Between RAPD samples only: $r = 0.04–0.20$,
  mean $r = 0.13$
Between AFLP samples only: $r = 0.55–0.73$,
  mean $r = 0.61$

The AFLP method evidently gave more consistent results.

### 3.3.5 The analysis of nucleotide fragments

The most valuable common attribute of nucleotide fragments or short sequences, by whichever method they are obtained, is that they can be treated as equivalent to genetic loci *provided that they are not linked* but inherited independently. A test should *always* be made of this important assumption. However, the method of analysis used will depend very much on the limitations of the method and the objective of the investigation (Brown 1996).

Once nucleotide fragments have been obtained they can be compared in a variety of ways. Most fundamentally, their composition can be determined by sequencing. When separated on gels, RFLPs with different lengths and linear patterns of distribution can be recorded and compared. However, detection of further variation requires Southern blotting using an appropriate probe. If, instead of RFLP, an amplification procedure is employed,

appropriate probes can now be tailored to particular sequences which may represent genes or regions with genes in either mtDNA or nDNA. The fragments obtained may differ in size, density or charge, and any of these properties can be used to separate them on appropriately selected gels. Their presence can then be visualized by appropriate staining. By using denaturing gels, pattern differences due to as little as a single nucleotide difference can be detected.

#### 3.3.5.1 Sequencing

Sequencing is, of course, only necessary when the nature of the amplified fragment is not known, so that it is not required when a primer of known sequence has been involved in producing the fragments. Thus the main use has been with RFLP fragments. Automatic sequencing is now common but is often based on the original dideoxynucleotide system devised by Sanger *et al.* (1977) for short lengths of DNA. The procedure is somewhat laborious if it is not automated.

The method depends on interrupting DNA synthesis of a copy of the fragment in a controlled manner. If the fragment is double stranded it is first denatured to single strands. One of these is then annealed with a short oligoncleotide primer which complements a known or probable region of it. The sample is divided into four subsamples, to all of which are added DNA polymerase, the four necessary deoxynucleotides, at least one of which is radio-actively or fluorescently labelled, and *one* of the four dideoxynucleotides. Under appropriate conditions, DNA is synthesized, but where an analogue dideoxynucleotide is incorporated into the growing chain, it terminates elongation. The four products are then separated on a denaturing polyacrylamide gel. The radio-active fragments can be visualized by autoradiography or if fluorescence has been used by laser stimulation of the gel. Since each chain will have been terminated by the equivalent dideoxynuceotide to the corresponding deoxynucleotide, i.e. a site of incorporation of ddGTP corresponds to that of normal dGPT, the sequence of the target DNA can be read off the autoradiograph or gel.

Sequences can be readily compared when printed out.

### 3.3.5.2 *Restriction fragment length polymorphisms (RFLPs)*

When first developed, RFLPs were immensely valuable for use with fungi because they could be generated at will and provided readily identifiable genetic markers in organisms with few natural phenotypic markers. Moreover, it was then seen as an advantage that, with test fungi, only 1 μg of DNA was routinely required and satisfactory results were obtained with as little as 0.1–10 ng when multiple copies of a target sequence were involved (Lee and Taylor 1990). However, the procedure did prove time consuming, since further analysis of variation involves the use of Southern blotting (see section 3.3.5.3).

The principal is very simple. DNA is extracted and aliquots exposed to one or more type II restriction endonucleases derived from prokaryotes. Each endonuclease recognizes a different specific short sequence of DNA, to which it binds. The sites are usually 4–6 bp long and subsequently the enzyme cleaves the DNA at a specific position within the sequence (Table 3.2).

Since recognition sites usually occur more than once along a DNA molecule, this results in a number of DNA fragments of different lengths according to the distribution of the cleavage sites along the molecule. The fragments can be separated by agarose gel electrophoresis; the shorter ones diffuse most rapidly. A small constraint is imposed by the fact that methylated cytosine bases are protected from endonuclease action (Antequera

et al. 1984). However, there are several positive advantages.

- The method can be applied to any part of a fungus, such as hyphae, spores, strands or rhizomorphs, and reproductive structures, and in some circumstances to symbiotic fungi in mycorrhizal roots (morphotypes) or lichens. It can also be used in combination with PCR techniques (see section 3.3.2).
- In general, RFLPs provide a ready source of multilocus genotypes which are, or are assumed to be, selectively neutral, behave as moderately polymorphic, usually co-dominant, markers and are inherited in a Mendelian manner and can therefore be followed in crosses.
- Identical samples of DNA exposed to the same restriction enzymes produce identical and highly reproducible fragment patterns of different lengths even in different laboratories, with different samples exhibiting different patterns. The position of each band can be calibrated against a range of marker fragments of known size or molecular weight, e.g. those generated by digestion of phage λ DNA by the *Hind*III restriction enzyme.
- Unlike isozymes, RFLPs are *not* confined to detecting transcribable regions only so that their numbers are far greater, being limited only by the number and kinds of endonucleases employed and their specific restriction sites. Moreover, they are more useful than isozymes because they do not suffer from the limitations of the latter due to the constraints imposed by the genetic code.
- Single-base substitutions (a cause of point mutations), insertions or deletions (indels) within a recognition sequence are detectable by a change in the length of the fragment and hence a change in the gel pattern owing to its altered position.
- Differences in the number of tandem repeats between restriction sites can also lead to differences in fragment lengths. If dispersed repetitive sequences with high variability occur, then each individual may exhibit a unique pattern of RFLPs. Therefore such regions can act as 'fingerprints' for individual genotypes, e.g. the shiitake mushroom *Lentinula edodes* [BA] where almost every strain examined had a unique RFLP pattern (Kwan *et al.* 1992) (see section 3.3.3.5).

**Table 3.2** The action of some commonly used Class II restriction endonucleases

| Name | Action |
|---|---|
| *Alu*I | 5-AG ▼ CT-3' |
| *Bam*HI | 5-G ▼ GATCC-3 |
| *Eco*RI | 5-G ▼ AATTC-3 |
| *Hae*III | 5'-GC ▼ CC-3' |
| *Hind*III | 5'-A ▼ AGCTT-3' |
| *Hinf*I | 5'-G ▼ ANTC-3' |
| *Pst*I | 5'-CTGCA ▼ G-3' |
| *Taq*I | 5'-T ▼ CGA-3' |

The symbol ▼ indicates where the endonuclease cleaves the nucleotide chain.

• The RFLPs—the gel phenotypes—mimic the underlying genetic polymorphism of the DNA.

However, there are a number of practical constraints and some sources of potential error.

• Preparation of RFLPs can be time consuming and a relatively large amount of high-quality uncontaminated DNA is required initially.
• Co-migrating fragments are *not* necessarily genetically homologous.
• If length mutations occur, i.e. structural variation in the material, fragments will be larger or smaller independently of the location of the restriction sites, so that fragments of comparable size will *not* necessarily represent shared restriction sites.
• Size and mobility of fragments are affected by the degree of methylation of the nucleotide bases and such methylation is not distributed evenly along the DNA molecule.

These sources of potential error should be tested. A further problem is that with some fungi, e.g. *Cladosporium fulvum* [FA], it has not proved possible to detect a great deal of variation and, since most of the fragments are repeated sequences, there is considerable ambiguity as to whether they are homologous or not (Brown 1996) although, with amplification methods (AFLP) (see section 3.3.4), genuine variation has proved to be detectable.

RFLPs have been applied to many situations. Different RFLP patterns can be recognized and used for identification of the genotypes of different species, e.g. *Aspergillus* [A] (Kozlowski *et al.* 1982), individuals or clones, e.g. *Blumeria graminis* f.sp. *hordei* [A] (Brown *et al.* 1990), and their frequency in different, or successive, populations can be determined. Therefore they can provide useful estimates of gene or genotypic diversity of populations and can be used to detect the location of particular clones in space and time as well as microgeographic differentiation. Their use in the estimation of genotypic diversity has a particular value in limiting cases in assessing the contributions of both sexually and asexually derived components of a population. Populations with active sexual reproduction will, through recombination, show a wide range of gel patterns which differ in subsequent generations, whereas gel patterns of completely asexual organisms are uniform over different generations except, of course, when rare mutations lead to change.

Because any part of the mycelium can be sampled, they can be used both to identify mycorrhizal fungi on roots (Egger 1995) and to demonstrate the common identity of aerial basidiocarps of known mycorrhizal species with their far more extensive underground mycelia, e.g. in a *Pinus muricata* forest (Gardes and Bruns 1992, 1996).

RFLPs can and have been restricted to mtDNA (Taylor 1986) or to a particular region of the DNA such as the rDNA region (Specht *et al.* 1984). Use of mtDNA is particularly attractive since the molecule is relatively small (25–175 kb (Brown 1987)), easy to purify and usually occurs in high copy numbers. It is often variable in composition in Eumycota, although less so in Straminipila and less than cDNA, and it shows considerable size variation even between supposedly closely related taxa, e.g. *Saccharomyces* [A] (Hoeben and Clark-Walker 1986) and *Suillus* [BA] (Bruns and Palmer, 1988). In contrast, rDNA is often conserved between taxa but is frequently highly polymorphic because of the large numbers of tandem repeats and variation in length of the ITS regions, e.g. *Pseudocercosporella herpotrichoides* [FA] (Nicholson *et al.* 1993). In some fungi, however, recombination is apparently suppressed in the ITS regions, e.g. *Coprinus* [BA] (Russell *et al.* 1984; Nicholson *et al.* 1993) and *Neurospora* [AP] (Russell *et al.* 1984).

*3.3.5.3 The use of probes with RFLPs: Southern blots*
Further information concerning the variation shown by RFLPs can be obtained by using specific probes, often supported by the Southern blot technique.

The most common procedure is to test the homology of a particular RFLP by hybridization with a specific DNA sequence probe, a considerable number of which have been employed. The exact nature of these probes has not always been ascertained but E9, for example, is a sequence of *B. graminis* f.sp. *hordei* [A] DNA which does not hybridize with barley DNA, is about 4 kb long and contains several repeated motifs. It hybridizes with about 30 different RFLP fragments after *Hind* III digestion of DNA in at least 100 different isolates of *B. graminis* f.sp. *hordei* [A] (Brown *et al.* 1990).

A number of the other probes detect and hybridize with dispersed repeated sequences in fungal nDNA, e.g. MGR586 in *Magnaporthe grisea* [AP] and RG57 in *P.infestans* [O], whereas others only hybridize with a small number of loci, e.g. STL40 with sequences at one or two loci in *Mycosphaerella graminicola* [AP].

The homology of specific restriction fragments with sequences in the probe, which may include a known marker gene located on a specific chromosome, can be determined most elegantly and rapidly using the Southern blot technique (Southern 1975).

The RFLP–DNA on the gel is denatured, transferred to a membrane or filter for which it has great affinity (e.g. nitrocellulose, nylon, Hybond etc.) and immobilized on it. This is generally achieved by denaturing the double-stranded DNA on the gel with alkali, neutralized in salt buffer to ensure that the strands then remain single. The single-stranded DNA is then eluted onto the membrane by capillary action induced by surrounding it by 50 per cent sodium chloride–sodium hydroxide (SSC) solutions of different strengths (Figure 3.5).

Once the DNA has been transferred, the membranes can be stored more or less indefinitely, providing a permanent record of the RFLPs. If a membrane is exposed to a denatured [32]P-labelled DNA probe, the probe will hybridize with those bands with which it possesses sequence homology and they can be identified by autoradiography of the treated membrane.

Several assays (up to 10) can be made of the same membrane, using different probes if desired. Great care needs to be taken with this technique but the successive steps and protocols are well documented (e.g. Weising *et al.* 1994). The technique has been applied with great success to *M. grisea* [AP]

using the MGR 586 probe. In fact, MGR 586 has a double value, for the 30–50 copies which occur in the rice blast fungus can be used to separate variants and lineages within that fungus, and supposedly related fungal strains found on other adjacent grasses can be distinguished because they only carry one or two copies of the probe (Hamer 1991).

An alternative method is to utilize a plasmid and ligate the separated DNA fragments to a site on it. For example, fragments of DNA of *Septoria trici* [FA] produced by exposure to *Sau*3A were separated, ligated to the *Bam*H1 site of the commercial vector pGEM4 (Promega), treated with intestinal calf alkaline phosphatase and transferred by plasmids to *Escherichia coli* HB101. Two hundred *E. coli* colonies from each transfer were retained for use as probes. These, labelled with [32]P by nick translation, could then be used to identify homologous sequences on the membrane in exactly the same way as just described. In this case they were also employed to localize sequences on particular chromosomes (McDonald and Martinez 1990b).

Although much use was made originally of RFLPs, probes and Southern blots, they are laborious and time-consuming procedures demanding very considerable care to ensure comparability of results. Hence they have been replaced by methods which produce amplified fragments.

### 3.3.5.4 *Amplified fragments*

Amplified fragments possess all the advantages of RFLPs plus the advantage that their nucleotide constitution is determined by the probe(s) used in their preparation. They can, of course, be analysed in much the same way as RLPs through differences in their size and gel patterns, although in many cases they provide greater discrimination. For example, single nucleotide substitutions can be detected more readily by SSCP or DGGE (see 3.3.3.3 and 3.3.3.4) with neutral or denaturing polyacrylimide gels because of differences in their mobility (Burt *et al.* 1996; Kurkela *et al.* 1999). They are also more effective with regions including repeated motifs, especially SSRs which are potentially a prolific source of information on variability as they are both frequent and widely distributed throughout the genome. Thus their discriminatory properties are greatly enhanced.

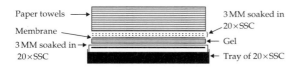

**Figure 3.5** Diagram to illustrate the arrangement for obtaining a Southern blot. Pressure is applied to the arrangement from above. There are various similar arrangements, and the values shown are typical. SSC, 1 : 1 sodium chloride–sodium hydroxide mixture. (Reproduced from Birt and Baker (2000) with permission from Blackwell Science).

## 3.4 The utility of different molecular screening techniques

Each of the methods described has its advantages and disadvantages for particular purposes and differs in its ease of manipulation, sensitivity and economy. Therefore there is no single ideal technique that meets all these requirements. Some ways in which the data obtained from gel patterns can be used and interpreted are described in Chapter 5.

RAPD is probably the least reproducible technique, especially when comparisons involving different laboratories are concerned, and the AFLP or SSR techniques are probably the most reproducible, but it has to be stressed that strict control of the conditions under which amplification takes place is the best way to ensure reproducibility with any of the amplification techniques. RAPDs and SSRs applied to tandem repeated regions, depending on their sizes, are probably the most effective approach when multi-locus data are required. SSRs are particularly valuable for discriminating between closely related taxa such as races within a species. AFLP has the immense advantage that it identifies by far the largest number of loci of any of these methods. This has obvious advantages for use in linkage studies and for single locus data also. At present, the procedure requires a considerable input of resources but no doubt this will be overcome as the technique is developed.

As far as different purposes and the material used are concerned, relatively conservative regions of nDNA, such as rDNA or enzyme proteins, are probably the best material for phylogenetic studies, although SSR is probably the most appropriate for taxa that are believed to have originated recently. This is because SSRs generate variation more rapidly as they are apparently more mutable.

## 3.5 Taxonomic identification using sequence markers

A popular objective of genetic analysis in fungi has been to provide markers to identify or distinguish different taxa. Considerable effort has been put into identifying specific sequences which will enable sterile mycelia, taxa which are difficult to separate taxonomically, difficult species or fungi of similar or different somatic morphology to be identified or distinguished.. A variety of techniques have been used, but a particularly rewarding method has been the use of the amplified PCR technique with the polymorphic tandemly repeated rDNA region whose basic organization is shown in Figure 3.6.

The four sequences 17 S, 5.8 S, 25 S and 5 S code for the ribosomal RNA units and are the most highly conserved regions. Both the ITS and the IGS regions contain variable non-coding regions. The IGS region is most variable in length (the 'variable-length spacer'), but the ITS1 and ITS2 regions are less highly conserved. The principal sources of variation between different taxa lie in the numbers of tandem repeats of the basic unit (17 S–5 S) and the variation in length of the IGS region. In addition, variation also occurs in the ITS regions. By using appropriate primers these can be treated either as a single 'ITS region' or as a variation in ITS1 distinguished from that in ITS2, e.g. *Colletotrichum* [FA] (Sreenivasaprasad *et al.* 1996). The 5.8 S region between these last two spacers is not usually used. A wide range of primers is available for preferential amplification of particular regions. A selection commonly used with ectomycorrhizal Basidiomycotina and Ascomycotina is given by Henrion *et al.* (1992). Since many of these code specifically for the fungal 'ITS region', they can be used with total DNA extracted from green plants associated in some way with one or more fungi. This is a great convenience in studying such material.

Amplified regions can be further analysed with restriction enzymes and the RFLPs obtained can be separated and compared. Which particular regions

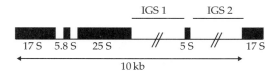

**Figure 3.6** Diagram to illustrate the structure of one unit of the rDNA of *Laccaria bicolor* [BA]. The RNA coding 17 S, 5 S and 25 S regions (rectangles) is relatively highly conserved but appreciable variation usually occurs in the length and composition of the ITS and IGS regions. The sequence ITS1–5 S–ITS2 is often used for phylogenetic comparisons and is often described as the 'ITS region'. (Reproduced from Selosse *et al.* (1996) with permission of Springer-Verlag.)

or combination of regions give the greatest discrimination depends on the fungus examined. For instance, in the genus *Laccaria* [BA], the ITS has a constant length although differing from other ectomycorrhizal genera and species. There is some variation in the length of the 17S unit, but large differences occur in both the 25S and IGS regions which were found to be the most useful for completely distinguishing between species of the genus *Laccaria* [BA]. By extending the variable 25S and IGS separations using derived RFLPs, complete inter- and intra-specific discrimination could be achieved (Henrion *et al.* 1992). However, in other fungi, the ITS region is the more variable and gives better discrimination, e.g. species and strains of *Colletotrichum* [FA] (Sreenivasaprasad *et al.* 1996).

In addition to the simple comparison of fragments and fragment patterns for analysing similarities and differences betweeen isolates of supposedly the same or different taxa, the whole region or individual fragments can be sequenced. These complementary kinds of approaches can often assist in the recognition of different causes of polymorphism to be detected in regions such as rDNA. In *Fusarium* [FA], for example, two ITS2 types occurred, one in high and the other in low copy number, but direct sequencing only provided evidence for the major type (O'Donnell and Cigelnik 1997). In *Hebeloma crustuliniforme* [BA] isolates, different ITS types occur within the same dikaryon and under these circumstances direct sequencing of ITS PCR products was ineffective because they were mixed (Aanen *et al.* 2001).

Therefore it is often wise to confirm any taxonomic or phylogenetic deductions drawn from the analysis of one region by comparison with those from another. Several other sequences have been used, either in addition or as alternatives to the rDNA region, for the purposes of taxonomic discrimination or evolutionary studies. They include those for mtDNA, the $\beta$-tubulin, hydrophobin genes and various enzymes, e.g. in *Aspergillus* Sect. *Fumigati* [A] (Geiser *et al.* 1998) and in *Pleurotus, Lentinula,* or *Ganoderma* [all BA] (Thon and Royse 1999). The enhanced ability to prepare specific primers (described earlier in section 3.3.2.1) from any known gene whose protein-encoding sequences in a fungus can be obtained from the GenBank database can now provide even greater opportunities to use further regions for these purposes.

## Further reading

A valuable comprehensive compendium of molecular techniques and related genetic and phylogenetic analyses used with such data is given by:

Hillis DM, Moritz C and Marble BK (1996). *Molecular systematics*. Sinauer Associates, Sunderland, MA.

*Shorter accounts of several methods are given in:*
Baker AJ (ed.) (2000). *Molecular methods in ecology*. Blackwell Science, Oxford.
Weising K, Nybom H, Wolff K and Meyer W (1994). *DNA fingerprinting in plants and fungi*. CRC Press, Boca Raton, FL.

*Isozymes are best dealt with in:*
Hames BD and Rickwood D (1990). *Gel electrophoresis of proteins: a practical approach*. IRL Press, Oxford.

*Recent critical accounts of PCR methods include:*
McPherson MJ, Hames BD and Taylor GR (1995). *PCR2. A practical approach*. Oxford University Press.
Newton CR and Graham GA (1997). *PCR* (2nd edn). BIOS Scientific, Oxford.

# CHAPTER 4

# Defining fungal individuals: ecological, biological and genetical aspects; sampling

However a population is defined, it is generally assumed that its individual members can be distinguished, except when dealing with clonal organisms. For many purposes the recognition of visible individual structures of a particular taxon is sufficient for ecological analysis and, indeed, fungi have been treated in this way, e.g. counts of taxonomically identical basidiocarps. But, in fact, the identification of a fungal individual by such inspection alone is usually impossible for the following reasons.

- In many cases, the mycelium is mostly submerged or embedded within a substrate so that external structures cannot readily be assigned to a particular mycelium.
- The growth habit of a fungal mycelium is indeterminate, so that its limits are usually not predictable.
- There is confusion over the limits of an individual because of the ability of hyphae to anastomose.
- A single mycelium can produce numerous visible structures which cannot always be assigned with certainty to a particular mycelium.

Because of its indeterminate growth, the spatial limits of a mycelium can only be recognized in a limited number of situations, e.g. the discrete radial mycelia of a *Penicillium* [A] or *Aspergillus* [A] developing on the skin of an infected fruit. Even here there can be ambiguity if adjacent spores germinate simultaneously, giving rise to a single mycelium that may have been produced by hyphal anastomoses so that it may be a single clone or derived from different clones. Inspection alone cannot usually determine whether adjacent colonies are genetically identical unless the advancing edges

of two or more mycelia interact in some way indicative of a recognition or non-recognition reaction (see Chapter 2, section 2.3). Even greater uncertainty exists when all that can be seen are numerous superficial structures on, or arising from, a substrate within which one or more mycelia may be growing. Such readily visible structures—asexual or sexual spore-bearing structures, lesions caused by fungal pathogens or, in the case of mycorrhizic forms, mycorrhizal roots—are often the item sampled. Yet, even if the embedded or subterranean mycelia from which they originate are well separated, this can rarely be detected by inspection. Moreover, as such structures often develop close together or are intermingled, further confusion can be caused unless there is an indication of whether one or more individuals are involved.

For example, two fungi which commonly grow in dead wood and produce clusters of external basidiocarps, namely the sulphur tuft *Hypholoma fasciculare* [BA] and a common bracket fungus *Coriolus versicolor* [BA], can represent entirely different conditions. The former is almost always a cluster of basidiocarps derived from a *single* mycelium, i.e. from one individual, whereas the latter is most frequently a cluster derived from *numerous genetically different* mycelia, i.e. from several individuals. When more than one such cluster of basidiocarps occurs on the same piece of wood, the uncertainty is increased even more. Have these all been derived from the same mycelium, from several clones of that mycelium, or from different mycelia?

One solution to the problem of identifying fungal individuals is to recognize that the functional unit is the mycelial clone and to adopt the term first coined to describe clonal higher plants, namely the **genet**, i.e. the product of a single zygote. However, in fungi the term is used somewhat differently and is applied to two different kinds of fungal organization. In Zygomycotina and Ascomycotina it is applied to each product of a zygotic meiosis, i.e. the mycelium derived from either a germ spore of the former group or an ascospore of the latter. In these fungi, a genet also includes all the mycelia derived from it by mitosis during asexual reproduction. In the Basidiomycotina, however, the term genet is applied to the dikaryotic mycelium derived from the fusion of two monokaryons, themselves derived from individual basidiospores, i.e. the products of a single zygotic meiosis. Similarly, if a dikaryon reproduces asexually, its products are part of the same genet.

In general, therefore, genets can only be identified for the purposes of considering the population genetics of a fungus after isolating and further investigating the properties of each isolate. This introduces a further inevitable constraint since it imposes appreciable secondary work on top of any initial sampling of a population. Indeed, what may appear to be a cluster of fungal individuals, such as a group of toadstools, may turn out to be only the easily visible multiple reproductive stages of a single genet!

Therefore sampling is of prime importance not only because it is rarely possible to investigate an entire fungal population but also to ensure that it is indeed a population, and not a single individual, that is being sampled. It generally requires some prior knowledge of the fungus under investigation, such as the attributes discussed in section 4.1 below. It often requires, in addition, a pilot survey. Once some idea of the spatial scale of the individuals and populations and their biology is known, a sampling procedure can be adopted to ensure that the appropriate components are detected most effectively. As a practical starting point, the simplest procedure is to describe a sample of the individuals of a particular taxon within a defined area, or volume, either at a particular instant or over a period of time. Ideally, any sampling procedure adopted should provide satisfactory measures of two main kinds of attribute:

- the physical extent and limits of individuals and populations in space, their numbers, their condition and/or duration in time and the numbers of individuals in each population
- the genetic attributes of the individuals and populations under examination, their breeding behaviour and their variability.

The former are described for individuals in section 4.1, the kinds of markers employed to determine their genetical attributes in section 4.2 and some sampling procedures in section 4.3. An indication of the range of methods available for describing and analysing populations is given in Chapter 5, and an account of the very diverse potential reproductive attributes which occur in fungi is provided in Chapter 8.

Lastly, the limits of variation within a population, whether determined from samples, or from the population as a whole, can be measured in various ways:

- by quantification of the kinds and frequencies of individuals or biologically significant groups within the sample
- by measuring the range of one or more, supposedly biologically significant, attributes of the sample and expressing it, or them, by some measure of variability
- by both kinds of description.

## 4.1 The physical and biological attributes of fungal genets

Effective sampling can only be achieved if something is known about the physical attributes of a taxon's mycelium and its general biology. Some fundamental considerations for all fungi are:

- the average size of a mycelium
- whether or not it is associated with a particular substrate or location
- the range of dispersal of its propagules
- its modes of persistence in a viable state other than the mycelium.

All these can vary with the biology of the particular fungus under investigation and the environments in which it occurs. Allowance needs to be made for these attributes and their variability when a fungus population is sampled.

## 4.1.1 Mycelial extent

The size and extent of the mycelial phase can differ between genera and species, between different genets of the same species or even between identical genets under different conditions. The common microscopic soil fungi demonstrate this clearly. Their ephemeral mycelial development may occupy only a fraction of a cubic millimetre, or a few cubic millimetres, but the form adopted can differ appreciably. For instance, *Cylindrocarpon* [FA] spp. develop as small patches on tap roots, *Penicillium* [A] and *Aspergillus* [A] often cover small pieces of substrate quite densely, *Mucor rammanianus* [Z] shows a similar behaviour but as nutrients are depleted the hyphae spread out for about 2–3 mm and chlamydospores develop, persisting when the hyphae die, and *Zygorrhynchus* [Z] hyphae appear to grow diffusely through the soil without any particular association with any substrate (Burges 1960). Where substrates are exposed to multiple infections, an apparently continuous mycelial region develops. Although discrete mycelia of a single species, of the same or different genotypes, are detectable for a brief period when numerous conidia germinate in close juxtaposition, their identity is rapidly lost as their mycelia grow, overlap or anastomose, e.g. germinating conidia of a powdery mildew such as *Blumeria graminis* [A] on the surface of a cereal leaf. Alternatively, an apparently continuous mycelial mat may be composed of an intricate three-dimensional mosaic of mycelia of different species or genets , e.g. fungal leaf-surface communities (Dickinson 1976) and Dutch elm disease (*Ophiostoma ulmi* [AP]) bark populations (Lea, cited by Brasier 1984) (Figure 4.1).

Hyphal intermingling or fusion of mycelia pose a particular problem which is exacerbated in unicellular fungi, such as yeasts, when adjacent discrete colonies readily coalesce as they grow and so the initial limits become lost. Nevertheless, it has proved possible to separate throat or vaginal yeast-like *Candida albicans* [A] ('thrush') infections, where different genotypes can often be distinguished although the scale of sampling procedures is, unavoidably, rather coarse. Intermingling is not unknown in hyphal fungi where, for example, confluent basidiocarps were found to belong to the

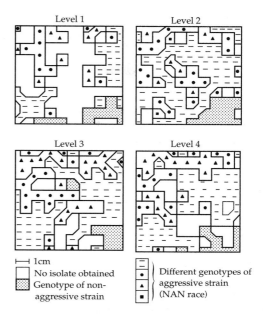

**Figure 4.1** Distribution of VCGs of *Ophiostoma novo-ulmi* [AP] in samples taken at four different levels of the bark at various sites from the outer to the inner region: level 1, conidiophores on inner bark surface; level 2, inner bark surface; level 3, within inner bark; level 4, within outer bark. Note the approximately corresponding positions of each genotype at the different levels. (Reproduced from Brasier (1984), with permission from the British Mycological Society.)

distinct intersterile groups I and II of the minute *Nidularia denudata* [BA] (Burnett and Boulter 1963). However, in many other cases where heterogeneous results are obtained, it is not always possible to determine whether this is due to multiple infections or mutation within an originally uniform colony.

In contrast with the microscopic mycelia of microfungi, perennial mycelia of Basidiomycotina may occupy several square or cubic metres of soil or other substrate, such as wood or roots, often as strands or as a spatially ill-defined mycelium. Mycelia may extend over several square centimetres of root surface, occupy an appreciable volume of the roots and produce numerous, apparently distinct, aerial basidiomata. The space occupied will depend upon the taxon involved, its age, its growth rate and the nature of its host in the case of a symbiotic mycelium. For instance, in Australia, patches of mycelia of *Armillaria luteobulbulina* [BA] apparently ranged from over 24 m$^2$ to 3.5 ha with a mean value of $5500\,m^2$ in association with *Eucalyptus* stands. Seventy-one percent of the areas consisted of single clones, with the maximum number found at any one locality being three (Kile 1983). The maximum rate of expansion of 1.3 m per year compared with 1.5 and 1.6 m per year for *A. mellea* agg. [BA] in stands of Norway spruce (*Picea excelsa*) and Douglas fir (*Pseudotsuga douglasii*) respectively in the UK (Rishbeth 1968), or 5.2 m per year in stands of slash pine (*Pinus elliottii*) in Zimbabwe (Swift 1958). As a genus, *Armillaria* [BA] apparently possesses some of the largest and most persistent mycelial genets known. For example, a single clone, clone I, of *A. gallica* [BA] in northern Michigan (USA) occupies at least 15 ha (Smith *et al.* 1992)! Other perennial pathogenic Basidiomycotina are almost as large and old. Over a dozen large genets of the wood-rotting bracket fungus *Phellinus weirii* [BA] were detected in an area of approximately 9 km × 5 km in a Canadian hemlock forest. In many cases the clones were dispersed over almost a kilometre, no doubt owing to the long history of fire in the region which resulted in the destruction and tree loss in the proximity of the mycelium. Genets ranged in estimated age from over 1100 to 380 years (Dickman and Cook 1989).

In contrast, genets of ectomycorrhizal Basidiomycotina are usually smaller than free-living or parasitic species. For instance, the largest detected genet of *Suillus pungens* [BA], an ectomycorrhizal fungus of Bishop pine (*Pinus muricata*) in California, was about $300\,m^2$ but had a growth rate of about 5 m per year; most free-living genets were smaller (Bonello *et al.* 1998). The size range of genets of different fungi is evidently very great and is clearly affected by the nature of the species, the individual's genome and the environment in which it occurs. Broadly speaking, average genet size increases from the ephemeral Zygomycotina through short-lived Ascomycotina to perennial Ascomycotina and the Basidiomycotina.

### 4.1.2 Substrate specificity and specific site locations

Mycelial distribution often reflects the heterogeneity of the substrate such as soil; for example, mycelia of nutrient-determined soil groups such as the 'sugar fungi' or other rhizosphere fungi are located close to sites of nutrient release and these, in turn, reflect the uneven disposition of roots and decomposing material in the soil. Any association with another organism, animal or plant, will obviously affect the distribution in space of the fungus, and while the basis may be nutritional, this is not always the case. For instance, clusters of chlamydospores or sclerotia of Eumycota or oospores of Oomycota occur in soil and represent all that is left of the sites of infected roots, now totally decomposed. In life, those pathogenic or symbiotic fungi which show host specificity are found most commonly, but not necessarily exclusively, in association with the appropriate host plant. For example, in the root pathogens *Pythium* [O] and *Phytophthora* [O], zoospores show positive taxes and develop in the vicinity of susceptible roots (Deacon 1993). In contrast, trap-bearing mycelia are only induced in nematode-trapping fungi in the vicinity of concentrations of nematodes, but undifferentiated mycelia of these fungi may occur elsewhere in the soil where there is a low concentrations of nematodes (Barron 1977). Any factor which promotes site localization is one possible cause of non-random distributions of either mycelia or reproductive bodies, and the sampling technique adopted should make allowance for such situations.

### 4.1.3 Persistence

There is remarkably little specific information about the persistence and life span of the mycelia of different fungal species, although many can be broadly classified as ephemeral or perennial. Ephemeral soil fungi, for example, may only persist in time for as little as a few hours and certainly no more than a few days, although their propagules can persist far longer in a dormant state. One of the few firm estimates for a perennial mycelium is that of *c.* 25 years for clones of *A. luteobubulina* [BA] in eucalypt forests (Kile 1983), while some of the *Phellinus weirii* [BA] (Dickman and Cook 1989) and *A. gallica* [BA] (Smith *et al.* 1992) referred to earlier are estimated to be more than 100 years and 1500 years old respectively!

There is more information concerning the viability and lifespan of sclerotia and chlamydospores (Sussman 1968), especially pathogenic fungi where such information is relevant to the design of protective treatments. For example, sclerotia of *Verticillium dahliae* [FA] can persist for 14 years in soil, and those of *Sclerotium* [FA] spp. can persist for 2–5 years. Even some spore types are remarkably persistent; for example, conidia of *Helminthosporium sativum* [FA] are viable for up to 20 months, teliospores of *Puccinia recondita* [BR] for 2 years, and oospores of *Pythium ultimum* [O] for months to years. In most cases the period of dormancy is very dependent upon environmental conditions.

A particular feature of many fungi that sporulate in soil is the widespread non-specific general fungistasis which inhibits their germination. A variety of explanations have been given for the cause of this phenomenon, but it is commonly supposed to be due to general energy deprivation in the substrate (Lockwood 1977; Lockwood and Filonow 1981). One consequence of this general fungistasis is that spores may remain dormant for several seasons before active growth recommences. During this period extraneous agents such as soil mites, earthworms or soil water movement can cause appreciable dispersal of dormant propagules, and as a result they come to be adjacent to populations of entirely different origins. However, the distances involved are likely to be of the order of centimetres, although water movement can occur over longer distances, especially under flood conditions. This phenomenon needs to be borne in mind when attempting to define such a soil population. The physiological effects of general fungistasis are ameliorated by the presence of highly organic matter, warmth and moisture, which usually reduce dormancy and longevity.

Different morphological states can often be differentially isolated depending upon the technique used and the viability of different structures. In some cases, propagules retain their viability beyond the life of their parent mycelium, so that sampling them is a measure of the selected descendants of the mycelial phase population, or of a potential new population, if and when they germinate. This kind of problem is particularly relevant to studies of submerged or soil fungi. Some of

the best cases studied are those of Oomycota, notably *Pythium* [O] spp. Figure 4.2 shows the origins of almost 500 dilution plate isolations of Pythiums from two soils over a 6-month period and the differential effects of the plating out media used (Ali-Shtayeh *et al.* 1986).

The majority were derived either from sporangia or from swollen hyphal bodies and very few from oospores, although the origins of nearly 30 per cent were unknown. However, it is possible to sample oospores of *P. ultimum* [O] and other species selectively by using their indifference to various fungicides and to heat. Such treatments enable them to be selected preferentially, germinated and analysed (Stasz and Martin 1988). Another class of structure which persists beyond the mycelial phase is the sclerotium, and this has been the basis of population sampling in *Sclerotinia sclerotiorum* [AD] (e.g. Kohli *et al.* 1992). In many cases such

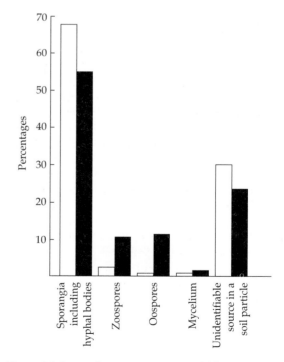

**Figure 4.2** Types and percentage occurrence of different kinds of propagule of *Pythium* [O] spp. in the soil. Isolations were from soil samples from a single site near Reading, UK, using dilution plates over a 6-month period. (Reproduced from Ali-Shtayeh *et al.* (1986), with permission from the British Mycological Society.)

structures can be selectively sieved out from soil or other substrates.

As indicated earlier, one important consequence of sampling persistent dormant bodies in substrates is that a sample can include propagules derived from different successive populations which were once actively growing. Therefore it is desirable to sample actively growing populations simultaneously but separately, together with dormant representatives of presumed earlier generations. These samples, respectively, provide information about the contemporary population and the kinds of genotypes that could contribute to future generations.

The determination of the opposite attribute to population persistence, namely population extinction, is often of importance. A 'population' may be subdivided into spatially separated components which recombine sporadically but which may individually increase or decrease at different rates or in different ways in growth or numbers, together with others which arise and become extinct—a **metapopulation**. Extinction is not always easy to observe unless individual genets or infections have been either marked or mapped. Even if this is done, mycelia may die down and regenerate later, while reproductive structures are neither always produced in successive reproductive seasons nor necessarily at exactly the same site. However, under favourable conditions such observations can be made, and a metapopulation structure has been described for natural populations of *Melampsora lini* [BR] in Australia (Burdon 1993). It is believed by many observers that metapopulations are probably a widespread pattern amongst fungi that are restricted to particular hosts or substrates and only capable of limited dispersal.

### 4.1.4  Propagule dispersal and cloning

Conidia and spores are responsible for relatively widespread dispersal compared with the range of dispersal of chlamydospores and sclerotia, although the latter, in particular, play an important role in persistence in time. The effectiveness of dispersal, either passive or active, can account for heterogeneity in populations. For example, in *F. oxysporum* [FA], genetically distinct localized clones occur in undisturbed natural soil, but distinct clones cannot be recognized in adjacent cultivated soil where a heterogeneous population seems to occur, perhaps reflecting the greater mixing of spores which occurs in the latter owing to agricultural operations (Gordon *et al.* 1992). Frequently a fungus propagates by both conidia and meiotic spores with different dispersal patterns, e.g. *Gremmeniella abietina* [AP] whose conidia are dispersed relatively short distances by rain splash whereas its basidiospores are wind dispersed for several tens of metres at least (Roll-Hansen F and H 1973). In such cases, an apparently single population may be a mixture of locally dispersed clonally propagated isolates and immigrant recombinant isolates which are more widely dispersed. This situation has been detected in the chestnut blight fungus *Cryphonectria parasitica* [AP] where the non-random aggregations of lesions on bark are due to genetically identical clones, probably derived through short-range dispersal of conidia from an existing infection on the tree. These are intermingled at random with genetically heterogeneous infections resulting from the more widely dispersed ascospores from ascomata on adjacent trees (Milgroom *et al.* 1991).

When new genotypes, hitherto unknown, suddenly appear in relatively isolated populations, mutation or rare recombinant events may be suspected, but long-range dispersal can also be a cause. The most notable example of this is probably the occurrence in Australia of East African wheat-rust pathotypes which were almost certainly introduced through very-long-range airborne dispersal across the Indian Ocean (Burdon *et al.* 1982).

## 4.2  Recognition of intraspecific groups, genets and fungal individuals

Specific or subspecific identification in fungi is still based mainly on morphological comparison, but this often proves difficult in practice. For instance, *Fusarium moniliforme* [FA] was originally recognized as a highly variable species which occurred on a wide range of hosts. Unsatisfactory attempts were made to subdivide it on morphological and host-range characters. Eventually, on the basis of confrontations between mycelia derived from single spores, it proved to be divisible into four mutually inter-sterile 'mating groups', a number now increased to seven with the possibility that others may be identified (Hsieh *et al.* 1977; Leslie 1991, 1995). The original four groups were initially described as varieties (Kuhlman 1982), but are now regarded as genetically and phylogenetically distinct **sibling** or **cryptic** species between which there is some host differentiation but whose morphological features overlap. Many similar cases are known and it has been evident for many years that morphologically similar fungi can conceal a variety of biological differences, further exacerbating the problem of recognizing genetically distinct individuals and species (see Chapter 11).

In general, it is often difficult to be sure how taxonomically homogenous a sample of isolates are since they may represent an apparent species, subspecies or variety, or some kind of intraspecific group. Attempts at subdivision may bring to light quite profound differences, as with *F. moniliforme* [A]. A valuable first approach with apparently similar isolates, once they have been cultured and their characteristics noted, is to pair them in all combinations. Any interactions observed may give an indication of the presence of different MTFs or VCGs. Failure of hyphal fusions on confrontations between mycelia of different isolates can also be an indication of the occurrence of different (sibling) mating populations, or even of incompatible MTFs, depending on the species. In Ascomycotina, monokaryotic anastomoses suggest an identity between VCGs but in Basidiomycotina they indicate potential sexual reproduction between species or sibling species. In both groups their absence suggests different VCGs or species.

### 4.2.1 Using the SI reaction

Within a single VCG, there is usually some quantitative, morphological and genetic variation between isolates, which presumably reflects the presence of different individual mycelia having in common only the same genes responsible for the SI reaction. In many fungal species this level of intraspecific discrimination is as far as individual recognition can be taken but it can be used for

limited population analysis. For instance, differences between populations can often be recognized on the basis of their VCG composition alone and these can be of real biological significance, e.g. samples of *F. oxysporum* [FA] along a transect running from cultivated to natural soils (see section 4.3.1.1 and Gordon *et al.* (1992)).

Sometimes, however, as in *Aspergillus nidulans* [AP], there is evidence for at least some additional genetic identity within a VCG. In *A. nidulans* [AP] all the individuals constituting a single VCG possess more features in common than they do with any member of a different VCG (Grindle 1963b). In this way some 19 VCGs have been recognized. Nevertheless, field samples usually included more than one VCG (Figure 4.3) so that they either did not differ in their habitat preferences or occupied spatially distinct microhabitats which had not been separated by the sampling technique used.

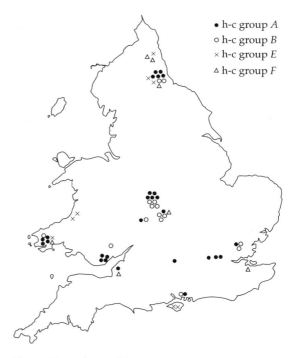

**Figure 4.3** Distribution of four VCGs (*A, B, E* and *F*) of *Aspergillus nidulans* [A] from soil samples in different regions of England. Note that different VCGs can either occur together in the same locality or uniquely and that some are widespread (e.g. *A* and *B*). (Reproduced from Croft and Jinks (1977), with permission from Academic Press.)

In *F. oxysporum* [FA], there is also evidence for considerable genetic identity between individuals of the same VCG since the RFLP patterns of the mtDNA are usually identical or very similar within each VCG but differ between them (Correll 1991). There may be a correlation of VCG type with other characteristics such as pathogen types, as in *F. oxysporum* f.sp. *apii* [FA] (Correll *et al.* 1986), although in other taxa, e.g. *F. oxysporum* f.sp. *lycopersici* [FA] (Elias and Schneider 1991), there is no obvious correlation. *Cryphonectria parasitica* [AP] is a good example of how VCGs have been used for a variety of purposes, namely:

- for analysing population structure (Milgroom and Cortesi 1999)
- to distinguish different populations, e.g. those on different species of oak (Nash and Stambaugh 1989)
- to infer from the occurrence of new VCGs in a population that this was due either to recombination or immigration from an adjacent population (Anagnostakis and Krantz 1987), hypotheses which could be tested by further genetic analysis
- to infer intercontinental spread of the fungus because of the occurrence of identical VCGs in different continents (Anagnostakis *et al.* 1986).

Originally, amongst the Basidiomycotina, the MTFs of diaphoromictic mating systems were used to characterize mycelial types, e.g. *Piptoporus betulinus* [BA] (Burnett and Partington 1957) and *Marasmius oreades* [BA] (Burnett and Evans 1966), and combined with other characters these are still useful markers. However, the character in these fungi that has proved most valuable for the recognition of individuals is the SI reaction between basidiomycotous dikaryons or equivalent mycelia. As indicated earlier, the SI reaction is not completely definitive. In some cases a misleading result may be given simply because the conditions for successful fusions are not right; for example, two compatible isolates of *Suillus bovinus* [BA] failed to fuse when inoculated at 10 mm apart but at a distance of 2 mm all reactions were unambiguous (Sen 1990). It also needs to be remembered that, even though incompatibility is generally a reliable guide to genetic non-identity, compatibility is not a *guarantee* of genetic identity; for example,

some somatically compatible *Suillus granulatus* [BA] isolates were not genetically identical as assessed by RAPD markers (Jacobson *et al.* 1993).

## 4.2.2 Using pathogenicity characters

As the genetic basis for virulence became known and specific virulence genes identified, pathotypes or races were identified by the presence of a particular virulence gene which was recognized by the use of the host cultivar carrying its 'matching' resistance gene (Wolfe and Schwarzbach 1975). This kind of information has many applications. In particular, pathotype composition has been used for estimating and comparing both phenotypic and genotypic variability in different populations of *P. graminis* [BR] in North America (Groth and Roelfs 1986) and the structure of European populations of *Blumeria graminis* [AD] (Limpert *et al.* 1990). However, it has to be borne in mind that the only certain genetic similarity between members of the same race or pathotype is for the virulence gene(s) carried; individuals may differ in numerous other ways genetically.

## 4.2.3 Using molecular markers

Although the methods described above to define biologically significant groups are useful, they only provide partial solutions to individual recognition. More information is needed to characterize a fungal individual than the use of a single phenotypic or genetic character. It can best be achieved by employing additional genetic information derived from isozymes, nDNA RFLPs or RAPD–PCRs, SSRs and mitochondrial data. For instance, VCGs, pathogenicity, isozymes and geographical distribution were used in combination to distinguish a number of biological groups within isolates of *Fusarium oxysporum* var. *vasinfectum* [FA] from crucifers (Bosland and Williams 1987). Even finer discrimination was achieved with *F. oxysporum* f.sp. *ciceris* [FA] on chick peas by combining RAPDs based on three different primers with pathogenicity data based on the differential reactions of 10 chickpea testers followed by principal co-ordinate analysis (Kelly *et al.* 1994). Not only could individual isolates be recognized, but the two kinds of pathotype—the wilt-inducing and leaf-yellowing—could be unequivocally distinguished.

Even finer discrimination can be made using SSRs, as with *Microbotryum violaceum* [BS]. Shykoff *et al.* (1999) used SSR polymorphism to distinguish otherwise morphologically indistinguishable individuals of this fungus on three hosts (*Silene vulgaris, Silene nutans* and *Dianthus carthusianum virginiana*) growing in the same area. When more than one isolate was inoculated onto the same host they could still be distinguished. Discrimination of a different kind was shown by Wang *et al.* (1997), who were even able to demonstrate the probability that successive isolates sampled in a population of *Gremmeniella abietina* [AP] would carry the same marker genes, although originating from different sexual events (*See 7A3*).

### 4.2.4 Summation

The object of the discriminatory procedures adopted must always be related to the reasons for undertaking the discrimination or description in the first place. If all that is required is sufficient information to ensure that disease-resistant crops are planted, then pathotype discrimination may well be sufficient. But if the object is to describe and compare population structures at different sites or under different regimes, a more detailed analysis of genotypes and individual genes present or absent may be required. When the identification of an individual, genet or genotype is essential, as many different kinds of available characteristics which can practically be included should be employed.

A classic example of the use of such a technique is the characterization and separation of very large, ancient and intermingled clones of *Armillaria gallica* [BA] and *A. ostoyae* [BA] (Smith *et al.* 1992, 1994). The species can be separated by their inter-sterility and identified positively using rDNA RFLPs or sequences. Morphology, SI, MTFs and both mtDNA and nDNA RFLPs were used together (Hodnett and Anderson 2000). The somewhat unexpected finding was that a characteristic but different genetic identity was demonstrated across the entire area of each genet sampled. Whether or not each was a single large intact mycelium or a population of long-lived clones (ramets) derived

from an initial infection was not determined. It might have been supposed that mutant sectors or mycelia would have arisen over the course of hundreds or even thousands of years, but this was not detected. Similar stability was indicated by the identity of MTFs around the periphery of intact fairy rings of *Marasmius oreades* [BA] whose ages were variously estimated to be several tens or hundreds of years (Burnett and Evans 1966; Mallett and Harrison 1988). Other characters were not studied. The cause of the MTF stability was not understood but could well reflect some form of stabilizing selection (see Chapter 9, section 9.4.3.2).

## 4.3 Sampling

Except in very unusual circumstances, any practical investigation of a population has to rely on sampling, which further increases the difficulties of recognizing the identity and spatial extent of individual mycelia. It raises questions such as what sample size and frequency of sampling is best, or even what fungal element is it best to sample—spore, sexual reproductive structure or mycelium? Therefore effective sampling must be on a scale related to the spatial spread of individual fungal mycelia, their morphology and their reproductive biology to allow for the distribution of propagules of all kinds. In practice, many of these are usually unknown quantities at the outset of an investigation. Unfortunately, therefore, identifying fungal individuals or populations and determining their breeding behaviour in nature is a time-consuming process.

In the vast majority of cases the extent of a population is not obvious on inspection, and even apparently superficial mycelia can be embedded in some substrate so that all the population may not even be visible! However, some knowledge of individual and population limits is an essential requirement to distinguish different genets, populations or subpopulations. Moreover, it is often desirable to know whether the individuals or genets are clustered, dispersed evenly, or randomly distributed. In addition, although not always necessary, it is frequently useful to be able to discriminate between individuals, for instance as a necessary step in investigating breeding behaviour or in determining the frequency of different genotypes in the population. To meet all these differing requirements, the sampling strategy needs to rest upon all available knowledge about the biology of the organism, particularly its growth habit and reproduction.

### 4.3.1 Spatial sampling procedures

Sampling involves the selection of the site from which the sample(s) will be taken, the number of samples taken, the frequency of sampling and the actual manner by which the sample(s) are obtained.

#### 4.3.1.1 Site sampling
The prime objective of site sampling is to determine the spatial location and limits of both individual genets and the population as a whole. As far as the mycelium and perennating structures are concerned, sampling methods based on those developed and used widely by ecologists and agriculturalists are applicable (e.g. Mueller-Dombois and Ellenberg 1974). When only minimal information is available, some form of hierarchical grid sampling, such as at the points of intersection of a rectangular grid at intervals of, say, 10 m, 1 m, 10 cm, 1 cm etc., is probably the best initial approach.

An alternative pattern of sampling is to examine sites within successive zones of concentrically divided circular areas. This may be appropriate if the fungus under study is known to originate from a particular site or limited region, or if the extent of spatial invasion by other species or individuals is to be detected. However, because most analyses of ecological or genetic population data assume that sampling has been at random, even in a hierarchical scheme, samples should be taken at random at each hierarchical level or in each concentric zone. Nevertheless, it has to be recognized that, inevitably, even the best sampling procedure will place some limitation on the information which can be gained.

For fungi embedded within a substrate, a grid can, of course, be extended to three dimensions by sampling upwards and downwards as well as laterally. In soil, sampling may well be related to different soil horizons, including the litter layer(s) and the general disposition and geometry of roots in the soil, or those of particular species. In woody substrates it may relate to softwood or hardwood regions, or to regions exhibiting different degrees of decay. One method of particular applicability to fungi whose mycelium is largely soil-borne or immersed in some substrate such as wood, is to plate samples from cores taken at different sites on a selective medium. Unfortunately, only a very few highly selective media are known. A good example is that developed for isolating *Phaeolus schweinitzii* [BA] from either wood or soil samples (Barrett 1978), or for soil Fusaria [AF] (Gordon *et al.* 1992).

Above ground level, samples may be taken at different heights in relation to features such as the nature and height of the vegetation canopy, or from different regions of the aerial shoot, i.e. basal leaves, stem leaves, older or younger leaves, inflorescence structures, fruits etc. In the case of pathogenic or symbiotic species, sampling sites are often defined by the distribution of lesions or mycelia on the hosts. In addition, it is often important to have some idea of both the extent and the effectiveness of the dispersal of fungal propagules. This involves using some kind of atmospheric sampling device followed by germination tests of the propagules so isolated.

In practice, the sampling scale will need to adjusted to match the probable scale of fungal individuals or structures in, or on, the substrate and its effective range of dispersal while, at the same time, satisfying statistical considerations. Six examples, two for subterranean fungi, two for fungi above ground and two for airborne propagules, will indicate the kinds of procedure used.

*Example 1* Gordon *et al.* (1992) wished to compare soil populations of *Fusarium oxysporum* [FA] in adjacent regions of cultivated and uncultivated land. Therefore they took six samples at 10-m intervals in each area along a linear E–W transect extending from the cultivated into the uncultivated land with a gap of 35 m between the main sampling areas. The soil samples were cores of dimensions 4 cm × 20 cm taken at depths of 2 cm and 10 cm both on the transect and within 5 cm on either side of it. The soil from each sample was homogenized, dilution plates were made on *Fusarium*-selective medium from each sample and the first 10 samples from each plate were used for genetic assay.

*Example 2*   Most toadstool populations have been sampled by isolating mycelium or basidiospores from the aerial basidiocarps only, but this neglects entirely the bulk of the mycelium buried in the substrate and those mycelia which are not producing basidiocarps.

Gardes and Bruns (1996) were particularly concerned to identify the population structure, both above and below ground, of a community of ectotrophic mycorrhizal and saprophytic fungi growing in a *Pinus muricata* forest. Therefore they took soil cores of dimensions 10 cm × 25 cm or 10 cm × 40 cm beneath, or within, a 50-cm radius around an aerial basidiocarp. Mycelium and mycorrhizal roots were then separated, and fungal material was isolated from the soil by a sieving, washing and dissection procedure. Although mycorrhizal roots were initially classified into differing morphotypes, DNA was extracted and two specific regions of nDNA and mtDNA known to be diagnostic for several agaric species were investigated using PCR primers and RFLPs. The nDNA region was that between the 18S and 28S regions and the highly conserved region of the mtDNA determining the large rRNA subunit. In this way basidiocarps, mycorrhiza and mycelium present in the samples were diagnosed and some semi-quantitative data obtained on numbers of basidiocarps and the frequency and abundance of different kinds of mycorrhizas (Figure 4.4).

Although the most common fungus in each core sample was usually also the most common above ground nearest to the centre of the core, there were remarkable differences below ground. For instance, although basidiocarps of *Suillus pungens* [BA] were amongst the most common, they only seemed to be associated with traces of subterranean mycelium, whereas basidiocarps of *Russula amoenolens* [BA], only detected twice in the three successive aboveground annual samples, were amongst the most abundant mycorrhiza detected annually under the soil. Individual clones were not recognized in this work, but it could have been further elaborated by identifying genotypes using RFLPs or the PCR technique. This study demonstrates clearly the

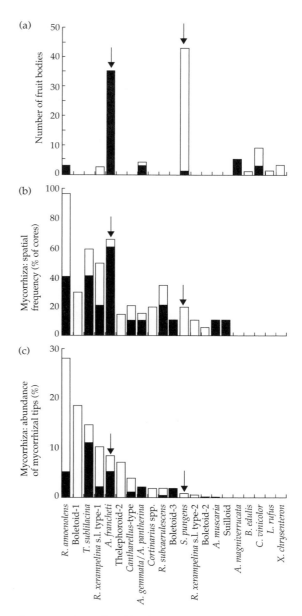

**Figure 4.4** The number of basidiocarps and both the spatial frequency and abundance of mycorrhizas based on soil core samples taken beneath basidiocarps of *Amanita francheti* [BA] and *Suillus pungens* [BA]: (a) numbers of basidiocarps within 1 m of the sampling sites recorded over a 3-year sampling period; (b) percentage number of cores in which mycorrhiza of a given species were present; (c) percentage abundance based on the number of mycorrhizal root tips recorded. (Reproduced from Gardes and Bruns (1966), with permission from the National Research Council of Canada.)

importance of sampling both above and below ground (or on the surface of, or within, any substrate) if the full extent of any population is to be determined. A comparable study (Dahlberg *et al.* 1997) in an old Norway spruce forest in Sweden gave similar results but employed a circular quadrat technique; but the subterranean mycorrhizal sampling was just outside the area to prevent interference with normal growth within it.

*Example 3*  A more elaborate sampling procedure was used to investigate Canadian populations of *Sclerotinia sclerotiorum* [AD] infecting canola (rape; *Brassica napus*) in four fields.

Two fields were sampled at each of two areas 450 km apart (Kohli *et al.* 1995). Fields 1 and 2, both of area 0.6 km$^2$, were 13 km apart and fields 3 and 4, of area 0.4 km$^2$ and 1.3 km$^2$ respectively, were 6 km apart. In each field an 800 m × 400 m grid was laid out to provide 128 quadrats of dimensions 50 m × 50 m, and these were sampled in two ways. Firstly, single samples, exactly 50 m apart, were taken from a fixed point in each quadrat. Secondly, one quadrat in each field was selected at random for intensive sampling at five sites on a diagonal transect which included the original sampling site. The actual samples collected were as follows:

- any apothecia at the sample sites
- sclerotia from eight to ten infected stems
- at full bloom, one petal selected at random from each of four flowers from five inflorescences.

Sampling was carried out in successive years at equivalent times. Once these various samples had been isolated, cultured and genetically identified, they provided basic information on the actual and potential infecting populations.

- Sclerotia provide a sample of the stem-infecting population and, since they are also responsible for asexual propagation and dispersal, indicate a measure of this infective potential.
- Petal samples compared with the sclerotial samples provide information on infection within the population as well as any air-borne infection from beyond the sampled area.
- Apothecial samples provide information about the potential air-borne infective population arising within the area.

*Example 4*  An example of a more elaborate, hierarchical but semi-randomized sampling procedure for a graminaceous foliar pathogen *Septoria tritici* [A] is shown in Figure 4.5. (McDonald *et al.* 1995).

This procedure enables genetic diversity to be partitioned:

- among locations in a field
- between leaves at a location
- among lesions on a leaf
- between pycnidia within a lesion.

*Example 5*  Special techniques are necessary to sample airborne spores. Indirect sampling of potentially mobile phases by trapping sedimenting spores or conidia using spore traps at a series of sites is probably the most common and simplest method of sampling the dispersion of airborne spores. Their profile can then be compared with that of local fungi. A variety of traps can be used—whirling arm samplers, cascade impactors and continuous volumetric slit traps (Gregory 1973). The abundance and diversity of basidiospore clouds of *Schizophyllum commune* [BA] in the Caribbean area was investigated by sampling five

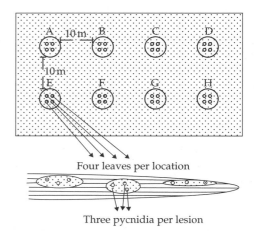

Four leaves per location

Three pycnidia per lesion

**Figure 4.5** Hierarchical sampling technique applied to different sites and pycnidial leaf lesions of *Mycosphaerella graminicola* [AP] on wheat: (a) location of samples in the field; (b) location of samples on each leaf. This technique enables comparisons to be made between field sites, between leaves at a site, between lesions on a leaf and between lesions on the same leaf. (Reproduced from McDonald *et al.* (1995a), with permission from the National Research Council of Canada.)

sites over a limited period using a range of genetically defined monokaryons as detectors. Compatible basidiospores could form dikaryons with the monokaryons in the traps and their genotypes could be determined. The clouds were shown to be genetically highly heterogeneous, as judged by their RFLP profiles, and largely reflected the local terrestrial populations. There was little evidence for extensive gene migration (James and Vilgalys 2001).

*Example 6*   During the growing season vast physically and genetically heterogeneous airborne clouds of barley or wheat mildew conidia (*Blumeria graminis* f.sp. *hordei* or *tritici* [A]) are dispersed over Europe. Their genetic structure was investigated.

The cloud composition varies with the crop grown, the season and climatic conditions. Knowledge of the cloud composition is important for detecting genetic changes and gene migration and, ultimately, for the application of control measures. Unfortunately, the concentration is largely below the detection level of the more usual spore traps, but a jet spore trap which can be mounted on the roof of a vehicle has been designed to overcome this limitation and, in effect, to provide an adequate air sample (throughput *c.* $400 \, l \, min^{-1}$) either at a fixed point or along a transect traversed by the vehicle (Schwarzbach 1978). Within the trap the conidia sediment onto living leaves of a host plant lacking resistance genes and can then be transferred to differential hosts whose infection patterns indicate the pathotypes present and their relative frequency.

It will be obvious from the examples described in this section that preliminary investigations to obtain some idea of the most appropriate scale on which to sample as well as the kinds of sample to take, will usually repay the extra effort involved.

### 4.3.1.2 Sample size
Obviously the number of samples actually taken will need to be related to any statistical procedures which may be employed for population description or genetic analyses. In general, increasing the numbers of distinct recognizable individuals is likely to be helpful to most kinds of analysis (Brown 1996). An absolute minimum of about 25–35 samples is desirable, but 100 or more is far more

satisfactory. However, sample size will always be constrained by the balance between the effort and expense involved in collecting and the potential gain in benefit to the intended analysis. Apart from determining the limits to any population, two other important properties which need to be investigated are the distribution patterns of individual mycelia or genets within a site, whether random, regular (uniform) or aggregated, and the scale of such patterns and its persistence. The latter can only be determined by direct observation.

### 4.3.1.3 Pattern recognition
In many cases the intrinsic pattern of a mycelium can be observed, such as 'fairy rings' in which the mycelium is relatively toroidal and compact in shape growing forward only at the outer margin, e.g. *Marasmius oreades* [BA]. In other cases, such as with many ectomycorrhizic Basidiomycotina, the external mycelium is more evanescent and a ring of basidiomata develops in the litter centred on an infected tree from whose roots the external mycelium radiates as strands or individual hyphae, e.g. *Amanita muscaria* [BA] around birch trees. The ring of basidiomata reflects a particular region of this external mycelium at which appropriate physiological differentiation occurs, resulting in the differentiation of reproductive structures. In others, such as the dry rot fungus *Serpula lacrymans* [BA] or the stinkhorn *Phallus impudicus* [BA], the mycelial pattern of strands, aggregations or basidiomata reflects the nutritional state of the substrate at the time that the growing edge of the mycelium reached it—strands are related to low nutrient status, whereas aggregations and basidiomata are linked to higher nutrient sites.

When the population is composed of numerous adjacent discrete mycelia, their degree of dispersion (random, uniform or clumped) may be of significance (Figure 4.6).

Clumping is of particular interest and various methods have been developed to detect it. Techniques based on sampling and comparisons of mycelial frequency in quadrats as applied to macroscopic organisms (Greig-Smith 1952; Dale 2000) can be used in some circumstances, and more elaborate measures have been developed to define both variability of point density within a prescribed

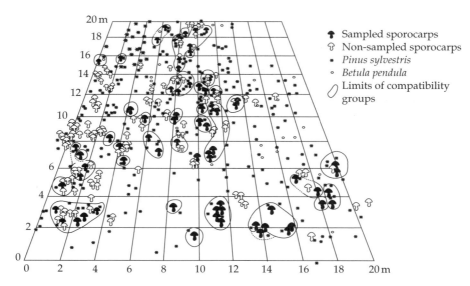

**Figure 4.6** Diagram illustrating the distribution of clones of *Suillus bovinus* [BA] in an old sandpit colonized by pine and birch at Björklinge, Sweden. Aggregations of basidiocarps belonging to the same clone are enclosed and the clones are disposed at random. (Reproduced from Dahlberg and Stenlid (1990), with permission from the Trustees of the *New Phytologist*.)

area and the relative position of the points within a wider enclosing area such as a field, e.g. incidence of diseased plants in a field or community (Ferrandino 1998). However, this is not always practicable nor sufficiently accurate to apply in other situations, especially on a small or microscopic scale.

The techniques of spatial autocorrelation, developed by geographers and applicable at any scale, have been adapted to biological situations in general (Sokal and Oden 1978) and by mycologists in particular (Milgroom *et al*. 1991; Milgroom and Lipari 1995a). In essence, patterns representing the Euclidean distances between mycelia belonging to identical genets, or genets having some genotypic element in common, which can be joined by 'Gabriel lines'* to form 'Gabriel networks', are compared with an equivalent pattern where the different genets are computed as though randomly arranged. A significant difference when the

patterns are compared indicates clumping of the genets composing the population. This method was used to investigate cankers of *Cryphonectria parasitica* [AP] on the bark of infected chestnut trees. Each canker was typified by determining its VCG or, in a second study, by using both nuclear and mitochondrial RFLPs. Milgroom and colleagues (Milgroom *et al*. 1991; Milgroom and Lipari 1995a) were able to demonstrate that VCGs, nuclear genotypes and mitochondrial haplotypes were all significantly clustered on individual trees. The proportion of DNA fingerprints and mitochondrial RFLPs shared between pairs of isolates within trees was correlated with the Euclidean distance between cankers. Therefore they suggested that this reflected the limited dispersal of conidia around cankers, a fact consistent with the observation that conidia are dispersed by water on the surface of trunks. However, since self-fertilization can occur in this species, localized dispersal of ascospores was also possible. In contrast with these observations it was found that VCGs were distributed at random between trees. The authors suggested that this reflected the potentially wider dispersal by ascospores.

* A 'Gabriel line' is a line which can be drawn between two contiguous points, A and B, and defined as the diameter of a circle which can be drawn to include only those points and exclude any other point C outside that circle, or when the angle ABC is greater than 90°.

*4.3.1.4 Practical techniques for isolating and recognizing fungi*

It is not really practicable to provide adequate guidance on the identification of fungi and methods for their isolation in this book. Few world monographs exist but some comprehensive aids to identification to the generic level are available for some widespread groups, usually of economic importance, e.g. Ascomycotina (Hanlin 1998a,b), Fungi anamorphici (Barnett and Hunter 1998) and Urediniomycetes (Cummins and Hiratsuka 1992), but local mycofloras are best consulted at specific and varietal levels. Similarly, practical techniques used to obtain samples and isolate a fungus will vary with the species and the associated substrate under investigation. General methods for some groups are available and useful compendia include the following: general methods (Gams 1992), soil fungi (Johnson *et al*. 1959; Grigorova and Norris 1990; Davet and Rouxel 2000) and air spora (Gregory 1973). Not infrequently, however, standard methods will need to be adapted to the particular situation under investigation and the most effective and up-to-date methods are best sought in individual papers published in appropriate journals. In many cases the precise identity of the fungus can only be made by inducing the reproductive phase, and this is not always easy. Few generalizations can be made but, not infrequently, the sexual reproductive phase is induced by employing low levels of nutrients.

Where the teleomorph is not available or absolute taxonomic discrimination is required, it is now evident that this can be achieved most reliably by using appropriate molecular markers (see Chapter 3, section 3.5, and Chapter 11).

## Further reading

Correll JC (1991). The relationship between formae speciales, races, and vegetative compatibility groups in *Fusarium oxysporum. Phytopathology*, **81**, 1061–4.

Dahlberg A, Jonsson L and Nylund J-E (1997). Species diversity and distribution of biomass above and below ground among ectomycorrhizal fungi in an old-growth Norway spruce forest in south Sweden. *Canadian Journal of Botany*, **75**, 1323–35.

Groth JV and Roelfs AP (1986). Analysis of virulence diversity in populations of plant pathogens. In *Populations of plant pathogens: their dynamics and genetics* (ed. MS Wolfe and CE Caten), pp. 63–74. Blackwell Scientific, Oxford.

Milgroom MG (1995). Analysis of population structure in fungal plant pathogens. In *Disease analysis through genetics and molecular biology: interdisciplinary bridges to improve sorghum and millet crops* (ed. JF Leslie and RA Frederiksen), pp. 213–29. Iowa University Press, Ames, IA.

Smith ML, Bruhn JN and Anderson JB (1994). Relatedness and spatial distribution of *Armillaria* genets infecting red pine seedlings. *Phytopathology*, **84**, 822–9.

Sokal RR and Oden NL (1978). Spatial autocorrelation in biology. I. Methodology. *Biological Journal of the Linnean Society*, **10**, 199–228.

Sussman AS (1968). Longevity and survival. In *The fungi: an advanced treatise*. Vol. II: *The fungal population* (ed. GC Ainsworth and AS Sussman), pp. 447–85. Academic Press, New York.

# CHAPTER 5

# Describing and analysing populations: basic genetic and phylogenetic aspects

The minimum requirement for describing and comparing populations in both space and time is to be able to identify and distinguish them from each other. Such discrimination is difficult in fungi because of the inherent problem of recognizing distinct individuals (see Chapter 4). By inspection alone, therefore, it is only possible to describe a population in territorial terms as a group of adjacent fungal mycelia or reproductive structures rather than as a group of individuals. However, if either the spatial extent of the mycelia or the genotypic constitution of the reproductive structures is known, the population can be described as:

a pattern of distribution of individuals with distinctive but comparable morphology or genotypes (however determined) in space and time.

This limited but practical definition is at least a starting point (see, as an example, Fig. 4.6). It enables patterns or changes in a contemporary population to be distinguished or described. However, further analysis must rest on inference and conjecture.

Further analysis of a population involves, minimally, some knowledge of the breeding behaviour of the individuals composing it and the range of dispersal of their spores or other propagules. Some idea of the limits of the population as a biological and genetic unit can then be obtained and some estimate made of how far the territorial entity is also a breeding entity. In addition, if it can be obtained, some general idea of the history of the population is also desirable. With these basic data, the definition of a population can be extended to:

a group of individuals possessing one or more integrative attributes which results in their functioning in some way

as an ecological and evolutionary unit, e.g. as an inter-breeding group, but distinguished in some way(s) from other such units in space, time or biologically.

Potentially, this definition extends investigation to the dynamics of populations in space and time, and hence to an analysis of the possible causes of population changes and, ultimately, evolutionary changes. However, to define a fungal population in terms adequate for a fully descriptive population genetic analysis, the following information is required:

- observations over several generations
- sampling of the visible and non-visible members who, nonetheless, may contribute genetically from time-to-time to the gene pool
- very often, additional genetic experimentation
- some information, preferably as far back as possible, about the origin and development of the population.

In the best-documented cases sufficient information is available to enable reasoned hypotheses and extrapolations to be made concerning the genetic make-up of a territorially defined population, its probable spatial and reproductive limits, and sometimes its probable history.

There can be considerable difficulties in getting even this far. The difficulty of defining a fungal individual has already been described. Even when the individuals making up a population can be defined phenotypically or genotypically, their breeding behaviour is rarely immediately obvious. For instance, the human pulmonary pathogenic fungus *Coccidiodes immitis* [FA] is known at two geographically distinct infection sites in California. At each site, isolates from groups of localized infected individuals can be defined genotypically. Each population is distinct but genetically highly variable, suggesting cryptic sexuality, yet the fungus apparently lacks any sexual reproduction (Burt *et al.* 1996). Of course, sexual reproduction may take place in the saprophytic stage in the soil where the fungus occurs sporadically (Greene *et al.* 2000) but, if so, it has never been recorded. Many other Eumycota (*c.* 18–20 per cent of all described species) are known only from their anamorphic state, and many more potentially sexual species are commonly found only as anamorphs in nature. Even if a teleomorphic phase is found, whether rarely or commonly, the range of possible sexual behaviour is still considerable and not immediately evident. Morphological evidence for regular sexual reproduction by itself adds little to an understanding of the behaviour of a population because obligate outbreeding, infrequent outbreeding, predominant inbreeding without the exclusion of outbreeding, obligate inbreeding or any combination of these can occur at different times and in different places, together with both facultative and obligate amixis! Thus a separate investigation into the reproductive biology of a fungus usually has to be made before its population behaviour can be fully understood. It is becoming increasingly apparent that in many supposedly asexual populations some recombination may occur, or has occurred in the past, by rare sexual recombination or even parasexual processes. How frequently such processes occur in nature is uncertain. Lastly, since present population structure can often be understood only in the light of its past history, further detailed knowledge and analysis is required, preferably from a wider area than that in which the population of immediate interest occurs.

As well as a knowledge of the phenotypes and genotypes of the individuals composing the population and its size and duration (see Chapter 4), the most important and useful genetic attributes of a contemporary population are as follows:

- measures of phenotypic or genetic diversity in the population
- assessment of the similarities and differences between individuals in the population

- assessment of any evidence for phenotypic or genetic clustering of individuals within the population and, if so, whether such clusters can be correlated with any other factor(s)
- evidence for the import or export of propagules of any type between populations, i.e. determinants of gene flow
- breeding behaviour—whether inbreeding or outbreeding is predominant, the frequency of outbreeding relevant to inbreeding, asexual or amictic behaviour.

The first three aspects will be discussed in this chapter, migration in Chapter 6 and breeding behaviour in Chapter 7.

## 5.1 Phenotypic and genetic diversity

Phenotypic diversity can often provide indirect information about the genetic attributes of a population and is usually easier to observe and quantify than genetic diversity itself. Nevertheless, a basic attribute of any population is its genetic diversity, which is made up of two components:

- gene diversity—the numbers and frequencies of different alleles at individual loci
- genotypic, or genotype, diversity—the numbers and frequencies of genotypically distinct individuals in a population.

The distinction is important. Gene diversity is *unaffected* by the mating system of the fungus, but reflects the size and age of the population and can be affected by selection. In contrast, genotypic diversity *is profoundly affected* by the mating system. Outbreeding promotes a great diversity of different genotypes, whereas inbreeding or clonal reproduction does not. Different genotypes are, of course, also subject to selection.

### 5.1.1 Phenotypic diversity

The provision of some measure of the diversity of a population is desirable as well as the possibility of comparing the similarities or differences between populations. Two thorough investigations of these properties as applied to fungal populations are the use of different measures of phenotypic diversity applied to the statistics collected annually by the US Department of Agriculture Cereal Rust Laboratory of the annual numbers of physiological races of wheat rust (Groth and Roelfs 1987) and data on virulence in barley powdery mildew in Europe (Müller *et al.* 1996). In the latter case, e.g. virulence in *Blumeria* (*Erysiphe*) *graminis* f.sp. *hordei* [AD/FA], the phenotypes recognized can be equated precisely with defined genotypic attributes, since they are determined by the resistance genes present in the barley host.

The simplest measure of phenotypic diversity $H$ is the ratio of phenotypes $r$, such as the number of pathotypes or VCGs in a population, to the total number $N$ of isolates sampled:

$$H = r/N.$$

However, this ratio is not very useful as it is very sensitive to sample size—the rarer the phenotype, the larger the number that needs to be sampled to detect it.

More elaborate indices exist to describe diversity in different populations. Some measures of diversity, developed initially by ecologists to compare the species richness of different communities, have been adopted to describe both phenotypic and genotypic diversity. One of these, a simple measure which is less sensitive to sample size, is the Gleason index:

$$H_G = (r - 1)/\ln N$$

where $r$ and $N$ are as defined for the previous measure. An improvement in sensitivity is achieved because as $r$ increases, the measure of the sample size diminishes through the use of its logarithmic value.

More sophisticated indices, i.e. the Shannon–Weaver index (Shannon and Weaver 1949) and Simpson's index of diversity (Simpson 1949), derived from ecological applications are commonly used for both phenotypes and genotypes. Both of these indices use the number and frequency of different genotypes or phenotypes. They are defined as follows:

$$H_{\text{Shannon}} = -\sum_{i=1}^{n} \ln p_i$$

or

$$H_{\text{Shannon}} = -\sum (n_i/N) \ln (n_i/N)$$

and

$$H_{\text{Simpson}} = 1 - \sum \frac{n_i(n_i - 1)}{N(N - 1)}$$

where $p_i$ is the frequency of the $i$th element, $n_i$ is the number of isolates of the $i$th element and $N$ is the size of the sample. When the sample size is variable, the following correction (Sheldon 1969) should be applied to the Shannon–Weaver index:

$$H_{\text{Shannon/corr}} = \frac{-\sum_i (p_i \ln p_i)}{\ln N}.$$

The complements of both indices are sometimes used, i.e. $1 - H_{\text{Shannon}}$ or $1 - H_{\text{Simpson}}$.

These broad indices of diversity have been used frequently, e.g. with rust populations (Groth and Roelfs 1987) and *Cochliobolus carbonum* [AP] (Leonard and Leath 1990). However, diversity is a complex notion. Two important components of the diversity of a population are its 'richness' and its 'evenness'. These concepts have not been applied very often to fungal populations but, stated most simply, **richness** is the number of different phenotypes divided by the total number in a sample, and **evenness** is the relative abundance of different phenotypes in a sample.

Sample size affects both the Simpson and Shannon–Weaver indices to some extent (Groth and Roelfs 1987; Müller *et al.* 1996), and also affects richness and evenness. In particular, small samples may fail to detect either.

The effect of sample size on the Shannon–Weaver and Simpson indices of diversity was clearly demonstrated by using large numbers of European-wide samples of *B. graminis* f.sp. *hordei* [AD] pathotypes collected at random as a number of regional sub-samples (Müller *et al.* 1996). It is evident that, although the Shannon–Weaver index is strongly affected by sample size below a figure of about 400, the Simpson index is remarkably unaffected above about 80–100 (Figure 5.1(a)). In this investigation richness $d$ and evenness $E$ were calculated as follows:

$$d = s_{\text{sub}}/n_{\text{sub}}$$

$$E = 1/\lambda_{\text{sub}} - 1/\exp H_{\text{sub}} - 1$$

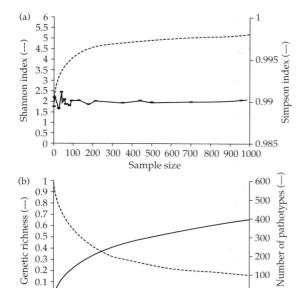

**Figure 5.1** Relationship of various attributes to sample size based on samples from the European population of barley mildew *Blumeria graminis* f.sp. *hordei* [A]: (a) relationship between sample size and two measures of diversity (the Shannon–Weaver index and the Simpson index); (b) the relationship between sample size and genetic richness. (Reproduced from Müller *et al.* (1996) with permission of Kluwer Academic.

where $s_{sub}$ is the number of different pathotypes, $n_{sub}$ is the size of the smallest sample, $\lambda_{sub}$ is the complement of the Shannon–Weaver index (i.e. $\lambda_{sub} = 1 - \lambda$) and $H_{sub}$ is the Shannon–Weaver index in random sub-samples of size $n_{sub}$. As might be expected, richness was inversely correlated with sample size in this series of samples, but this was not true of evenness (Figure 5.1(b)). There was also a high correlation (0.96) between the Simpson index and evenness, although the correlation between this index and richness was only 0.55. In addition to richness and evenness, two further measures—complexity and dissimilarity (see section 5.1.2) were also investigated.

The Simpson index is probably better than the Shannon-Weaver index as a general measure of genetic diversity for the following reasons:

• the Shannon–Weaver index may be more strongly affected by the sample size than the Simpson index

• the Simpson index is generally highly correlated with the other measures described above

• in practice, samples or populations are rarely of the same size, but correlations between the Simpson index and both complexity and dissimilarity are usually high (0.73 and 0.81 respectively in this example).

## 5.1.2 Phenotypic similarity and dissimilarity

It is often useful to know just how much similarity or dissimilarity there is between two or more individuals or populations. A number of indices, borrowed from ecological analyses, have been proposed for such comparisons.

### 5.1.2.1 Similarity

Numerous coefficients of similarity $S$ are in use. The most common are Jaccard's coefficient of similarity $S_J$ (Jaccard 1908), the simple matching coefficient $S_{SS}$ (Sneath and Sokal 1973) and Dice's index $S_D$ (Lynch 1990) which, with binary data, i.e. presence or absence scored as 1 or 0 respectively, is equivalent to the nucleotide diversity measure $S_{NL}$ developed by Nei and Li (1979). These measures of similarity are given by

$$S_J = \frac{n_{xy}}{n_x + n_y} - n_{xy}$$

$$S_{SS} = n_{xy} + \frac{n_{-xy}}{n_x + n_y}$$

$$S_D = \frac{2n_{xy}}{n_x + n_y}$$

$$S_{NL} = \frac{n_{xy}}{(n_x + n_y)/2}$$

where $n_x$ and $n_y$ are the total numbers of attributes associated with populations $x$ and $y$ respectively, $n_{xy}$ is the number of attributes common to both $x$ and $y$, and $n_{-xy}$ is the number absent from both $x$ and $y$.

It is not immediately obvious which measure of similarity is the most appropriate to use in any particular case. However, a common situation is one where the similarities between RFLP–PCR-derived bands on gels derived from different isolates, considered as phenotypic attributes, are being compared. Lamboy (1994a) showed that, with such material, the Nei–Lei measure of similarity was the least susceptible to bias. Such comparisons are not always technically easy to make. Although, ideally, DNA extraction and amplification protocols should be rigorously optimized, if that has not been done, some bands may be faint, some may be missing or spurious bands may be present. The average of completely replicated treatments can then be used or a correction can be applied (Lamboy 1994b). However, good methodology is far preferable to this or any other type of correction.

### 5.1.2.2 *Dissimiliarity*

Dissimilarity is often an informative measure of the differences between populations. In the case of the European populations of *B. graminis* f.sp. *hordei* [AD] cited in section 5.1.1 (Müller *et al.* 1996), dissimilarity for virulence was measured by the mean number of specific virulence differences between the different pathotypes in the sample. It was estimated as the mean dissimilarity

$$D_{\text{mean}} = \sum_{i<j} \frac{d(a_i a_j)}{n(n-1)/2}$$

where $d(a_i a_j)$ is the number of different virulence characters for each pair $(a_i, a_j)$ and $n(n-1)/2$ is the number of pairwise comparisons between $n$ genotypes.

This measure of mean dissimilarity can be estimated in the same way for any other phenotypic character.

### 5.1.3 Genetic and genotypic diversity

The measures used for expressing phenotypic diversity can equally well be applied to expressing genetic diversity. The Dice similarity index is used most commonly when comparisons are being made either between nucleotide fragment profiles (bands) on gels, or between the average number $\Pi$ of pairwise nucleotide differences in sequences. This is, effectively, the same as the mean dissimilarity, i.e.

$$\Pi = \sum_{i<j} \Pi_{ij} \frac{1}{n(n-1)/2}$$

where $\Pi_{ij}$ is the difference between the $i$th and the $j$th sequences and $n$ is the number of sequences in the sample.

When using genetic diversity data, **genetic richness** is defined as the number of alleles potentially found at a locus, and **genetic evenness** is a measure of whether the number of alleles at a locus occur equally frequently. For example, if three alleles are available at a locus, do they occur equally frequently $(1:1:1)$ or not, such as one predominantly and the other two rarely (e.g. $100:2:1$) (Brown and Weir 1983)? Another useful attribute is the mean number of loci per isolate or **complexity**. However, when genotypic diversity is being considered, the contributions of genetic richness and genetic evenness to the sample are neither distinguished, nor do they account for similarities or dissimilarities among these elements in a sample.

Another useful concept of genetic diversity, applicable to diploids and therefore dikaryons, is that of the **average heterozygosity**, defined in terms of gene frequency. If there are $m$ alleles in a randomly breeding population and $x_i$ is the population frequency of the $i$th allele at a locus, the heterozygosity of the locus can be defined as

$$h = 1 - \sum_{i=1}^{m} x_i^2.$$

It follows that the average heterozygosity $H$ is the average of this quantity over all loci. So, for $r$ loci

$$H = \sum h_j/r$$

where $h_j$ is the value of $h$ for the $j$th locus.

Even though heterozygosity is inapplicable to either haploids or polyploids, the probability for such populations that two randomly chosen genes are different is the equivalent quantity and can be defined in the same way (Nei 1973). He has also called the quantity $(1-h)$ the **expected homozygosity** $J$. These quantities and concepts can be used when comparing populations regardless of whether they are diploid, dikaryotic, haploid or polyploid (see section 5.1.4).

Software packages such as POPGENE (version 1.2) (Yeh *et al.* 1997) exist for calculating various indices, both phenotypic and genotypic.

## 5.1.4 Comparing genetic variation within subdivided populations or between populations

Populations are often dispersed as sub-populations which may or may not be connected by gene flow. The description of variation in subdivided populations is not only useful in its own right but is relevant to migration between populations (see Chapter 6, section 6.3) and to inbreeding within small populations (see Chapter 7, section 7.1). The earliest approach to describing their genetic variation was that developed by Wright (1943, 1951, 1965) for diploid organisms. He defined a quantity $F$, the inbreeding coefficient of a population, as 'the departure from the amount of homozygosis under random mating toward complete homozygosis'.

In a subdivided population he partitioned $F$ into three further quantities $F_{ST}$, $F_{IS}$ and $F_{IT}$, which are measures, respectively, of:

- the genetic differentiation of the subpopulations, i.e. inbreeding due to population subdivision (always positive)
- inbreeding within subpopulations
- inbreeding in the total population as a whole.

They are related as follows:

$$F_{ST} = (F_{IT} - F_{IS})/(1 - F_{IS})$$

or, in terms of the frequency $q$ of a homozygous allele $a$,

$$F_{ST} = V(q)/q(1 - q)$$

where $V(q)$ is the variance in the frequency of $a$.

The average values of the three coefficients were shown by Nei (1977) to be

$$\overline{F}_{IS} = \frac{\overline{H}_S - \overline{H}_0}{\overline{H}_S}$$

$$\overline{F}_{IT} = \frac{\overline{H}_T - \overline{H}_0}{\overline{H}_T}$$

$$\overline{F}_{ST} = \frac{\overline{H}_T - \overline{H}_S}{\overline{H}_T}$$

where $\overline{H}_0$ is the average observed heterozygosity within a subpopulation over loci, $\overline{H}_S$ is the average expected heterozygosity within subpopulations over loci and $\overline{H}_T$ is the average of the expected heterozygosity in the total population over loci.

These measures, in one form or another, can readily be applied to diploid populations such as the Straminipila or to dikaryotic fungi such as the Basidiomycotina.

Nei (1973) also approached comparisons between populations in a somewhat different way, which can be used more readily with nucleotide fragment data, by determining a **coefficient of gene differentiation** $G_{ST}$ analogous to $F_{ST}$. He derived this from a consideration of his two widely used indices, the **genetic identity** $I$ and the **genetic distance** $D$ (Nei 1972, 1987). These indices presuppose that the populations or samples compared have been genetically separated for sufficient time for some nucleotide differences to have accumulated and for nucleotides to have been substituted at a constant rate, i.e. a 'molecular clock' is assumed to be operating. Each is expressed as a single numerical value.

For a single locus with $n$ alleles, where $J$ is a measure of the probability that two genes chosen at random from a population are identical, Nei's normalized genetic identity is

$$I_N = J_{xy}/(J_xJ_y)^{1/2}.$$

If, for populations $x$ and $y$, the frequencies of the $i$th allele are $p_{i,x}$ and $p_{i,y}$ respectively, $J_{xy}$, $J_x$ and $J_y$ are given by

$$J_{xy} = \sum_{i=1}^{n} p_{ix}p_{iy}$$

$$J_x = \sum_{i=1}^{n} p_{ix}^2$$

$$J_y = \sum_{i=1}^{n} p_{iy}^2.$$

Nei's genetic distance can then be defined in terms of $I$ or $J$ as

$$D = -\ln I_N$$

or

$$-\ln J_{xy} - (\ln J_y)/2 - (\ln J_x)/2.$$

In the case of multiple alleles, the various values of $J$ are calculated by summing over all alleles at all loci and determining the average value per locus by dividing this sum by the number of loci. The genetic identity and distance are then calculated in exactly the same way as before using the average values $\hat{J}'_{xy}$, $\hat{J}'_x$ and $\hat{J}'_y$:

$$I'_N = \frac{\hat{J}'_{xy}}{(\hat{J}'_x \hat{J}'_y)^{1/2}}$$

and

$$D'_N = -\ln I'_N.$$

Nei's basic assumption of a constant rate of nucleotide replacement is not necessarily correct. If it is not, a similar relationship is used but the Js are the *geometric* means over loci $j'_{xy}$, $j'_x$ and $j'_y$ respectively.

The values of genetic identity, $I$ and $I'$, range from zero, when *no* alleles are shared between the two populations, to 1, when they are *all* identical. Similarly, the values of $D$ and $D'$ range from infinity, when the populations have *identical* allele frequencies, to zero when they have *none* in common.

Nei's measure can be used with enzyme polymorphisms or nucleotide marker data. It will be recalled that the frequencies of enzyme polymorphisms derived from electrophoretic data are less discriminatory because only about a quarter to a third of amino acid substitutions in proteins are detected (see Chapter 3, section 3.2). Nevertheless, Nei's measure has often been used with such data. One example of its use is its application to 13 isozyme loci in 24 different populations of *Atkinsonella hypoxylon* [AP] occurring on four different host grasses, one *Stipa* species and three *Danthonia* species (Leuchtmann and Clay 1989) (Table 5.1). The value of $I$ between isolates from *S. leucotricha* and those from the *Danthonia* spp. was low, and isolates from *D. spicata* and *D. compressa* showed the greatest similarity. While there was considerable variation amongst the *Danthonia* isolates, those

from the four populations of *Stipa* sampled were distinct and identical. Subsequently, the *Stipa* isolates have been recognized as a distinct species, *A. texensis* [AP] on the basis of the electrophoretic comparison, their incompatibility with other isolates and a re-examination of their morphological characters (Leuchtmann and Clay 1989).

Nei's coefficient of gene differentiation $G_{ST}$ can also be defined in terms of $D$ as the ratio of the average gene diversity $D_{ST}$ between populations to the gene diversity $H_T$ in the total population:

$$G_{ST} = D_{ST}/H_T = H_T - H_S/H_T$$

where

$$D_{ST} = \sum_k \sum_l D_{kl}/s^2,$$

there are $l$ and $k$ sub-populations in a total of $s$ sub-populations and the average gene diversity within subpopulations is $H_S = (1 - J_S)$.

For instance, using allozyme data derived from a world sample of *Schizophyllum commune* [BA], it is possible to distinguish two groups of isolates, one confined to the eastern hemisphere and the other to the western hemispheres (James *et al.* 1999) (Table 5.2).

For small samples of unequal size, which is not uncommon in practice, the values of $H_S$ and $H_T$ are biased estimates and are replaced by

$$\hat{H}_S = \frac{\tilde{n}}{2\tilde{n} - 1}(1 - J_S)$$

and

$$\hat{H}_T = 1 - J_T + \hat{H}_S/2\tilde{n}$$

**Table 5.1** Nei's genetic identity $I$ among isolates of *Atkinsonella hypoxylon* [AP] from four grass hosts, three *Danthonia* species and *Stipa leucotricha*, based on the frequency of 36 alleles at 13 enzyme loci

| Host species | D. spicata | D. compressa | D. sericea | S. leucotricha |
|---|---|---|---|---|
| D. spicata | 1.000 | 0.883 | 0.647 | 0.208 |
| D. compressa | | 1.000 | 0.818 | 0.279 |
| D. sericea | | | 1.000 | 0.308 |
| S. leucotricha | | | | 1.000 |

Nei's $I$ varies from complete identity at $I = 1$ to non-identity at $I = 0$.
Reproduced from Leuchtmann and Clay (1989), with permission from the National Research Council of Canada.

**Table 5.2** Pairwise estimates of genetic structure and gene flow amongst pan-global *Schizophyllum commune* [BA] populations

| Populations | $G_{ST}$ | $\chi^2$ | Df | P | $D_m$ |
|---|---|---|---|---|---|
| Eastern and Western hemispheres | 0.180 | 609.2 | 36 | <0.001 | 0.159 |
| Northeastern and Southeastern USA | 0.020 | 25.8 | 37 | NS | 0.024 |
| Australasia and Asia | 0.000 | 3.6 | 28 | NS | 0.015 |
| Australasia and North America | 0.212 | 375.2 | 47 | <0.001 | 0.203 |
| Asia and North America | 0.279 | 548.0 | 43 | <0.001 | 0.232 |

Df, degrees of freedom; NS, not significant.
Summed $\chi^2$ values test allelic heterogeneity over loci. $D_m$ is Nei's genetic estimator of minimum genetic distance between populations.
Reproduced from James *et al.* (1999).

where $\bar{n}$ is the harmonic mean of the sample sizes ($n_i$) from each of the $s$ sub-populations. For haploids, $2\bar{n}$ is replaced by $\bar{n}$ in the above equations (Brown 1996).

If it is suspected that there are subdivisions nested within those first recognized, it is possible to partition $D_{ST}$ further (Chakraborty 1980). Smaller regions may exist within wide geographical regions, and smaller homogenous groups can be recognized by partitioning $D_{ST}$ into $D_{RT}$ and $D_{SR}$, where $D_{RT}$ represents the variation between the regions and $D_{ST}$ is the variation between populations within the regions. Then

$$D_{ST} = D_{RT} + D_{SR}$$

and therefore

$$H_T = H_S + D_{RT} + D_{SR}$$

and

$$G_{RT} = D_{RT}/H_T$$

where $H_T$ is the total variation due to variation amongst populations within regions. Table 5.3 illustrates these parameters applied to a wide range of variously distributed fungal species.

Many fungal species are characterized by high levels of variation and, not surprisingly, those with wide distributions show high levels of genetic diversity even though in many cases the populations are partially isolated geographically.

The use of $G_{ST}$ has been criticized on the grounds that it does not allow for differences in sampling, thereby introducing a bias (Slatkin 1985). However, such a bias can be overcome either by dividing the gene diversity in each population by the sample size (Boeger *et al.* 1993), or by a method described by Nei and Chesser (1983) which allows for unequal sample numbers, as was the case with the *S. commune* [BA] data referred to above.

Another measure that can also be used is Weir and Cockerham's $\theta$, which is formally equivalent to Wright's $F_{ST}$. This measure has the advantage that it allows for unequal samples and sample sizes. Full details of how it can be estimated are given by Weir and Cockerham (1984) and Weir (1996b).

## 5.2 Character association: the recognition of distinct genotypic groups and their relationships

It is often useful to investigate whether isolates are clustered or associated in some way and, if so, how such clusters can be interpreted. One common question is whether genotypic identity reflects taxonomic identity and whether phylogenetic relationships can be deduced between differing genotypes.

A range of procedures are available to investigate these questions. The detailed protocols are often complex. Since most are now available as computer programs, this account will be limited to the kinds of procedure that can be adopted.[*] The principles involved are discussed by several authors (e.g. Sokal and Sneath 1963; Sneath and Sokal 1973; Bridge 1998; Page and Holmes 1998). In addition, working details can be found in the appropriate software manuals and in a variety of publications (e.g. Excoffier *et al.* 1992; Randerson 1993; Swofford *et al.* 1996; Weir 1996a).

Three different approaches can be used:

• ANOVA (**an**alysis **o**f **var**iance) or AMOVA (**a**nalysis of **mo**lecular **var**iance), the modification

---

[*] The computer programs referred to in this book are the most recent available at the time of writing but they are frequently updated. In practice, particular care should be taken to ensure that the most recent versions are used. Access to the latest version may often be obtained through the World Wide Web.

**Table 5.3** Genetic variation in natural fungal populations based on molecular markers

| Reference | Species | Data type | Geographic scale | No. of populations | No. of isolates | No. of genotypes | $H_T$ | $H_S$ | $G_{ST}$ |
|---|---|---|---|---|---|---|---|---|---|
| 1 | Pyricularia grisea | RFLP | Local | 3 | 1516 | 130 | 0.65 | 0.53 | 0.19 |
| 2 | Mycosphaerella graminicola | RFLP | Local | 3 | 512 | 472 | 0.425 | 0.421 | 0.010 |
| 3 | Phaeosphaeria nodorum | RFLP | Subregional | 9 | 432 | 426 | 0.51 | 0.49 | 0.03 |
| 4 | Crumenulopsis sororia | Allozyme | Subregional | 6 | 1181 | ? | 0.26 | 0.25 | 0.041 |
| 5 | Mycena rosea | Allozyme | Subregional | 8 | 112 | ? | 0.457 | 0.425 | 0.07 |
| 6 | Suillus tomentosus | Allozyme | Subregional | 4 | 43 | 37 | 0.228 | 0.162 | 0.289 |
| 7 | Lycoperdon pyriforme | Allozyme | Subregional | 7 | 310 | 49 | 0.287 | 0.252 | 0.122 |
| 8 | Cryphonectria parasitica | RFLP | Regional | 9 | 449 | ? | 0.43 | 0.34 | 0.20 |
| 9 | Atkinsonella hypoxylon | Allozyme | Regional | 6 | 200 | 16 | 0.262 | 0.103 | 0.607 |
| 10 | Agaricus bisporus | RFLP | Regional | 4 | 214 | 188 | 0.539 | 0.431 | 0.201 |
| 11 | Armillaria gallica | RFLP | Regional | 4 | 272 | 110 | 0.528 | 0.516 | 0.023 |
| 12 | Neurospora intermedia | Allozyme | Inter-continental | 4 | 145 | ? | 0.485 | 0.420 | 0.145 |
| 13 | Rhynchosporium secalis | Allozyme | Inter-continental | 3 | 150 | 34 | 0.278 | 0.230 | 0.164 |
| 14 | Mycosphaerella fijiensis | RFLP | Inter-continental | 4 | – | – | – | – | 0.32 |
| 15 | Agaricus bisporus | RFLP | Inter-continental | 2 | 342 | 272 | 0.561 | 0.539 | 0.039 |
| 16 | Schizophyllum commune | Allozyme | Inter-continental | 7 | 136 | 130 | 0.428 | 0.336 | 0.214 |
| 17 | Pythium ultimum | Allozyme | Inter-continental | 6 | – | – | 0.139 | 0.123 | 0.115 |

$H_S$ and $H_T$ describe within-population and total heterozygosity; differences between $H_T$ values are *not* comparable because of varying techniques used. $G_{ST}$ represents the proportion of genetic diversity due to subpopulation differentiation in allele frequencies.

**References**: 1, Chen et al. (1995); 2, Chen et al. (1994); 3, Keller et al. (1997); 4, Ennos and Swales (1991); 5, Boisselier-Dubayle et al. (1996); 6, Zhu et al. (1988); 7, Huss (1996); 8, Milgroom and Lipari (1995); 9, Leuchtmann and Clay (1996); 10, Xu et al. (1997); 11, Saville et al. (1996); 12, Spieth (1975); 13, Goodwin et al. (1993); 14, Carlier et al. (1996); 15, Xu et al. (1997); 16, James et al. (1999); 17, Barr et al. (1996).

Based on James et al. (1999) with permission of the Society for the Study of Evolution.

developed specifically for haploid molecular genetic data
- non-hierarchical ordination methods such as principle component or principle co-ordinate analysis
- hierarchical/cladistic methods, especially when taxonomic similarity or phylogenetic relationships are sought.

ANOVA and AMOVA are used to demonstrate differences between groups of individuals in relation to a single variance that is distributed normally or when the data can be so distributed after a suitable transformation. In contrast, non-hierarchical methods can be used with multivariate data which need not necessarily be normally distributed. They are probably most effectively applied to discriminate within populations showing a more or less continuous range of variances either *within a species*, or *between very closely related taxa*. However, if it is suspected that the populations studied include *different taxa of specific or higher rank having some phylogenetic connection(s)*, then hierarchical cladistic methods are likely to be more appropriate (Bridge 1998). A variety of techniques, the majority using sequence-derived data, are available to determine hierarchical groupings, i.e. potentially phylogenetically related groups. Even though the basic data used with all these methods are, increasingly, either DNA sequences or patterns of nucleotide fragments derived in various ways, such data can also be combined with other attributes such as morphological, physiological or isozyme data.

Another question is whether there is any correlation between such associated or clustered isolates possessing the same or very similar genotypes and specific ecological, regional, geographical or historical features. These issues are discussed in Chapter 6, section 6.3, and Chapter 12.

## 5.2.1 ANOVA and AMOVA

The application of the analysis of variation to partitioning haploid or diploid genotypic data within and between populations is useful, and full protocols are given by Weir (1996b).

Increasing use is now being made of AMOVA. The method was first used in the analysis of human haplotype mitochondrial data (Excoffier *et al.* 1992), but it can equally well be applied to any haplotype data including, of course, data from haploid fungi. Its approach is analogous to that of normal ANOVA and a software package is now available (Schneider *et al.* 1997). The basic data are first converted into a matrix of squared distances between all pairs of the haploid isolates, from which it is possible to obtain an estimate of the various variance components. A good example is provided by the analysis of variation of North European *Gremmeniella abietina* [AD] (Hellgren and Högeberg 1995).

Samples from *Pinus sylvestris*, *Pinus contorta* and *Picea abies* which differed in their reproductive behaviour were obtained throughout the southern and northern regions of Fennoscandia, i.e. the southern regions of Norway, Sweden and Lapland and northern Sweden and northern Finland respectively. Those from the northern region produced apothecia and pycnidia relatively abundantly on the parts of the trees covered by deep snow, mainly in winter. Trees scattered throughout both regions, but predominantly in the south, produced apothecia infrequently, but pycnidia developed on current-year shoots throughout the crowns of the trees. The question arises as to whether these two phenotypically distinct disease manifestations are caused by two genetically different fungal taxa.

Amplified fragments of nuclear DNA (RAPDs) were obtained from the ITS region of rDNA and the M-13 minisatellite region of each isolate and separated to give banded patterns on agarose gels. Only reproducible bands were treated as alleles and were scored as either present (1) or absent (0); covarying bands were excluded. The measure of difference in banding patterns between all combinations of isolates was estimated as the square of the Euclidean distance $\delta_{xy}$ between two isolate banding patterns $x$ and $y$, where

$$\delta_{xy}^2 = \sum_{i=1}^{s} (x_i - y_i)^2$$

and $x$ and $y$ are scores for band $i$ (1 or 0) and $s$ is the number of band positions employed.

Incidentally, $F_{ST}$ can also be estimated from

$$F_{ST} = \frac{\sigma B}{(1 - B)B}$$

where $B$ is the mean frequency of a band in all populations and $\sigma B$ is its variance.

Fragment patterns derived from the M-13 primer showed some variation but fell into two clear groups separated by a constant diagnostic band. Regardless of host species, one group had been obtained from isolates from large trees—the large tree type (LTT)—and the other group from isolates from small trees—the small tree type (STT). Variance components could then be calculated for the following comparisons:

- populations of LTTs and STTs causing disease on large and small trees
- populations from northern and southern regions
- populations on different host species.

The results obtained are shown in Table 5.4 where it can be seen that there are:

- significant differences between the STT and LTT populations
- no significant differences between LTT populations from the two geographical regions (the STT are almost entirely confined to the northern region)
- a significant difference when all the *P. sylvestris* plus *P. abies* isolates, whether showing STT or LTT features, were compared with the *P. contorta* isolates, although the proportion of the variance was only decreased from 37.7 to 33.5.

Therefore the most significant differentiating factors appear to be the size of the tree and climatic conditions affecting snow cover, and a far smaller but appreciable influence of the effect of host preference between *P. contorta* and the other conifers.

## 5.2.2 Ordination methods

Ordination methods are potentially useful when more than one variate is under consideration, and ordination can be employed when variates are not distributed normally. Therefore ordination methods can be used for any such data, in particular those derived from nucleotide fragments since these data are often not distributed normally owing to selective bias which may be introduced by the use of particular primers or probes (Bridge 1998).

The object of the methods is to detect patterns correlated with one or more variables such as host preferences, environmental factors or geographical factors. Principle co-ordinate and principle component analysis are potentially useful when the data relate to populations of a single taxon or intraspecific taxa, i.e. where the variation is more likely to be continuous or semi-continuous, without wide breaks, as occurs if different species are involved.

**Table 5.4** Analysis of variance (AMOVA) of banding patterns of amplified DNA fragments obtained by the PCR using 150 isolates of *Gremmeniella abietina* [AD] from Northern and southern Fennoscandia

| Source of variation | Df | Sum of squares | Mean squares | Variance component | Percentage total variance | P |
|---|---|---|---|---|---|---|
| *Groups of STT and LTT populations* | | | | | | |
| Between groups | 1 | 106.9 | 106.9 | 1.39 | 37.7 | <0.001 |
| Populations within groups | 1 | 47.9 | 3.4 | 0.16 | 4.4 | <0.001 |
| Within populations | 131 | 278.9 | 2.1 | 2.13 | 57.8 | <0.001 |
| *Groups of large tree types (South Sweden, North Sweden, South Norway)* | | | | | | |
| Between groups | 2 | 14.4 | 7.2 | 0.20 | 14.4 | 0.06 |
| Populations within groups | 7 | 10.2 | 1.4 · | 0.09 | 6.3 | <0.02 |
| Within populations | 61 | 68.3 | 1.1 | 1.12 | 79.3 | <0.02 |
| *Groups of isolates from Pinus sylvestris and Pinus contorta* | | | | | | |
| Between groups | 1 | 95.4 | 95.4 | 1.21 | 33.5 | <0.001 |
| Populations within groups | 14 | 59.5 | 4.2 | 0.27 | 7.4 | <0.001 |
| Within populations | 131 | 278.6 | 2.1 | 2.13 | 59.1 | <0.001 |

Df, degrees of freedom.
Reproduced from Hellgren and Högeberg (1995), with permission from the National Research Council of Canada.

In essence, ordination methods extract successive components from a matrix of similarities or distances between the variates, using so-called R statistics for comparing all possible pairs of attributes, or Q statistics for similar comparisons between individuals. Typically, the summed squares and cross-products between the individuals or their attributes are analysed to provide eigenvectors and the results are generally plotted as an ordination diagram. The procedure, which is usually carried out using a suitable statistical program (e.g. Minitab, MVSP, PCA, NTSYS) is designed to optimize a sequence of axes so that each progressively minimizes the variation left over by its predecessor(s). A clear account of both the principles and procedures used is given by Randerson (1993) who comments: 'In ordination, each individual is placed on one or more constructed axes so that its geometrical position relative to its fellows reflects its similarity to them'.

An excellent example of principle co-ordinate analysis is that employed to investigate the genetic structure of populations of the northern corn blight fungus *Setosphaeria turcica* [AP] (Borchardt *et al.* 1998). Seventy RAPD markers were identified from 264 isolates from four continents. Many of the markers were restricted to a single continent: nine to Europe, eight to Mexico, five to China and eleven to Africa (Kenya). Pairwise comparisons were made using the Dice index (see section 5.1.2) and, after a scalar transformation and computation of eigenvectors the principal co-ordinates, were plotted as shown in Figure 5.2.

The first two co-ordinates account for 34 per cent of the total variation in the sample but, as can be seen, a clear separation of the principal regions was obtained and the physical distance between different samples on the diagram gives some indication of their affinities. Some populations showed closer affinity than others; for example, those from North and South China were close but remote from those from Mexico, although these last were quite close to European isolates. This suggests two working hypotheses: the introduction of Mexican

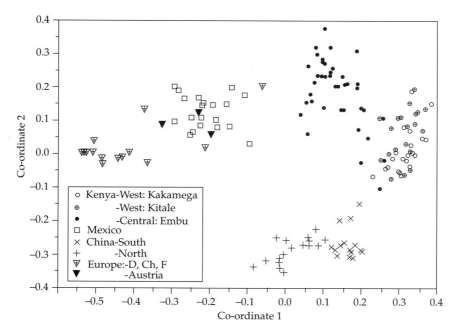

**Figure 5.2** Diagram illustrating a principal co-ordinate analysis of RAPDs of haplotypes of 264 isolates from four continents of *Setosphaeria turcica* [FA], the cause of northern leaf blight of maize. Comparisons were based on pairwise similarities using the Dice coefficient of similarity. The first two co-ordinates account for 34 per cent of the variation. Note the correlated clumping of much of the data to demonstrate clear regional groupings of most of the isolates and the apparent similarity of some of the European and Mexican isolates tested. (Reproduced from Borchardt *et al.* (1998), with permission of the American Phytopathological Society.)

pathotypes into Europe, and the differentiation within China of two main races of the pathogen. Of course, these hypotheses need to be tested further, but these findings and inferences indicate the potential value of this method of analysis.

### 5.2.3 Hierarchical methods

Hierarchical methods can be used with morphological or sequence data, or combinations of such data. The use of nucleotides in a DNA sequence, or of amino acids in a protein sequence, as a *character* is, on the whole, less subjective than is the use of morphological characters.

When molecular data are employed the first essential, having obtained a number of sequences, is to compare and align them. This is now most easily done using a multiple alignment program such as CLUSTAL V (Higgins *et al.* 1991) or CLUSTAL X (Thompson *et al.* 1997) (also at http://inn.prot.weizmann.ac.il/software/ClustalX.html). The object is either to compare, as characters, comparable positions along nucleotide sequences of a conserved gene (such as rDNA, $\beta$-tubulin, etc.) from different isolates, or to compare some matrix derived from them in order to infer their presumed phylogenetic relationships. Basically this involves determining the most probable sequence of changes to explain the observed differences between the sequences being compared. The measure of difference adopted is usually the number of character changes that would be required to convert one sequence to another. Neither determination *necessarily reflects* either the *actual* historical sequence of events that have taken place, or the *actual* number of events that have occurred, since there is no means of knowing whether a change has occurred more than once in the course of evolution. Reversals or parallel changes almost certainly do occur, so that the amount of sequence divergence observed is almost certainly an underestimate of that which has actually taken place, a phenomenon called **homoplasy**. Therefore, if sequence data are used in any way to determine phylogeny, an appropriate correction needs to be applied.

In addition, two important assumptions underlying such methods are that the characters employed are *independent* and *homologous*. Therefore,

in general, restriction fragment data are *not suitable* because they may well violate the former assumption. However, they have been used, with appropriate safeguards, with AFLPs when comparing very similar taxa or isolates of the same taxon (Bakkeren *et al.* 2000).

The outcome of the various methods is to determine which arrangement of the data best expresses their potential phylogenetic relationships by arranging them in some kind of meaningful hierarchy. Finally, it is necessary to test the reliability of the arrangement or, if more than one possible arrangement is found, to determine which best fits the data.

An arrangement so derived is usually represented as a branching structure—a **dendrogram**. There are various kinds of dendograms. The simplest—a **cladogram**—only demonstrates the *relative* relationships in a group of individuals, with adjacent branches depicting the greatest similarity or closest ancestry. The **branch points**, or **nodes**, reflect divergence from a previously common sequence and the **length of a branch**, or **distance**, is a measure of the mean number of estimated character changes (substitutions) required to convert one sequence to another.

If the cladogram purports to demonstrate presumed phylogenetic relationships with some indication of relative affinities, it is described, technically, as a **tree**. An important measure is the total number of estimated changes needed to construct a tree, i.e. the **tree length**. Frequently, alternative changes can account for the observed nucleotide data but, in general, the object is to determine the shortest tree length that best fits the data available. Trees may be rooted or unrooted. The former are assigned a point of origin, either subjectively or by comparison with an 'outgroup', i.e. data derived from a supposedly unrelated isolate or organism, often presumed to be ancestral in phylogenetic comparisons. Unrooted trees merely indicate relationships by distance from a common origin. Rooted trees are generally preferred (Figure 5.3). Various kinds of rooted tree differ in their properties (Figure 5.3). **Additive trees** include some measure of the closeness or difference between individuals, e.g. branch lengths based on the number of substitutional steps

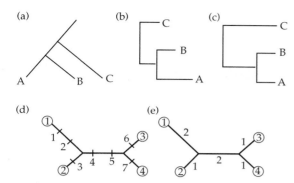

**Figure 5.3** Trees. (a) A basic cladogram, interpreted as (b) an additive tree, where the horizontal axis indicates the amount of genetic change, or (c) an ultrametric tree, where the horizontal axis represents time. (d), (e) Identical trees based on the same nucleotide sequence but derived by either (d) parsimony or (e) distance methods. In the former, the individual contributions to each branch are shown.

**Table 5.5** Tree-building methods

| Data type | Method of tree building |
|---|---|
| Distances | Clustering algorithm |
| | Fitch–Margoliash |
| | Neighbour joining |
| | UPGMA |
| Nucleotide/amino acid sites | Optimality criteria |
| | Maximum parsimony |
| | Maximum likelihood |

UPGMA, unweighted pair-group method with arithmetic means.

between them, or possibly some measure of evolutionary change. If an additive tree is drawn so that the tips of the branches are equidistant from its root, it is described as an **ultrametric tree** and its branches can be thought of as representing the relative degree of divergence through evolutionary time determined by some kind of molecular clock.

A variety of different procedures are available to achieve the optimum hierarchical arrangement. Many use computerized algorithms which reduce the burden of computation. Details of these procedures can be found in compilations such as Hillis *et al.* (1996). A clear and simple guide to procedures, including computerized computations, is provided by Hall (2001). Here, only an outline will be given of the two alternative approaches adopted to convert data derived from population samples into trees, i.e. **distance methods** and **discrete character methods**.

In distance methods, **distance** is the number of nucleotide (or amino acid) substitutions that have occurred along a branch of a treee. The aligned sequences, for example, are first converted into a pairwise **distance matrix** (or **similarity matrix**) of differences (or similarities) betwen the sequences which are then converted into a tree (see section 5.2.3.1). In contrast, **characters** are represented by the sequence of individual nucleotides, and in discrete character methods, these raw data, whatever their nature, are converted directly into

a **data matrix** from which a tree can be obtained (see section 5.2.3.2).

There are also two main methods of constructing the tree, either by seeking to cluster the data, whether as a distance matrix or as raw data, or by relating the data to the form of the tree in the most optimal way, using optimality criteria. Of course, if the same data are used to start with, the length of the tree should be the same regardless of whether it has been estimated by a difference method or a discrete method, since these are simply different methods of estimating the tree. In practice, however, this is not necessarily the case because of errors in the data or the corrections and approximations which may be used in the estimations. The ways in which the type of data used and the method of tree building adopted can be combined are summarized in Table 5.5.

Whatever the method employed, it is essential to recognize that *any* method is capable of generating a tree from *completely random data* as a result of entirely chance events! Thus, even when a tree has been generated, it is essential to estimate the likelihood of how far it fits the basic data and how far it reflects a genuine departure from a random pattern, which is the basic null hypothesis (see section 5.2.3.3).

### 5.2.3.1 Distance methods

The **Fitch–Margoliash method** (Fitch and Margoliash 1967) was developed as one of the earliest methods for producing an unrooted tree from a distance matrix. It is based on first comparing the two most similar sequences, then comparing them with a third, the least different to either, and so on. It is assumed that each pairwise distance measurement is independent, but this is not necessarily true and so corrections are necessary. Therefore an optimization routine is followed to bring the calculated branch lengths, reflecting the numbers of hypothetical substitutions separating branches, as close as possible to those expected. The method is included in the PHYLIP package (Felstenstein 1993) (also at http://evolution. genetics.washington.edu/phylip.html), but the optimization routine takes some time even when using a computer. Therefore it is not used as often as either of the following methods.

A much more common method is the **neighbour joining method** which is based on an algorithm (Saitou and Nei 1987). It *assumes* that the data can be represented by an additive tree. In essence, the algorithm arranges and successively rearranges items in a distance matrix into clusters by an iterative process, based on minimum distances between sequences, designed to minimize the total length of the final tree, i.e. it determines heuristically the minimum evolutionary tree necessary to connect items. Three computer packages which include the necessary algorithm are PHYLIP, PAUP* (Swofford 1998) (also at http://www.sinauer.com) and MEGA (Kumar *et al.* 1993).

Originally, the simplest, most common and most popular method relying on clustering based on similarity was the **unweighted pair group method with arithmetic means** (UPGMA), although neighbour joining methods are now often preferred. UPGMA involves making successive approximations to find the best least-squares ultrametric tree. Therefore, unlike the two previous methods, it includes the *assumption* that there is a molecular clock which determines a regular rate of nucleotide substitution. This assumption, that the evolutionary rate is regular, may or may not be correct (Gillespie 1991). In essence, distances are calculated from the proportion of character differences between isolates relative to the total number of characters used. If DNA or protein (enzyme) sequences are used, corrections are usually made for the estimated percentage of repeated substitutions. The three computer packages mentioned above also include the elements of UPGMA necessary both to correct sequence data and to determine the relationships of the sequences.

Despite the questionable assumption concerning a constant evolutionary rate of nucleotide substitution, this is still a widely used method because its final form is an ultrametric tree.

Distance methods, in general, suffer from two kinds of disadvantage: first the method used to estimate distance (or similarity) may be unsatisfactory, and second the method used for actually building the tree based on these distances may not be optimal.

### 5.2.3.2 *Discrete character methods*

Discrete character methods employ two main techniques to determine tree structure—**maximum parsimony** and **maximum likelihood**. When using parsimony, sequences are first aligned either manually or using a computer program such as CLUSTAL V (Higgins *et al.* 1991) and then compared. It is often theoretically possible to convert one sequence into another by a number of different alternative changes, each representing the equivalent minimal number of possible changes necessary to effect the conversion. The method of parsimony selects the smallest number of changes required to effect the conversion, i.e. the least number of nucleotide substitutions. The *implicit assumption* underlying parsimony is that changes in the nature of the substitutions are both relatively rare and minimal. A possible source of error is that not all nucleotide substitutions are equally likely. For instance, transversions, i.e. pyrimidines $\leftrightarrow$ purines (e.g. $A \rightarrow C$), are less common than transitions, i.e. purines $\leftrightarrow$ purines (e.g. $A \rightarrow G$) or sometimes pyrimidines $\leftrightarrow$ pyrimidines. However, allowance for this or other kinds of substitution can be built into the program (so-called 'weighted parsimony'). Another problem is that as the number of sequences compared increases, the number of possible trees also increases, but at a much faster rate (Felsenstein 1978). Although the program PAUP* (Swofford 2000) is available, considerable computation is involved if several sequences are being compared. This puts a practical limit on the number of comparisons that can be made. Moreover, inevitably, parsimony cannot always provide a single unambiguous tree because the observed changes can sometimes be accounted for in a number of equally valid ways, each involving the same number of steps. However, even if a unique arrangement cannot always be determined, further discrimination can be achieved by applying the maximum likelihood method.

When more than one tree is available, the one which is most likely can be inferred using the maximum likelihood method. In essence, the method provides two kinds of information—an indication of which set of branch lengths is the most likely, and which of all possible trees is the most likely. The method has to assume some model of sequence evolution on which the likelihood can be based. A number of models exist. The Jukes–Cantor model is the simplest and earliest (Jukes and Cantor 1969). It assumes that the four DNA bases are *equal* in number in the molecular sequence, that substitutions follow a Poisson distribution and that every kind of substitution is *equally likely*. However, there is evidence that these assumptions are not always true, and Kimura (1980) modified the hypothesis by allowing different ratios of transitions : transversions while Felsenstein (1981) suggested that allowance should be made for variation in base numbers. Clearly, all these suggestions are valid for some sequences in some organisms, and a further model has been developed which allows for variation in both base composition and the ratio of transitions to transversions (Hasegawa *et al.* 1985).

The computation of maximum likelihood also involves computing all possible ancestral nodes within a tree because they are, of course, unknown. Therefore the process is time consuming because the probability of any situation at any site has to be estimated given all possible combinations of immediately ancestral sites. Thus, as a simple example, for two terminal bases connected by a single node, four possible ancestral combinations, corresponding to the four bases of DNA, need to be calculated, while for four connected terminal bases such as A–T, with two connected nodes X–Y, there are 16 possible alternative ancestral arrangements at the nodes as shown in Figure 5.4. Thus computer time and storage can be genuine limitations on the method (Tehler *et al.* 2000) which, in practice, has to be restricted to relatively small data sets.

**Figure 5.4** Possible alternative ancestral arrangements at the nodes for four connected terminal bases A–T with two connected nodes X–Y. X–Y can be any of the pairs A–A  A–C  A–G  A–T, C–A  C–C  C–G  C–T, G–A  G–C  G–G  G–T, or T–A  T–C  T–G  T–T.

*5.2.3.3 Estimating the likelihood of any tree and discriminating between alternative trees*

Trees are often constructed on quite limited data and it is helpful if some kind of estimate can be made of their reliability. Bootstrapping gives a rough but indicative answer to the reliability of each branch but not of the tree as a whole. Nevertheless, it is useful and usual to include bootstrap values in any hierarchical analysis. A **bootstrap** in this context is a large number of pseudo-replicated sets of data (1000 + are commonly used) on which each branch is based (Felsenstein 1985). They are generated by sampling the original character matrix with random replacement to give a sample of pseudo-replicates the same size as the original data set. Thus some data sets are not replicated at all, others may be replicated more than once and many will differ from either. The percentage occurrence of the *original* character matrix in the bootstrap sample is a measure of the reliability of the branch concerned. It is usual to label each node in a tree with the corresponding bootstrap value. As a rough rule of thumb, less than 50 per cent occurrence is regarded as providing *no* support for a branch, while 95 per cent or above is regarded as showing *strong* support; intermediate values are regarded as only providing *limited* support.

Parsimony can be applied to discriminating between the two best alternative trees; in particular, it can be used to investigate whether the differences are simply due to random error. If it is assumed that all nucleotide sites are independently and identically distributed, then it is possible to compare the trees by investigating the difference $D$ in their total length, where $D = \sum D_i$, i.e. the sum of the minimum numbers of nucleotide substitutions at the $i$th informative site where $n$ is the number of such sites. The null hypothesis is that there is no difference between the trees, and this can be tested with a paired $t$-test with $n-1$ degrees of freedom, where

$$t = \frac{D/n}{s_D/\sqrt{n}}$$

If there is no advantage in one tree over the other $D = 0$, but if one is better than the other $D \neq 0$ (Kishino and Hasegawa 1989). The necessary programme is available in both PAUP* and PHYLIP.

Increasingly, one tree is obtained from one set of sequences and others from either a different sample of the same sequence, or from sequences of other genes in the same organism. Similarly, in other situations where both nucleotide and morphological data are available from the same organism, the alternative trees may not agree. The question such situations pose is: Are the discrepancies due entirely, or largely, to stochastic variation, or do they have some phylogenetic significance? One possible course of action is to combine the data sets on the supposition that a more reliable result can be obtained if the widest possible range of data is used. However, this is not always the best course to take so that a choice has to be made between this and testing the separate estimates. The best course to be adopted is rarely immediately obvious! However, an excellent discussion on how such choices can be made and, if combining the data is ruled out, how best to proceed thereafter has been provided by Huelsenbeck *et al.* (1996).

## 5.3 Population history: the coalescent and nested clade analysis

The contemporary structure and genetics of any population inevitably reflect its history, in part at least, so that any method which provides historical information could be of use in interpreting current situations. One particularly useful approach to this

problem is to investigate gene genealogies, where each node represents a coalescence event, on the assumption of neutral evolution (Kingman 1982; Hudson 1990). Several methods are now available and a useful summary is given by Emerson *et al.* (2001). They all have three features in common:

- the recognition that a phylogenetic or gene tree embodies in a concise form the history of a species or a population, and that gene genealogies using haploid or haplotype data are most useful for this purpose
- any set of lineages can be traced back, i.e. inwards from terminal branches, to where they shared a common ancestry, i.e. the node where they *coalesce*
- such lineages are assumed to have arisen through neutral evolution, i.e. changes are entirely due to mutation at a statistically constant rate (a molecular clock), and genetic drift, there is no selection or recombination and both mating and sampling are random.

It is important to be able to extrapolate back through a tree to the node representing the most recent common ancestor (MRCA). Figure 5.5 demonstrates this and, incidentally, shows a test can be made of whether a molecular-clock hypothesis is valid. X is the MRCA of the sequences (or species) A, B and C, although A and B have developed through the intermediary common ancestor O. The evolutionary distance $d_{AB}$ between A and B is equal to the sum of their branch lengths

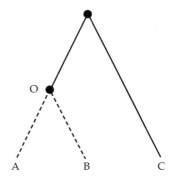

**Figure 5.5** Diagram to illustrate an MRCA and a basic test for the occurrence of a 'molecular clock' (see text for explanation). (Based on Avise (1994).)

from a common ancestor at node O, i.e. $d_{OA} + d_{OB}$. Consider now the relations of A, B and C. It follows that

$$d_{AC} = d_{OA} + d_{OC}$$

and

$$d_{BC} = d_{OB} + d_{OC}$$

or, rearranging,

$$d_{OA} = d_{AC} + d_{OC}$$

and

$$d_{OB} = d_{BC} + d_{OC}.$$

Hence, by subtraction,

$$d_{OA} - d_{OB} = d_{AC} - d_{BC}.$$

However, if a molecular clock is operational, $d_{OA}$ and $d_{OB}$ should be equal since their difference is zero, and hence

$$d_{AC} = d_{BC}$$

and these two distances can be measured empirically. If a molecular clock is operational, these distances should, statistically, be equal.

Coalescence theory assumes the neutral theory of evolution (Kimura 1983), and the coalescent point is a function of population size. Thus individuals in a small population compared with those taken from a large population will differ by fewer substitutions. If the cumulative distribution of coalescent events, logarithmically transformed, is plotted against time, adopting as a molecular-clock hypothesis the percentage base substitutions, for example, inferences can be made about the population size over the period investigated. Such plots, known as lineage through time (LTT) plots can be computed using such program packages as LAMARC (available at http://evolution. genetics.Washington.edu/lamarc.html). This package includes several programs, all of which use non-recombining haplotype sequence data, namely COALESCE, which estimates $N_e$ representing a constant population, FLUCTUATE, which estimates both $N_e$ and the growth rate of an exponentially growing or declining population, and MIGRATE, which estimates $N_e$ and the migration rates of $n$ constant populations, i.e. it is a measure of gene flow. This last program can use isozyme or

microsatellite data in addition to nucleotide sequence data (Beerli and Felstenstein 1999).

Another useful technique, dependent on coalescent theory, is nested clade analysis. It is described in a series of papers by Templeton and colleagues (Templeton *et al.* 1987, 1988, 1992; Templeton and Sing 1993; Posada *et al.* 2000). The results of such an analysis can demonstrate various genetic events such as mutation or hybridization. If nucleotide data can be correlated with spatial and geographical distribution, it is possible to make inferences about migration and gene flow (Templeton 1995, 1998; Posada *et al.* 2000).

In essence, the approach is simple; although the details are somewhat complex, the principles are reasonably clear. An algorithm is applied to haplotype sequence data from many well-dispersed individuals either for one locus or for several combined loci, and an unrooted tree is constructed with at least 95 per cent probability that all linkages are plausible. This tree is then converted into a nested design in which changes due to a *single* mutational change are grouped together from the outermost tips of the tree inwards, i.e. 'one-step clades' (Figure 5.6).

From these one-step clades, two-step clades due to two changes can be grouped together, then three-step clades and so on. Rules for establishing these nested groups are provided. Each reflects a

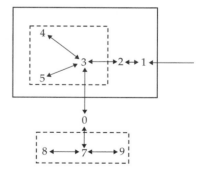

**Figure 5.6** Diagram illustrating the principle of nested clade analysis. The diagram represents part of a cladogram derived from non-recombinant haplotypes obtained as described in the text: 1–5 and 7–9, mutations; 0, missing inferred mutation. Lines joining each number represent a single successive mutational step, so that each pair of mutations joined in this way is a one-step clade (dashed-line box). Successive one-step clades form a two-step clade (solid-line box), and so on to higher clades.

historical change in the sequence. The basic common-sense assumptions underlying these rules are as follows.

• The clades at the tips of the tree represent the historically most recent changes and are the most recently derived; the more distal clades, located increasingly further into the tree, inevitably have more mutational derivatives and are older.
• Older sequences are more geographically dispersed than younger sequences.

The statistical significance of each clade level can be estimated using a NANOVA (nested ANOVA) program developed to analyse the data and, indeed, a NANOVA can be used further to analyse the significant differences between clades at the same level. In this way the historical importance of particular kinds of change can be determined.

Two measures of geographic distance are used, the clade distance and the nested clade distance. **Clade distance** is a measure of the geographical spread of individuals within a clade and is determined by obtaining the geographic centre of all the individuals within a clade. The **nested clade distance** is the distance of clade from the geographic centre of their particular nested category, i.e. it is

a measure of how far individuals in closely related clades are to each other. A series of predictions based on coalescent theory and simulations have been developed which enable various population parameters to be tested from the accumulated data, since they give rise to different expected patterns as set out in Table 5.6 (Crandall and Templeton 1996; Posada *et al.* 2000).

The only published fungal application to date using the coalescent approach and nested clade analysis is that of Carbone and Kohn (2001) (Figure 5.7 and Table 5.7). Otherwise, these methods have not yet been been widely used with fungi.

In Carbone and Kohn's analysis it was possible to avoid using the few isolates in which recombination was involved, so that the techniques could be applied to gene trees derived from sequences of several genes in a wide sample of isolates of *Sclerotinia sclerotiorum* [AD] and also to a smaller sample of various *Sclerotinia* [AD] species (see Chapter 12). Six sequences were determined from the sample of 341 isolates of *S. sclerotiorum* [AD] from different hosts with a wide geographical spread. The most informative sequences in the sample were nucleotide regions covering the IGS of the rDNA repeat unit and of the elongation

**Table 5.6** Expected patterns under the different models of population structure and historical events using nested clade analysis

**Pattern 1. Restricted gene flow**
(a) Tip clades narrowly geographically distributed. Some interior clades broadly distributed
(b) Average clade distance of interior clades—average clade distance of tip clades is significantly large
(c) Average clade distance should increase with increasing clade level in a nested series of clades
(d) The same patterns hold for nested clade distances unless some gene flow is due to long-distance dispersal events, when significant reversals of the above patterns can occur with nested clade distances

**Pattern 2. Range expansion**
(a) Significantly large clade distances and nested clade distances for tip clades, and sometimes significantly small for interior clades under contiguous range expansion, but some tip clades should show significantly small clade distances under long-distance colonization
(b) Interior clade distance—tip clade distance is significantly small for contiguous range expansion; interior nested clade distance— tip nested clade distance is significantly small for long-distance colonization
(c) The above patterns are not recurrent in the cladogram or are geographically congruent

**Pattern 3. Allopatric fragmentation**
(a) Significantly small clade distances at higher clade levels
(b) The pattern of distances in (a) should represent a break or reversal of the distance pattern established by the lower-level nested clades
(c) Clades showing patterns (a) and (b) should be connected to the remainder of the cladogram by a larger than average number of mutational steps
(d) The above patterns are not recurrent in the cladogram or are geographically congruent

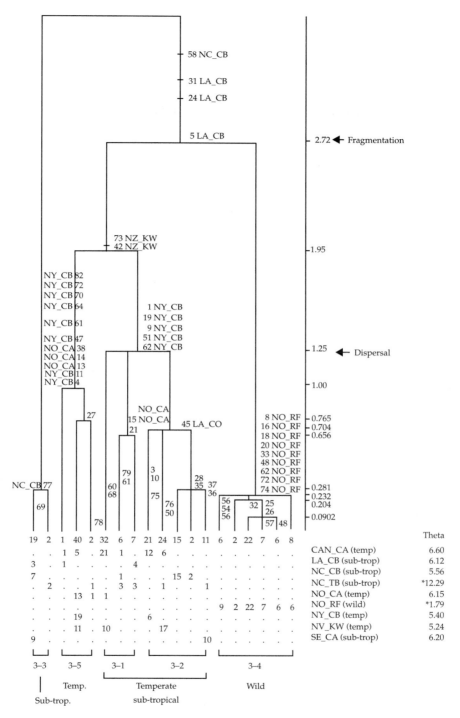

**Figure 5.7** An illustrative genealogy based on the IGS regions of 341 haploid isolates of *Sclerotinia sclerotiorum* [AD/FA] using a multilocus approach and based on an application of coalescent theory and nested clade analysis. The diagram represents the highest root probability based on a migration matrix and coalescent simulations. The time-scale on the right is the estimated time to the MRCA in units of $N_e$ generations. Base substitutions are indicated by a dot and the site number. The geographical sampling areas are shown at the base. The original paper should be consulted for further details. (Reproduced from Caborne and Kohn (2001), with permission from Blackwell Science.)

**Table 5.7** Estimates of the effective numbers of migrants exchanged between sampling areas each generation

| Sampling area | Estimates of $2N_e m$ | | | | | | | | |
|---|---|---|---|---|---|---|---|---|---|
| | 1,*x* | 2,*x* | 3,*x* | 4,*x* | 5,*x* | 6,*x* | 7,*x* | 8,*x* | 9,*x* |
| 1 CAN_CA | – | 0.0 | 0.0 | 0.3 | 2.8 | 0.0 | 0.0 | 0.0 | 0.0 |
| 2 LA_CA | 0.0 | – | 1.0 | 0.0 | 0.0 | 0.0 | 0.0 | 0.0 | 0.0 |
| 3 NC_CB | 0.0 | 0.0 | – | 1.2 | 0.1 | 0.0 | 0.0 | 0.0 | 0.0 |
| 4 NC_TB | 0.0 | 3.6 | 0.0 | – | 0.0 | 0.0 | 0.0 | 2.0 | 0.6 |
| 5 NO_CA | 0.0 | 0.0 | 0.0 | 0.0 | – | 0.0 | 0.9 | 1.4 | 2 |
| 6 NO_RF | 0.0 | 0.0 | 0.0 | 0.1 | 0.0 | – | 0.0 | 0.0 | 0.0 |
| 7 NY_CB | 0.4 | 0.0 | 0.0 | 0.0 | 0.7 | 0.0 | – | 0.0 | 0.0 |
| 8 NZ_KW | 0.4 | 0.0 | 0.0 | 0.0 | 0.8 | 0.0 | 0.0 | – | 0.0 |
| 9 SE_CA | 0.0 | 0.4 | 0.0 | 0.3 | 0.0 | 0.0 | 0.0 | 0.0 | – |

$2N_e m$ is derived from a sample of 241 haplotypes of Sclerotinia sclerotiorum [FA/AD] from nine geographically distinct sampling areas using the MIGRATE program.
The migration parameter is read as with the following example: sampling area 6 (NO_RF)/4,*x* indicates the estimated migration rate from sampling area 4 (NC_TB) into sampling area 6 as 0.1, which is a very low level of migration.
Sampling areas and crop source of sample: CAN_CA, Canada on canola (rape); LA_CA, Louisiana on canola (rape); NC_CB and NC_TB, North Carolina on cabbage and tobacco respectively; NO_CA and NO_RF, Norway on canola (rape) and *Ranunculus ficaria*; NY_CB, New York on cabbage; NZ_KW, New Zealand on kiwi fruit; SE_CA, SE states, USA on cabbage.
(Reproduced from Carbone and Kohn (2001), with permission from Blackwell Science.)

factor-1$\alpha$ (EF-1$\alpha$). After alignment, the evolutionary history was reconstructed for both samples to enable nested clade analysis to be carried out on the population sample. It was extended further by the use of GEODIS 2.0 (available at http://bioag.byu.edu/zoology/crandall_lab/geodis.html) (Posada *et al.* 2000). This enables significant associations between clades for geographical position and haplotype and for host and haplotype to be determined. Indels and recombinant data were excluded from this analysis, as required by coalescence theory, although they were reintroduced later in the program. The nested clade was also rooted to enable the sequences to be ordered. Once the nested clade network had been rooted, successive sequences were extrapolated back in terms of the generations to the MRCA, thereby arranging them along a relative coalescent time-scale in units of $N_e$ generations. Associations of haplotypes were established with clear geographical areas, as demonstrated at the termination of the genealogy (Figure 5.7). MIGRATE was then used to determine the effective numbers of migrants exchanged between the sampling areas and the magnitude of $\theta$

(equivalent to $F_{ST}$) between different areas estimated. Its magnitude appeared to depend on the number of populations in a particular sampling area. Finally, the various sequences of events were determined from migration matrices employing the GENETREE program (Griffiths and Marjoram 1996) (available at http//www.maths.monash.edu.au/-mbahlo/mpg/gtree.html).

In summary, the *S. sclerotiorum* [AD] population sample indicated an initial fragmentation event of its common ancestor. This was followed by dispersal in space and to different hosts, resulting in recurrent restrictions on gene flow because of isolation by distance and, eventually, to speciation. The migration matrices made it possible to determine the effective number of migrants exchanged per generation ($2N_e m$) between the three areas that were recognized—temperate, subtropical and wild (=Norwegian=*S. sclerotiorum* [AD] isolates on *Ranunculus ficaria*). Within this species, the temperate agricultural and Norwegian clades were seen from the species sample to be fully differentiated as species (see Chapter 12) but the other clades were not. The temperate subtropical populations in clades 3–1 and 3–2 are evidently more recently evolved from the common *S. sclerotiorum* stock.

## Further reading

Bridge P (1998). Numerical analysis of molecular variability: a comparison of hierarchic and non-hierarchic methods. In *Molecular variability of fungal pathogens* (ed. P Bridge, Y Couteaudier and JM Clarkson), pp. 291–308. CAB International, Wallingford.

Groth JA and Roelfs AP (1987). The concept and measurement of phenotypic diversity in *Puccinia graminis* f.sp. *tritici*. *Phytopathology*, **77**, 1395–9.

Hall, BG (2001). *Phylogenetic trees made easy: a how-to manual for molecular biologists*. Sinauer Associates, Sunderland, MA.

Hillis DM, Moritz C and Mable BK (1996). *Molecular systematics*, Sinauer Associates, Sunderland, MA.

Hudson RR (1990). Gene genealogies and the coalescent process. In *Oxford surveys in evolutionary biology*, Vol. 7 (ed. D Futuyama and J Antonovics), pp. 1–44. Oxford University Press.

Nei M (1987). *Molecular evolutionary genetics*. Columbia University Press, New York.

Sneath PHA and Sokal RR (1973). *Numerical taxonomy* WH Freeman, San Francisco, CA.

# PART III

# Processes in populations

Processes that determine genetic change or stability are going on at all times, many of them simultaneously, in populations of fungi. Figure P3.1 illustrates those of most significance for their population genetics.

New variation is generated by mutation and disseminated through spore dispersal and migration, by heterokaryosis, by recombination, by hybridization and by introgression. Population numbers wax and wane, new populations arise and others become extinct. Populations may comprise homogenous clones or be genetically heterogeneous; they may occur as large single perennial populations, smaller discrete populations which rarely exchange genetic material or complex metapopulations. Their constituent genotypes are constantly exposed to random gains and losses and to natural selection which results in altered gene frequencies. All these processes ensure a constant and dynamic response to the continuously varying environment in which the populations either flourish or decline and, ultimately, evolve through the creation of new species and the extinction of others. In this section the various processes, other than speciation, which occur in fungal populations will be described.

Mutation and migration is considered first since the ultimate source of all new variation is mutation. Experimental methods for generating and isolating mutants and studies of their properties have occupied much of fungal genetics. Therefore it is surprising that so little attention has been paid to spontaneous mutation in fungi either in the laboratory or in nature. It is an area which would benefit greatly from more attention in the future. Once mutants have arisen it is not clear how far purely stochastic fluctuations, i.e. genetic drift, ensure their maintenance either in small fungal populations or in those which show large numerical fluctuations. Migration is a further complicating

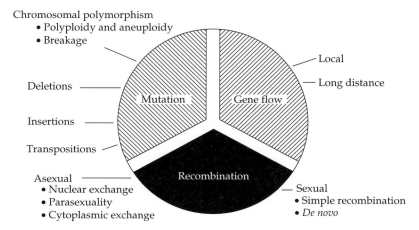

**Figure P3.1** Diagram to illustrate the principal genetic processes (mutation and recombination) and gene flow which determine variation in fungal populations (Based on Burdon (1987), with permission of the American Phytopathological Society).

factor. Mechanisms which bring about spore release and dispersal in fungi have received much attention, but their relevance for the general population genetics of all fungi, especially gene flow between populations, has recieved less attention. Because it is such an important agent in introducing new genetic material into populations, sometimes more so than mutation, gene flow is considered next.

A consideration of the generation of variation through meiotic recombinant processess follows naturally, since the most effective way in which new multigene combinations are generated in populations is by recombination. In many fungi the consequences of recombination may be distorted by gene conversion, and therefore its effects on populations are considered. Recombination is also affected by mating systems. Their actual operation in nature must play a significant but complex role in determining succesful fertilizations. Although the majority of fungi are probably self-fertile, this does not usually preclude out-crossing to a greater or lesser extent. Similarly, even some self-incompatible species, apparently mandatory outbreeders, are capable of some inbreeding through a variety of different mechanisms. Thus many fungi in natural situations can, potentially, combine in- and outbreeding in various ways which are not always immediately obvious. Indeed, it is not always obvious when a fungal population is clonal or whether recombination has taken place recently or in the past.

Fungi have been shown to undergo novel recombination processes in the laboratory but it is not clear how signifcant these are in nature. Nevertheless, without adequate knowledge of these processes the interpretation of patterns of variation within or between populations could be difficult or inadequate. They include heterokaryosis, when whole nuclei exchange and association occurs so that nuclear genotypes are recombined intact. Parasexuality—transient diploidy followed by mitotic recombination and, eventually, the restoration of haploidy—although probably uncommon, is another complicating possibility. A particular aspect of recombination in other organisms, namely hybridization, even if infrequent in fungi as is generally supposed, is significant for population biology through combining genotypes, initiating intogression of novel genes into other races or species and providing the possibility of allopolyploid formation. Little attention has been paid to this process in fungi, and there is a widespread belief that, except for a few well-authenticated examples, it is rare and of little practical significance. Therefore its significance is re-assessed.

Having considered how variation is generated in fungal populations, attention inevitably shifts to how chance and natural selection sieve the multiplicity of genotypes constantly created. Much of the available information on fungi, whether from experimental or natural situations, deals with selection for increased or altered pathogenicity in phytopathogenic fungi in agricultural situations. How natural selection acts on other traits and in other ways in fungi has only just begun to be investigated. Adaptation to habitats and changes in adaptation are an important outcome of selection, and intraspecific competition, which has not received much attention from mycologists, is another important agency in determining the success or failure of competing populations. Competitive effectiveness is clearly an important component of the general fitness of a population or species but fitness involves more than that. Finally, therefore, genetic fitness is briefly considered.

# CHAPTER 6

# The generation of variation—I Mutation and migration

Strictly, mutation is the sole source of new variation, although it is recombination (see Chapter 7) which mainly generates novel multilocus genotypes. Mutations can occur in both nDNA and mtDNA, and the latter are quite common in fungi. Before the structure of DNA had been elucidated, mutations were recognized by the phenotypic changes they caused. Thus, a complete range from point mutations to structural chromosomal alterations and aneuploidy was known from fungi, but there was little evidence for the origin or frequency of either auto- or allopolyploidy in fungi, although both kinds of change undoubtedly occur. Migration between populations must also occur frequently because dispersal in many fungi is highly efficient. Therefore, in a natural population, it may be difficult in practice to distinguish the appearance of a novel gene or phenotype derived in this way from the consequences of mutation. Hence it is convenient to consider mutation and migration together. Migration is also important in its own right as a process which promotes gene flow between existing populations or results in the replacement of extinct populations by similar or different ones.

In this chapter, consideration of how mutants become established in populations and the frequency of detectable mutations precedes a consideration of the kinds of mutation that occur and have become established. Subsequently, modes of gene flow in fungi are considered together with methods for its detection.

## 6.1 The establishment of mutants, random genetic change and the frequency of detectable mutations

Mutation is due to major or minor changes in DNA, but the detection of a specific mutation can only be inferred in a population after several generations have elapsed and the mutant character becomes apparent as the result of random fixation or selection. In a few favourable situations phenotypic detection can be relatively rapid in fungi, as in the case of new virulent mutants and fungicide-resistant mutants of pathogens of cultivated crops, or under selective situations such as occur in industrial situations. Chromosomal mutants can sometimes be detected through their effects on fertility, and some aneuploids are detectable because they affect the normal growth pattern almost immediately. But, in general, immediate mutant detection is not possible, especially in natural or semi-natural situations. Much the same is true of random genetic change, although fixation in small populations can, potentially, be inferred.

The increased understanding of DNA and of changes in its sequence and structure now provides far more detailed knowledge of the types of molecular change that underlie mutation. Many nucleotide substitutions, changes or even losses result in no change in the proteins they determine because of the redundancy of the genetic code. They are phenotypically indetectable. An important and widespread belief is that the majority of mutations are neutral in effect at the molecular level, although some significant molecular changes are undoubtedly disadvantageous and are selectively eliminated as demonstrated by Kimura (1968, 1983) and, to a lesser extent, by Ohta (1992), who considered most mutants to be mildly deleterious. Incidentally, it was implicit in the earlier pre-molecular considerations of Fisher (1930) and Wright (1931) that the majority of significant mutational changes were initially neutral in effect. Deleterious mutants are lost most rapidly in haploid fungi, but more slowly from diploids or dikaryotic forms where they may be masked by the presence of a viable allele.

Whether or not a new mutation persists and is ultimately beneficial to the organism depends on the mutation rate, the genome in which it is located and the size of the population. If the mutation rate $\mu$ is high and the rate of reverse mutation is low, even in a small population, a mutant or rare allele has a better chance of eventual survival than if the opposite conditions prevail. Thus the rare potentially advantageous mutants, regarded as the most significant by some at even the molecular level (Gillespie 1991), probably become established in a population only through recurrent mutation. This is because any gene present in a fungal population at a low frequency can be lost or fixed as the result of the inevitable random sampling of conjugating gametes or individuals in sexual reproduction, or, if the population is maintained predominantly by asexual spores, when these are dispersed and either germinate and persist or perish. Such stochastic changes occur regardless of whether or not the gene confers a potential selective advantage. For neutral genes, on average, a little over a third of all alleles present at a low frequency are lost in this way in each generation and their probability of fixation is only equal to the frequency of the gene in the population.

Therefore the size $N$ of the population is important. First, the probability of a favourable gene being generated is proportional to the size of the population. Second, the smaller the population the more significant are stochastic events, resulting in **genetic drift**, i.e. change in gene frequency due to chance fluctuations in numbers of individuals, as first described by Wright (1931). Both selection and drift are affected by the **effective population size** $N_e$, i.e. the population restricted to those individuals which have an equal expectation of being the parents of any sexual or asexual progeny formed. Most commonly $N_e$ is smaller than the actual numbers of individuals observed apparently forming the population. There are many reasons for this in fungi.

First, and most simply, some individuals may not breed at all or only breed after the mycelium has reached a certain maturity, or the fungus is a strict inbreeder.

Second, in outbreeding fungi the numbers and kind of mating types present affect the size of $N_e$. Therefore, for a dimictic fungus having both *mat A*- and *mat a*- carrying individuals

$$N_e = \frac{4N_{mat\cdot A}N_{mat\ a}}{N_{mat\ A} + N_{mat\ a}}$$

Therefore, if the numbers of the two MTFs are equal $N = N_e$, but if they are *unequal*, the size of the breeding population is somewhat reduced. For example, if the ratio is $2:1$ then the effective population size is about 10 per cent less than the actual numbers of individuals.

Third, if the numbers of offspring contributed by individuals to the next generation is unequal, i.e. if the variance in offspring numbers exceeds the mean number of offspring, this may have a substantial effect in reducing numbers.

Fourth, an even more drastic factor reducing population size is an impairment of the breeding abilities of some of the population, e.g. as a consequence of 'female sterility', as occurs in many Ascomycotina (see Chapter 7, section 7.2.6.1). It can then be shown that

$$N_e = \frac{4N^2N_h}{(N + N_h)^2}$$

where $N$ is the total number of individuals and $N_h$ is the number of normal, i.e. hermaphrodite, individuals in the population (Leslie and Klein 1996). Equal numbers of normal and female sterile individuals reduce the size of the breeding population by just over 10 per cent, but as the number of female sterile individuals in the population increases $N_e$ declines rapidly. Moreover, in populations which propagate asexually for an appreciable period, the numbers of normal individuals will tend to decline because female sterility can arise from a variety of mutational events. Therefore when a sexual reproductive phase does intervene, female sterile mutants may have accumulated, reducing further the numbers of normal individuals and hence $N_e$.

Fifth, the breeding system in diploids such as Straminipila can affect population number if there is a high degree of inbreeding. The effective population size is $N/(1 + F_{IS})$, where $F_{IS}$ is the inbreeding coefficient. Therefore if the population is completely inbred, as might occur with a homothallic Oomycete, $N_e$ is effectively halved.

Finally, many fungi undergo wide seasonal or annual fluctuations in numbers and it can be shown that the value of $N_e$ is then equal to the harmonic mean $\tilde{N}$ of the fluctuating population size. Hence

$$\tilde{N} = \frac{t}{\Sigma(1/N_i)}$$

where $t$ is the time interval and $N_i$ is the smallest population in the series (Wright 1938). In such cases $N_e$ will be very small, approaching $N_i$.

Strictly, the probability of selection of a favourable allele is proportional to $N_e$ and the effective number of alleles maintained in a population $n_e$ is also related as follows (Kimura and Ohta 1971):

$$n_e = 4N_e\mu + 1$$

for a diploid and

$$n_e = 2N_e\mu + 1$$

for a haploid.

The most significant consequence of conditions which result in $N_e$ being less than $N$ is that the effective population may be so reduced that drift becomes a significant factor in determining the fate of vulnerable, i.e. rare, or newly mutant alleles. If $N_e < 10$–$100$ a neutral allele may be lost quite quickly, perhaps by as much as 0.1 per cent per locus per generation, but if $N_e < 1000$–$10\,000$ the rate of loss is progressively reduced to as little as $10^{-4}$ per generation. It will then take far longer for an allele either to be lost or to go to fixation, so that new or rare alleles will, on balance, be exposed to natural selection for much longer.

There are three kinds of situation, all of which occur with fungi, when populations might become so small that genetic drift becomes significant.

First, drift could occur discontinuously in those fungi whose population size fluctuates and where asexual predominate over sexual generations. Such fluctuations are commonly associated in pathogenic fungi with epidemic conditions. Reduced populations commonly occur as the epidemic phase declines or when growing conditions deteriorate, but both of these conditions are often associated with the onset of a sexual phase. Therefore new genotypes could be injected into the population as it increases in numbers even if the variability of an asexual surviving population has been reduced by drift. No data are available for fungi, but it should be possible to investigate such situations.

Second, viable small populations, which are geographically and ecologically isolated, could occur in which drift would be effective. Meta-populations, i.e. subdivided populations, of asexual Eumycota occur in which extinction of some sub-populations is followed by recolonization of the sites. In such a case the effective population size is

$$N_e = \tilde{N} + n(2\lambda) + n\tilde{N}\mu/\lambda$$

where there are $n$ different genotypes, $\tilde{N}$ is the harmonic mean of the effective population size of each genotype per unit, $\lambda$ is the rate of genotype extinction and $\mu$ is the mutation rate. Consequently, $N_e$ is appreciably smaller than $N$ (Maruyama and Kimura 1980). Populations of sexually reproducing fungi can show similar behaviour and they provide some evidence for the minimal size of population necessary to ensure continued survival. For example, in Kosciusko National Park, Australia, the probability of survival for infections by the naturally occurring *Melampsora lini* [BR] on plants of the native *Linum marginale* is in the range 0.017–0.005. For a 50 per cent chance of survival, 50 to 150+ hosts need to be infected, which gives a lower estimate for the value of $N_e$, assuming a single infection per host. The figure for *Triphragmium ulmariae* [BR] on *Filipendula ulmaria* is much larger at 350+ hosts (Burdon 1993). Smaller populations of either of these annual pathogenic rusts become extinct. These minimal estimates for pathogenic fungi are somewhat above those where drift might play a significant role. However, it does seem possible that saprophytic species, especially those with perennial mycelia which probably require less stringent conditions for establishment, might sustain smaller population numbers and so be liable to genetic drift.

Third, drift could occur in isolated populations which have arisen either as founder populations after long-range dispersal, for example, or in relict populations which have become isolated as a consequence of a major environmental change such as a glacial epoch. The known long-range dispersal of rust fungi from southern Africa to Australia is a situation which probably resulted in a small founder population, but no evidence is available for either the initial population size or for drift.

Two well-documented examples of isolated populations whose origins are probably due to environmentally imposed isolation are the dikaryotic bracket fungus *Fomitopsis rosea* [BA] in Northern Europe and the chestnut blight fungus *Cryphonectria parasitica* [AP] in Switzerland.

*F. rosea* [BA] occurs infrequently on dead spruce in small isolated populations, often widely separated, in the fragmented unmanaged taiga forest of northwest Russia and Fennoscandia (Hogberg and Stenlid 1999). PCR preparations were made from three primer-defined regions (one representing the M13 sequence and another a repetitive sequence $(CAGA)_4$) from several single-spore isolates taken from each basidiocarp sample and the bands obtained were compared. Since several spores were obtained from each basidiocarp, it was possible to detect homozygous, heterozygous and polymorphic loci and so to estimate heterozygosity. The basic data were corrected for under-estimation of heterozygosity by increasing the observed numbers by $(0.5)^3$ and by reducing the number of homozygous samples by $(0.5)^{3/2}$. The nucleotide regions detected which represented loci all segregated as $1:1$, but the observed and corrected heterozygosity was appreciably lower than expected (Table 6.1). In particular, the lowest values were found in those regions of Fennoscandia subjected to the longest periods of forest fragmentation where the mean observed heterozygosity was 0.09 compared with an expected value of 0.19. Two explanations which would account for these observations are:

(a) that they indicated drift in the small isolated populations resulting from forest fragmentation;
(b) that the populations resulted from random selection due to ancient founder effects dating from the period of woodland spread followed by sporadic scattered infections in the Holocene.

The only directly observed evidence for reduced variability in founder populations comes from *C. parasitica* [AP] in Switzerland (Hoegger *et al.*

**Table 6.1** Mean values of observed ($H_O$) and expected ($H_E$) heterozygosity in isolates of *Fomitopsis rosea* [BA] in four sites plus a pooled sample from southern Fennoscandia

| Location | Mean heterozygosity | | P values | Standard error |
|---|---|---|---|---|
| Syktyvar | $H_O$ | 0.25 | | |
| | $H_E$ | 0.30 | 0.06 | 0.002 |
| Lyali | $H_O$ | 0.18 | | |
| | $H_E$ | 0.27 | 0.0001* | 0.0001 |
| Yaksha | $H_O$ | 0.18 | | |
| | $H_E$ | 0.31 | 0.0002* | 0.0001 |
| Suobatt | $H_O$ | 0.29 | | |
| | $H_E$ | 0.24 | 0.28 | 0.005 |
| South central | $H_O$ | 0.09 | | |
| Fennoscandia | $H_E$ | 0.19 | 0.000* | 0.000 |

* Observed heterozygosity significantly different from expected at $P = 0.01$. (After Högberg and Stenlid (1999), with permission from Blackwell Science.)

2000). Four populations were compared. Three were recently established populations at Choëx, Weggis and Murg, all north of the Swiss Alps which form a barrier between them and older established populations of chestnut blight fungus in northern Italy. The fourth was a population at Claro which was newly established but was within an older previously infected area south of the Alps. The mating type ratio and numbers of VCGs were determined for each population and, in addition, isolated DNA was probed using pMS5 and the data were subjected to a similarity comparison. In the three populations north of the Alps, mating type ratios differed markedly from the normal $1:1$ and the Shannon index for the VCGs ranged from zero to 0.58, with two of the three populations carrying only a single VCG. At Claro, although the MTF ratio deviated from $1:1$, the Shannon index for the VCGs was 1.64, with nine VCGs being identified. Three and five fingerprints occurred at Choëx and Weggis, but in both populations a single multilocus genotype predominated; three genotypes occurred at Murg, one of which predominated. Thus these populations are virtually clonal in respect of genotypic variation. This compares with the Claro population with 33 genotypes, none being predominant. Evidently the three new isolated founder populations are appreciably less variable than that at Claro, which was not isolated from the main infected areas in Italy, although even here mating was unlikely to be at random.

A third possible example is the numerous effectively self-fertile and therefore genetically isolated populations of *Gaeumannomyces graminis* [AP] in South Australia. It is claimed that these populations are periodically reduced in size as a result of cereal crop rotations and seasonal effects, so that the genetic structure of the surviving founder populations is largely determined by drift (Harvey *et al.* 2001). There is certainly not much evidence of gene flow, but although the variation in each population is small, the genetic variance of the cereal region as a whole is not inconsiderable. Until the difficult problem of determining the actual size of such populations has been solved, there is bound to be uncertainty about potential bottlenecks and subsequent genetic drift.

Whether new mutant genes are actually lost due to drift in small populations is not known, but such evidence as is available suggests that the majority of fungal populations are large enough for loss not to occur for that reason. Therefore selection almost certainly plays a dominant role in establishing new mutations in most fungal populations. However, an indirect selection mechanism, genetic hitch-hiking, can result in the chance increase and establishment of a gene.

### 6.1.1  Genetic hitch-hiking

When a new mutant which is neutral or slightly deleterious in effect is tightly linked to a gene which already confers a high selective advantage on the individual, it will increase in parallel with its positively selected neighbour (Maynard-Smith and Haigh 1974; Thomson 1977). It is not known how frequently such 'hitch-hiking' events occur in fungi in connection with new mutants but, statistically, they must occur by chance from time to time.

Hitch-hiking is favoured in those fungi that propagate asexually for several generations, since this maintains the genotype unchanged. Thus, if the necessary conditions exist for hitch-hiking at the outset, genes will remain associated together. A possible example is provided in barley mildew by 'unnecessary virulence' genes, i.e. virulence genes carried at a far higher frequency than the matching

resistance genes in the crop. During 1985 in the UK, the most common strains of *B. graminis* f.sp. *hordei* [AD] carried *V-a7* and *V-(Ab)*, virulence genes matching the principal resistance genes present in the contemporary barley crop, plus three 'unnecessary virulence' alleles, *V-a6*, *V-k* and *V-(CP)*. The latter occurred in 60 per cent or more of all mildew isolates sampled, although the frequencies of the matching resistance genes in the barley crop ranged only from zero to 1.7 per cent. The virulence genes are unlinked, but it was suggested that the unnecessary selectively neutral virulence genes had been maintained as a consequence of hitch-hiking (Brown and Wolfe 1990).

### 6.1.2  The frequency of mutation in fungal chromosomes

There are very few estimates of spontaneous mutation rates in fungi. One of the earliest measurements was for recessive lethal mutations in air-dry conidia of *Neurospora crassa* [AP] (Auerbach 1959). She found that their number increased by about 0.3 per cent per week at 32 °C but more slowly, 0.1 per cent per week, at 4 °C. Presumably, these mutants reflected mutations at a number of unspecified loci in non-replicating nuclei. Mutation rates at specific loci have also been examined. By screening uninucleate microconidia, a single spontaneous auxotrophic mutant was found amongst 3000 tested, suggesting an upper mutation rate of about $10^{-6}$ (Tatum *et al.* 1950). In keeping with this was the total absence of mutants at the *ad-3* locus in a sample of $7.6 \times 10^{-6}$ microconidia, even though, when irradiated, 180 mutants were obtained from far fewer microconidia ($10^{-6}$) (de Serres and Kolmark 1973). Most other estimates have been of back mutation rates from a mutant to the wild-type condition. More recently, a number of loci in *N. crassa* [AP] and *Saccharomyces cerevisiae* [A] have been examined. The forward mutation rate at the *MTR* locus in uninucleate conidia in liquid medium of *N. crassa* [AP] was $2 \times 10^{-7}$ conidia and the back mutation rate was an order of magnitude smaller (Stadler *et al.* 1991). Drake (1991) recorded rates of $2.77 \times 10^{-8}$ ascospores for spontaneous mutation at the *URA 3* locus in *S. cerevisieae* [A], and $1.76 \times 10^{-7}$ and $2.7 \times 10^{-7}$ uninucleate conidia for *ad-3AB* and *mtr* respectively in *N. crassa* [AP]. He suggested that the average mutation rate was proportional to the reciprocal of the genome size, giving a rate of 0.003 per genome per replication. Because Drake's sample is so small it is difficult to assess whether or not his generalization is valid, but there is, at least, a reasonable degree of similarity between the various mutation rates per conidium reported for *N. crassa* [AP].

### 6.1.2.1 Fungal mutation in nature

Even if more reliable data were available, it is debatable how useful knowledge of the average rate of mutation under natural conditions would be.

First, the rate of mutation alone is insufficient to bring about change very rapidly. As already described, the chances of a mutant persisting depend on whether it confers an advantage or disadvantage on the individual or is neutral, and the statistical chance that it will survive. This is affected by population size and the forward/back mutation rates. The smaller the population the less likely it is that a mutant will survive. In fact, most mutations are recurrent and most can back mutate; therefore if an allele $A$ of frequency $p$ mutates to $a$ with frequency $u$, and the back mutation rate is $v$, then the frequency of $a$ due to mutation alone is

$$\Delta q = up - vq$$
$$= u - q(u + v).$$

At equilibrium, setting the relationship to zero,

$$q_{eq} = u/(u + v)$$

If, as seems reasonable, $v$ is small compared with $u$, after one generation

$$p_1 = p_0 - up_0.$$

Hence

$$p_1 = (1 - u)p_0$$

or, in general, after $t$ generations

$$p_t = (1 - u)^t p_0.$$

Therefore change in gene frequency as a result of mutation alone is very slow indeed. With mutation frequencies of the order found in fungi ($10^{-6}$ or $10^{-7}$), mutation alone would take many tens of thousands of generations to reduce the frequency

of a gene to a tenth its original value, and this would probably be an over-estimate since the forward rate $\mu$ would be opposed by the frequency of back mutation. In the event that the mutant was disadvantageous, it would be selected against. However, combined with positive selection, the frequency of a mutant in a population could be appreciably increased.

Mutation frequency is affected by environmental conditions as well as by selection. For instance, Auerbach's (1959) observation of the differential frequency of recessive lethals at 32 and 4 °C adds support to Drake's suggestion, following Koch (1971), that the rate might be very different under conditions of starvation or stress. Stress is claimed to increase the mutation rate in both bacteria and plants, but there is very little information about whether it affects the frequency of mutation in fungi. Indeed, only one direct demonstration has been provided of environmental stress on mutation in wild isolates in natural conditions (Lamb *et al.* 1998).

Lamb and his colleagues isolated mycelia of the self-fertile *Sordaria fimicola* [AP] from soil and dung (on which latter it regularly forms perithecia) from the north- and south-facing slopes of 'Evolution Canyon' in Lower Nahal Oren, Mount Carmel, Israel. This canyon runs east-west, is about 120 m long and 100 m deep, and is 400–500 m wide at the top, reducing to 100–200 m across the bottom. The geological basis of the canyon, limestone, is uniform throughout. The 25° north-facing slope is thickly clothed with lush temperate 'European' vegetation, whereas the highly insolated 35° south-facing slope whose climate fluctuates considerably, often being 300 per cent hotter than the north side, is an arid open savannah. Eleven isolates were taken, three from three sites at 60, 90 and 120 m above sea level on each side and two from the canyon bottom. They formed perithecia and some ascospores were germinated, grown on to a further perithecial stage and scored for two further generations. Normally this fungus has black ascospores although, experimentally, a number of ascospore colour mutants (grey, pale grey or colourless) have been isolated and behave as single-gene mutations (Olive 1956). Ascospores from wild isolates were examined for similar spontaneous mutants and also for spontaneous mutants resistant to acriflavine. Significantly *higher* mutation rates both to non-black ascospores and to acriflavine resistance were found in the isolates taken from the *highly insolated south-facing slope*. In the two subsequent perithecial generations this differential mutation rate was shown to be, in part, heritable (Table 6.2).

This study demonstrates the importance of fully recording the climatic and ecological conditions in which fungi are sampled for population analyses if an adequate analysis is to be made. Further investigations of this kind are desirable since, in nature, spores and mycelia must often be exposed to prolonged periods of stress of various kinds such as low nutrition, high insolation or exposure to ultraviolet radiation, dehydration and extremes of temperature, and, increasingly, environmental mutagens.

In contrast to this situation there is little evidence for mutation in the mycelia of two long-lived

*Armillaria gallica* [BA] clones in Michigan (Hodnett and Anderson 2000). Samples were taken from 22 sites in two clones. Judged from the linear distance between the furthest points apart on the mycelium of clone I, it had undergone an estimated minimum of $6.5 \times 10^7$ cell divisions over a time period estimated to be of the order of 1500 years (Smith *et al.* 1992). Using the SSCP technique, 41 regions were screened in the nuclear DNA isolated from the samples, the equivalent of screening $10^4$ bp. *No variation was detected between isolates from either individual or in the MTFs.* Assuming mutation rates of the order given in section 6.1.2, at least 65–650 mutations should have been detected; their total absence was an astonishing finding. It is suggested that either the mutation rate was lower than might have been expected or the estimated number of cell divisions appreciable fewer, but both processes will need further investigation since no explanation can be given for these findings at present, apart from the constancy of the MTFs.

**Table 6.2** The spontaneous mutation rate in homothallic isolates of *Sordaria fimicola* [AP] from different aspects of 'Evolution Canyon' in Israel

| Wild-type strains | Octads isolated | | Asci with 4 + :4m ascospores | | Limits of three repeats[a] |
|---|---|---|---|---|---|
| | **Octads** | **Ascospores** | **Mean %** | **Standard error** | |
| *South-facing slope* | | | | | |
| S1 (dung) | 61 | 375 | 3.49 | ±0.92 | 3.40–3.81 |
| S2 (dung) | 74 | 734 | 2.37 | ±0.06 | 2.14–2.38 |
| S3 (soil) | 51 | 077 | 4.41 | ±0.24 | 4.08–4.89 |
| *Bottom of valley* | | | | | |
| B | 7 | 522 | 2.14 | ±0.08 | 2.00–2.32 |
| *North-facing slope* | | | | | |
| N5i (soil) | 68 | 750 | 1.25 | ±0.03 | 1.19–1.31 |
| N5ii (soil) | 7 | 262 | 1.23 | ±0.06 | 1.08–1.31 |
| N6 (soil) | 66 | 077 | 1.27 | ±0.00 | 1.21–1.30 |
| N7i (dung) | 59 | 969 | 0.89 | ±0.05 | 0.79–1.00 |
| N7ii | 8 | 883 | 1.18 | ±1.18 | 0.81–1.36 |
| *Slope totals* | | | | | |
| South-facing | 187 | 145 | 3.25 | ±0.03 | 3.02–3.57 |
| North-facing | 210 | 941 | 1.13 | ±0.03 | 1.09–1.21 |

Wild-type ascospores are uniformly black and spontaneous mutations in the laboratory are rare; *m* represents any colour mutant, white or grey.

[a] The three repeated tests took place over a 5-month period during which, although in store, there was no substantial change in the mutant frequency.

(Reproduced from Lamb *et al.* (1998) with permission of the Genetical Society of America.)

There is evidence for balancing selection maintaining the MTFs of *Neurospora* [AP] and *Coprinus cinereus* [BA] relatively unchanged genetically over geologically long periods of time (millions of years) (see Chapter 9, section 9.4.3.2). However, that situation is very different from maintaining a whole genotype genetically unchanged for a period of minimally hundreds to thousands of years, even if based on a relatively small but random sample of sites.

## 6.2 Natural mutants of fungi

Fungal genetics was developed using induced mutations, but a variety of established spontaneous natural mutants are known. However, the actual nature of the mutation is known in only a few cases. The vast numbers of induced mutations do not provide a useful guide to the origins, frequencies, distribution or persistence of spontaneous mutations in populations. Extensive tests using *A. nidulans* [A] demonstrated that spontaneous mutants of various kinds were comparable in their expression to induced mutants (Teow and Upshall 1983).Unfortunately, because enrichment techniques were used to isolate the mutants, no estimate could be made of the mutation rates involved. Spontaneous mutants isolated included prototrophic white, yellow and pale green conidium colour mutants and 15 auxotrophs—five requiring *p*-amino benzoic acid, two requiring adenine, three requiring methionine, and one each requiring aneurin, biotin, lysine, nicotinic acid and pyridoxine. A considerable number were resistant to toxic substances including potassium chlorate, sodium selenate, *p*-fluorophenylalanine, acriflavine, actidone, 8-azaguanine and the fungicide benomyl. All appeared to be single-gene mutations and *no* structural changes were detected in any of them. This contrasts with comparable induced mutants where the mutagenic agents can often cause structural chromosomal changes as well.

### 6.2.1 The nature of spontaneous natural mutants in fungi

There is a dearth of biochemical information about the nature of spontaneous mutations. Now that

genes can be cloned and base sequences more readily determined, some information is becoming available. For example, it is now known that spontaneous mutation of the avirulence gene *Avr4* in *Cladosporium fulvum* [FA], the cause of leaf mould in tomatoes that carry the resistance gene *Cf-4*, can be due to as little as a single amino-acid substitution in the elicitor protein it produces, i.e. a single base in a codon. In one case the change, $(G \rightarrow A)$, resulted in the replacement of cysteine by tyrosine. In another, two changes, $(U \rightarrow C)$ and $(U \rightarrow A)$, resulted in two amino acid substitutions—histidine for tyrosine and isoleucine for threonine. In yet another mutant, the loss of a single nucleotide resulted in a frame shift reducing the N-terminal end of the protein from 18 to 13 amino acids (Joosten *et al.* 1994). The consequences of any of these changes in the elicitor protein was to make the mutants virulent to tomatoes carrying the *Cf-4* resistance gene. In other mutants of *C. fulvum* [FA] virulent to tomatoes carrying the resistance gene *Cf-9*, the avirulence gene *Avr9* appeared to have been lost entirely.

Differences between sequences conferring virulence or non-virulence are appreciably larger in other fungi. In *Magnaporthe grisea* [AP], for example, mutant virulent strains carrying the *avr-Mara* gene differed from the wild type carrying *Avr-Mara* by two deletions about 13.5 and 25 kb long.

Thus very small changes in the genetic material—nucleotide substitutions or deletions, or frameshift mutations due to insertions or deletions—can all result in biologically significant spontaneous mutants. Since a gene-for-gene relationship is probably widespread in natural populations of fungal pathogens and their host plants (e.g. Clarke 1997; Burdon 1997) (see Chapter 9, sections 9.4.1 and 9.4.2), it is not surprising that mutations to virulence have been detected relatively frequently in various non-pathogenic strains of fungi, e.g. *F. oxysporum* [FA] (Gordon and Martyn 1997).

The detection of apparently new successive virulence mutations over time have characterized most plant pathogenic fungi of economic importance. This situation is shown clearly in the potato blight *Phytophthora infestans* [O]. Prior to the early 1980s populations worldwide were clonal and lacked sexual reproduction, except in Mexico. The numbers of clones circumscribed by isozyme and RFLP analysis were quite few in number. A single clone, genotype US-1, was introduced from the USA to Western Europe in 1844–1845 and subsequently exported to the rest of the world (Goodwin *et al.* 1994). Nevertheless, more than 10 specific virulence genes have arisen in it and combinations of these have given rise to some 40 or more different pathotypes, i.e. a new pathotype about every 2.3 years since 1845.

A comparable example amongst the Eumycota is *Puccinia striiformis* f.sp. *tritici* [BR], which is an entirely clonal fungus despite producing nonfunctional basidiospores. It was introduced to Australia in 1979, and in the 10 years to 1988 some 15 new mutant pathotypes were identified, almost all certainly due to single-gene mutations (Wellings and McIntosh 1990). This is an appreciably more rapid generation of pathotypes than by *Phytophthora infestans* [O], but this is not really surprising. Estimates of daily uredospore production by *P. graminis tritici* [BR] vary from $2.5 \times 10^{-12}$ per hectare per day with 10 per cent infection of the crop, to $10^{-11}$ per hectare per day with 1 per cent infection (Rowell and Roelfs 1971; Parlevliet and Zadoks 1977). *P. striiformis* [BR] is likely to be comparable to black rust of wheat so that, if the mutation rate is of the order of $10^{-5}$ or $10^{-6}$, even allowing for uredospores being dikaryotic, there is clearly an immense potential for many mutants of all kinds to be produced daily over the several weeks in which uredospores are shed! Although prolific, spore production by *Phytophthora* is not comparable with that of the rusts.

No other cases of spontaneous mutation are as clearly documented as those for virulence and certainly none are documented so extensively or for so many different species. Nevertheless, since a novel virulence gene is likely to be highly positively selected, these cases give some idea of the upper likelihood of *persistent* mutations under natural or quasi-natural conditions.

*P. striiformis* f.sp. *tritici* [BR] illustrates another feature concerning spontaneous mutations, namely their uneven distribution in space. For instance, Tainshui and other sites in Gansu Province, China, have long been notable for the generation of large numbers of pathotypes, presumably due to new mutant virulence genes (Li and Shang 1989). In fact, it appears that the region is a site of general high mutability since DNA fingerprinting analysis using a moderately repetitive specific sequence, PSR311, has demonstrated that the area also has a high level of general genetic diversity (Shan *et al.* 1998). The region is mountainous with cool winters and summers, favourable to continuous growth of the fungus. There are also a diversity of climatic niches. Thus there is a longer period and diversity of habitats in which new mutations can accumulate and be maintained than elsewhere in China. Whether that accounts entirely for these observations is not known. Alternatively, episodes of high spontaneous mutability occur in other fungi (see section 6.2.2) which, if they occurred in this species, could account for the observations. Unfortunately, no Chinese data are available on the nature and frequency of spontaneous mutants in *P. striiformis* [BR].

In contrast, some genes, notably the complex MTFs of Basidiomycotina, are remarkably resistant to attempts to induce new functional mutations and spontaneous mutations are unknown. For instance, although MTFs are either allelic or idiomorphic, numerous attempts by Raper and his colleagues to induce a new *functional* MTF in the tetrapolar *Schizophyllum commune* [BA] were never successful. Those that were obtained were usually due to self-fertility (Raper 1966). Nevertheless, mutations at the molecular level can occur in MTFs since different isolates which show identical mating type behaviour are now known to differ in the nucleotide sequences of the *A-1* factor (T.Y. James, unpublished data).

However, the absence of mutation to a new functional MTF condition is probably the norm. In addition to the experiments with *S. commune* [BA], extensive treatments by a variety of mutagenic agencies, both physical and chemical, of many millions of uninucleate conidia of the bipolar *Vararia granulata* [BA] carrying the *A1* MTF entirely failed to detect a single new functional mutant, although reconstruction experiments suggested that a frequency of at least $1 \times 10^{-8}$ should have been detected (J.H. Burnett, unpublished data). A similar stability has been found in nature in fairy-ring fungi such as *Marasmius oreades* [BA] whose mycelia persist for at least several decades. All basidiocarps derived from the same ring and tested over successive seasons were found to have their MTF constitution unaltered (Burnett and Evans 1966 and unpublished data).

In *Coprinus cinereus* [BA] there is now evidence that a specific sequence in the *B* MTF is not only very ancient but has been maintained virtually unchanged as a consequence of stabilizing selection (May *et al.* 1999). Whether new mutants arise but are selected against, or whether no new functional mutant MTFs are ever produced by these basidiomycotous fungi is not known. However, it is potentially useful for population studies to know that these loci are effectively immutable.

An exceptional situation occurs in one strain of *Pleurotus ostreatus* [BA] where new *B* factors can apparently be generated by crossing over within the *B* MTF locus (Larraya *et al.* 2001).

Cytoplasmic elements can also alter the phenotype of a few fungi. Many dsRNA mycoviruses are now known (Koltin and Leibowitz, 1988), most of which have no obvious phenotypic effect on their host fungi other than senescence. However, some such as the dsRNA viruses responsible for the 'killer' factors in *S. cerevisiae* [A] (Bevan and Makower 1963) and *Ustilago maydis* [BS] (Wood and Bozarth 1973) have a clear effect. These viruses are transmitted cytoplasmically by cell fusion. Whether their ability to kill other sensitive genetic strains of their own species, which they do, confers any advantage is unknown.

Cytoplasmic DNA plasmids also occur. They also appear to be associated with toxin production such as the linear plasmids in some yeasts (Cong *et al.* 1994) or those in *Alternaria alternata* [FA] which encode a peptide, tentoxin, which is toxic to some higher plants (Shepherd 1992). Transmission is cytoplasmic in the same way as the dsRNA viruses.

## 6.2.2 Gene and chromosomal changes associated with recombination and gene conversion

Inherited changes in genes and chromosomes are often associated with either meiosis or mitosis. There are a variety of reasons for this association, but many situations involve some kind of unequal or aberrant pairing, the formation of hybrid nDNA or, sometimes, the consequences of strand repair mechanisms.

One mutational situation is of sporadic occurrence, namely episodes of high mutability. Various causes are known for this situation. One that has been described from yeast in laboratory populations is the presence of a mutator gene or genes. These result in deficiencies in chromosomal DNA repair mechanisms and consequently give rise to high spontaneous mutability (Gottlieb and von Borstel 1976; Hastings *et al.* 1976). They are expressed in connection with meiosis and, in some cases, mitosis (Golin and Esposito 1977). Similar mutations have neither been detected in natural populations of yeasts nor isolated from other fungi, but they might well be expected to occur in any fungus.

Transposable elements of various kinds which actively affect gene structure and result in mutability, nucleotide excisions and chromosomal rearrangements are widespread in fungi (Daboussi 1996; McClure 1999). In some species they make up an appreciable fraction of the nDNA, e.g. *Fusarium oxysporum* [FA] where clusters of various DNA transposons occur (Hua-Van *et al.*, 2000). Retrotransposons, originally described in yeast (*Ty* elements) have also been found in a number of other fungi including *Neurospora* [AP] (*Tad1*), *Cladosporium fulvum* [FA] (*CfT-1*), *Fusarium oxysporum* [FA] (*skippy*) (Anaya and Roncero 1995; Daboussi and Langin 1995) and *Botrytis cinerea* [FA] (*flipper* and *Boty*) (Levis *et al.* 1997).

The *Ty* family of retrotransposons is the most fully studied (Liebman and Picologlou 1988). Each is a 6-kb length of DNA dispersed at various sites within the chromosomes. About 30 copies of *Ty*1 or 6–12 copies of *Ty*2 occur in each nucleus of Baker's yeast, and rather fewer copies (4–20) in wild yeasts. They are capable of transposition to other sites where they are associated with recombination of both reciprocal and non-reciprocal homologous nuclear DNA, resulting in a variety of aberrations, deletions, duplications, inversions or reciprocal translocations (Codón *et al.* 1997). In yeast, the *Ty* elements appear to be preferentially integrated upstream of transfer RNA (tRNA) loci (Eigel and Feldmann 1982), and in some cases there appear to be distinct 'hotspots' where integration occurs preferentially (Ji *et al*. 1993). Here phenotypic expressions involve a variety of multisite mutations of various genes (Liebman *et al*. 1979).

Instabilities at the *am* locus in strains of *N. crassa* [AP] are caused by the retrotransposon *Tad1*, and it has been shown that it can also be transferred between nuclei in a heterokaryon from the nucleus that carries it initially to other nuclei from which it is absent (Kinsey and Helber 1989; Kinsey 1990).

Apart from the possibility of inducing gene mutations, the fact that retrotransposons can cause chromosome breakage may be related to the widespread structural chromosomal variation in fungi such as that in *M. grisea* [AP]. A number of transposons and repetitive elements have been identified from this fungus (Talbot 1998). Some, such as *grasshopper*, are characteristic of pathogenic isolates from a specific host, in this case millet (*Eleusine* spp.). Others appear to occur as commonly in isolates pathogenic to rice as in those pathogenic to other grasses, e.g. *MG-SINE* which occurs in up to 100 copies in its host (Kachroo *et al*. 1995). A number of repeated elements, such as *MG-SINE*, contain 5' ends and can act as sites for recombination during mitosis. As a consequence, chromosome breakage and translocations, both homologous and non-homologous, may occur and give rise to the extensive variation in chromosome length and karyotype reported from USA lineages by Talbot *et al*. (1993). It is not clear whether the great variation in size and number of mini-chromosomes (500–2000 kb) found in largely infertile rice isolates from a world sample can be accounted for in this way (Orbach *et al*. 1996). However, since many mini-chromosomes contain repetitive elements, they could have arisen from translocations.

Despite these known or potential effects on fungi, many carry retrotransposons with no detectable phenotypic effect. For example, the retrotransposons *Boty* and *flipper* occur in about 75 per cent of all *Botrytis cinerea* [FA] isolates collected in the Champagne district of France, but elsewhere only *Boty* occurs. They do not appear to affect the fungal phenotype. *Boty* is a 6-kb *gypsy*-like retrotransposon and *flipper* is a 1842-bp class II element, so they clearly arose independently but presumably successively, unless there has been horizontal

transfer of *flipper* to a *Boty*-containing strain (Giraud *et al*. 1997). This is unlikely since, using RFLP markers, it was shown that the strains were genetically very different and recombination even between two sympatric strains was very restricted.

Chromosomal rearrangements are probably more common in fungi than has been realized or recorded. Translocations have been detected in *M. grisea* [AP] (Skinner *et al*. 1991) and *Septoria tritici* [FA/AD] (McDonald and Martinez 1991), and many have been found in *Coprinus cinereus* [BA] (Pukkila and Casselton 1991), while insertional translocations are said to be common in *Neurospora* [AP] (Perkins 1985), amounting to about 8.5 per cent of all recorded rearrangements (Perkins 1997). Of course, the causes of chromosomal mutations are various, and no doubt the reasons for such high chromosomal polymorphism are likely to be equally varied. Spontaneous chromosome breakage and reunion undoubtedly occur, as already described for *M. grisea* [AP]. However, the chromosomal rearrangements in *Nectria haematococca* (*F. solani*) [AP] undergoing experimental transformation with an alien plasmid *pLD* (derived from *pFOLT4R4*) are apparently due to the incorporation of this autonomously replicating alien plasmid (Kistler and Benny 1992). This is unusual.

There is no doubt that some plasmids confer phenotypically detectable effects and that transposons bring about changes to the karyotype and some site specific mutations but how far these effects contribute to selectively significant change cannot at present be accurately assessed.

Chromosomal length polymorphism is widespread in fungi, e.g. *Candida albicans* [A] (Magee and Magee 1987), *S. cerevisiae* [A] (Ono and Ishino-Arao 1988), *Ustilago hordei* [BS] (McCluskey and Mills 1990), *U. maydis* [BS] (Kinscherf and Leong 1988), *Cochliobolus heterostrophus* [AP] (Tzeng *et al*. 1992), various forma speciales of *Fusarium oxysporum* [FA] (Kim *et al*. 1993; Boehm *et al*. 1994) and *C. cinereus* [BA] (Pukkila and Casselton 1991). Many length polymorphisms appear to be associated with either mitosis or meiosis. The human pathogenic basidiomycotous yeast *Cryyptococcus neoformis* [BT] exists in a number of karyotypes. It is meiotically unstable and many different karyotypes are found after meiosis has occurred. There is also considerable chromosome-length polymorphism. Surprisingly, much karyotype variation appears to be genetically neutral and is not associated with any noticeable phenotypic changes (Boekhout and van Belkum 1997). Individual strains are more or less stable, only rarely giving rise to mutant karyotypes, but sequential isolations from the same parent often show small karyotypic changes. It has been suggested that these may be the result of rare mitotic recombination within a clone. Karyotype changes associated with mitosis are also known from the strictly anamorphic ascomycotous human pathogenic yeast *Candida albicans* [A] (possibly the type I strain of *C. stellatoidea* [A] (Kistler and Miao 1992)), where as many as 1.4 per cent of mitotic progeny showed altered phenotypes and karyotypes, although $1 \times 10^{-3}$ or $1 \times 10^{-4}$ are more usual frequencies. Size, colour and texture of fungal colonies were altered and, although the origin and nature of the chromosomal changes were not determined unambiguously, they appeared to be either large deletions or insertions which could have been caused by mitotic crossover associated with chromosome duplication and non-disjunction (Rustchenko-Bulgac *et al.* 1990; Rustchenko-Bulgac 1991).

Meiotically unstable fungi are far more common, e.g. *Leptosphaeria maculans* [AP] (Plummer and Howlett 1993, 1995) and *Ustilago hordei* [BS] (McCluskey and Mills 1990; McCluskey *et al.* 1994; Gaudet *et al.* 1998). In the latter fungus most of the polymorphism is associated with changes in the numbers of tandem repeats in the rDNA sequence on chromosome 4. Indeed, it seems probable that many chromosome-length polymorphisms in fungi result from gains or losses in such rDNA regions, or in telomeric DNA, as in *S. cerevisiae* [A] (Gaudet *et al.* 1998).

A situation which may be mistaken for a mutation is that caused by changes in very closely linked genes or by changes in intragenic recombination frequencies due to specific *rec* factors. In *Neurospora crassa* [AP] it proved possible to demonstrate that *rec* genes occurred in several wild strains and it seems likely that polymorphism for this kind of control is widespread in natural populations. However, these are normal recombination effects. Of more interest for populations are the cases

of abnormal recombination resulting in gene conversion.

Gene conversion occurs when one of a pair of alleles is converted into the other, usually at meiosis. Therefore it is most easily recognized in linear tetrads such as those found in ascomycotous fungi (Lamb 1996). In *Sordaria fimicola* [AP], for example, black (wild-type) or white (mutant) pigmented ascospores determined by a pair of alleles normally segregate in a ratio of $4+ : 4m$, but aberrant segregation ratios also occur such as $2+ : 6m$, $2m : 6+$, or $3+ : 5m$, $3m : 5+$, or $4+ : 4m$ asci where the segregating ascospores are not equally divided between the upper and lower halves of the ascus (Kitani and Olive 1967, 1969).

Conversion arises from mismatching of strands in heteroduplex chromosomal DNA. This is normally corrected by the repair system when excising and replacing mispaired bases in one or both strands as a single replacement event, i.e. when the sites are very close together. Gene conversion may occur in 0.2–10 per cent of all meioses in Ascomycotina but higher frequencies have been recorded, e.g. 14.6 per cent in the apple scab fungus *Venturia inaequalis* [AP] (Boone and Keitt 1956) and 33–55 per cent in yeast [A] (Detloff *et al.* 1991).

A further complication is gene conversion disparity, i.e. when the relative amounts of conversion, $+m$ or $m+$, differ. It can be detected by measuring the overall frequency of the +allele in aberrant tetrads (Lamb 1996). Conversion frequencies are also affected by specific conversion control factors (*ccfs*) and in some case their effects are large. For example, in *Ascobolus immersus* [AP] conversion frequencies of the gene *w1* can range from 0.9 to 23.6 per cent depending on which of three unlinked alleles, *ccf2*, *ccf3* or *ccf4*, is present as only *ccf2* is closely linked to the *w1* locus (Helmi and Lamb 1983).

While the effects of gene conversion may be immediately apparent in a cross, its effects on allele ratios in natural populations have not been observed. A number of theoretical assessments have been made and the consequences for populations and evolution explored (Lamb and Helmi 1982, 1989; Lamb 1984, 1986). Gene conversion will have its most significant effect when two neutral alleles at any one locus are in equilibrium. Change in their frequency will then be determined more by the rates and direction of gene conversion than by spontaneous mutation rates, as the magnitude of the former is so much greater. Even when there is some selective disparity between alleles, the effect may still be to modify the rate of selective change.

Gene conversion has only been studied in detail in a few fungi, but there seems little doubt that it occurs widely in fungi although its quantitative extent is not known. Therefore how far it contributes to change in gene frequency in natural populations is, at present, conjectural.

## 6.2.3 Mitochondrial mutation in mtDNA: high-frequency specific transfer

Circular mtDNA can undergo gene mutations and rearrangements analogous to those described for nDNA. The mitochondrial genome is evidently appreciably more uniform in most Oomycota (Straminipila ) than in the eumycotous Ascomycotina and Basidiomycotina (McNabb and Klassen 1988). However, there are exceptions, and the mitochondria of *Phytophthora megasperma* [O] (Förster *et al.* 1989) and *P. capsici* [O] are very diverse, as are isolates of *P. citricola* [O] from avocado (Förster *et al.* 1990). In contrast, there is a high frequency of length mutations and a relatively low frequency of mutation due to base substitutions in the mtDNA of the Eumycota, e.g. *Atkinsonella hypoxylon* [AP] (van Horn and Clay 1995) and *Agaricus* [BA] spp.(Hintz *et al.* 1985).

Linear mitochondrial DNA plasmids appear to be particularly common in fungi (Meinhardt *et al.* 1990). A global survey of *Neurospora* [AP] spp. suggests that about 50 per cent of all isolates carry a mitochondrial plasmid (Griffiths *et al.* 1995). A common cause of mtDNA disruption is the presence of linear or circular plasmids which are capable of being inserted into the mtDNA, resulting in either dysfunction or mutation of mitochondrial genes. The mycelium usually becomes impaired in growth or senescent, e.g. *petite* mutants in *S. cerevisiae* [A] and senescent mutants in *Neurospora* [AP] spp. and *A. nidulans* [A] (Kempken 1995). In general, the senescence and plasmids are co-inherited maternally although, very exceptionally, paternal inheritance has been detected (cited

in Griffiths and Yang (1993)). However, some linear mitochondrial plasmids are known which have no detectable phenotypic effect even though they may be very abundant, e.g. those demonstrated by fluorescent labelling in the take-all fungus *Gaeumannomyces graminis* [AP] var. *tritici* and var. *avenae* (Henson and Caesar-TonThat 1995).

Many plasmids are highly mobile and appear to be transferred between species in nature, e.g. seven different plasmids in *Neurospora* [AP] spp. (Arganoza *et al.* 1994). Such transfer can occur intraspecifically at heterokaryon formation, e.g. senescence plasmids in *Neurospora* [AP], or even during transient hyphal fusions between different species (Griffiths *et al.* 1990; Collins and Saville 1990). In addition, there is now increasingly strong evidence for plasmid transfer between both species and strains in *Neurospora* [AP] by introgression following limited sexual recombination (Bok and Griffiths 1999).

Two species have been investigated in some detail, *N. intermedia* [AP] and *Podospora anserina* [AP]. The *kalilo* and *maranhar* senescent strains of *N. intermedia* [AP] when first investigated, were restricted to Hawaii and a region of India respectively. Both are 8.6-kb linear plasmids, *kal*DNA and *mar*DNA, which replicate autonomously in the mitochondria but can become inserted in the mitochondrial genome, resulting in senescence and death of the mycelium (Griffiths 1992). These plasmids appear to be genetically unrelated. However, four variants of the *kalilo* plasmid are known in various species of both *Neurospora* and *Gelasinospora* [both AP] and occur in various regions, e.g. Louisiana and Moorea, Tahiti, as well as Hawaii. Introgressive transfer has been achieved experimentally between several different species, e.g. *N. intermedia* to *N. crassa* and *N. tetrasperma* [all AP] (Bok and Griffiths 1999).

The other plasmid which has been investigated fully is a small (0.8 μm) circular plasmid in *P. anserina* [AP] which can exist as either an amplified autonomous circular plasmid or an integrated intron of the gene for subunit I of cytochrome oxidase in the mitochondrial DNA (Kück 1989). As the cells of the fungus age, multiple rearrangements occur in the mtDNA and the leading hyphae of the colony become senescent. Exceptionally, however, some recombinants become stabilized and even induce longevity (Hermanns and Osiewacz 1996)! An increase in longevity has not been recorded elsewhere and it is not clear how it arises.

Senescence is a widespread phenomenon in all fungi, and although these kinds of mutants have often been found to be responsible, this is not always the case, nor is a plasmid always involved. For example, in a world survey of 132 strains of *N. intermedia* [AP], 26 cases of senescent phenotypes were found, none of which appeared to be associated with a mutant mitochondrial genotype or with a mitochondrial plasmid.

The importance of the role played in nature by either senescence-inducing or, if it turns out to be a common phenomenon, longevity-inducing plasmids is not known. It has been suggested that senescence is of little significance in natural populations since some mycelia, such as those of *Neurospora* [AP] spp., only persist for a short period *in nature* before undergoing sexual reproduction, during which senescence can be suppressed (Griffiths and Yang 1993). Whether this is the case or not, mutations associated with increased longevity, if widespread, become of even greater interest and relevance because of the existence of the long-lived perennial mycelia of many species. Unfortunately, the longevity of mycelia in nature is a subject on which virtually no information exists (see section 4.1.3).

A remarkable phenomenon which could be mistaken for mutation is the high frequency transfer of specific mtDNA sequences between different strains following the establishment of a heteroplasmon/heterokaryon. It has been identified in several Ascomycotina, e.g. *Saccharomyces* [A] (Sanders *et al.* 1976), *Neurospora* [AP] (Bernard and Küntzel 1976) and *Aspergillus nidulans* [A] (Earl *et al.* 1981). It has usually been detected as constant size difference in the mitochondrial genome between two strains owing to the insertion of quite specific sequences at specific positions. The inserted sequences do not appear to be necessary for the activities of the recipient strain and, indeed, there is some evidence that they are not translated. Insertion is evidently due to a recombinational event of very high frequency approaching or equalling 100 per cent. Its genetic significance is obscure.

## 6.2.4 Aneuploidy and polyploidy (see also Chapter 8)

Transient aneuploidy persists for variable times and arises from various causes such as non-disjunction (e.g. *N. crassa* [AP]) or during haploidization of exceptional diploid nuclei in a normal haploid (e.g. *Aspergillus nidulans* [A]). Persistent aneuploidy can also occur but too little is known, either experimentally or in nature, to comment reliably on its significance. However, it does appear that haploid fungi will not normally tolerate anything other than $n+1$ aneuploids for any great length of time. However, anamorphic hybrids between *Epichloë* spp. and related anamorphic *Neotyphodium* [FA] spp. are usually heteroploid and apparently persist in nature (Kuldau *et al.* 1999).

Nothing is known about the spontaneous incidence of polyploidy in the Eumycota. There is no doubt that polyploids occur sporadically in fungi (see Chapter 8, section 8.2.3). An increasing number of diploid species are known, e.g. haploid *Verticillium dahliae* [FA] and its stable diploid or near-diploid descendant *V. longisporum* [FA], originally described as a variant of *V. dahliae* [FA] (Ingram 1968). In general, most ascomycotous diploid forms investigated are unstable, as are the derived aneuploids in the reversion phase to the haploid condition e.g. *Aspergillus nidulans* [A] (Pontecorvo and Käfer 1958).

Both autoploidy and allopolyploidy are well documented in species of *Allomyces* [C] (Wilson 1952; Emerson and Wilson 1954). Apart from the occurrence of variable numbers of bivalents and a small number of quadrivalents at meiosis in *Cyathus stercoreus* [BA] (Lu 1964) and *Xylaria curta* [AP] (Rogers 1968), suggesting segmental allopolyploidy, there is no indication of whether autopolyploidy or allopolyploidy predominates amongst polyploid Eumycota. The polyploid series in domesticated *S. cerevisiae* [A] is autopolyploid (Leupold 1956).

Polyploidy has been found more frequently in the Straminipila, e.g. Pythiaceae and Saprolegniaceae (Win-Tin and Dick 1975) and *Phytophthora* [O] (Sansome 1965), and therefore may be more common in this group than in the Eumycota. Only diploid *P. infestans* [O] occurs in Mexico, although diploids, triploids and tetraploids have been isolated from Peru (Tooley *et al.* 1989) and amongst US and European strains, from which aneuploids have also been recovered (Tooley and Therrien 1991). *Phytophthora megasperma* [O] includes isolates ranging from $n = 10/12$ to $n = 30+$ (Sansome and Brasier 1974; Hansen *et al.* 1986).

The nature and origin of these polyploids is not known with certainty and there are various possibilities. Autoploidy is certainly possible. Nuclei of different ploidy can occur within the same mycelium (Sansome 1977), so if a $3n$ or $4n$ nucleus gave rise to gametangia, the result of a cross would be polyploid. Secondly, an aberrant meiosis of a rare self-fertile $2n$ oogonium (see Chapter 7, section 7.2.9.) could give rise to a polyploid clone. Lastly, since several interspecific crosses are known from a number of *Phytophthora* [O] spp., it seems possible that all polyploidy could occasionally arise after hybridization. Indeed, the existence in Sri Lanka of both a sterile diploid form of *P. meadii* [O], with abnormal gametangia and non-viable oospores, and a fertile normal tetraploid form is highly suggestive of an allopolyploid situation (Sansome *et al.* 1991).

### 6.2.5 Mini-chromosomes

Mini-chromosomes are a regular element in the chromosome complement of several fungi. They are often lost at meiosis but, exceptionally, in *Cochliobolus heterostrophus* [AP] they actually increase (Jones 1995). However, most mini-chromosomes are mitotically stable. Their effects are very varied. In many species they appear to have no phenotypic effect at all, but in others, e.g. *Magnaporthe grisea* [AP] (Orbach *et al.* 1996), they are associated with 'female' sterility although a causal connection has not been demonstrated. Mini-chromosomes are common in pathogenic non-fertile rice isolates, but are largely absent from grass isolates that are mostly fertile. In the pathogenic species *Nectria haematococca* [AP] the genes that detoxify a host's phytoalexin are carried on a mini-chromosome (Van Etten *et al.* 1994). If this is lost, the fungus also loses its pathogenicity.

At present it is difficult to generalize on the significance of mini-chromosomes for fungi.

Obviously they can arise or be lost in different populations and in some cases, e.g. *M. grisea* [AP], are characteristic of particular populations. Since many mini-chromosomes include a mixture of high- or low-copy repetitive sequences they can arise from the activity of transposons and so appear to have a spontaneous origin.

## 6.3 Migration and gene flow

Novel genotypes are found in all fungal populations from time to time. While mutation may be responsible, migration, i.e. the dispersal and establishment of individuals resulting in gene flow, can never be ruled out without further testing. Migration can have a number of important consequences.

- Migration into an isolated population is a source of novel genetic material, equal to or even exceeding any increase in novel genotypes as a result of mutation, and, if the population is small, can also overcome any tendency to genetic drift.
- In populations between which migration is common, it tends to increase the genetic homogenization of both populations by reducing the genotypic differences between them.
- Migration followed by establishment in a new locality can result in the foundation of an entirely new population, often genotypically different from the parental population (the 'founder effect').

The principal agents of gene flow in fungi are spores, both asexual and sexually produced, and fertilizing agents such as microconidia and spermatia. Of course, gene flow cannot occur unless dispersal is followed either by establishment of a sporing mycelium, ideally in the vicinity of another population, or by somatic or sexual fusion with a pre-existing different mycelium. Mycelial establishment is a consequence of normal spore dispersal, but successful sexual fusions can only occur when the spore is capable of acting as a fertilizing agent. Therefore one of the important steps in ensuring gene flow through sexual reproduction is the final transfer of fertilizing agents, such as conidia or, more usually, microconidia, to a receptive gametangial or progametangial structure. It is regrettable therefore that there is almost complete ignorance on how far fertilizing agents, other than conidia acting in this role, can be dispersed while retaining their vitality.

A considerable theoretical literature has developed in connection with migration in other Eukaryota, but despite innumerable studies, both theoretical and practical, of spore dispersal, data on migration and gene flow in fungi are still sparse except for a few pathogens (McDermott and McDonald 1993). Recently a preliminary general approach has been developed (Rogers and Rogers 1999).

Gene flow has not been measured directly in natural fungal populations. The experimental introduction of genetic markers and their subsequent detection elsewhere is technically feasible but has not been investigated thoroughly. Attributes such as virulence or fungicide resistance, usually naturally occurring and determined by a single gene, once detected or introduced, can, in principle, be followed as they spread through a single widespread population or between adjacent populations (see Chapter 9, section 9.4.1). However, if this procedure is to be effective, the following prerequisites are essential.

- Gene frequencies in the populations investigated should be at or near equilibrium.
- There should be some idea of the *spontaneous* mutation rate in each population at the marker gene locus or loci, since migration can only be assumed after allowing for mutation.
- The markers should be selectively neutral and, if more than one is used, they should be inherited independently because most indirect methods for estimating migration assume independent assortment of alleles. This requirement is of particular importance when nucleotide fragments or sequences such as RFLPs, RAPDs or AFLPs are being used as markers. A check should be made that the markers are derived from different chromosomes.

These conditions have rarely been ascertained. Consequently, reliance on the range of dispersal of propagules in nature and their establishment is the best practicable available measure, together with indirect methods for the estimation of the magnitude of gene flow.

## 6.3.1 Fungal dispersal in nature: airborne dispersal

The vast majority of fungal spores, however liberated, are dispersed through the air. Dispersal is common over relatively short distances. Fungal spores are rarely distributed horizontally more than 100 m from their source of origin and frequently for far shorter distances; for example, spores dispersed in splash droplets are rarely dispersed more than 20 cm and there is a very steep decline in the number of spores transported over this distance. An essential condition for more widespread dispersal is that the spores should rise above the laminar boundary layer when they may then be carried higher by atmospheric turbulence (Ingold 1971; Gregory 1973) (Figure 6.1; see also Figure 1.7). The effect of the height of liberation and the degree of turbulence on spore dispersal is dramatic, as shown by Gregory (1966) (Table 6.3).

Even under optimal conditions, as studies on the spread of epidemics show very clearly, individual dispersal paths may be relatively short, (thousands of metres or less), but cumulatively successive dispersal–infection–dispersal sequences can result in appreciable distances being covered quite rapidly. This may well be a common mode of dispersal by many fungi even over shorter distances.

One example is dispersal by the boletaceous fungus *Suillus grevillei* [BA] (Zhou *et al.* 2001). Two stands of *Larix kaempferi* at the foot of Mount Fuji, Japan, are about 700 m apart, with two further small and rather scattered *Larix* populations, roads and other trees in between. Fortunately, *S. grevillei* [BA] forms mycorrhiza exclusively with the *Larix*. Individual fungal genets were identified using PCR amplification at a single locus of three SSR regions and, in this way, 22 genotypes were recognized in the stands. Identical genotypes were located mostly in clusters, indicating a somewhat limited range (appreciably less than 700 m) of basidiospore dispersal and establishment.

Once spores have been transported upwards through the laminar boundary layer, gusty intermittent winds are a particularly effective transport agent because eddies promote both horizontal and vertical diffusion (Aylor 1990). For example, although individual conidia are not widely

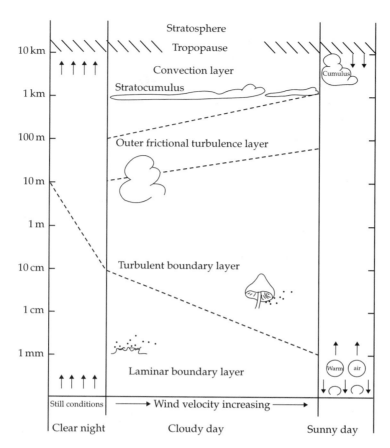

**Figure 6.1** Diagram illustrating the layers of the atmosphere which affect spore dispersal by fungi. Note that the vertical scale is logarithmic. The boundaries of the atmospheric layers, notably that of the laminar boundary layer, vary by day and night and according to the weather. (Modified from Gregory (1973).)

**Table 6.3** Estimates of the effect of atmospheric turbulence on the dispersal of spores of different sizes

| Spore size, turbulence | Probable flight range (m) | |
| --- | --- | --- |
| | Liberated at 0.1 m | Liberated at 1.0 m |
| Large spores | | |
| Turbulence low | 4 | 47 |
| Turbulence high | 50 | 50 |
| Medium spores | | |
| Turbulence low | 37 | 180 |
| Turbulence high | 200 | 850 |

Reproduced from Gregory (1966), with permission from the Colston Research Society.

dispersed, there is no clustering but widespread mixing of clones of *Sclerotinia sclerotiorum* [AD] in fields of oilseed rape (canola) in Canada, indicating effective dispersal of its spores (Kohli *et al.* 1995).

Spores approaching the boundary with the convection layer, about 1–2 km above the ground, depending on the nature of the ground cover and the degree of turbulence, may enter this layer where they are liable to convective uplift and can be transported by high-altitude winds. It is such spores which, potentially, can migrate naturally over very long distances, in some cases even between continents.

Transport in aeroplanes, on humans or as a result of unsupervised or unregulated human activity, such as being attached to exported items, is, of course, an increasingly probable mode of fungal migration. For example, *Puccinia polysora* [BR], a minor pathogen of maize in the USA, became an important pathogen throughout Africa following its introduction into West Africa in about 1949, almost certainly by air transport (Cammack 1959). Similarly, the recent epidemic of Dutch elm disease probably entered the UK on imported elm logs through failure of quarantine control.

There is a rather broad relationship between the nature of the dispersal unit, the agent responsible for dispersal and the viability of the dispersal unit. In general, small pigmented air-borne spores or conidia are likely to be dispersed most widely; dark pigmentation has often been shown to contribute to resistance to ultraviolet radiation (Pugh and Buckley 1971). In contrast, the eventual germination of a long dormant sclerotium, followed by spore formation and localized dispersal, can be an equally unexpected source of inoculum that cannot be entirely ruled out.

The consequences of migration are most likely to be confused with mutation in those cases where dispersal occurs either over a sufficiently long distance for it not to be immediately obvious where the source of the migrants is located, or where no obvious source of the new genotypes is apparent. The origin of several new and genetically distinct pathotypes of *Puccinia graminis* f.sp. *tritici* [BR] in Australia appear to be unambiguously due to spectacular long-distance migration. The evidence comes from virulence and isozyme analysis and is consistent with knowledge of high- altitude wind-circulation patterns. Isozymes (Burdon *et al.* 1982) were identified from eight enzyme systems and the isozyme data are given in Figure 6.2. Identical isozyme patterns in samples from Southern Africa and the new pathotypes which appeared suddenly in Australia in 1954 confirmed that they had migrated rather than arisen *de novo* in the latter continent. The most striking feature of this case is not that the migrant spores actually became established in Australia but that it is known precisely how this occurred. From meteorological data and high-altitude balloons a circulation pattern had been detected in the southern hemisphere which could have been responsible for this long-distance dispersal from Southern Africa to Australia. This

and other established long-range dispersal paths have been described and illustrated by Nagarajan and Singh (1990) (Figure 6.3)

Significant but far shorter migrations are far more frequent. The spread of new genotypes of *B. graminis* f.sp. *hordei* [A] across Europe has been followed by sampling over many kilometres at a large number of sites (Limpert 1987). An important factor is the amount of inoculum being produced, so that as this declines so does the amount dispersed. Both quantities decline in the UK from spring to mid-June when the amount dispersed is virtually negligible (O'Hara and Brown 1996). Similarly, the annual northward migration of new genotypes of *P. graminis* [BR], derived by recombination in Mexico and southern North America, to the wheatlands of Canada in the spring and in the opposite direction in late summer has been described as the 'Puccinia path' (Stakman 1933). In some years this migration is mediated by successive dispersal flights, but in others the long-range winds are sufficiently strong to disperse the spores over the entire distance.

A combination of human transport followed by natural long-range dispersal can be immensely effective in bringing about vast migrations. The most striking example is that of *P. polysora* [BR] in which examination of herbarium specimens in Africa and eastwards, following the probable human introduction of the pathogen from the Americas, has demonstrated the presumed windborne spread of *P. polysora* [BR] across the African continent, from West Africa to the southeast, at a rate of about 1000 km per year (Figure 6.4) (Cammack 1958; Gregory 1973).

Suitable sampling techniques, such as isolations from adjacent sites, soil and airborne-spore collections, or the comparison of genotypes from different regions may enable dispersal to be inferred as the most probable explanation of the appearance of novel genotypes, but ambiguous situations often occur. An excellent example is asexual *Aspergillus flavus* [A] in cotton fields in the USA. Samples taken from soil and cotton bolls in a single field over a 3-year period (1987–1989) showed the presence of a few 'new' VCGs in 1988 and 1989 (Bayman and Cotty 1991). Four possible origins exist for the 'new' VCGs. A mutational origin is possible but,

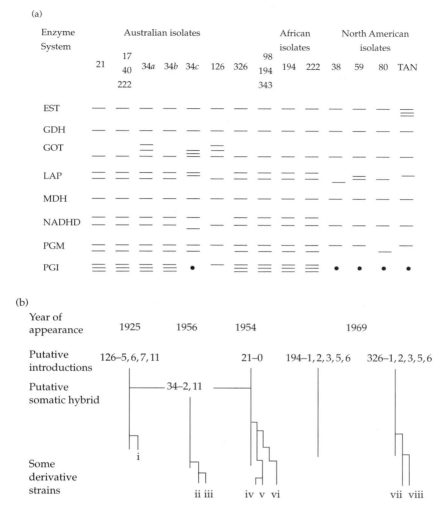

**Figure 6.2** The probable origin of races of *Puccinia graminis* f.sp. *tritici* in Australia from Southern African races. (a) Diagrammatic comparison of gels of isozymes of some Southern African, Australian and North American rust races. Note the similarities between the Southern African and Australian gel patterns. (b) Diagrams of probable phylogeny of the Australian races based on the data in (a). (Reproduced from Burdon *et al.* (1982), with permission of CSIRO Publishing.)

alternatively, the dry conidia of *A. flavus* [A] are certainly aerially disseminated. A third source is the persistence of sclerotia formed earlier on soil mycelia which had germinated. Each sclerotium can develop 3–20 conidiophores from which conidia are dispersed locally (Wicklow and Donahue 1984; Wicklow and Wilson 1986). Cultivation, even 'minimal' cultivation, could expose different dormant sclerotia on the soil surface and, if these had germinated, potentially different genotypes could

be released. 'New' genotypes derived in this way from earlier populations would then represent 'migration in time' rather than in space.

The least likely possibility is an origin from parasexual recombination within the existing population. However, since in this investigation no samples were taken from adjacent fields, discrimination between migration, mutation, sclerotial germination or parasexuality as the source of new VCGs cannot be made with certainty. The situation

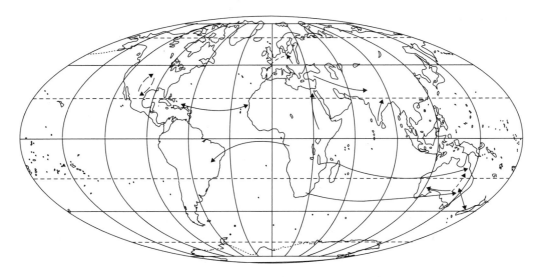

**Figure 6.3** Principal identified routes for airborne long-distance dispersal of propagules such as fungal spores. (After Nagarajan and Singh 1990, with permission.)

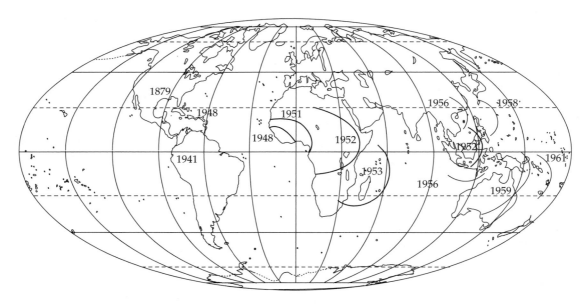

**Figure 6.4** The spread of *Puccinia polyspora* [BR] after transport to Sierra Leone in 1949 from the Americas. Its mode of origin in Africa is unknown. High-altitude winds are a possible source, but transport of infected material to Africa by aircraft is thought more probable (Cammack 1958). (Modified from Gregory (1973).)

demonstrates that even when migration in space can be ruled out, 'migration in time' should always be considered.

Virtually nothing is known about the aerial dispersal or duration of viability of fertilizing agents such as microconidia which are usually exceedingly light and small. It is often assumed that their viability is of limited duration so that their biologically effective dispersal distances must be small.

## 6.3.2 Fungal dispersal in nature: dispersal by animals

A number of fungi disperse their spores, including fertilizing microconidia, through the intervention of animals, and a variety of adaptations promote such dispersal. The majority are probably transferred by small and microscopic soil animals—nematodes, mites and insects of various kinds—but some are dispersed by larger animals.

The simplest mode of animal dispersal is the incidental ingestion of fungal spores of all kinds and their deposition, in viable condition, in faeces. In many cases, such as soil animals, the distance traversed is of the order of centimetres, millimetres or less. Viable spores have been recovered from the faecal pellets of many mammals and birds, and the coprophilous fungal flora bears witness to the very considerable numbers of fungi that are dispersed in this way, ranging from the Zygomycotina to the Basidiomycotina. Viable spores have even been recovered from the droppings of birds, e.g. the semi-aquatic soil fungi *Rhizophlyctis rosea* and *Hyphochytrium catenoides* [both C] from blackbird droppings. In such cases the distance covered may be considerable and, because of the feeding habits of the bird, the faeces may well carry genetically distinct spores which can germinate in close juxtaposition, facilitating gene flow even more effectively.

Many ascomycotous conidia and microconidia are slimy so that they stick to the bodies of passing small or microscopic animals such as nematodes, mites, and insects. In other cases the spores are released in a slime or viscous droplet which attracts insects by either its smell or its sugar content. Consequently the spores are either ingested or adhere to the body or mouth-parts of the visiting animal, ensuring dispersal. The stinkhorns (*Phallus* spp. [BA]) and their relatives are perhaps the best-known example of this adaptation, but it is common in many Ascomycotina, such as conidia of *Ophiostoma ulmi* [AP]. The same adaptation also serves to promote the transfer of potential fertilizing agents. In the case of *O. ulmi* [AP], mites transfer conidia on their bodies over a few millimetres to the protoperithecia developing in the inner bark (Brasier 1978) and this pattern may well be widespread amongst fungi that develop a stromatic mycelium under the bark of trees. For instance, *Daldinia loculata* [AP] forms an extensive stroma just within the bark of infected birch twigs on which its perithecia develop. There is evidence that perithecia at different sites on the same stroma can be fertilized by at least three genetically different conidia. This could well be achieved by crawling mites or other insects in much the same manner as in *O. ulmi* [AP] (Johannesson *et al.* 2001). Such a pattern is analogous to gene flow brought about by pollen in higher green plans.

Another group in which transfer of fertilizing pycnidiospores is almost always due to insects which have been attracted is the heterothallic rust fungi and other Basidiomycotina (Brodie 1936). Homothallic rust fungi usually lack pycnidiospores. The prime attractants in hetrothallic species are 'nectar' droplets exuded by the developing pycnidium and, in the case of some rusts (e.g. *Puccinia suaveolans* [BR]), the attractive odour accompanying the exudation which is easily detected by humans. The effective dispersal distance for pycnidiospores and other fertilizing agents is presumably determined by their viability as effective fertilizing cells.

One of the must bizarre adapations to insect dispersal also occurs in the rust fungi. In various crucifers, such as *Arabis* species, pseudoflowers are induced as a consequence of infection by the rust fungus. The sites of the pseudoflowers coincide with the development of uredosori and so spore dispersal is promoted (Roy 1993).

## 6.3.3 Establishment of dispersed fungal propagules

It is an obvious truism that the vast majority of dispersed fungal propagules do not become established, otherwise there would be an even greater abundance of fungal mycelia than is actually observed! Of course, vast numbers are lost during dispersal, some are inviable and, for yet others, environmental conditions are unsuitable even if a propagule reaches an appropriate substrate. For instance, it is estimated that many Agaricomycetidae produce $10^9$–$10^{10}$ basidiospores per year but only about $10^{-9}$ become established (Buller 1931).

Limited data exist for the numbers of spores necessary to ensure the minimal probability of one infection or of establishment (equivalent to the dilution end-point). Most of the data relate to the infection efficiency of pathogenic fungi. In general, the figure is minimally in the range of tens to thousands of spores, e.g. *c*.10 for short-range infection of tomatoes by *Fusarium* [FA] (Haymaker 1928) or 15 000 for wind-borne infection of wheat by the smut fungus *Tilletia* [BS] (Heald 1921). Thus the probability of establishment by a successful migrant, and hence of effective gene flow, is relatively small.

However, the efficiency of gene flow may be appreciably higher than suggested by such data since in many fungi the probability of germination of a spore is greatly increased if it is in the vicinity of a mycelium of the same species. Hyphal fusion would then be more probable provided that spore and mycelium were compatible. Similarly, the chances of a successful fertilization between spores, micro-conidia and similar structures which act, or can act, as fertilizing agents deposited in the vicinity of trichogynes or other receptive hyphae are greatly increased by tropistic growth of the potential recept-ive structure towards the fertilizing agent. For instance, most of the Ascomycotina which develop trichogynes show such behaviour, producing attractant pheromones (Bistis 1983), and amongst the Basidiomycotina, e.g. *Coprinus* spp. [BA], monokar-yotic hyphae show tropistic growth towards oidia leading to fusion and dikaryotization (Kemp 1975).

### 6.3.4 Estimating gene flow indirectly

The actual structure of any natural population in nature is rarely known. Different models have been developed to mimic situations believed to occur and to measure the supposed gene flow between populations (Slatkin and Barton 1989). Historically, the first model considered was unrestricted exchange between equal-sized sub-groups of a single sub-divided population—the 'infinite islands' model (more usually, the 'islands' model) (Wright 1951). Of course, this is not a very realistic model. Other models are the 'stepping-stone' model, where migra-tion is only between adjacent populations arranged linearly in a one- or two-dimensional matrix, or the more complex 'lattice' and 'continuum' models. It is doubtful whether these latter two are any more com-parable with reality than the first two simpler models.

Migration can be measured in two ways—either as the probability of migration, or as the number of individuals, gametes or spores which have migrated. Thus, in the island and stepping-stone models, migration is measured either as the prob-ability $m$ that each propagule or individual in a population is an *immigrant*, or as the number $Nm$ of individuals that have migrated, where $N$ is the population size (more usually, $N_e$). In the lattice and continuum models, the measures are either the standard deviation $\sigma$ of the distance moved by each gamete or propagule in any one direction, or alternatively the neighbourhood size $N_p$ within which individuals migrate, which is roughly com-parable with the area from which the parents of an individual can be drawn. These last two models will not be considered further here since they do not greatly improve the estimates of migration.

In the islands model, consider an allele $a$ that occurs in the populations with *average* frequency $p_t$ in generation $t$ when the gene flow between popu-lations has a probability $m$. Any $a$ allele chosen at random from any one population could have one of two origins: *either* from that population in the pre-vious generation $(t-1)$ where it would have occurred with a probability of $(1-m)$ at a frequency of $p_{t-1}$, *or* from another population with a prob-ability $m$ and therefore with frequency $\bar{p}$. If other processes, such as mutation, drift or selection, are ignored $\bar{p}$ will remain constant over all generations, since it is the average frequency, and

$$p_t = p_{t-1}(1-m) + \bar{p}m$$

whose solution from generation 0 is

$$p_t = \bar{p} + (1-m)^t(p_0 - \bar{p}).$$

In other words, regardless of the initial frequency of $a$ in any population, under migration it con-verges to $\bar{p}$.

In the special case where the gene frequencies $p_D$ and $p_R$ in a donor and receptor population are known, as well as the gene frequency in the receptor population *after* migration has occurred, $m$ can be determined from

$$m = \Delta p/(p_D - p_R).$$

However, these values will be known only rarely. A possible situation is when a phenotypically well-expressed character originates in one part of a widespread population and the allele migrates to a distant location through the population, e.g. a new, or newly selected, advantageous virulence allele migrating through a population such as the virulence factor $V_g$ in *B. graminis* f.sp. *tritici* [A] which arose in Germany in 1931 and subsequently spread throughout Europe (Wolfe and McDermott 1994).

The most common method of estimating gene flow is by using Wright's *F*-statistics to compare differentiation between effective populations of similar size. Wright (1943) demonstrated that the number *Nm* of individuals which migrated was related to $F_{ST}$ in diploids as follows:

$$F_{ST} = 1/(1 + 4Nm).$$

Hence in diploids (and also in dikaryotic fungi)

$$Nm = [(1/F_{ST}) - 1]/4$$

and in haploids

$$Nm = [(1/F_{ST}) - 1]/2$$

However, in practice, the populations compared are not usually of equal size or, more commonly, are of unknown size, so that an alternative often used is to employ Nei's $G_{ST}$ (Boeger *et al.* 1993) or Weir and Cockerham's $\theta$ (see Chapter 5, section 5.1.4) instead of $F_{ST}$.

A rather different method of estimating *Nm* utilizing 'private alleles' has been proposed by Slatkin (1985) and gives comparable results to the previous methods (Slatkin and Barton 1989). A 'private' or 'rare' allele is one that occurs in only *one* population. Slatkin showed that $\log Nm$ decreases approximately as a linear function of the average frequency of private alleles and derived the following empirical formulae:

$$\ln[\bar{p}(1)] \approx a \ln(Nm) + b$$

for a diploid/dikaryon or

$$\ln[\bar{p}(1)] \approx [a \ln(Nm) + b]/2$$

for a haploid where $\bar{p}$ is the average frequency of private alleles and $a$ and $b$ are constants whose value is determined by simulations and varies with the size of the sample examined. 'Private' alleles may be more sensitive indicators of migration because they are rare and will not be dispersed unless dispersal is quite considerable. This is a possible advantage of this method.

These methods draw attention to two other aspects of gene flow. The first is that *Nm* falls off rapidly as the value of $F_{ST}$ declines (Figure 6.5). Assuming diploid or dikaryotic populations,

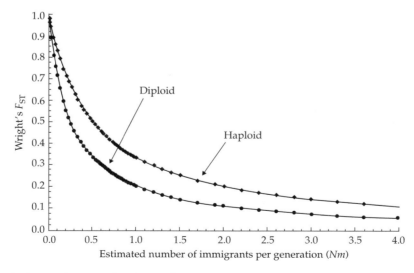

**Figure 6.5** Relationship between the estimated number *Nm* of migrants per generation and Wright's $F_{ST}$. Curves represent populations of diploid/ dikaryotic and haploid individuals respectively. (Reproduced from Rogers and Rogers (1999), with permission from Kluwer Academic Publishers.)

a frequency of 0.5 for the allele in the migrant individuals and any islands model, Wright (1931) demonstrated that if $Nm > 1$, it would override any tendency to drift in the receptor population. This figure, which clearly only applies to a particular set of circumstances, has been used as a rough yard-stick to evaluate the significance of migration. Values greater than 1 are accepted as evidence that gene flow is significant, and values below 1 indicate that little gene flow has occurred. While this is a useful indicative yardstick, allowance should always be made for any substantial difference between the conditions under investigation and Wright's assumed conditions.

One further property of the islands model is of some value. If the populations involved are only short distances apart and they are at equilibrium, the logarithm of the average number $M$ of migrants per generation is negatively correlated with the logarithm of the geographic distance between them (Slatkin 1993). This correlation is illustrated in Figure 6.6(a) based on the distribution of *Crumenulopsis soraria* [AD] on Scots pine in Scotland (Ennos and Swales 1991), while the absence of a correlation is evident in Figure 6.6(b) based on the distribution of *Cryphonectria parasitica* [AP] in eastern North America (Milgroom and Lipari 1995b). Populations of the former species have existed for a long time and presumably gene exchange is at equilibrium, whereas *Cryphonectria parasitica* [AP] populations have either not reached equilibrium or isolation by distance is prevented by long-range migration (Milgroom and Lipari 1995b). The former suggestion seems more plausible since *C. parasitica* [AP] is a relatively recent migrant to North America.

Gene flow of significance in the past history of populations or hybridization between species can be of importance in understanding their present structure. The methods based on the nested clade analysis of enzyme genealogies or phylogenies developed by Templeton and his colleagues (Templeton *et al.* 1995; Templeton 1998; Posada *et al.* 2000) (see Chapter 5, section 5.3) can be used for this purpose. When such data are associated with biogeographical data they can be used to 'identify and localize the effects of restricted gene flow and historical events on geographic associations of haplotypes'.

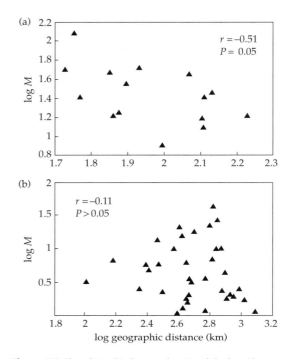

**Figure 6.6** The relationship between log $M$ and the logarithmic geographical distance $C$ between sub-populations in (a) *Crumenulopsis soraria* [AD] and (b) *Cryphonectria parasitica* [AP]. The *C. soraria* data are for Scotland (Ennos and Swales 1991); the *C. parasitica* data are for nine sites in eastern North America. Note that the correlation is only significant for the former. (Reproduced from Milgroom and Lipari (1995b), with permission of the American Phytopathological Society.)

The appropriate general inferences concerning gene flow were set out in Chapter 5 (Table 5.5). A special case of gene flow of particular phylogenetic importance is species hybridization. Phylogenetic data involving the species concerned are subjected to nested clade analysis. The absence of hybridization is demonstrated if the haplotypes from each species are found to be *nested distinctly* before they are nested with those of another species, but if the species are *separately nested* before the nesting of all haplotypes from a single species, then hybridization is indicated.

## Further reading

*Mutation*

Daboussi M-J (1996). Fungal transposable elements: generators of diversity and genetic tools. *Journal of Genetics*, **75**, 325–39.

Drake JW (1991). A rate of spontaneous mutation common to DNA-based microbes. *Proceedings of the National Academy of Sciences of the United States of America*, **88**, 7160–4.

Lamb BC and Helmi S (1989). The effects of gene conversion control factors on conversion-induced changes in allele frequencies in populations and on linkage disequilibrium. *Genetica*, **78**, 167–81.

## Migration

Gregory PH (1973). *Microbiology of the atmosphere*. Leonard Hill, Aylesbury.

Nagarajan S and Singh DV (1990). Long-distance dispersion of rust pathogens. *Annual Review of Phytopathology*, **28**, 139–53.

Rogers SO and Rogers MM (1999). Gene flow in fungi. In *Structure and dynamics of fungal populations* (ed. JJ Worrall), pp. 97–121. Kluwer Academic, Dordrecht.

Slatkin M and Barton NH (1989). A comparison of three indirect methods for estimating the average levels of gene flow. *Evolution*, **43**, 1349–67.

# CHAPTER 7

# The generation of variation—II
# The importance and diversity of inbreeding and outbreeding

Recombination in fungi can involve the reassortment of genes, chromosomes, extra-chromosomal elements, or even whole nuclei. The first two occur most frequently at meiosis but mitotic recombination is possible. The reassortment of extra-chromosomal elements and whole nuclei consequential on hyphal anastomoses, leading to heterokaryosis and heteroplasmon formation, is also possible and can be detected. The basic processes concerned in sexual and parasexual recombination have been outlined earlier (see Chapter 1, section 1.4).

In this chapter meiotic recombination only will be considered. Consideration of heterokaryosis, parasexuality, hybridization and polyploidy will be deferred to Chapter 8.

In nature, meiotic recombination is regulated and affected by a variety of environmental and genetic factors which determine the balance between sexual and asexual reproduction as well as that between inbreeding and outbreeding. Therefore the potential maximum sexual recombination that a fungus can exhibit may not be achieved; indeed, sexual recombination may be completely inhibited for long periods. Clearly, this will affect the genetic variation detectable in natural populations. Nevertheless, even infrequent sexual episodes or sexual reproduction involving very few individuals is sufficient to generate appreciable recombination.

An indication of the order of magnitude for sexual reproduction to be effective in a population is provided by a supposedly clonal population of *Coccidioides immitis* [FA] in California. Unexpectedly, no two of the 30 isolates examined were genetically identical which suggested that recombination was

occurring, or had occurred, in the population, and the question arose as to how frequent this would need to be to generate such variation. Burt *et al.* (1996) considered a predominantly clonal population in which, nevertheless, recombination and drift occurred. Assuming that *C. immitis* [FA] was, in fact, an ascomycotous fungus and that the probability that two isolates taken at random are derived from the same ascospore is $f$, in the next generation the same probability $f^1$ will be

$$f^1 = (1/N)(1-s)^2 + (1-1/N)f(1-s)^2$$

where $N$ is the population size and $s$ is the proportion of isolates that are derived sexually. At equilibrium,

$$f_{eq} = \frac{(1-s)^2}{1 + 2s(N-1) - s^2(N-1)}$$

and, approximating,

$$f_{eq} = \frac{1}{1 + 2sN}$$

or

$$sN = 1 - f_{eq}/2f.$$

Assuming that the Californian population was at equilibrium and since no two isolates were identical, $f = 0$. However, the sample was small and so, supposing that in fact $f < 0.05$ in the population as a whole, $sN \geq 9.5$. Thus, regardless of the size of the population of *C. immitis* [FA], only a few individuals need be sexually derived each generation to ensure very appreciable variation in the population. Evidently, even a single round of sexual reproduction is sufficient to generate considerable genetic variation whose effects can persist through many generations. Indeed, it is now possible to detect ancient hybridizations using gene genealogies

In natural situations, especially where the loci involved are not well marked and their structure is not understood, recombinational events readily distinguished under experimental conditions can sometimes be mistaken for mutation. One situation is where allelic recombination is subject to genetic control by *rec*-genes such as has been described in *Neurospora* [AP] (see Chapter 6, section 6.2.2) (Smith 1966). Another is where cross-overs occur between repetitive sequences, thus generating deletions, duplications, inversions, translocations or dicentric and acentric fragments, depending on which repeats pair and their orientation. These chromosomal abnormalities may be detected, e.g. by electrophoresis, but in the absence of other information, their origin may not be realized. Therefore combined genetic and cytological investigations are worthwhile.

Heterokaryosis and parasexuality can usually be demonstrated easily in experimental situations, but have rarely been demonstrated in nature. The presently accepted wisdom is that neither of these processes plays much part in bringing about recombination in natural populations, but supporting evidence is limited and, in fact, the true situation is largely unknown (see Chapter 8).

## 7.1 Sexual reproduction in nature

Self-fertilization is probably the most common mode of sexual reproduction in fungi but many species are obligatorily cross-fertile. Since most species produce spores or conidia asexually as well, which often also act as fertilizing agents, there is a wide spectrum of breeding behaviour. Broadly, populations may be:

• exclusively asexual and therefore clonal (e.g. *Fusarium oxysporum* [FA])
• predominantly asexual but with a single interpolated sexual phase on a regular (seasonal) cycle (e.g. *Blumeria* (*Erysiphe*) *graminis* [A])

• predominantly clonal, with sexual recombination being relatively infrequent (e.g. *Stagnospora nodorum* [FA])
• exclusively sexual—either exclusively outbreeding (e.g. *Piptoporus betulinus* [BA]) or exclusively inbreeding (e.g. *Neurospora africana* [AP])
• potentially sexual but secondarily homothallic—hence generally inbreeding but rarely outbreeding to a small extent (e.g. *Neurospora tetrasperma* [AP]).

Even when a species is potentially capable of sexual reproduction, it may only occur extremely rarely in nature. There are many reasons for this: adverse environmental or nutritional conditions, the differential distribution of interfertile strains in cross-fertile species or a predominance of sterile, or partially sterile, strains. Season, climate and nutrition provide important triggers for sexual reproduction and their absence or alteration may have profound inhibitory effects on the process. In the limit, the fungus may remain wholly mycelial but more often periodic responses are the result. Thus, in the north temperate zone, many agarics ('toadstools') normally develop basidiomata only in the autumn, but this can be delayed if conditions are dry and the trigger provided by autumnal rains is delayed. Edaphic and nutritional conditions have equally profound effects so that, for example, the agaric *Omphalia maura* [BA], although evidently present as soil mycelia, had never been known to produce basidiomata on a mor soil near Reading, UK, until it was limed (Hora 1959).

The absence of a compatible partner in outbreeding fungi normally prevents sexual reproduction but there is ample evidence for many fungi that compatible partners are usually available in nature. Direct demonstrations have been made with *Coriolus versicolor* [BA] and *Mycena galopus* [BA]. In both cases monokaryotic mycelia, one to a compartment in petri dishes divided up into several chambers, were exposed to the spore rain in their natural habitat. *C. versicolor* [BA] was exposed for 6, 12 and 24 h in a woodland, and *M. galopus* [BA] was exposed for 14 h overnight at five different localities in woodlands where it grew. For *C. versicolor* [BA] almost as many dikaryons were detected in mycelia exposed for 6 h as in those exposed for 12 h, and for *M. galopus* [BA] 68–100 per cent of the exposed mycelia were dikaryotized. In a number of cases lines of antagonism were present between the dikaryons so formed, indicating that more than one kind of genetically distinct dikaryon had developed through fusion of the common tester monokaryon with naturally wind-dispersed basidiospores which were genetically different but compatible (Williams *et al.* 1984; Frankland *et al.* 1995). In an experiment using a similar technique, several populations of the cosmopolitan *Schizophyllum commune* [BA] gave similar results at each site but the spore clouds differed in their MTF composition, i.e. the spore clouds reflected local populations and only showed limited dispersal (James and Vilgalys 1996). It must also be realized that under natural conditions the probability of achieving a compatible fusion will be much less than these experimental results. Even when compatible MTFs are present, not only will acceptor monokaryons be less accessible than the exposed plate cultures but climatic or infertility phenomena may prevent sexual reproduction. The latter situations appear to be the case in the straminipilous gametangia-forming populations of *Bremia lactuca* [O] in California, in contrast with the importance of sexual reproduction for this fungus in Europe (Hott *et al.* 1987).

Sexual reproduction can also fail for genetic reasons. Infertile or partially sterile strains occur quite widely. About 25 per cent of all the dimictic Mucoraceous species isolated by Blakeslee from soil over many years were sterile and failed to cross either between themselves or with fertile tester strains (Blakeslee 1904b; Blakeslee and Cartledge 1927; Blakeslee *et al.* 1927). A similar situation is found across the geographical and host range of the potentially dimictic *Magnaporthe grisea* [AP] (Leung *et al.* 1988; Notteghem and Silué 1992). In this species the mating types are neither distributed evenly across the range nor numerically equal: 32 per cent are *Mat*-1, 20 per cent are *Mat*-2 and the remaining 48 per cent are incapable of reacting with either of the tester strains. Even the presence of both compatible mating types is not necessarily sufficient to guarantee successful sexual reproduction. For example, in two rice pathogen samples, one from the Ivory Coast and the other from Cameroon, with both mating types present as

assayed by the tester strains, all crosses within the populations were intersterile. Strains from different hosts also differ in their fertility; thus isolates from wheat or millet are highly fertile and those from rice are predominantly 'female' sterile, apart from one fully fertile 'female' isolate from French Guiana. Regardless of the host plant, there is a high correlation between the presence of mini-chromosomes and infertility in this species but whether this correlation has a causal basis is not known.

Variations from normal sexuality and biological features which bear on the success of sexual reproduction in nature will be described later, but first, evidence for its fundamental importance for fungi in natural situations will be described.

### 7.1.1 The genetic consequences of sexual reproduction

The outcome of sexual reproduction in promoting and maintaining genetic diversity in a population is self-evident. In fact, there are only a few investigations among the fungi which directly demonstrate its consequences and importance, although a good deal of circumstantial evidence for the contribution and significance of sexual reproduction to fungal populations has been provided by investigations of their phenotypic and genotypic variation using a wide variety of markers.

The most convincing investigation of the significance of sexual recombination on a broad scale was that made with *Puccinia graminis* f.sp. *tritici* [BR] in the USA (Roelfs and Groth 1980; Groth and Roelfs 1982; Burdon and Roelfs 1985a,b; Roelfs *et al.* 1997). By the mid-1920s a barberry eradication programme east of the Rocky Mountains in the Great Plains area virtually eliminated the alternative host for *P. graminis* f.sp. *tritici* [BR], which is necessary for the sexual stage to develop. Therefore populations have been effectively asexual since then (Roelfs 1982). In contrast, barberry eradication was *not* practised in the Pacific Northwest, and so here the populations certainly include sexually produced progeny. The genetic diversity, pathotype composition and isozyme phenotypes of these two regional populations were investigated and compared in 1975. Virulence diversity was determined by the set of differential lines used to distinguish pathotypes in the USA and Canada. In the Great Plains area, 17 pathotypes were detected prior to barberry eradication and, indeed, the numbers of virulence phenotypes dropped steadily from 1918 to 1975 over the period of barberry eradication (Figure 7.1). By 1975, the numbers had fallen to six distantly related pathotypes which could be grouped into nine clonal lineages, each including a few pathotypes differing in only one or two specific virulence genes. In contrast, in the sexual populations of the Pacific Northwest 100 pathotypes and at least 80 electrophoretically determined enzyme phenotypes were detectable in 1975. In these populations, isolates with the same multilocus enzyme phenotype always differed in their combination of virulence genes. Thus, by 1975 there was a notable difference in the numbers of virulence phenotypes in the two populations: 23.5 per cent (100/426 isolates) in the sexual populations and only 0.7 per cent (17/2377 isolates) in the asexual populations.

Because each differential race used to classify the pathotypes carried a single unique resistance gene,

it was possible to employ a modified form of the Shannon index of diversity $D_i$, namely a linear function of the number $I$ of differentials used, i.e. $D_i = -\sum \ln 0.5$. The overall departure $D_f$ of the individual departures from the value of 0.5 for each differential can be estimated (Roelfs and Groth 1980). If $d_i$ is the increment of diversity detected by each differential line which shows a departure from a frequency of virulence of 0.5,

$$\Delta d_i = -[(0.5 + d_i) \times \ln(0.5 + d_i) + (0.5 - d_i) \times \ln(0.5 - d_i)]$$

and the total loss in diversity is

$$D_f = \Delta D_i.$$

By using the Shannon index in this way, the genetic diversity of the sexual populations was found to be 39 per cent of the total potential diversity and that of the asexual populations was only 12 per cent. Moreover, the degree of genetic association in the asexual populations was much greater at 39 per cent than that in the sexual population, where it was only 14 per cent (Figure 7.2).

The position on the Great Plains has not remained static since 1975. In 1989 a new pathotype, QCCJ(12), was detected which differed from all the races previously known in these 'asexual' populations. It was capable of infecting some wheats but 95 percent of the samples of this pathotype were isolated from barley (*Hordeum*). Its virulence pattern differed from all former pathotypes;

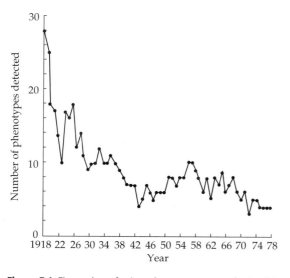

**Figure 7.1** The numbers of unique phenotypes amongst the *Puccinia graminis* f.sp. *tritici* [BR] populations of the Great Plains region of the USA between 1917 and 1978. Standard differential hosts were employed for detection. Note that the barberry eradication programme was virtually complete by 1928 so that the population was effectively asexual over this period. (Reproduced from Roelfs and Groth (1980).)

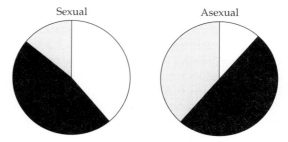

**Figure 7.2** Virulence diversity of the populations of *P. graminis* f.sp. *tritici* [BR] in the USA in 1975: (a) the sexual population from Washington and Idaho; (b) the asexual population of the Great Plains region. White, detected diversity; black, loss of diversity due to departure of virulence from a frequency of 0.5; stippled, loss in diversity due to association of virulences. (Reproduced from Groth and Roelfs (1987), with permission from Blackwell Science.)

in particular, it was avirulent to wheat cv McNair 701. Since none of the characters shown by this pathotype were known previously in this region, it seemed unlikely that a rare sexual event had produced QCCJ(12) on the Great Plains. However, avirulence to cv McNair 701 was known from the sexually reproducing populations of the Pacific Northwest. The most plausible hypothesis is that the pathotype first arose through recombination in the sexually reproductive populations of the Pacific Northwest and was then, exceptionally, wind dispersed eastwards to the region of the Great Plains (Roelfs *et al.* 1997). If this is, indeed, the case, in future the new pathotype will be expected to vary on the Great Plains only as a consequence of mutation, since these populations lack sexual recombination. Therefore this situation provides a dramatic demonstration of the importance of:

• a thorough knowledge of the genetic attributes of pre-existing populations
• a detailed knowledge of the incidence of sexual reproduction in all populations
• rare, relatively long-distance, dispersal and gene flow between populations.

Perhaps the most notable and dramatic example of the global importance of sexual reproduction for previously asexual populations is that of *Phytophthora infestans* [O]. The potato blight pathogen is believed to have originated in Mexico. In the early nineteenth century it spread to the USA and from there to Europe and the rest of the world. In the late twentieth century. Seven allozymes were used initially to identify genotypes of *P. infestans* [O], although both nDNA and mtDNA markers are now available. In neither Canada nor the USA were there large genotypic differences nor great variation. Subsequent studies showed that contemporary genotypes of *P. infestans* [O] from the rest of the world resembled and were clearly related to those in the USA. It became evident by the mid-twentieth century that, apart from Mexico, the fungus was probably a single asexual clone, modified only by mutation. The original clone, US-1, had spread to Europe in the 1840s (possibly with others) and from thence throughout the world. However, populations from Mexico were known to be more variable than those in the rest of the world

and the alleles for glucose phosphate isomerase-1 (*Gpi-1*) and peptidase-1 (*Pep-1*) were found to be in Hardy–Weinberg equilibrium. This suggested that sexual recombination occurred there, but it was only in 1950 that both the *A1* and the *A2* mating types were actually identified in Mexico (Niederhauser 1956). The clone in the rest of the world was *A1*. A dramatic change occurred in the early 1970s–1980s as a result of the export of strains with the *A2* mating type from Mexico (Table 7.1). Although the first published *A2* record was from Switzerland in 1984 (Hohl and Iselin 1984), it had probably been introduced to Europe some years earlier (Drenth *et al.* 1994) and subsequently to Israel and Egypt, followed by China, Japan and Brazil. Worldwide, sexual reproduction became potentially possible. Completely new, hitherto unknown, genotypes have been detected in The Netherlands, Poland and the UK, where the *A2* mating type has also been present, as well as recombinants between the new and older clones (Figure 7.3). As a further development, some of the new genotypes are displacing the older clonal genotypes (Spielman *et al.* 1991; Drenth *et al.* 1994) Table 7.1). Thus new strategies for resistance breeding will have to be developed to protect Solanaceous crops.

The situations observed in *P. graminis* f.sp. *tritici* [BR] and *P. infestans* [O] both developed over a

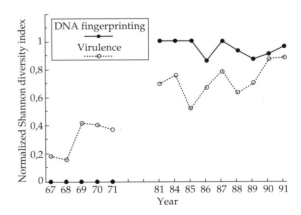

**Figure 7.3** Normalized Shannon diversity index for virulence and DNA fingerprint analysis of populations of *Phytophthora infestans* [O] in The Netherlands between 1967 and 1991. The *A2* MTF was probably introduced to The Netherlands about 1979–1980. (Reproduced from Drenth *et al.* (1994) with permission of Kluwer Academic.)

**Table 7.1** The number of isolates of *Phytophthora infestans* [O] analysed for virulence and DNA fingerprint patterns and the number of RG57 genotypes identified in isolates obtained prior to the introduction of the *A2* mating type factor to The Netherlands

| | Virulence | | Numbers of isolates with different DNA fingerprint patterns | |
|---|---|---|---|---|
| | Isolates | Races | Isolates | RG-57 genotypes |
| 1966 | | | 1 | 1 |
| 1967 | 39 | 5 | | |
| 1968 | 63 | 3 | 1 | 1 |
| 1969 | 13 | 4 | | |
| 1970 | 13 | 4 | 1 | 1 |
| 1971 | 20 | 4 | | |
| 1974 | | | 1 | 1 |
| 1978 | | | 1 | 1 |
| Totals (pre-*A2*) | 148 | 8[a] | 5 | 1 |
| 1981 | 46 | 19 (12)[b] | 5 | 5 |
| 1984 | 7 | 5 (5) | 4 | 4 |
| 1985 | 16 | 6 (4) | 13 | 13 |
| 1986 | 14 | 8 (4) | 9 | 7 |
| 1987 | 23 | 15 (2) | 17 | 17 |
| 1988 | 32 | 15 (8) | 30 | 25 |
| 1989 | 56 | 25 (14) | 57 | 40 |
| 1990 | 44 | 29 (18) | 30 | 24 |
| 1991 | 15 | 12 (9) | 14 | 12 |
| Totals (post-*A2*) | 253 | 73 (26) | 179 | 134 |

[a] Races containing virulence factors 1, 2, 3, 4 and 10.

[b] Races in parentheses that only contain the same virulence factors 1, 2, 3, 4 and 10. The RG57 genotype arose in the USA but became widespread there after *A2* had been introduced and presumably came to Europe with strains carrying *A2*. (Reproduced from Drenth *et al.* (1994) with permission of Kluwer Academic.)

number of years. Selection obviously played a significant part, confirming the importance of the generation of new genotypes by sexual reproduction on which selection can act.

## 7.2 Mating systems and their variations

Mating systems are often treated as if rigid in their expression, and it is these situations which have received most attention. Different mating systems often occur in closely related species. The most common situation is when a genus includes both heteromictic and homomictic species, e.g. *Mucor* [Z] (di- and homomictic), *Neurospora* [AP] (di-, homo- and homoheteromictic) and *Coprinus* [BA] (diaphoro- and homomictic), with some of the species being taxonomically very close. Often a particular mating system is typical of a genus, or of sections within a genus, e.g. *Marasmius* [BA] where

unifactorial and bifactorial diaphoromictic groups occur as well as some homoheteromictic species (Gordon and Petersen 1991; Gordon *et al.*, 1994). Whether, and if so how, one outbreeding system can evolve into another is still an area of great ignorance, but two cases are known.

The Sordariaceae, e.g. *Neurospora, Sordaria, Gelasinospora* spp. [all AP], include dimictic, homo-dimictic and homomictic species, and these last have been shown to lack one of the two MTFs (*mat a*), although how this condition arose is unknown. Homodimixis could have arisen by selection for broader and shorter asci (see Chapter 10, section 10.1.2).

Secondly, *Ustilago maydis* [BS] has a bifactorial diaphoromictic system determined at two unlinked loci. One locus has a pair of alleles, *mat A1/A2*, but there are several effectively allelomorphic *mat B* MTFs. Most other species are dimictic, e.g. *U. hordei*

[BS]. In such species, nucleotide sequences equivalent to those in *U. maydis* [BS] occur but the *mat A1* and *mat A2* factors are closely linked to the numerous *mat B* factors. Therefore the mating system appears to possess only two MTFs with a number of alleles: *mat A1Bx* and *mat A2By* (Bakkeren and Kronstad 1993). Since most of the Ustilaginales are dimictic, the condition in *U. maydis* [BS] presumably arose as a consequence of a breakage of the linkage between the *mat A* and *mat B* factors.

A mating system is not, of course, detectable by inspection but neither is sexual reproduction in the many Basidiomycotina that lack any morphological sexual differentiation. For example, many Homobasidiomycetes lack dikaryons, clamp connections or both, and possess multinucleate hyphal compartments. Nevertheless, they show analogous behaviour to fungi possessing regular dikaryons with clamps. Species of *Stereum* [BA] possess whorls of multiple clamps and have multinucleate compartments on both their primary and secondary mycelia, yet they appear to have a mating system similar to a typical dikaryon-forming unifactorial diaphoromictic fungus. However, confusion can be caused because, while compatible matings between basidiospores differing in one pair of factors, *C1/C2*, produce a normally growing mycelium, an abnormal mycelium develops if another pair of factors, *B1/B2*, are present and differ (Coates *et al*. 1981; Coates and Rayner 1985a,b).

The potential outbreeding ability of different mating systems varies with the numbers of MTFs (Table 7.2). Where the number of MTFs in a population is below the maximum for the species, the degree of outbreeding will be reduced but, in practice, in many Basidiomycotina the numbers present in a population are still very large so that random mating is approached (Table 7.3).

In addition, a number of situations are known where mating systems can either be modified to give some degree of both outbreeding and inbreeding via the sexual cycle, or the fungus possesses alternative mating systems.

## 7.2.1 Mating systems in Chytridiomycotina

Apart from the aquatic genus *Allomyces* [C], very little is known about the sexuality or mating systems of the Chytridiomycotina. In a classic study performed 60 years ago, Emerson (1941) demonstrated different patterns in the sub-genera Euallomyces and Brachyallomyces. In the former, meiosis in thick-walled resistant sporangia on a sporophytic generation give rise to zoospores that develop into morphologically similar gametophytic mycelia. These develop well-defined spatially separated male and female gametangia that release motile gametes (Emerson and Wilson 1949). Prior to fertilization, male gametes tend to collect in the vicinity of undischarged female gametangia which

**Table 7.2** Examples of the effects of the numbers and symmetry of mating type alleles on in- and outbreeding in Eumycota

| System | MTFs | Component loci | | | | Inbreeding potential (%) | Outbreeding potential (%) |
|---|---|---|---|---|---|---|---|
| | | A | A | B | B | | |
| Dimixis | A/a | 2 | | | | 50 | 50 |
| Unifactorial diaphoromixis | ∞As | 4 | | | | 50 | 75 |
| | | 10 | | | | 50 | 90 |
| | | 20 | | | | 50 | 95 |
| Bifactorial diaphoromixis | ∞As and ∞Bs | 2 | 2 | 2 | 2 | 25–36 | 56 |
| | | 3 | 2 | 3 | 2 | 25–36 | 69 |
| | | 5 | 5 | 5 | 5 | 25–36 | 92 |
| | | 8 | 8 | 2 | 2 | 25–36 | 74 |

Reproduced from Burnett (1976).

**Table 7.3** Out- and inbreeding potential in some natural populations of Basidiomycotina

| Organism | Sample size | Area or distance between limits of population | Outbreeding potential (%) | Inbreeding potential (%) |
|---|---|---|---|---|
| Bifactorial species (tetrapolar) | | | | |
| *Pleurotus ostreatus* [BA] | 35 | $16 \times 32\,km^2$ | 96.2 | 25 |
| | 26 | $64\,km^2$ | 93.9 | 25 |
| *P. sapidus* [BA] | 7 | $64\,km^2$ | 82.5 | 25 |
| | 13 | $20\,km^2$ | 74.3 | 25 |
| | 18 | 5 km (one tree) | 66.7 | 25 |
| | 8 | *c.* $1\,m^2$ | 65.0 | 25 |
| *Polyporus obtusus* [BA] | 24 | ? | 96.9 | 25 |
| *Coriolus versicolor* [BA] | 37 | *c.* 6.5 km | 91.6 | 25 |
| | 12 | $50\,cm^2$ (one stump) | 80.5 | 25 |
| *Schizophyllum commune* [BA] | 12 | 15.8 ha | 98.5 | 25 |
| | 15 | 11.7 ha | 96.8 | 25 |
| *Ustilago maydis* [BS] | 23 | *c.*2 ha | 38.5 | 25 |
| | 10 | *c.*5 ha | 31.4 | 25 |
| Unifactorial species (bipolar) | | | | |
| *Piptoporus betulinus* [BA] | 100.5 | British Isles | 96.4 | 50 |
| | 33 | 3.8 ha | 78.3 | 50 |
| | 29 | 0.76 ha | 76.4 | 50 |

Reproduced from Burnett (1976).

release the pheromone sirenin (Machlis 1958a,b). Sirenin attracts male gametes specifically and induces a topotaxic response to any source of the pheromone; it elicits no response in either zoospore or female zoogametes (Carlile and Machlis 1965a,b). Thus, potentially, both self- and cross-fertilization can occur since autopolyploids that can cross with the haploids are known (see Chapter 8, section 8.2.1.1). However, the relative frequency of the crosses is not known. The polyploid forms are fully fertile and show a similar alternation of generations. *Allomyces anomala* [C], the only species in the Brachyallomyces, possesses a mycelium comparable with that of the sporophytic Euallomyces but totally lacks a sexual phase. Zoospores released by its resistant sporangia simply propagate the parental form. Exceptionally, a very few gametophytes have been produced experimentally either by germinating sporangia in 0.02 M $KH_2PO_4$ solution, or by cutting off hyphal tips before the resistant sporangia develop and subculturing the hyphae for several transfers on agar. The gametophytes so produced sometimes resemble those of *A. arbuscula* [C] or *A. macrogynus* [C], two

species of Euallomyces! Therefore *A. anomala* [C] may be a sterile hybrid or derivative of a hybrid between these two species.

### 7.2.2 Mating systems in the Zygomycotina

Most information is available from the Mucorales amongst the Zygomycotina. Blakeslee (1904a) was the first to recognize and investigate sexuality in various species of *Mucor* [Z], *Rhizopus sexualis* [Z] and other species of Mucorales. He realized that, although the 'pin moulds' developed gametangia, they did not exhibit sexual differentiation. Some occurred as two compatible strains and others as a single self-fertile strain. As a consequence he coined the terms 'heterothallism' and 'homothallism' to describe outcrossing and selfing species respectively. The notion that fungi show physiological rather than morphological sexual differentiation unlocked subsequent understanding of all fungal sexuality. Nevertheless, the genetical basis of heterothallism in the Zygomycotina is still not known, mainly because their zygotes are so difficult to germinate, e.g. less than 1 per cent in

*Mucor hiemalis* [Z] (Gauger 1965), and there are very few genetic markers. Indeed, considerable uncertainty still exists as to the location and occurrence of meiosis and segregation. Numerous nuclei enter the zygote after gametangial fusion and at least one, probably many, fuse. At least, the numbers of nuclei decline as the zygote ages and four patterns have been described.

- In the *Mucor* [Z] pattern, supposedly fused nuclei decline for 6 days after which no more fused nuclei can be detected.
- In *Rhizopus* [Z] and *Absidia* [Z] both supposedly fused and unfused nuclei co-mingle in the zygote only and do not occur in either the germ tube or the sporangium.
- In *Phycomyces* [Z] both supposedly fused and unfused nuclei occur in the zygote, germ tube and germ sporangium.
- In *Syzygites megalocarpus* ( = *Sporodina grandis*) [Z], a homothallic species, neither paired nor apparently fused nuclei have been observed.

These observation by Cutter (1942a,b) were largely confirmed by Sjöwall (1945). Neither chromosomes nor meiotic configurations have ever been seen (Callen 1940; Cutter 1942a,b; Laane 1974), but segregation data exist which demonstrate that meiosis has occurred. In *Phycomyces nitens* [Z] multinucleate spores isolated from 84 germ sporangia showed both di- and tetra-type segregation of four markers, apparently from a *single meiosis per sporangium* although, exceptionally, some segregants from a single sporangium had been derived from more than one meiosis. In addition to normal segregants, some mating type heterokaryons were obtained and many germ sporangia were sterile (Eslava *et al.* 1975).

In contrast, in *Mucor mucedo* [Z] and *Rhizopus stolonifer* [Z], the spores from a germ sporangium are *all* the same mating type, either *mat +* or *mat −*, and the derived mycelia are either sterile or azygosporic (Köhler 1935; Gauger 1977). In contrast, two factor crosses in *M. hiemalis* [Z] produced multinucleate spores, although normal segregation and meiosis are apparently followed by the migration of a single segregant nucleus into each germ sporangium. Sometimes these were larger than usual, uninucleate and heterozygous for

mating type. The resultant mycelia were azygosporic and were presumed to be derived from diploid fusion nuclei since their mycelia sectored to give either normal or sexually functional *mat +* or *mat −* sectors which retained their characteristics on subculture.

Finally, in *Choanephora cucurbitarum* [Z] the spores in the germ sporangia are uninucleate and *mat +* or *mat −* nuclei segregate normally provided that a pair of alleles which control fast and slow growth rates are also segregating. However, if two strains of similar growth rate are crossed, the zygote can germinate directly to give a 1:1:2 segregation of *mat +* : *mat −* : *mat + /mat −* heterokaryons (Yu and Ko 1999) Thus the isogenicity of the strains determines their segregational outcome.

From the limited investigations performed to date it is evident that recombination can occur in the dimictic Mucorales but that its outcome is often distorted for unknown reasons. Moreover, natural isolates of mucoralean fungi are often sexually inactive, frequently as many as 14–15 per cent of all isolates (Blakeslee 1904b; Gauger 1966), and azygosporic forms are not uncommon. The single homothallic species investigated, *Sporodinia grandis* [Z], does not appear to undergo recombination and is presumably amictic. In addition, for many species, zygotes are rarely found in nature. Therefore, how far recombination contributes to the variability of populations of Mucorales under natural conditions is problematical.

## 7.2.3 Variations on basic mating systems in Ascomycotina

A variety of variants on the usual dimictic or homomictic alternatives are known.

### 7.2.3.1 Multiple fertilization
Multiple cross-fertilization is possible as was first shown experimentally with *Neurospora* [AP] (Nakamura and Egashira 1961). It also occurs naturally in other fungi. Isozyme markers and naturally occurring ascomata were used to show that outcrossing is the normal means of fertilization in the dimictic ascomycotous *Crumenulopsis soraria* [AD] (Ennos and Swales 1987). However, in a small

percentage of cases a single ascoma had been fertilized by more than one genetically distinct, presumably splash-dispersed, microconidium.

A similar phenomenon is known in addition to the normal dimictic system in *Cryphonectria parasitica* [AP]. The protoperithecium contains at least *two* nuclei ('female'), and if more than one microconidium fuses with the trichogyne then, subsequently, each penetrating 'male' nucleus can fuse with a different 'female' protoperithecial nucleus. A further complication is that if more than one 'male' nucleus is present, genetic exchange and segregation can occur between them prior to fusion with a resident 'female' nucleus. Thus, potentially, the resulting sexually produced progeny can carry genes derived from one 'female' and both 'male' parents (Anagnostakis 1982)! The complex situation in *C. parasitica* [AP] is, to date, unique and it is not known how widespread the occurrence of multiple fertilization is in the Ascomycotina. If genetic results appear to be anomalous, this possibility is always present and needs to be investigated.

### 7.2.3.2 Inbreeding in normally dimictic Ascomycotina

Various degrees of inbreeding can occur in some dimictic fungi. In addition to the modes of outbreeding behaviour shown by *C. parasitica* [AP], it can also self *provided that* normal outcrossed perithecia have developed first. A selfed perithecium arises through the fusion of the genetically identical resident 'female' nuclei in unfertilized protoperithecia *adjacent* to normal developing crossed ones. Therefore in this fungus a final complexity is the possibility of a single perithecium giving rise to both half-selfed (two 'female' nuclei) and half-crossed (one 'female' nucleus and one 'male' nucleus) progeny within the same perithecium. In nature, both types of outcrossed selfed or half-selfed–half-outcrossed perithecia have been known to occur in adjacent positions on the trunk in addition to clonal isolates derived from asexual conidial infections. As Anagnostakis remarks 'Genetic information can be gained from analyses of progeny of this kind, but a great deal of data is needed for meaningful interpretations' (Anagnostakis 1988).

Somewhat different behaviour is shown by the dimictic *Sordaria brevicollis* [AP] and *Ophiostoma novo-ulmi* [AP]. In *S. brevicollis* [AP] only the *mat A* strain is capable of producing a few self-fertile perithecia when grown alone, especially if incubated in the dark. All the progeny are *mat A*, but meiosis appears to have occurred for, if the parental *mat A* culture is heterokaryotic for spore colour genes, these segregate in the ascus. Presumably adjacent protoperithecial nuclei, whether genetically identical or not, can fuse and undergo meiosis, as can occur in *C. parasitica* [AP]. However, *S. brevicollis* [AP] differs from that species since the development of normal crossed perithecia *inhibits* rather than promotes the development of self-fertile ones in a *mat A* culture (Robertson *et al.* 1998). In *O. novi-ulmi* [AP] the situation differs from both these species. A small number of scattered perithecia develop after prolonged incubation of *B*-mating type cultures alone on elm twigs. These smaller malformed perithecia form fewer ascospores than in normal crosses but segregate for *both A*- and *B*-mating types. This 'pseudo-selfing' is thought to be due to a mutation of the *B* to the *A* mating type in some nuclei, thus rendering the mycelium self-fertile. Less frequently, possible cases of the reverse mutation $B \to A$ have been found (Brasier and Gibbs 1975; Brasier 1984). However, another mechanism, mating type 'switching' comparable with that found in yeast, could account for such selfing phenomena.

### 7.2.3.3 Mating type switching

A special mechanism exists for converting the normal dimictic system of yeast (*S. cerevisieae* [A]) into a homomictic system through gene switching, provided that the gene *HO* (for homothallic) is present. This gene is active only in cells which have produced a bud, i.e. are binucleate, and acts by switching the mating type of one nucleus into the compatible form *a* to $\alpha$ in *a/a* cells or $\alpha$ to *a* in $\alpha/\alpha$ cells. The cells thus acquire both *mat a* and *mat* $\alpha$ MTFs and fertilization can occur. The switch depends on the occurrence of the two alternative but functionally inert sets of MTF information carried in every cell in addition to the functional *mat*-locus. In the presence of *HO*, either a functionally inert *mat a* or *mat* $\alpha$ gene is preferentially inserted into the *mat*-locus where it becomes functional and is translated (Ahmad 1953; Hicks *et al.* 1977).

A superficially similar condition (Perkins 1987), whose genetic basis is not understood in all cases, appears to occur sporadically amongst the Ascomycotina. This is the regular production of asci with 50 per cent self-sterile cross-fertile ascospores and 50 per cent self-fertile ascospores, e.g. *Chromocrea spinulosa* [AP] (Mathieson 1952), *Sclerotinia trifoliorum* [AD] (Uhm and Fujii 1983a,b) and *Ceratocystis coerulescens* [AP] (Harrington and McNew 1997). The first of these exhibits an additional feature, an extra mitotic division in the ascus. This results in a 16-spored ascus with eight large self-fertile ascospores which produce scattered perithecia on their mycelium and eight small sterile ascospores. The latter produce abundant perithecia only at the junction where one of their germ mycelia is opposed to one derived from a large ascospore. It has been suggested that all these fungi are basically dimictic but regularly undergo mutation at meiosis, rendering half the progeny self-sterile. The difference in spore size is explained as a pleiotropic expression of the MTFs. However, in the apparently homothallic *C. coerulescens* [AP] species complex (*C. virescens, C. pinicola, C. fimbriata* etc.), the 'switching' mechanism originally suggested (Harrington and McNew 1997) is now known to be quite complex. Self-fertile isolates carry both MTFs, but self- fertilization results in the selective deletion of the *Mat-2* locus, so producing self-sterile but cross-fertile *Mat-1* isolates only. However, cross-fertilization between a self-fertile isolate and the slower growing *Mat-1* isolates results in self-fertile progeny, i.e. carrying both *Mat-1* and *Mat-2* loci. Thus in this species complex both selfing and crossing are possible and the system is potentially maintained by the selective deletion mechanism.

Convincing evidence has not been provided for either a 'switching' or mutational cause for the behaviour of any of the other cases, such as *O. novo-ulmi* [AP]. Moreover, most accounts of unusual mating behaviour to date have been based on laboratory studies and it is not known whether any of them show this behaviour in nature. However, if this is the case it suggests that, in nature, limited selfing in supposedly obligatory cross-fertile fungi may be more common than is usually supposed. Therefore breeding systems may be more flexible in nature than has been assumed.

### 7.2.3.4 Preferential crossing between siblings

Some sibling crosses are possible between the meiotic products of a single meiosis in all fungal mating systems: 100 per cent in dimictic species, and 50 per cent and 25 per cent respectively in unifactorial and bifactorial diaphoromictic fungi. The amount is reduced if meiospores are widely dispersed so that sexual fusions occur predominantly between non-sibling progeny. Even so, a variety of causes promote preferential crosses between sibling meiospores. Mechanical conditions which, potentially, favour the restoration of parental genotypes through such fusions are not uncommon. Selfing from such causes is less effective than the kinds of genetic situations just described. The enclosure of sibling asci within a single dispersal structure will clearly increase the chances of preferential fusion of sibling meiospores. This occurs in hypogeous Ascomycotina such as the truffles *Tuber* [A] and *Elaphomyces* [A], in which the subterranean sporocarps enclose the meiospores which germinate and fuse *in situ*.

In many Ascomycotina, there is a bias towards sibling fertilization as a consequence of the ascospores being discharged simultaneously, or in very close succession, so that they tend to be deposited and collect in a drop of fluid. As this dries the ascospores are drawn even closer together. The germ tubes of compatible spore pairs may then fuse at germination, thus restoring the parental genome.

### 7.2.3.5 Outcrossing in normally self-fertile species

Outcrossing has been detected both experimentally and in natural isolates of normally selfed Ascomycotina. Experimentally, if hyphae of two genetically different but somatically compatible mycelia of *Sordaria fimicola* [AP] come into contact, crossed perithecia develop in the contact regions (Olive 1956). In other cases the presence of compatible but genetically different conidia can fertilize a normally self-fertilizing protoperithecium. For example, using *Gaeumannomyces graminis* [AP], a self-fertile

Pyrenomycete, it proved comparatively easy to produce crossed ascospores in experimental conditions (15 per cent, (9/60), crossed perithecia) (Blanch *et al.* 1981), although no evidence has yet been found for outbreeding in nature (O'Dell *et al.* 1992; Harvey *et al.* 2001). Nevertheless, both these situations might well occur naturally; indeed, a naturally occurring case (although the precise mechanism is unknown) is that of the wild British isolates of *Emericella* (*Aspergillus*) *nidulans* [A] used for genetic analyses in the laboratory. There is now clear evidence that, historically, they were derived from outcrossed strains before being brought into experimental use (Geiser *et al.* 1994, 1996).

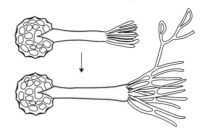

**Figure 7.4** Germination of a teleutospore of *Tilletia tritici* [BS]. The spore germinates to form a promycelium bearing terminal haploid basidiospores which can produce further secondary spores. Fusion frequently occurs between compatible spores, thus effecting self-fertilization.

### 7.2.4 Variations in basic mating systems in Basidiomycotina

Variations in mating systems comparable to those described above for Ascomycotina also occur in Basidiomycotina. Multiple fertilizations have not been recorded in Basidiomycotina, and where SI systems are present it is probably precluded, nor have any phenomena comparable with mating type switching been reported.

A bias towards preferential fusion between basidiospores derived from the same tetrad is common in some groups. Smut fungi are the most notable examples. For example, in some species of *Tilletia* and *Urocystis* [both BS], each diploid teliospore (the basidial equivalent) forms a germ tube at the tip of which haploid sporidia develop that often fuse in pairs via a conjugation tube before becoming detached. Thus the parental genotype is almost always restored (Figure 7.4).

In *Microbotryum violaceum* ( = *Ustilago violacea*) [BS], a common pathogen of campions (*Silene*) and related Caryophyllaceae, the potentially outbreeding dimictic system is effectively reduced to self-fertilization (Hood and Antonovics 2000). In the laboratory, where high nutrient media and temperatures of 15–25 °C tend to be used, uninucleate sporidia are budded off from the meiotic tetrad and dikaryons form between compatible sporidia derived from the same or different dikaryons. However, at 15 °C and under low nutrient conditions resembling natural conditions, postmeiotic fusion occurs almost entirely between cells of the developing tetrad. Since the MTFs are linked to the centromere, these intra-tetrad matings are almost entirely between nuclei that have separated at the first meiotic division (automixis) and re-establish the parental dikaryon. However, if such matings are delayed, adjacent cells of the linear tetrad develop sporidia; in many cases these carry haplo-lethals closely linked to the centromeres and therefore die. In these circumstances, only one MTF is recovered. A widespread bias, usually in favour of the *mat A1* MTF occurs in both North American and some European populations (Oudemans *et al.* 1998; Kaltz and Shykoff 1997). In Europe, the condition varies on different host plants (Kaltz and Shykoff 1999). In effect, therefore, in nature this fungus is probably predominantly self-fertile with occasional wide outcrosses.

Other situations merely increase the likelihood of sibling crosses. Structural bias occurs in another group, the Nidulariales. In the bird's nest fungi (*Nidularia, Cyathus, Sphaerobolus* [all BA]) the aerially borne basidiospore-forming tissue (gleba) is enclosed within a wall and dispersed as a single unit—a peridium. Although the basidiospores probably germinate only when the peridial wall eventually decomposes, the contiguity of sibling meiospores increases the likelihood of the parental dikaryon being restored. Sibling fusion is less common with tetrads of basidiospores although, since in most Basidiomycotina the spores of each tetrad are discharged in rapid succession, they can be deposited in close proximity if the spore path is short; otherwise they become more widely dispersed.

### 7.2.5 Secondary inbreeding: homoheteromixis (secondary homothallism)

Genetical and cytological mechanisms exist which result in inbreeding predominating in basically heteromictic Ascomycotina and Basidiomycotina. An appreciable number of these secondarily homoheteromictic (secondarily homothallic) fungi are known.

#### 7.2.5.1 Homoheteromixis in Ascomycotina

The two mechanisms that have been described for homo-dimictic Ascomycotina are illustrated in Figure 7.5. In these, the consequence of segregation and ascospore formation is that each of the four ascospores formed is heterokaryotic for mating type and the mycelia which develop from them are self-fertile.

In a small proportion of cases in both species, some spores homokaryotic for MTFs are produced. In *P. anserina* [AP], rarely the MTFs segregate at the *first* meiotic division, so that adjacent nuclei carry the same MTF, or, if differentiation is abnormal, two or more small uninucleate ascospores are formed. The latter situation occurs in *N. tetrasperma* [AP] but there are additional requirements. In heterothallic Neurosporas the MTFs act as *het*

alleles, preventing somatic fusion between *Mat A* and *Mat a* mycelia, but in *N. tetrasperma* [AP] the presence of an allele of the *tol* gene, *tol^T*, suppresses their heterokaryon incompatibility function (Jacobson 1992).

In two strains of *N. tetrasperma* [AP], for example, there were as many as 2–10 per cent uninucleate ascospores, and in addition as many as 12–16 per cent uninucleate homokaryotic conidia were produced on the same mycelium (Raju 1992). The uninucleate ascospores from an ascus behave exactly like the ascospores of normally eight-nucleate species, i.e. the spores and conidia are self-sterile but can also cross either with mycelia derived from uninucleate ascospores of compatible mating type or with mycelia from binucleate self-fertile ascospores. If a bisexual heterokaryon re-forms, its nuclear behaviour reverts to that typical of a four-spored ascus. However, most matings between homokaryotic progeny, unless they are from the same lineage, are often partially or wholly sterile. When all possible crosses were made between uninucleate ascospores derived from 10 strains from three localities (Louisiana, Hawaii and New Zealand) only 27 per cent of them were completely fertile with four germinable ascospores. In a further 24 per cent, one to three ascospores were normal, in 15 per cent they were completely sterile, not even producing perithecia, and the remaining 36 per cent showed partial sterility. The morphological and cytological features associated with the partially sterile crosses suggest the involvement of protoplasmic incompatibility, although its nature has not been defined (Jacobson 1995). It appears that one cause for the sexual dysfunction observed on crossing derived monokaryons of *N. tetrasperma* [AP] is if they are heterogenic for the heterokaryon incompatibility allele *het-c*. It appears likely that, to function normally, *N. tetrasperma* [AP] should be homoallelic at the *het* loci (Saenz *et al.* 2001). Thus limited outbreeding is not entirely lost in such homoheteromictic species but can clearly be disrupted from this and other causes.

A number of other species with four-spored asci are known in genera normally possessing eight-spored asci and their origins resemble one of the two just illustrated. In most of them, as in the two

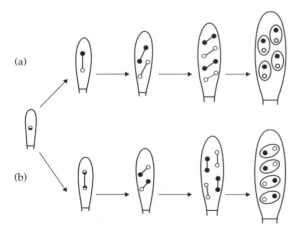

**Figure 7.5** Two different ways in which the homodimictic condition is maintained in Ascomycotina: (a) *Neurospora tetrasperma*; (b) *Podospora anserina*. The condition is determined by the orientation of the meiotic spindles, and by first-division segregation of the MTFs in *N. tetrasperma* and second-division segregation in *P. anserina*. (From Burnett (1975).)

already mentioned, some uninucleate ascospores are usually produced. Thus some outbreeding is probably also retained here.

### 7.2.5.2 Homoheteromixis in Basidiomycotina

A post-meiotic mitosis occurs in many basidia, but its location and consequences are varied. In some it occurs in the basidium and all or only four nuclei enter the basidiospore. In others mitosis occurs in the sterigmata followed by back-migration into the basidium, and in yet others mitosis does not occur until the meiotic products have entered the spore when the proximal nuclei may migrate back into the *basidium*, e.g. *Lentinus tigrinus* [BA] [Hibbett *et al.* 1994). Consequently, in some species the basidiospores are binucleate but, depending on the migration pattern of the mitotic products, the spores may carry a single or complementary meiotic products. Lange (1952), Kennedy and Burnett (1956) and Petersen (1995a) have reviewed the variety of ways in which compatible MTFs, derived ultimately from the same meiosis, may be brought together in a single basidiospore. Of course, there is always a 50 per cent chance of sibling meiospores crossing in normal unifactorial diaphoromictic fungi and a 25 per cent chance in normal bifactorial diaphoromictic species (see Table 7.2) but in some species the percentage is much higher. The simplest situation is that in which only two-spored basidia are developed so that, inevitably, two of the four meiotic products enter each basidiospore. This situation has been described in species of *Mycena*, such as *M. epipterygia* [BA], where both two- and four-spored basidial forms occur. Basidiospores from the former give rise to dikaryons on germination, whereas those from four-spored basidia are monokaryotic. Some of the various possible situations are illustrated in Figure 7.6.

One of the best understood examples of a homoheteromictic two-spored species is the cultivated edible mushroom *Agaricus bisporus* [BA], where the majority of basidiospores (88 per cent or more) are heterokaryotic for MTFs although the expectation on random segregation would only be 67 per cent. In this case, there is cytological evidence to suggest that this results from the orientation of the second meiotic division spindles (Evans 1959). Superficially similar situations occur in a number of other agarics. Spores homokaryotic and heterokaryotic for MTFs arise on a single basidium if a mitotic division occurs before the four meiotic nuclei migrate into the basidiospores, it results in the development of eight-spored basidia. These are not uncommon in the Agaricales-Leucosporae (Kühner 1945) and occur sporadically elsewhere in the Basidiomycotina, e.g. the bolete *Suillus grandis* [BA] (Jacobson and Miller 1994), where 5–10 per cent of the basidiospores are binucleate and potentially could give rise to dikaryons. It has been suggested that, where this condition occurs, heterokaryotic spores could be at a competitive advantage over monokaryotic spores since the dikaryon is ready formed in the former at germination (Tommerup 1990).

In other cases, however, four of the eight nuclei degenerate selectively, and segregation then appears to be normal, e.g. *Peniophora ludoviciana* [BA] (Biggs 1938), or the number of basidiospores which develop, sometimes in successive crops, equals the number of basidial nuclei, e.g. *Clitocybe aurantiaca* [BA] (Maire 1902). However, if all the nuclei migrate into the basidiospores, by chance, some may carry identical MTFs and others different MTFs, and on germination these behave as monokaryons or dikaryons respectively.

In the minute dikaryotic bird's nest fungus *Mycocalia denudata* [BA], the four- or eight-nucleate condition is determined by an allelic pair of genes *Pd* and *pd*. The former is dominant and determines the precocious division of the four meiotic nuclei within the basidium (either *Pd* + *pd* or *Pd* + *Pd*), whereas in isolates carrying *pd* alone, only one of the four nuclei migrates into each basidiospores and undergoes a further mitotic division in each spore. Therefore all basidiospores in this fungus are ultimately binucleate, but whether they are genetically *Pd/Pd, Pd/pd* or *pd/pd* can only be determined from their subsequent behaviour on germination (Figure 7.7) (Burnett and Boulter 1963). Apparently similar facultative homoheteromixis (sometimes called amphithallism) occurs sporadically elsewhere in Basidiomycotina (Kennedy and Burnett 1956), but its genetic basis is not known.

One feature common to all homoheteromictic Basidiomycotina is the effect of the breeding system on heterozygosity in the progeny. Whether

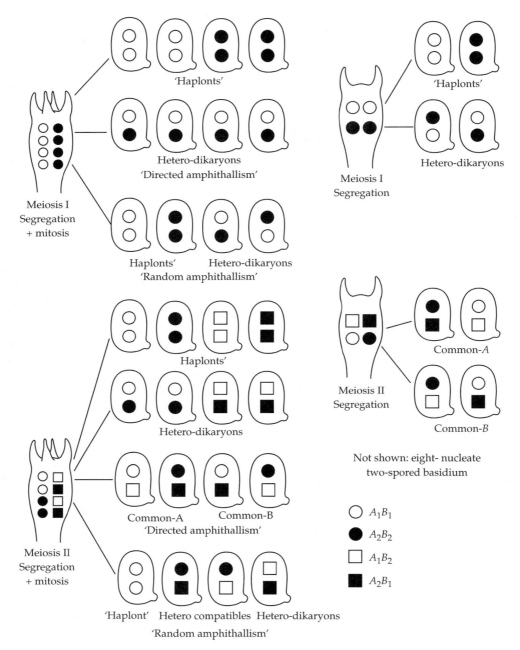

**Figure 7.6** Diagrammatic representations of how homodiaphoromixis can be maintained in Basidiomycotina, assuming a tetrapolar condition. The left-hand diagrams show basidia where the post-meiotic mitosis occurs before migration of the nuclei into the basidiospores to give an eight-nucleate basidium. They also illustrate the consequences for MTF loci (*A* and *B*) which segregate at the first division (upper left) and second division (lower left) of meiosis. The segregation in the basidium may be genetically determined ('directed') or at random. The right-hand diagrams show two spored basidia, where nuclear migration precedes mitosis with first-division segregation (upper right) and second-division segregation(lower right) at meiosis. (Reproduced from Petersen (1995), with permission from the Mycological Society of America.)

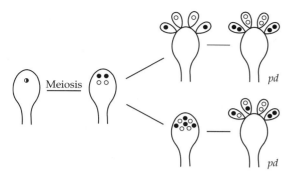

**Figure 7.7** The action of the *Pd* and *pd* alleles on segregation in the basidium of *Mycocalia denudata* [BA]. Homodikaryotic recessive *pd pd* strains show normal post-meiotic behaviour, but *Pd Pd* or *Pd pd* strains undergo a precocious mitotic division after meiosis to give an eight-nucleate basidium. Since the eight nuclei migrate at random into the basidium, on average the basidiospores are approximately 50 per cent homokaryotic and 50 per cent heterokaryotic for MTFs. (From Burnett (1975).)

as a result of the nuclear dynamics in the basidium or purely by chance, intra-tetrad pairing, i.e. pairing of the haploid products of the *same* meiosis, can occur. The consequences of this on the heterozygosity of the progeny are variable but it results, in general, in an increase in heterozygosity compared with the parental types or dikaryotic parent. The extent depends on various factors such as the linkage between various loci and that of the MTFs (Kirby 1984); for example, in *Agaricus bisporus* [BA] the expected heterozygosity of the progeny is greater than 50 per cent of that in the parental dikaryon and can be even higher than the 67 per cent obtained at normal segregation (Langton and Elliott 1980).

## 7.2.6 Polymorphic and partial sexual expression

Some fungi exhibit genetic polymorphism in respect of the balance between sexual and asexual reproduction. For example, isolates of *Glomerella cingulata* [AP] from apples, grapes or peppers can develop both conidia or perithecia, whereas those from *Ipomoea* or okra produce few or no conidia but abundant perithecia. In addition, some isolates are known which can only produce conidia (Wheeler 1956). The range of behaviour is set out in Table 7.4

An even more surprising development is that, apparently, four conidial isolates of this species (as its anamorph *Colletotrichum gloeosporoides* [FA]) from pecan (*Carya* spp.), fell into a typical dimictic pattern, *Mat A* or *Mat a*, but a fourth isolate was cross-fertile with *both* types. This suggests a unifactorial diaphoromictic system, i.e. a single locus with multiple alleles, comparable in mating behaviour to similar systems in the Basidiomycotina. Thus *Mat a* is designated *Mat A1*, *MatA* as *Mat A2* and the fourth isolate as *Mat A3*. Moreover, isolates from different hosts (e.g. *Medicago*) would not cross with pecan isolates, while those from *Mangifera* only produced sterile asci (Cisar and TeBeest 1999). This suggests that isolates from different hosts may be different mating populations, and it is not clear whether a range of different mating systems occurs in *G. cingulata* [AP] or whether more than one species is involved with different mating systems.

Another species, *G. graminicola* [AP], the cause of anthracnose of maize, also has an unusually complex mating system. Mating in heteromictic strains is determined by multiple alleles at a single locus but, in addition, homomictic strains occur which give rise to heteromictic strains with high frequency (Vaillancourt and Hanau 1991, 1992). Evidently, the genus *Glomerella* [AP] appears to have an unexpectedly complex mating system for a member of the Ascomycotina which, to date, is not paralleled elsewhere in the group. However, the mating systems of only relatively few Ascomycotina have been studied and similar complexity may well occur elsewhere.

### 7.2.6.1 Female sterility

Polymorphism for sexual expression is quite widespread amongst Ascomycotina and, indeed, loss of 'female' sexuality is a common situation. Protoapothecia and protoperithecia are morphologically and physiologically quite complex, and many different mutations can bring their dysfunction and hence induce 'female' sterility. This may account for its relatively common occurrence. Loss of these structures in part or in whole, e.g. physical or functional loss of trichogynes, does not necessarily prevent sexual reproduction. In some species, normal conidium/trichogyne fusions are replaced by hyphal fusions and the nuclei

**Table 7.4** A complex mating system: morphology, behaviour and fertility of isolates of *Glomerella cingulata* [AP] which differ at their *A* and *B* loci when paired in plate culture

| Genotypes of cultures | Perithecial production | | | | | | | |
|---|---|---|---|---|---|---|---|---|
| | **A + B +** | **A2B +** | **A + B1** | **A + B2** | **A2B1** | **A2B2** | **A1B +** | **A1B1** |
| A + B + | − | | | | | | | |
| A2B + | **1** | | | | | | | |
| A + B1 | 2 | X | | | | | | |
| A + B2 | 2 | **X** | **X** | | | | | |
| A2B1 | 2 | 2 | 1 | **1** | | | | |
| A2B2 | 2 | 2 | **1** | X | 0 | | | |
| A1B + | 1 | **1** | 2 | 2 | 2 | 2 | | |
| A1B1 | **3** | **3** | **X** | X | X | X | 0 | |
| A1B2 | **3** | **3** | X | X | X | X | 0 | 0 |

The *A* locus determines protoperithecial production and the *B* locus determines the degree of *self*-fertility. Each locus has three known allelomorphs, +, 1 and 2. The loci are linked with a recombination frequency of 44.7%.
Key to reactions: **bold type**, cross-fertility; 1, light but distinct ridge of perithecia; 2, heavy ridge of perithecia; 3, very heavy ridge of perithecia; X, weak or irregular interaction; 0, no reaction.
(Based on Wheeler and McGahen (1952).)

introduced then migrate to potential protoperithecial or protoapothecial sites where reproduction occurs, e.g. *Humaria granulata* [AD] (Gwynne-Vaughan and Williamson 1930). How widespread this is in nature is unknown, but it is possible that this process could occur facultatively, even where 'female' sterile strains are the norm.

In *Fusarium solani* f.sp. *cucurbitae* race 1 [FA/AP] the situation is even more complex. In addition to the two MTFs, sexuality is affected by two other genes, *M* and *C*, which affect the development of protoperithecia, trichogynes and conidia. Eight genotypes have been identified:

- *MC mat +* and *MC mat −*: 'hermaphrodite' with both conidia and trichogynes
- *mC mat +* and *mC mat −*: 'male', i.e. lacking trichogynes but with conidia
- *Mc mat +* and *Mc mat −*: 'female', i.e. with trichogynes but lacking conidia
- *mc mat +* and *mc mat −*: 'neuter', i.e. without trichogynes or conidia.

These genotypes are distributed unequally over the range of the species; for example, in race 2 in California the *Mc mat +* genotype prevails, in The Netherlands the *MC mat −* genotype prevails and in Australia the *MC mat +* genotype prevails. As a consequence, although sexual reproduction is unknown in nature in *F.solani* f.sp. *cucurbiteae* [FA/AP], it can readily be achieved experimentally by bringing sexually compatible strains together (Snyder *et al.* 1975). If migration of a compatible strain were to occur naturally, fertility would be restored to such a region but it has not yet occurred outside the laboratory.

A similar situation has been recorded in *F. sulphureum* (*Gibberella cyanogena* [AP]) although here the majority of isolates are *mat −* (Gordon 1954). In several Ascomycotina other kinds of mutation can bring about female sterility by impairing either the normal development or the functioning of apothecia or perithecia.

Quite often 'female' sterile strains coexist with hermaphrodite strains. In that event the 'female' strains may contribute unequally at sexual reproduction, for both contribute 'male' gametes equally but the 'female' gametes are contributed entirely by the hermaphroditic parent. Thus the 'female' sterile parent is selected against (Leslie and Klein 1996). Therefore, on first consideration, it is somewhat surprising that the condition is so widespread. However, the selective disadvantage of 'female' sterile strains could be compensated for in various ways. 'Female' sterile strains would be favoured if either the strains show unequal conidial production, with that of the 'female'sterile strain being the greater, or the conidia differ physiologically in their effectiveness in contributing 'males ' at sexual reproduction. Whether either of these conditions ever obtains in the laboratory or in nature is not known.

Another factor affecting the situation is the overall mutation rate to 'female' sterility, since the condition can be caused by a variety of different mutations. If this rate is relatively high and a large number of non-sexual generations intervene between each round of sexual reproduction, both of which are very probable circumstances in nature, 'female' sterility is likely to persist (Leslie and Klein 1996). Whether, in fact, such situations occur in nature in any 'female' sterile fungus requires further investigation.

### 7.2.7 Sterility

Partial sterility can arise as a consequence of mutants affecting meiosis, resulting in biased segregations and ultimately affecting gene frequency in a population. The best known are the 'spore-killer' genes which occur in the Ascomycotina *Neurospora*, *Podospora*, *Gibberella* and *Cochliobolus* [all AP] (Raju 1994). In crosses between spore killers and sensitives, a condition determined by a pair of alleles ($Sk^k$ and $Sk^s$ or $Spk$ and $spk$, depending on the species) the asci contain only four normal ascospores; the other four fail to develop. Homozygous crosses between either spore killers or sensitives produce normal eight-spored asci. Spore killers are found in natural populations and are fairly common in *N. sitophila*, *Gibberella fujikuroi* and *Podospora anserina* [all AP]. For example, at least seven different 'spore-killer' genes are known from *P. anserina* [AP] (van der Gaag *et al.* 2000). However, since it is secondarily homothallic, the effect is that only half the spores, those that carry *spk*, abort, giving rise in this case to two-spored asci. These ascospores germinate and the mycelium develops normal four-spored asci in the perithecia.. The distribution of spore killers in populations of *Cochliobolus heterostrophus* [AP] is uneven. It does not occur in race T (the cause of the southern corn blight epidemic) at all, and only in about half the strains of the most common race O (Bronson *et al.* 1990). In crosses between the two races evidence was provided that the *spk* gene was

linked with the dominant *Tox1* gene, associated regularly with a translocation and responsible for determining a toxin to maize produced by race T isolates (Leach *et al.* 1982; Tzeng *et al.* 1989).

Sterility can arise from other mutations which impair post-fertilization development, a not uncommon condition in Ascomycotina. A variety of situations has been demonstrated in species of *Cochliobolus* [AP]. In the dimictic *Cochliobolus* (*Helminthosporium*) *carbonum* [AP], a recessive allele *s* causes sterility if both partners carry it but crosses involving the *S* allele, i.e. *mat A S* × *mat a s* or *mat a S*, or *mat A s* × *mat a S*, are fertile. The *s* allele occurs sporadically throughout the population, e.g. it occurred in 17 out of 75 *mat A* isolates and nine out of 73 *mat a* isolates. A similar situation is known in *C. heterostrophus* and *C. spicifer* [both AP] (Nelson 1964; Webster and Nelson 1968). *C. spicifer* [AP] demonstrates further complexity. A pair of alleles, $i_1$ and $i_2$, if present in otherwise compatible partners, can inhibit the mating reaction, e.g. *mat A* $i_1$ × *mat a* $i_1$, or *mat A* $i_2$ × *mat a* $i_2$ or *mat A* $i_1$ $i_2$ × *mat a* $i_1 i_2$ are all ineffective crosses (Webster and Nelson 1968).

Even when crosses are possible, their fertility can vary quantitatively, as in *C. carbonum* [AP] (Nelson, 1970a). When compatible partners of this species carry only the dominant allele *I* (unrelated to $i_1$ or $i_2$ mentioned above), the crosses are fertile. However, some 58 *mat A I* isolates showed a range of fertilities when crossed in all combinations with 64 *mat a I*. *Mat A I* segregants ranged from 22 to 98 per cent fertile and *mat a I* segregants ranged from 28 to 98 per cent. When 10 isolates of each genotype were crossed in all combinations, fertility ranged from zero to 100 per cent, with the former resulting from crosses from the those originally with the lowest fertility, and the latter from those with the highest fertility (Table 7.5).These results suggest that in this fungus fertility is not only determined qualitatively by major genes but also has a polygenic or multigenic basis.

### 7.2.8 Amixis

Homomictic fungi are often assumed to be self-fertilized and this is undoubtedly the predominant situation. However, a few fungi are known where

**Table 7.5** Multigenic control of fertility in *Cochliobolus carbonum* [AP]: the results of crossing isolates which differed in fertility when crossed against the total sample of 122 isolates

| Mating type *a* isolates of differing % fertility | Mating type *A* isolates of differing % fertility | | | | | | | | | |
|---|---|---|---|---|---|---|---|---|---|---|
| | 28 | 35 | 43 | 50 | 57 | 64 | 75 | 87 | 92 | 96 |
| | Percentage fertility of crosses | | | | | | | | | |
| 22 | 0 | 0 | 20 | 34 | 47 | 40 | 53 | 67 | 80 | 80 |
| 33 | 13 | 27 | 27 | 34 | 47 | 47 | 47 | 13 | 87 | 80 |
| 42 | 20 | 47 | 34 | 47 | 40 | 60 | 53 | 80 | 80 | 80 |
| 50 | 40 | 47 | 47 | 60 | 60 | 60 | 73 | 80 | 80 | 80 |
| 58 | 40 | 60 | 53 | 67 | 73 | 80 | 80 | 80 | 87 | 80 |
| 64 | 47 | 53 | 53 | 67 | 67 | 73 | 80 | 87 | 80 | 87 |
| 75 | 60 | 60 | 73 | 73 | 80 | 73 | 87 | 87 | 93 | 87 |
| 86 | 73 | 73 | 73 | 80 | 80 | 93 | 93 | 100 | 100 | 100 |
| 91 | 80 | 87 | 73 | 80 | 87 | 93 | 100 | 100 | 100 | 100 |
| 98 | 80 | 80 | 73 | 93 | 87 | 100 | 100 | 100 | 100 | 100 |

Based on data from Nelson (1970a), with permission from the National Research Council of Canada.

meiosis is replaced by mitotic divisions even though spores then develop which resemble normal meiospores, e.g. *Podospora arizonensis* [AP] (Mainwaring and Wilson 1968). Similar situations are known from Basidiomycotina such as *Volvariella volvacea* [BA], the Chinese straw mushroom. The basidiospores and subsequent monokaryotic mycelium are haploid, as estimated by intensity of DAPI staining, but two identical nuclei enter each basidium, fuse and undergo meiosis, with one haploid nucleus entering each of the four basidia and basidiospores (Chiu 1993). These amictic fungi are, in effect, obligatory inbreeders. It is not known how many fungi behave in this way.

### 7.2.9 Variations on basic mating sytems in Straminipila

Outcrossing in normally self-fertile Straminipila is also known, e.g. *Pythium ultimum* [O] and other *Pythium* spp. Most self-fertile mycelia are monoclinous, i.e. they develop terminal oogonia and, immediately beneath them, antheridia which fertilize them. However, some isolates develop additional antheridia elsewhere, and these may fertilize oogonia on either the same or another mycelium. So-called hyphal swelling (HS) strains only produce globose asexual spores and are apparently sterile, but when paired with hermaphrodite strains they develop antheridia and behave as functional males, with sexual reproduction taking place as normal. Therefore HS strains are basically female-sterile (Saunders and Hancock 1994). Even more complex behaviour is known from *Achlya heterosexualis* [O], a predominantly homomictic species, and *Pythium sylvaticum* [O], which is predominantly dimictic, but in both sexuality appears to be relative. Thus isolates can be arranged in a hierarchical series based on whether they act towards other isolates as male or female, i.e. from 'strong' females at one extreme to 'strong' males at the other, with those between being intermediate in their behaviour (Barksdale 1965; Pratt and Green 1973) (Figure 7.8). The behaviour of these fungi has been attributed to the interaction between their phenotypic reaction to pheromones released by the different conjugating strains. As far as the breeding system is concerned, it not only provides greater versatility but also confuses the genetic interpretations of crosses and the genetics of the population.

The majority of species of *Phytophthora* [O] appear to be homothallic. When heterothallism occurs it is regularly associated with a complex reciprocal translocation involving four chromosomes with which the mating type factors are associated. Although two mating tpes, *A1* and *A2*, appear to be involved, their genetic architecture

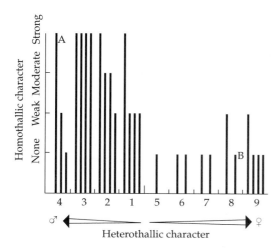

**Figure 7.8** Relative sexuality amongst 36 isolates of *Pythium sylvaticum* [O]. Each vertical bar represents the behaviour of a single isolate. Parental types are A and B. The so-called 'homothallic character' is assessed by oospore production in an isolated culture, i.e. a measure of selfing. The 'heterothallic' character relates to the intensity of oogonium production in crossed cultures. (Reproduced from Pratt and Green (1973), with permission from the National Research Council of Canada.)

has not yet been elucidated. Unequal numbers of *A1* and *A2* progeny are not uncommon, and *P. dreschleri* [O] behaves as if secondarily homothallic, evidently being trisomic for a mating type chromosome. Thus its progeny include *A1, A2* and *A1A1A2* individuals (Mortimer *et al.* 1977).

Sexual reproduction was unknown in *Phytophthora infestans* [O], except in Mexico, since the *A2* MTF was absent from other populations worldwide for about 140 years until both mating types were detected in European and North American populations in the 1980s (Hohl and Iselin 1984; Deahl *et al.* 1990). However, sporadic development of oospores had been noted prior to the introduction of isolates carrying the *A2*, and this effect was enhanced in both mating types when grown together but separated by a porous polycarbonate membrane (Shattock *et al.* 1986). On germination, only *A1* progeny were obtained but some were homozygous for the peptidase allele, although the parental strain was heterozygous. Later investigations have shown that oospores can be induced in *A2* cultures of *P. infestans* [O] by exposure to volatile materials produced by avocado roots, *Trichoderma* species or some fungicides

such as chloroneb (Brasier 1992), and that selfing ability varied with different clones according to the composition of the growth medium, with some *A2* clones being particularly susceptible (Smart *et al.* 1998, 2000). Thus it is possible that selfing can occur in nature under some circumstances, and that when both mating types are juxtaposed the oospores formed include some self-induced oospores, although the majority are crossed oospores.

## 7.3 The balance of outbreeding and inbreeding

The outcomes of various breeding systems are determined by a complex interplay between mycelial compatibilities of all kinds, the metabolic and nutritional status of individual mycelia and environmental factors. No cases are known where all the determinants have been identified, and only a beginning has been made in describing the different determinants involved, much less their interplay. Consequently it is not surprising that the balance of in- and outbreeding is affected differently across the distribution of a species. Some indication of the range of variation possible is illustrated in the following examples.

### 7.3.1 Environmental effects on sexual reproduction

Climatic and seasonal factors affect mating as well as mutation rate (see Chapter 6, section 6.1.2). Samples of *Setosphaeria turcica* [AP], the cause of some maize diseases, illustrate the effect of climate. Ascomata are not known in nature, although two MTFs have been identified from the field. Analysis using the RAPD technique showed that the genotypic diversity $H_S$ of isolates from Mexico, Kenya and southern China was extremely high, whereas $H_S$ values for European isolates were distinctly lower and there was evidence for a single clone extending throughout Germany, Switzerland and France. The index of multilocus association $I_A$ indicated that the tropical populations are virtually wholly outbreeding; indeed, in one Kenyan sample and the Mexican sample there is no evidence for clonal lineages at all. However, this was not the case in temperate samples: only one mating type was present in some regions and in others their numbers were very unequal. Clearly, there are significant differences between the two geographical regions. It was suggested that, in addition to the lack of the necessary genetic conditions, that climatic conditions for the induction and development of ascomata were rarely met in temperate climates. In contrast, in the tropics, sexual reproduction accounts for virtually all the genetic diversity (Borchardt *et al.* 1998).

Many temperate-zone pathogens demonstrate regular seasonal effects. It is common for a single annual autumnal sexual reproductive stage to be interpolated between several successive asexual conidial stages. Successive crops of conidia of barley mildew, *Erysiphe (Blumeria) graminis* f.sp. *hordei* [FA/AP], occur between October and July in north temperate Europe, and sexual reproduction is confined to June and July, with ascospores being released in the autumn. Values of gametic disequilibrium (GD – *see* 7A.4) for seven unlinked virulence alleles and one for fungicide resistance compared in July and October populations gave an indication of how far ascospores released at the end of summer affected the composition of the total spore population. Approximately 24 per cent of the spores from the autumn population were derived from sexual reproduction, i.e. a regular injection of new recombined genotypes occurs annually (Brown and Wolfe 1990). Thus, every year in the following months a novel population representing the previous year's selected but asexually propagated genotypes plus the new recombinants becomes available for a new round of selection during the asexual phase.

*Leptosphaeria nodorum (Stagnospora nodorum)* [AP] and *Mycosphaerella graminicola (Septoria tritici)* [AP] have a similar cycle to each other and to *B. graminis* [A]. The GD was estimated in North American populations of both the former species (McDonald *et al.* 1994, 1995). Seven RFLP markers were used in *L. nodorum* [AP] and, correcting for sampling identical clones, only 7 per cent of the comparisons showed significant allele association, indicating that the isolates tested were almost entirely the result of outcrossing. A similar situation was found in *M. graminicola* [AP] where one population was sampled in Oregon and another in California and

their GD was estimated using DNA fingerprints. In the Californian population 27 per cent showed significant disequilibrium until corrected for sampling of identical clones on the same leaf when it fell to 0.5 per cent, but in the Oregon population, where the sampling procedure avoided repetitious sampling of clones, all associations were random. Thus, in both populations virtually 100 per cent of the isolates were derived from outcrosses.

Another common life-cycle pattern in perennial fungi is when sexuality is more or less seasonal but both phases often coexist for an appreciable period so that both sexual and asexual propagules contribute to the populations. Two such fungi, *Crumenulopsis soraria* [AD] (Ennos and Swales 1987) and *Gremmeniella abietina* [AD] (Wang 1997), show a relatively simple reproductive pattern of this type.

In Scottish populations of *C. soraria* [AD], isozymes were used to investigate population structure and non-random associations between alleles at two loci by using the standardized gametic disequilibrium parameter $D$, but this never exceeded 10 per cent. Therefore this species is largely outbreeding, despite the coexistence of an appreciable asexual phase.

Populations of *G. abietina* [AD] in northern Sweden develop apothecia abundantly so that sexual reproduction might be expected to be of importance. Thirty-two RAPD nDNA fragments obtained from a total sample of 126 pycnidia were used to estimate the GD for all combinations of pairwise loci in two ways. First the entire collection of 126 pycnidial isolates was used, and then only a single representative of each genotype was used. The latter procedure was adopted in order to correct for the possibility that some pycnidia were so close together that they could have developed from the same mycelium. When an exact test of significance was used, 147 of the possible 496 combinations (29.6 per cent) showed significant non-random association, but when only single genotypes were tested the number decreased to 73 of 496 (14.7 per cent). Thus just over half the initial associations were spurious, resulting from sampling clones more than once. Therefore in this species a high degree of random outbreeding, as was expected from the biological situation, clearly contributed to the range of genotypes in the populations. This was borne out by the finding that 95.5 per cent of the genotypes differed by more than five RAPD fragments, i.e. were largely unrelated, and only 1.6 per cent of the genotypes differed by one or two RAPD fragments. Nevertheless, the populations occurred as patterns of local intermixed genotypically distinct clones as a consequence of the contribution of the asexual pycnidiospores. Unfortunately, comparable data from southern Sweden, where the sexual phase is relatively uncommon, are not available.

## 7.3.2 More complex in- and outbreeding patterns

The situations in *Cryphonectria parasitica* [AP] (Milgroom *et al.* 1992, 1993; Milgroom 1995) and *Ophiostoma novo-ulmi* [AP] (Brasier 1978, 1984, 1986a,b,c, 1988; Brasier and Gibbs 1975) are much more complex.

*Cyryphonectria parasitica* [AP], an oubreeder, is also capable of self-fertilization or partial fertilization (see Chapter 7, section 7.2.3.2). Thus it can reproduce clonally by both self-fertilized ascospores and conidia. A number of different populations in different geographical regions have been investigated. The marker loci studied were unlinked nDNA RFLPs, and DNA fingerprints were used as well as VCGs. Thirty-nine isolates were sampled in one series and 22 perithecia in another from a population at Mount Lake, Virginia (Milgroom *et al.* 1992, 1993). In the first sample, the genotypic diversity was much less than expected on a hypothesis of random mating although, when single genotypes only were tested, to avoid bias due to sampling clones, it could not be ruled out. In the second investigation a multilocus estimate of the proportion of outcrossed perithecia in which only a single fertilization had occurred, i.e. those with segregating progeny, was used. Sixteen of the 22 perithecia sampled (74 per cent) segregated. The number of actual outcrosses observed was then compared with the number expected under random mating using a method developed for diploid plants but modified for haploids (Shaw *et al.* 1981; Shaw and Brown 1982). The values of the estimate observed were 0.74 for RFLP data and 0.73

for fingerprint data, indicating a significant amount of outcrossing (over 70 per cent), assuming that the population was not subdivided and mating was random. The VCG data gave appreciably lower estimates of outcrossing, possibly because the diversity of VCGs was low in this population, or, perhaps because mixed mating had occurred. Four other populations sampled in North America confirmed this pattern of both in- and outbreeding in the same population; self-fertilization was found in 19–32 per cent of these populations. Because of the unusual mating system of *C. parasitca* [AP], the question arises as to how self-fertilization can occur even with biparental fertilization. In an attempt to answer this question, the number of DNA fragments actually segregating from a 'crossed' perithecium were compared with the mean number expected under fully random mating. In three further populations in North America there were significantly *fewer* fragments than would be expected by chance. Thus some of the inbreeding observed is due to partially induced self-fertilization in potentially outcrossed cases.

Very different results were obtained in populations sampled in Grimacco, northern Italy, and in Switzerland. In Italy the small sample of 14 only contained two VCGs, and the 11 DNA fingerprints highly correlated with them showed significantly greater similarities within the VCGs than between them. This suggests strongly that recombination was infrequent in this population, but the sample was too small for certainty (Liu *et al.* 1996). In contrast, only 3 per cent and 11 per cent self-fertilization were found in the Swiss populations where outbreeding obviously predominated (Bissegger and Milgrom 1994).

This valuable series of investigations demonstrates that, even in a single species, mating behaviour in one population is not necessarily a guide to behaviour in another, as well as indicating the complexities which can arise when a fungus is capable of both out- and inbreeding, however unusual either pattern may be. *C. parasitica* [AP] is not unique in this respect.

The fungus responsible for the recent epidemic of Dutch elm disease, *Ophiostoma novo-ulmi* [AP], now recognized as a new species, is one of the few fungi where the diversity of factors which affect sexual reproduction has been thoroughly investigated. The situation here is exceedingly complex. An important role is played by beetles (*Scolytus* spp.), which bring about the initial infection, and mites, especially *Tyrophagus putrescentiae*, which feed on mycelium in bark but leave protoperithecia and perithecia untouched and inadvertently transport conidia and ascospores on their bodies. As a result mites play an important role in fertilization (Brasier 1978). If the activity of the beetles responsible for initially infecting the elms is reduced and the activity of mites in bringing somatically and sexually compatible mycelia together is absent or reduced, mycelial growth in the bark is limited and sexual reproduction is either absent or decreased. Temperature certainly affects the beetles' movements profoundly, and mites are similarly affected by environmental conditions and parasitic attack. Thus both the extent of the disease and the effectiveness of different modes of reproduction and dispersal are climatically limited by indirect environmental effects and biotic interactions with other bark-dwelling organisms.

Outbreeding is determined not only by the MTFs but also by their interaction with the SI system as well as the presence or absence of dsRNA d-elements (Brasier, 1984). Although in the original *O. ulmi* [AP] strains the two MTFs (A and B) are present in equal numbers, this is not the case in *O. novo-ulmi* [AP] where the B type usually predominates and the A type only accounts on average for 15 per cent, although it can range from 4.1 to 55.6 per cent. Theoretically this should reduce the probability of compatible matings to about 25 per cent, but any such deficiency is compensated for by the much greater production of protoperithecia by the A type compared with their sparse production by the B type. Moreover, the absence of one MTF in any population can be compensated for by the 'pseudo-selfing' phenomenon described earlier (see section 7.2.3.2). Although that involves one round of inbreeding, it does restore both MTFs to the population when the ascospores germinate and, with that, the opportunity to outbreed again.

Mating efficiency is also affected by the rather unusual somatic incompatibility system in this fungus which appears to show a graded response related to the considerable numbers of VCG factors it carries. A range of compatible interactions is shown depending on the numbers of common VCG factors in the interacting mycelia (see Chapter 2, section 2.3.1.5). Full incompatibility (**w** reaction) is associated with the interpenetration of opposing hyphae into the incompatible mycelium, where the degree of penetration reflects the mycelial vigour of each mycelium and other unknown factors (Brasier 1984). If the interacting mycelia carry different MTFs, however, effective interpenetration will promote sexual reproduction. Indeed, 10 times as many mature perithecia were produced in fully somatically incompatible **w** reactions than in fully somatically compatible ones! Far from restricting cross-breeding, through confining sexual reproduction to somatically compatible strains, the VCG interactions in this species can positively promote it and so reduce a potential reduction in outbreeding. VCG interactions also affect the transfer of the cytoplasmic d-factors, which occurs with increasing efficiency as somatic intercompatibility increases, e.g. only a 4 per cent transfer with incompatible **w** reactions compared with almost 100 per cent in fully compatible **c** reactions. Another aspect of d-factors is their ultimate effect on reproduction. The presence of d-factors impairs growth rate, mycelial vigour and perithecial production and hence the numbers of viable ascospores. On the other hand, d-factors are apparently *not* transmitted through ascospores, so that the resultant mycelia, on germination, are fully restored to vigour, lacking d-factors. Thus an anomalous situation arises where it may actually be advantageous for two d-factor-carrying strains to cross, with the reduced ascospore output being compensated for by the improved vigour of the progeny!

Finally, the breeding behaviour differs in different regions. The advancing front of the epidemic now spreading across Europe is largely clonal, being characterized by the presence of a predominant VCG (a 'super VCG') and only one MTF, usually the B type (see Chapter 10, Table 10.2). In the areas behind the front, heterogeneity of both VCGs and mating types is reasserted. Therefore a further variable to be considered, when investigating the breeding behaviour of any fungus, is its ecological status in the communities where it occurs.

Despite the unravelling of this complex series of interactions affecting reproductive ability, it has not proved possible even in *O. novi-ulmi* [AP] to assess the relative contributions to population diversity of sexual and asexual reproduction or of outbreeding and inbreeding. Nevertheless, it provides a clear example of just how complicated it may be to unravel apparently simple breeding issues under natural conditions.

How far these two examples, *C. parasitica* and *O. novo-ulmi* [both AP], are exceptional in the complexity of their mating behaviour and how far it simply reflects the intensive studies made on them is not known. It could well be that such complexity is the norm simply awaiting investigation, rather than the exception.

### 7.3.3 In- and outbreeding: conclusions

What clearly emerges from these examples is that, although the breeding system of a fungus is the major determinant of whether it will be outbreeding or inbreeding, the genotypic composition of a population will also be influenced by other factors. The limiting case is where an isolated population of an outbreeding species is entirely clonal. Its genetic composition will not be affected by the breeding system of the species because parental genotypes will simply be re-formed at every round of mating. In addition to the genetic composition of the population, its size and distribution will similarly affect the degree of inbreeding or outbreeding, as will the dispersal of the species' spores and its mycelial growth rates. Nevertheless, the evidence available to date suggests that most fungal populations are genetically diverse or, even if locally genetically homogenous, are within dispersal

distance of genetically different populations. Therefore the breeding system is of considerable significance for population genetics, but is not necessarily fixed throughout the range of the species. Considerable variation is often possible within any one system. The relative balance between outbreeding and inbreeding is a complex matter and can vary between populations. How it is achieved will depend on the balance of sexual to asexual reproduction, the mating system, whether rigid or variable, and its expression modulated by environmental and climatic variables.

In sexually breeding Eumycota it looks as if one of the most common population patterns is likely to be reproductively mixed, where the individuals are derived both sexually, including some selfed derivatives, and clonally through asexual reproduction. The techniques used by Wang and coworkers (Wang 1977; Wang *et al.* 1977) to investigate the sexual–asexual, crossed–selfed balance in populations of *Gremmeniella abietina* [AP], described in section 7.3.1, enable quantitative estimates to be made of selfing and outcrossing. This method, described in section 7A.3 below, is equally applicable to other fungi, including dikaryotic and diploid forms, so that its extended use would provide a much clearer picture of how this aspect of fungal population structure differs both between species and in response to varying environmental conditions.

One further point needs to be made, namely that it is far from easy to know or discover in a population whether recombination is occurring or has taken place in the past. A range of tests, both observational and statistical, are now available to investigate the situation. These are described in the Appendix to this chapter. However, even when evidence is found for recombination having been involved in a population's history, its time, place and mechanism still remain to be elucidated.

## Appendix: Distinguishing between clones and recombinant populations

It is often desirable to know whether a population is clonal or whether recombination is occurring or has occurred. Unfortunately, this is not always obvious. One reason for this in fungi is the common life history pattern of a single recombinant

episode that is sandwiched between many tens or hundreds of asexual generations. Consequently, the numbers of asexually produced individuals far exceed those produced by recombination, and successive samples taken in the asexual phases may appear to show a population with an unchanging genetic constitution. In fungi exhibiting this kind of behaviour, the recombinant phase often occurs just before, or at the onset of, adverse conditions for somatic growth. Therefore at the start of a new growing period, populations of many Ascomycotina and Basidiomycotina are derived from two different sources:

• non-recombinant mitospores from individuals that have persisted from the previous growth phase
• recombinant meiospores that now germinate.

For instance, in *B. graminis* f.sp.*hordei* [AD] in the UK and most of Western Europe, early spring populations are derived mostly from over-wintered ascospores that now germinate plus a smaller number derived from mycelia which have persisted on over-wintering 'volunteer' hosts. Masses of conidia develop rapidly and initiate new asexual generations. Thus the spring and summer populations are derived from a mixture of non-recombinant genotypes representing survivors from the previous growth phase plus new recombinants derived from the ascospores. Of course, where adjacent spring and summer barley crops are grown, the genetic constitution of the mildew populations will obviously be even more complex because of the different kinds of potential inocula available.

In other cases, ascospores can germinate at any time. For example, in Britain, *Mycosphaerella graminicola* releases ascospores from October through to the following July. Recombinant genotypes can then occur virtually throughout the growing season.

The ratio of recombinant to non-recombinant genotypes in a population is, of course, also affected by the mating system of the species concerned, whether in- or outbreeding. Cases occur where the amount of genetic variation detected suggests that mutation is unlikely to be the sole cause and that recombination has occurred, although there may be no contemporaneous evidence for it. This may be a reflection of its occurrence in the past, even in contemporary anamorphic or amictic fungi (see Chapter 11). Indeed, with any taxon, whatever its apparent present state, it is difficult to demonstrate either that it has not undergone some recombinant process in the past, or that it has always been entirely clonal.

Even if it is neither possible to tell whether the genetic variation is due to contemporary or relict activity, nor to establish whether the recombinational event was mitotic or meiotic, it is sometimes possible to obtain an insight into the possible circumstances. Suggestive information concerning the period of recombination can be obtained as described later (see section 7A.6).

The likelihood of recombination can be broadly assessed, although there is a dearth of evidence for parasexuality in natural fungal populations. Indeed, details of parasexuality rely heavily on experimental investigations with laboratory strains of *Aspergillus nidulans* [FA]. If parasexuality follows the same pattern in other fungi, it is probable that there is a great difference between the frequency of mitotic and meiotic crossing.

This expectation is based on observations of mitotic recombination in a very few Ascomycotina. The only available quantitative data, using colour and auxotrophic mutants of *A.nidulans* [A : FA] (Pontecorvo 1959) gave a meiotic recombination index of $2 \times 10^{-2}$ and a mitotic index of $2 \times 10^{-5}$, i.e. mitotic recombination was a thousandfold *less* frequent. Another possible indicator of parasexuality is the occurrence of partial recombinant diploid or aneuploid nuclei, since in *A.nidulans* diploid nuclei are unstable and progressively loose chromosomes until the haploid condition is restored. Here too, the number and range of observations on the process of de- diplodization are extremely small.

Therefore, if there is evidence for recombination, it is strongly indicative of sexual reproduction at some stage of the history of the species unless there is positive support for parasexuality.

A number of properties of populations can be examined to determine whether a population sample demonstrates clonality, recombination or some evidence for both conditions (Milgroom 1995, 1996; Taylor *et al.* 1999). They include:

• direct observation of functioning reproductive structures
• the genotypic diversity *G* of a population or populations
• the extent of gametic disequilibrium *GD* in a population

• the extent of multilocus association in a population measured by the index of association $I_A$
• the length, number and resolution of trees derived from multilocus data—clonal populations show least reticulation, i.e.shorter, fewer and better resolved trees than those derived from recombining populations (the parsimony tree length permutation test (PTLPT)).

In general, no single property, taken alone, provides a decisive resolution since each property can be affected by various population processes, e.g. different causes can give rise to identical situations. Thus, if different loci are shown to be associated, this could be due to close physical linkage, cloning or the chance selection for more than one locus.

Therefore a consensus between different tests is highly desirable before favouring any specific hypothesis.

## 7A.1 Direct observation of reproduction in fungi

Observation is a qualitative measure that is not always easy to make in nature. A variety of situations make direct observation of reproduction difficult, even when it does occur.

**1. Pleomorphism when the teleomorphic phase has gone unrecognized** A classic example is *Septoria tritici* [FA] which was first described in 1843, although its teleomorph *M. graminicola* [AD] was not recognized and correctly named until 133 years later, in 1976, on cereal stubble in New Zealand (Sanderson 1972, 1976). In the UK, the significance of the sexual phase as a source of inoculum was only appreciated in the mid-1980s (Shaw and Royle 1989), and in the USA recombinants were only identified, using RFLPs, in 1989 (McDonald and Martinez 1990a).
**2. A totally unknown sexual phase although evidence may exist suggesting a recombination phase** *Coccidioides immitis* [FA], which is responsible for a human lung infection, was assumed to be entirely asexual. However, each of 30 isolates derived from 25 patients in a single outbreak in Tucson, Arizona, were genotypically distinct and

the probability that each was unique was very high. This suggested that some kind of 'cryptic' sexuality was involved (Burt *et al.* 1996). The likelihood became even more probable when isolates from two further outbreaks in California and Texas, some 400 miles from Tucson, provided a further range of distinct genotypes except that one that was common to California and Texas. (Later it was found that this Texan patient had, in fact, contracted the infection in California!) In contrast, successive isolations from individual patients always possessed the same genotype as that originally recorded. These observations suggest that this fungus does indeed undergo recombination, possibly in an unknown saprotrophic soil phase, although there is little migration between recombining populations, but once a patient is infected it remains clonal until released into the soil again.
**3. When a fungus appears clonal although potentially outbreeding** An excellent example of this situation is the rice blast fungus *Magnaporthe grisea* [AP] where populations in many areas appears to comprise a few clones only (Levy *et al.* 1993; Chen *et al.* 1995; Roumen *et al.* 1997). Moreover, sexual recombination is improbable over much of its natural range because of widespread male and female sterility and inequality of compatible mating types. Nevertheless, sexual recombination can readily be demonstrated under laboratory conditions and almost certainly occurs in some natural populations in Uttar Pradesh in the Indian Himalayas, the probable centre of diversity of its hosts. Genetic evidence of recombination followed by clonal propagation was demonstrated in these populations and sexually fertile and hermaphrodite strains were isolated (Kumar *et al.* 1999). Elsewhere, evidence has also been found of mitotic recombination in some natural strains (Zeigler *et al.* 1997).

Observation of apparently gametangial or other reproductive structures are, by themselves, at best suggestive but not conclusive, although they do suggest that sexual reproduction is possible or has been in the past. For example, *Sclerotium cepivorum* [FA] appears to be clonal but produces abundant microconidia comparable with 'male' fertilizing agents in other Ascomycotina. It was probably capable of sexual reproduction in the past (Couch and Kohn 2000) (see also section 7A.6).

An excellent example of the identification of sexual reproduction by direct observation is provided by the straminipilous fungus *Phytophthora infestans* [O]. From the original introduction of this pathogen to Europe over 150 years ago, when existing strains showed great similarity, specific matching pathotypes arose as new potato varieties were introduced. All carried the same MTF, *mat A1* (Fry *et al.* 1992) and no good evidence existed for recombination in nature. In contrast, within a few years of the introduction into Europe of *mat A2*, probably in the late 1980s, populations were found with both MTFs, fertile oospores and a great variety of novel genotypes (Drenth *et al.* 1994) (see section 7.1.1).

## 7A.2 Comparisons of genetic diversity

A test for the randomness of mating can be made by comparing gene frequencies, although allowance has to be made both for the independent or non-independent inheritance of the loci used and for the even or uneven distribution of alleles being compared. Multilocus haplotypes would be expected to occur with *equal frequencies* if a population is *mating at random*, provided that the loci which are being compared are inherited independently and the allele frequencies are evenly distributed at each locus. The former requirement can be tested for with known independent loci, but the latter requirement is less likely to be met. In that case, the probability $P$ that two isolates selected at random will share the same multilocus haplotype is given by (Keller *et al.* 1997)

$$P = \prod_{l=1}^{n} J_{il}$$

where $J_i$ is Nei's genetic identity, I (see section 5.1.4) for each of $l$ loci. Any departure indicates a departure from random mating. This of itself does not discriminate between recombination and clonality but is a useful indicator which needs to be associated with other tests.

## 7A.3 Comparisons of genotypic diversity

In fungi as in other organisms, the greater the degree of outbreeding in a population, the greater is the number of multilocus genotypes present, so that, in the limit, no two genotypes would be identical. Inbreeders, in contrast, would be expected to show more limited genotypic diversity, whilst clonally multiplied or amictic fungi would be expected to show none at all. The simplest way to investigate the situation is to determine and compare the index of genotypic diversity $G$ of different populations (Stoddart and Taylor 1988), where $G = 1/p_i^2$ and and $p_i^2$ is the frequency of the $i$th multilocus genotype. The significance of the variance of $G$ can be tested either by estimating a $t$-statistic (Stoddart and Taylor 1988, or by randomization of the data to generate a null statistic against which the value of $G$ can be compared (Peever and Milgroom 1994).

The following slight modification is necessary if anonymous RFLP markers or repetitive sequences are used:

$$G = \frac{1}{\sum_{x=0}^{n} [f_x(x/n^2)]}$$

where $n$ is the size of the sample and $f_x$ is the number of genotypes observed $x$ times in the sample, i.e. either the number of different fingerprints in the sample or the number of multilocus haplotypes. The maximum possible value for $G$ is the number of individuals in the sample, and to obtain the percentage maximum diversity, $G$ is divided by $n$ (McDonald *et al.* 1994). However, the use of genotypic diversity is at best indicative, since a contemporary population can be clonal although having had a history involving recombination, e.g. *Sclerotium cepivorum* (Couch and Kohn 2000).

Even when identical genotypes occur in a population in close proximity, suggesting clonality, it may be desirable to test whether they have arisen by selfing or simply represent the same genotype produced by different sexual events at different times. Such a situation arose in an isolated population of *Gremmeniella abietina* [AD]. This species is dimictic and produces conidia (Uotila 1992), and so both sexual and asexual progeny can be produced

(Wang *et al.* 1997). There was strong prima facie evidence for the occurrence of clones. Genotypes which had occurred more than once in the population were either on the same tree or on the same branch; none occurred on trees more than 5 m apart. This spatial pattern of repeated genotypes suggested that they were asexually produced clones, with dispersal of conidia being limited to a few metres at most. However, conceivably, identical genotypes could have resulted from sexual reproduction or inbreeding due to the limited size of the population.

These possibilities were tested using two measures $P_{gen}$ and $P_{se}$ to determine the origin of identical genotypes (Wang *et al.* 1997). $P_{gen}$ provides an estimate of the probability that consecutive samples which originated from different sexual events would, by chance, have the same multilocus genotype; $P_{se}$ estimates the probability that different sexually produced isolate samples somewhere in the population would, nevertheless, posses the same multilocus genotype. Then

$$P_{gen} = \prod p_i$$

where $p_i$ is the frequency of the $i$th allele in the population and

$$P_{se} = 1 - (1 - P_{gen})^G$$

where $G$ is the number of distinct genotypes in the population.

In this particular case the values obtained for both these quantities were exceedingly low in virtually every case. Since *G. abietina* [AD] is potentially outbreeding and the evidence for the occurrence of VCGs, which could reduce outcrossing to some extent is unconvincing (Furnier *et al.* 1984), the repeated genotypes are, indeed, most likely to be clones derived through asexual reproduction.

## 7A.4 Gametic disequilibrium

In a randomly outbreeding population, alleles at unlinked loci are *randomly assorted*, because of independent assortment. This contrasts with clonal or a strictly inbreeding sexual population; the latter is an unusual situation in fungi despite the numerous homothallic species. When clonality does occur, alleles at all loci, even those on different chromosomes, remain *closely associated* because of the lack of recombination. Thus the degree of non-random association of alleles at different loci can be used as an indicator of clonality. In diploids this can best be determined at the gametic phase so that this measure has been termed 'gametic phase disequilibrium' or, more simply, **gametic disequilibrium** (GD). Less precisely, GD is also termed 'linkage disequilibrium' because the rate of approach to random association in a sexually reproducing organism is reduced by linkage. The degree of deviation $D$ from non-random association between loci is used to measure GD, with the null hypothesis being that $D = 0$, i.e. that the loci under test *are* associated. Methods for calculating $D$ and testing its significance are described by Brown *et al.* (1980), Milgroom (1995) and Weir (1996b).

For diploid fungi, such as *Phytophthora* [O] (Tooley *et al.* 1985), the simplest way to test for a departure from random mating is to test for a departure from Hardy–Weinberg equilibrium at individual loci. The extent of any deviation from equilibrium, in particular a reduction in heterozygotes compared with those expected, can be used to estimate the inbreeding coefficient $F$. Prior to the spread of *mat A2*, populations from Central Mexico tested using allozyme markers were consistent with the Hardy–Weinberg equilibrium. Those from Canadian and North American populations were not. Thus the sexual nature of the Mexican populations, possessing both MTFs, was confirmed as was the asexual nature of those populations that possessed only the single MTF *mat A1*.

Haploid Eumycota, which lack heterozygosity, have to be treated differently. Suppose that the frequencies of any two alleles at different loci are $u$ and $v$. Then $D$ is given by

$$D_{uv} = p_{uv} - p_u p_v$$

where $p_{uv}$ is the gametic frequency and $p_u$ and $p_v$ are the observed frequencies at the two loci. Its significance can be tested by using a chi-square test with one degree of freedom:

$$\chi^2_{uv} = \frac{n D_{uv}}{p_u(1 - p_u)p_v(1 - p_v)}$$

where $D_{uv}$ is the maximum likelihood estimator for the disequilibrium coefficient between alleles $u$ and $v$, and $p_u$ and $p_v$ are the allele frequencies observed at the loci. Fisher's exact test of independence (McDonald et al. 1994; Weir 1996b) should be applied to any pairs of alleles that show a significant departure from random expectation.

Estimates of $D$ are not without their technical problems, however, because its value is affected by both statistical and biological constraints (Lewontin 1988). The value of $D$ alone can only be taken as an indication of clonality or sexual recombination.

Obviously, linked loci will bias the estimate of $D$ and the absence of linkage between the loci under test should be established prior to estimating $D$. The statistical significance of $D$ is also affected by sample size (Brown 1975) so that testing the null hypothesis that $D = 0$ is rather ineffective with small samples, and with large numbers of loci or alleles there are several estimates of $D$ and hence ambiguity. In practice, intermediate allelic frequencies and a sample size greater than 100 are probably optimal for estimating $D$. One way of avoiding these problems, since the value of $D$ depends on the actual gene frequencies, is to use an alternative measure, $D'$, namely

$$D' = D/D_{\max}$$

where $D_{\max}$ is the maximum $D$ possible for a set of allelic frequencies at the two loci (Lewontin 1988). This variant plus the use of Fisher's exact test of significance can give very different results from the simple test of the value of $D$.

An excellent example is provided by *Stagnospora nodorum* [FA] (McDonald et al. 1994), where both $D$ and $D'$ were estimated using pairs of loci derived from RFLPs of isolates from a single field of infected wheat. When all the isolates were used, but only one example from any identical clone was included, the non-random proportion was 7 per cent, but when Fisher's exact test of significance was applied none were significant. Since *S. nodorum* has a teleomorphic phase, *Leptosphaeria nodorum*, this is not entirely surprising. However, identical clones often occurred close together although scattered throughout the field and this, presumably, reflects asexual multiplication through

conidia. However, even when $D$ was estimated for the whole sample, the non-random proportion, using the exact test, only rose to 6 per cent. Thus the majority of the population (94 per cent) was derived by sexual reproduction, presumably by ascospore infection.

GD has been used to determine the relative contributions of recombinant and non-recombinant reproduction in a variety of fungi, including *Mycosphaerella graminicola* [AP] (McDonald et al. 1995; Zhan et al. 1999). These studies are highly informative because they demonstrate the importance of the sampling techniques employed. Ninety-three isolates were taken from 19 infected wheat leaves in a Californian population, an average of nearly five isolates per leaf. Using RFLPs as markers, many turned out to be genetically identical and 22 genotypes were recognized in all. Genetic disequilibrium was estimated between all 66 possible pairwise comparisons of alleles at each locus and for the 718 allele-by-allele comparisons in two ways:

- for all 93 isolates as sampled
- for 22 'clone-corrected' samples.

In the latter, genotypes found twice or more were only included *once* in the analysis in order to remove any possible bias due to spurious associations between loci resulting from sampling the same clone several times.

In the non-corrected sample, 50 (76 per cent) of the pairwise allele comparisons and 298 (42 per cent) of the allele-by-allele comparisons were statistically significant, suggesting a high degree of *asexual* reproduction. Even when Fisher's more exact test of significance (Weir 1996b) was applied, 27 per cent of the latter comparisons were still significant. However, when the clone-corrected data were used, only 12 per cent and 9 per cent were found to be significant and using Fisher's test this reduced to 0.5 per cent, i.e. the population was, in fact, *predominantly outbreeding* and only a very small proportion could have been derived from non-recombinant asexual conidia. This result compares favourably with the outbreeding behaviour found in isolates from Oregon where only single samples were taken from each infected plant. Therefore it is always desirable when estimating

GD to take only single samples from each site or, more generally, always to use clone corrected samples.

In addition to sampling and statistical problems, estimates of GD have to be used with caution for biological reasons, since clonality is not the only possible interpretation. For instance, if the sample examined is derived from several geographically or ecologically isolated sub-populations between which recombination occurs only very rarely or not at all, it may show GD even if there is appreciable panmixia within each sub-population. Selection which increases fitness, such as that in fungi due to epistatic interactions between loci, or genetic drift may also result in apparent GD. It has also been suggested that if samples are taken during a sudden localized epidemic situation, the data may be interpreted as demonstrating an apparent GD. This will be the case if a particular recombinant genotype is first highly selected and the organism then increases rapidly and asexually. At sampling, the asexually produced progeny are most likely to exceed by far the numbers of sexually produced genotypes until normal equilibrium is restored and, indeed, recombinants may never be detected (Maynard-Smith *et al.* 1993). The predominance of one or a few genotypes found in summer populations in Britain of *B. graminis* f.sp.*hordei* [A] is a possible example of this and there is a drop in the value of *D* between July and October estimates. This is correlated with an estimated increase to about 24 per cent of the isolates derived from ascospores and hence recombination in the autumn, when perithecia have formed (Brown and Wolfe 1990).

One final twist to the interpretation of 'clonal' reproduction is provided by data from *Cryphonectria parasitica* [AP] where some 'clonal' reproduction in the field is due to self-fertilization in a sexual but homomictic fungus. This was demonstrated by isolating the progeny of 22 perithecia. Only 16 (73 per cent) showed any segregation; the rest (27 per cent) showed no segregation and were genotypically identical with their progenitors. Since clonal reproduction through conidia can also occur in this fungus, if the population had been sampled without this knowledge it might have been concluded that it had a more complex origin including both recombinant ascospores and asexually produced conidia.

## 7A.5 The index of association

An alternative approach, originally applied to wild populations of barley (Brown *et al.* 1980) is a summary statistic, the index of association $I_A$. It avoids problems of sample size and any ambiguities arising from large numbers of pairwise comparisons. It has been adapted for use with haploid bacteria and fungi, e.g. bacteria (Maynard-Smith *et al.* 1993), *Pyrenophora teres* [AP] (Peever and Milgroom 1994), and *Coccidioides immitis* (Burt *et al.* 1996). $I_A$ uses a measure $K$ of the amount of multilocus association in a haploid population assessed by the number of loci that are different when two individuals are compared. In a diploid or dikaryotic fungus, $K$ is the number of heterozygous loci. The differences between a recombinant population and a clonal one are as follows.

- In a **recombinant** population genetic distance is expected to be distributed normally with its variance $s_K^2$ being equivalent to the variance of a normal distribution, i.e. $s_K^2 = \sigma K^2$.
- In a **clonal** population the distribution will be skewed and show an excess of extreme distances so that $s_K^2 > \sigma K^2$.

The **observed variance** is the distribution obtained by comparing each individual with every other individual in the sample. The values of the **expected variance** are derived by obtaining a $1000+$ times bootstrapped value of the observed data to provide a null value with which the observed variance is compared. A difference is regarded as significant when

$$s_K^2 = 100(1 - \alpha)\% \, \sigma_K^2.$$

where $\alpha$ is an acceptable type I error. Alternative methods for testing the significance of a difference are given by Brown *et al.* (1980), Milgroom (1995) and Maynard-Smith *et al.* (1993). In general, in the *absence* of recombination $I_A > 0$.

## 7A.6 The parsimony tree permutation test

A further method, originally developed to test whether bacteria were recombinant (Guttman and Dykhuizen 1994), is to construct phylogenetic trees based on different genes and see if the trees are congruent. This method has been further developed by Taylor and colleagues (Burt *et al.* 1996; Taylor *et al.* 1999) as a parsimony tree length permutation test (PTLPT) using data from *unlinked* polymorphic loci. The basic assumption is that *all* regions of the genome will be inherited similarly in clones, i.e. the gene genealogies from two unlinked loci will be *congruent*, but in populations which show recombination they are unlikely to be congruent. Therefore, using sequence data, a tree is calculated by applying parsimony to the sample data.

If the sample is *clonal* a number of expectations will be realized:

• either a *single* tree or a few short *well-resolved* trees will be obtained since clones will, in effect, appear to evolve cladistically
• a tree derived from the combined data sets of two or more loci will *not* be longer than the sum of the lengths for each locus alone.

However, if *recombination* has occurred the expectations are as follows:

• many *longer poorly resolved* trees are likely to be obtained because of the reticulating effect of recombination
• a consensus tree derived by combining the data will appear 'comb-like' not branching.

These latter expectations were met when Burt *et al.* (1996) investigated the supposedly asexual *Coccidioides immitis* [FA]; the consensus tree based on the 62 most parsimonious trees was almost completely unresolved, with only a single bifurcated branch. This was the first unequivocal demonstration of the occurrence of recombination in a presumed exclusively anamorphic fungus.

Another excellent example is provided by some Canadian populations of *Sclerotium cepivorum* (Couch and Kohn 2000). Using VCGs, sequence polymorphisms and cDNA fingerprints, only three clones appeared to be present and each field population was predominantly a single clone. However, when the procedure just described was applied using six variable loci and the VCG groups, clear evidence was obtained of recombination (Figure 7A.1). The five most parsimonious,

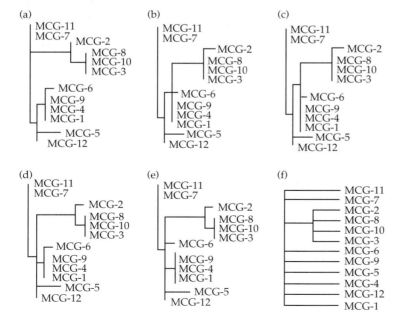

**Figure 7A.1** Diagrams to illustrate the five most parsimonious trees (a–e) based on five loci and the combined tree (f) derived from all the loci examined in *Sclerotium cepivorum* [FA] Trees a–e represent the five clonal, most parsimonious trees relating the 12 mycelial compatibility groups (MCG) found in the samples. Tree f is the derived strict-consensus combined tree and its 'comb-like' appearance suggests that recombination has occurred at some time in the genealogies of the five clones. (Reproduced from Couch and Kohn (2000), with permission from APS).

non-congruent trees (A–E) derived from the data are shown together with the 'comb-like' strict consensus tree inferred from combining all the data (E). The interpretation based on these findings was that onions infected with a few clones of *S. cepivorum* had recently been introduced to the Canadian sites and, since the contemporary fungus lacks sexuality, they had persisted in their clonal state. However, at some time in the past, presumably elsewhere, the fungus had undergone recombination, hence the evidence from the phylogenetic trees based on the consensus genealogical tree.

## 7A.7 Summary

It will be apparent that the determination of the breeding behaviour of a population is difficult to investigate directly and frequently both sampling and statistical tests for in- or outbreeding, or the balance between them, is complex to carry out and interpret. Highly suggestive evidence for recombination is found with increasing frequency in fungi where, hitherto, it was not expected. In many cases the recombination detected is likely to have been a past event, e.g. the *S. cepivorum* [FA] clones in Canada referred to above, or even a contemporary, but undetected, event as was possible with *C. immitis* [FA/A?]. When recombination is detected, none of the available methods, apart from resampling at time intervals, provides unambiguous evidence to determine whether or

not it is a *contemporary* process. Even this, as discussed in section 7A.1, may be difficult to demonstrate with certainty.

In general, therefore, the more that is known or can be discovered about the current physiology and ecology of the reproductive biology of a fungus, the easier it is to interpret statistical tests on samples designed to investigate the genetic basis for recombination and its significance.

## Further reading

Burt A, Carter DA, Koenig GL, White TJ and Taylor JW (1996). Molecular markers reveal cryptic sex in the human pathogen *Coccidioides immitis*. *Proceedings of the National Academy of Sciences of the United States of America*, **93**, 770–3.

Elliott CG (1994). *Reproduction in fungi: genetical and physiological aspects*. Chapman & Hall, London.

Leslie JF and Klein KK (1996). Female fertility and mating type effects on effective population size and evolution in filamentous fungi. *Genetics*, **144**, 557–67.

Maynard-Smith J, Smith, NH, O'Rourke M and Spratt BG (1993). How clonal are bacteria ? *Proceedings of the National Academy of Sciences of the United States of America*, **90**, 4384–8.

Nei M (1987). *Molecular evolutionary genetics*. Columbia University Press, New York.

Petersen RH (1995). There's more to a mushroom than meets the eye: mating studies in the Agaricales. *Mycologia*, **87**, 1–17.

Raper JR (1966). *Genetics of sexuality in higher fungi*. Ronald Press, New York.

# CHAPTER 8

# The generation of variation—III Recombination through heterokaryosis, parasexuality, hybridization and polyploidy

Hybridization and polyploidy are not uncommon amongst diploid Eukaryota, although the significance accorded to them as evolutionary processes continues to be debatable, but these processes are poorly documented in fungi. In contrast, heterokaryosis and parasexuality are much less common in diploid Eukaryota but well documented in Eumycota where they are seen to be a reflection of the structure and biology of the fungal mycelium. However, all four of these processes in fungi are far better known and studied in the laboratory than in natural conditions where the question of how they affect populations of both Eumycota and Straminipila is still far from clear.

In this chapter, because heterokaryosis and parasexuality are essentially intraspecific phenomena they will be considered before hybridization and polyploidy which effectively broaden the recombinational base through being mainly interspecific processes.

## 8.1 Heterokaryosis and parasexuality

Heterokaryosis—the occurrence of genetically different nuclei within a mycelium—can arise either by mutation in a resident nucleus or as a result of hyphal anastomosis followed by one way migration or exchange of genetically distinct nuclei. A variety of different kinds of heterokaryon occur in fungi reflecting both the structure of the mycelium and its compartmentation and the genetically different kinds of nuclei that can occur in a heterokaryon (Figure 8.1).

Once established, a heterokaryon is more effective than heterozygosis in 'sheltering' genetic variation since, potentially, all possible nuclear ratios can occur from zero to 100 per cent, rather than simply three alternative genotypes (*AA*, *Aa* and *aa*) as in diploids. Moreover, a heterokaryon can dissociate to give homokaryons, so 'releasing' the hidden variation.

Dikaryons of outbreeding Basidiomycotina are the one example where heterokaryosis, in the form of the dikaryon, occurs regularly; in this case the nuclei differ minimally in their MTFs. Dikaryons also demonstrate clearly that heterokaryosis does not necessarily result in parasexual recombination, although it provides a necessary condition for it to take place.

Parasexuality further requires the chance formation of a diploid nucleus which subsequently undergoes, equally by chance, mitotic recombination and haploidization when the chromosomes segregate (Pontecorvo 1959). What determines the formation of diploid nuclei within hyphae and for how long they persist is simply not known. Although the mechanics of mitotic recombination is becoming increasingly explicable in the light of increased understanding of the recombination process, the causes of haploidization—the dissociation and successive loss of chromosomes until the haploid condition is restored—are still obscure.

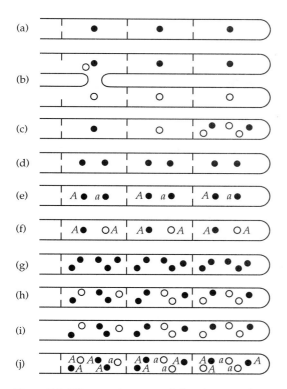

**Figure 8.1** Different nuclear patterns in homokaryons and heterokaryons: (a) monokaryotic homokaryon; (b) transient heterokaryon (e.g. *Verticillium dahliae* [FA]); (c) apical heterokaryon, in which distal homokaryotic compartments are isolated by complete septa (e.g. *Fusarium oxysporum* [FA]); (d) dikaryotic homokaryon; (e) MTF dikaryon, in which each compartment carries two nuclei each with one MTF; (f) heterokaryotic dikaryon, homokaryotic for MTFs; (g) multinucleate homokaryon (e.g. *Neurospora crassa* [AP]); (h) multinucleate heterokaryon (e.g. *N. crassa* [AP]); (i) coenocytic multinucleate heterokaryon (e.g. *Mucor* spp. [Z]); (j) multinucleate heterokaryon, also heterokaryotic for MTFs (e.g. *Neurospora tetrasperma* [AP]). (From Burnett (1975).)

At one time, heterokaryosis and parasexuality appeared to provide a plausible explanation for variation in Fungi Anamorphici or other fungi with impaired sexual reproduction. Clearly, if they were frequent phenomena in fungal populations, they would increase the potential amount of recombination although it was thought that the contribution of these mechanisms was likely to be minor when compared with meiotic recombination. Possible exceptions to this are when sexual reproduction is apparently absent or greatly reduced, as in some Fungi Anamorphici and the Glomales, or where

either genetical or environmental conditions have affected sexual reproduction adversely.

The best direct evidence for natural heterokaryons depends upon their isolation and laboratory investigation, but good circumstantial evidence comes from cases where parasexuality has been demonstrated since, inevitably, it involves prior heterokaryosis.

A current widespread view is that heterokaryosis and parasexuality play little part in natural situations, but this belief needs to be accepted with caution for three reasons.

**1.** The fact that both heterokaryosis and parasexuality can be demonstrated so easily in *experimental* situations in almost *any* fungal species suggests that they also occur in nature, the limiting factor being inadequate methods for their detection. Therefore their occurrence should not be discounted when assessing natural situations until more sensitive methods for detecting these rare events have been developed.

**2.** Mitotic crossing-over, when detected in *experimental* situations, occurs at far lower frequencies than meiotic crossing-over, so that such rare recombinants can easily be missed in naturally occurring situations, or, if only some recombinants are recovered, mistaken for mutants. If the spontaneous mutational pattern of a fungus under investigation were known, allowance could be made for this possible confusion. Unfortunately, there is still a dearth of such information from natural populations.

**3.** It is relatively easy to detect heterokaryosis and parasexual processes in experiments using strains with well-defined markers and selective methods which favour the detection of progeny that have undergone recombination. Prior to the introduction of linked molecular markers it was difficult to detect parasexuality in natural isolates with the kinds of markers available because its detection requires precise knowledge of both gene location and allele frequencies in populations. In addition, adequate information concerning the causes and range of a population's variation, such as its natural mutability, the presence, behaviour and consequences of transposons and the occurrence of dispersed repetitive DNA sequences, is usually not known. The situation should improve with increased use of molecular markers, with relatively detailed and accurate genetic maps including information on genome organization and with reliable data on allele frequencies in natural populations.

### 8.1.1 Heterokaryosis

In a technical sense, heterokaryosis is inevitable in virtually every fungus at some time. Mycelia contain thousands, perhaps millions, of nuclei in perennial mycelia, in some of which a gene or genes must, by chance, undergo mutation and establish, even if in one fungal compartment only, a heterokaryotic condition. This situation may arise by chance in any fungus. Of course, a mutant nucleus could be selected against and lost but, provided that the mutant is selectively neutral or even advantageous, its numbers could remain low or even increase slightly. A heterokaryotic region would then develop in the mycelium and if perpetuated could come to occupy hyphal apices or sporangial initials. In such cases the perpetuation and dispersal of an heterokaryon would be greatly increased. Evidence from synthetic heterokaryons demonstrates that apical regions are, indeed, potentially heterokaryotic. Both diploid and haploid nuclei in *Aspergillus nidulans* [A], revealed by fluorescent staining, occur in the multinucleate apical cells of synthetic heterokaryons. Indeed, both types of nuclei also occur in the other compartments distal to the apex (Clutterbuck and Roper 1966).

In contrast, the components of a heterokaryon become dissociated if an apex becomes occupied predominantly or wholly by a single nuclear type. This could reflect either chance or selection acting differentially on the nuclei (see Chapter 9, section 9.3.3).

#### 8.1.1.1 Heterokaryosis in Mucorales
Heterokaryosis is a possible but underrated source of variation in Mucorales. By isolating sporangiospores over many (more than 40) successive asexual generations from a single strain of *Mucor mucedo* [Z], seven morphologically distinct strains

were isolated which retained their characteristics over subsequent subcultures (Köhler 1935). Since somatic hyphal fusions are unknown in the Mucorales, presumably spontaneous mutants had arisen and persisted in the coenocytic mycelium to form a multiple heterokaryon. The heterokaryon possessed the normal appearance of this species so that the mutants were recessive. They could be propagated by multinucleate sporangiospores.

Another possible origin for the heterokaryotic state in mucoraceous fungi is abnormal nuclear segregation after meiosis. When the initially multinucleated zygote germinates to form a germ sporangium, more than one nucleus can enter the sporangium, and subsequently the germ spores, in some species. Consequently, heterokaryons could and do develop. In an appreciable number of cases germ spores give rise to mycelia heterokaryotic for mating type, readily detectable by their bright orange colour due to abnormal carotene production, e.g. *Phycomyces nitens* [Z] (Burgeff 1924). These heterokaryons are usually unstable, and normal sectors, homokaryotic for mating type, develop. Other undetected morphological or physiological heterokaryons could be generated and perpetuated in the same way.

It does not seem improbable that heterokaryosis is a common and important feature in the Mucorales where sexual reproduction appears to be, at best, erratic (see Chapter 7, section 7.2.2).

#### 8.1.1.2 Heterokaryosis in Glomales
Heterokaryosis is believed to play an important role in determining the variability of the Glomales. More than one ITS sequence has been obtained from a single multinucleate spore. This suggests strongly that some of the spores, at least, are heterokaryotic (Sanders *et al.* 1995; Lloyd-Macgilp *et al.* 1996; Hijri *et al.* 1999). However, for several reasons, it is technically difficult to work with certainty on a single glomalean mycelium. Therefore caution is necessary in interpreting the claims for heterokaryosis.

The Glomales are the most widespread obligately endotrophic mycorrhizic fungi in the world. They can only be cultured independently for a very short period and cannot easily be maintained in pure culture. They occur in the underground parts of Bryophyta and Pteridophyta, in the roots of Gymnosperms and in at least 80 per cent of all contemporary Angiosperms (Smith and Read 1997). Indeed, they have been described as 'universal' mycorrhiza because of their virtual lack of host specificity, something that has extended apparently throughout their fossil and recent record, i.e. 450–500 myr (Pirozynski and Dalpé 1989).

Their hyphae are coenocytic and multinucleate and, although the mycelium is characteristically intra-radical, it also extends into the surrounding soil as extra-radical hyphae. Unusually for Zygomycotina, intra-mycelial hyphal fusions are not uncommon between hyphae in either location (Tommerup 1988). They are particularly abundant in the extra-radical fan-like mycelium of *Glomus mosseae* [Z] in culture (Giovannetti *et al.* 2001). Anastomoses between hyphae derived from spores from different mycelia of the *same* species at germination have been recorded under laboratory conditions but with much lower frequency (Giovannetti *et al.* 1999). However, it is not known if any mycelia are heterokaryotic. There is no evidence for inter-specific anastomoses.

With one possible exception, *Gigaspora decipiens* [Z], most Glomales reproduce solely by means of relatively large terminal or lateral spores, although some species produce sporocarps in which the spores are borne. The spore wall, often of several distinguishable layers variously sculpted and usually thick, is frequently characteristic of the species. Depending on the species and spore size, each spore possess hundreds or even thousands of nuclei, not all of which enter the germ hyphae at germination (Burggraaf and Beringer 1989). Spores can remain dormant for long periods and germination can be induced by the presence of a compatible host root.

The sole claim for sexual reproduction is in *G. decipiens* [Z]. It has only been described briefly and has never been confirmed (Tommerup 1988). 'Suspensors' and a 'zygote', somewhat reminiscent of those developed by Mucorales, were found on mycelia of two strains in pot cultures, in natural habitats and under aseptic experimental conditions *in vitro*. In addition, 'azygospores' were described, morphologically different from the normal spores of this species, and these are claimed to occur

naturally on other species of *Gigaspora* [Z] and *Acaulospora* [Z] (Tommerup and Sivasithamparam 1990). None of these structures have been described by other workers. The current prevailing view is that glomalean fungi are entirely *asexual*, as far as morphological evidence is concerned (Sanders 1999).

With the application of molecular markers, considerable intra-specific genetical variation has been detected that is most readily explained by heterokaryosis. However, there are two major experimental difficulties.

**1.** Many host plants have been shown to be infected by more than one glomalean species or isolate and successful infection is usually only achieved by exposing roots to several spores. Thus it is uncertain whether isolated spores all come from the same mycelium and whether all spores are genetically alike, even if originating from one mycelium.

**2.** Because of the complexity of wall structure and location, they are often colonized by other fungi whose DNA may contaminate the supposedly glomalean sample during isolation. However, this can be detected in isolated DNA.

Isozyme polymorphisms have been detected in several glomalean fungi, although care needs to be taken to compare samples taken at the same developmental stage, e.g. allozymes detected in intra-radical hyphae differ from those in spores (Hepper *et al.* 1986). They have enabled the Glomales to be characterized unambiguously as haploid fungi (Rosendahl and Sen 1992). Altogether, isozyme polymorphisms have proved useful for demonstrating intraspecific variation, but intraspecific variation has not been detected.

More success in detecting intraspecific variation in Glomales has resulted from the use of DNA techniques. The number of copies of rDNA in the Glomales is low compared with other fungi having a similar sized genome, e.g. 10–75+ copies in *S. castanea* [Z] compared with *c*.140+ copies in *Saccharomyces cerevisieae* [A] or 220 in *Neurospora crassa* [A] (Hosney *et al.* 1999). The entire rDNA sequence has been determined for *Glomus mosseae* [Z] and *Scutellospora castanea* [Z] (Franken and Gianninazzi-Pearson 1996), and it is evident that the Glomales possess unique sequences enabling them to be separated from other fungi. This helps to exclude contaminant DNA.

*Two* distinct homologous sequences were detected from a *single* spore of *G. mosseae* [Z] (Sanders *et al.* 1995). All the 22 differential sites in the sequence could have arisen by a combination of substitutions or insertion/deletion changes in a nucleotide sequence. This is presumptive evidence for the spore being heterokaryotic and, since several nuclei enter each spore, for the mycelium being heterokaryotic also. Striking support for this possibility has been provided from other glomalean fungi.

First, starting with the contents of a single spore of *Scutellospora castanea* [Z] (*c.* 800 nuclei), successive dilutions reduced the volume to a level where, on a Poisson distribution, the majority would be expected to include material from only a *single* nucleus. PCR preparations of the ITS1 and the ITS4 regions were made, in all, from the dilutions from 24 spores. ITS isolates of five different lengths segregated among the different fractions, indicating the likely occurrence of genetically different nuclei *within the same spore*. The differences detected with *G. mosseae* [Z] suggested nucleotide substitutions, insertions or deletions. Comparison of the results from different single spores showed that ITS length variants were present or absent at different frequencies in each spore and that presences or absences were apparently independent of each other, i.e. there was no evidence for linkage between them (Hijri *et al.* 1999).

Second, ITS polymorphism within a single spore, similar to that demonstrated in *S. castanea* [Z], was also demonstrated in *Gigaspora margarita* [Z] (Lanfranco *et al.* 1999). In addition, using a specific M13 mini-satellite PCR primer, intra-sporal genetic variation in mini-satellite numbers was demonstrated in *Gigaspora margarita* [Z] (*c.* 2000 nuclei) by comparing the patterns obtained in first-generation spores following infection of clover by a single spore (Zézé *et al.* 1997). This suggests either multiple mutations in a mycelium derived from an originally homokaryotic spore or, more plausibly, segregation from an originally heterokaryotic spore and its derived mycelium.

Lastly, high levels of variation were detected in the rDNA sequences from single spores of *Acaulospora colossica* [Z] isolated in close proximity in a small soil sample (Pringle *et al.* 2000). Using ITS primers 1 and 4 followed by reamplification, the ITS regions were sampled and, in most cases, single DNA bands differing in size were recovered. Because of their heterogeneity, they could not be sequenced until they had been cloned. They then showed great variation, but it is not clear whether the heterogeneity of such single-spore rDNA is due to the presence of different copies in a nucleus, the presence of genetically different nuclei in the spore or even both conditions. However, rDNA consists of sequences in tandem repeats, and such sequences rarely undergo independent evolution but usually evolve in concert. Differences between sequences are generally 'ironed out', either by limited meiotic recombination or through a similar result achieved by 'concerted evolution', i.e. a combination of biased gene conversion and unequal crossing-over (Zimmer *et al.* 1980). Therefore these data suggest that *neither* meiosis *nor* concerted evolution occurs in this fungus and supports the view that these fungi are exclusively asexual. The observed variability could reflect the persistence of chance uncorrected mutations in the rDNA sequences, whether in the same or in different nuclei within a spore. If the Glomales are heterokaryotic, its origin is unknown although accumulated mutations, just as in *Mucor mucedo* [Z] (see section 8.1.1), are always possible and intramycelial anastomosis should not be discounted. Because several nuclei enter spore initials in the Glomales, populations of genetically different nuclei could enter spores at differentiation if the mycelia are heterokaryotic. Further segregation could then occur at spore germination, since it appears that not all spore nuclei enter the germ hyphae.

Therefore a contemporary working hypothesis is that glomalean populations appear to be maintained as a mixture of homokaryons and constant but potentially changeable heterokaryons. Segregation in the latter is determined by the chance distribution of nuclei at spore formation and germination. If, indeed, their populations have been maintained effectively in this asexual manner over hundreds of millions of years, the Glomales will occupy a unique position in the fungi, but, on the evidence available at present, rare stochastic sexual or parasexual episodes, however unlikely, cannot be wholly excluded.

### 8.1.1.3 Heterokaryosis in Ascomycotina and Fungi Anamorphici

The most common example of natural heterokaryosis in septate filamentous fungi of the Ascomycotina and Fungi Anamorphici is the 'dual phenomenon'. Predominantly mycelial and predominantly conidial strains can regularly be isolated by culturing a normal mycelium; the strains are believed to have arisen as spontaneous mutants (Hansen 1938). This causes problems in maintaining culture collections since 'mycelial' isolates may gradually predominate, but this is probably not significant under natural conditions. For example, various soil Fusaria only retain their sporulating ability and characteristic morphology if maintained in soil culture (Miller 1946). Under such conditions, selection apparently acts to suppress biologically undesirable traits becoming manifest.

In most Ascomycotina and Fungi Anamorphici, an equally probable origin for heterokaryosis to mutation is through hyphal fusions. Thus, with successive anastomoses between different mycelia, multiple heterokaryons could arise.

Complementation, much studied in experimental situations, is an equally characteristic advantage of the heterokaryotic state. Genetic deficiencies in one or more nuclei can be complemented by other nuclei carrying effective alleles. In some cases heterokaryons may even exceed the expression of either homokaryon in some respect, e.g. the enhanced growth rate shown by some heterokaryons of *Neurospora tetrasperma* [AP] compared with those of its component homokaryons (Dodge 1942)—so-called 'heterokaryotic vigour'.

Dissociation of heterokaryons can occur either by sectoring or through the production of uni- or multinucleate conidia, provided that only one nucleus enters the conidial initial (Pontecorvo 1946). Early model experiments with *Aspergillus* [A] indicated that the most common cause of heterokaryon sectoring is that component nuclei determine unequal growth rates, and heterokaryon stability is associated with little or no difference in growth rate (Pontecorvo and Gemmell 1944).

Nevertheless, highly unstable transient natural heterokaryons may arise quite often but remain undetected. Such transient heterokaryons provide an opportunity for the 'horizontal' transfer of nuclei and certainly for the transfer of non-nuclear elements, such as the *kalilo* plasmids which cause senescence (Griffiths *et al.* 1990).

Whether nuclear balance within a heterokaryon is important has not been fully resolved despite several experimental investigations of synthetic heterokaryons. Heterokaryons with greatly disparate ratios of nuclear components are well documented in Ascomycotina and Fungi Anamorphici, although nearly all the cases are from experimental situations (Ramsdale 1999). Assays of heterokaryosis are liable to error because nuclear ratios are generally sampled by isolating spores which, initially at least, are uninucleate. If there is any bias at spore formation in favour of the incorporation of a particular kind of nucleus, then the determination of the nuclear ratios will also be biased. Alternative procedures, such as chemical treatments which cause the breakdown of the mycelium and its rounding off into uninucleate or few-nucleate fragments that can be further treated, can sometimes get round this problem.

A heterokaryotic condition can persist virtually undetected. In synthetic stable dual heterokaryons of *Neurospora* with large differences in the numbers of their component nuclei, it proved difficult even with selection to eliminate the minority nuclei completely (Pittenger and Atwood 1956). Similarly, initially uninucleate spores isolated from a single mycelium of *Fusarium graminearum* [FA] grew uniformly when plated, but in a selection experiment it proved possible to demonstrate that, in fact, $5 \times 10^{-6}$ per cent of all the spores carried a mutant determining the colonial growth habit (Wiebe *et al.* 1993). Sometimes, if conditions are altered to favour a minority component of such heterokaryons, the nuclear balance can be shifted and they become unstable and dissociate. Therefore, if heterokaryons with numerically discrepant nuclear ratios occur in nature, the minority type can provide a small but potentially significant reservoir of genotypically distinct nuclei which could be preferentially selected by changed conditions. However, whether the results obtained with model selection experiments reflect the behaviour of naturally formed heterokaryons is not known, although the results obtained with natural heterokaryons like *M. mucedo* 40+, *Penicillium cyclopium* [FA] or *F. graminearum* [FA] (see Chapter 9, sections 9.3.2 and 9.3.3) suggest that they would behave in a similar manner.

Evidence against the natural occurrence of heterokaryons is strong for some species. One of the earliest and strongest arguments against its widespread occurrence in nature was the finding that even in Ascomycotina and Fungi Anamorphici, where heterokaryons could easily be forced under laboratory conditions, they appeared to be rare or absent from wild isolates (Caten and Jinks 1966). The improbability of natural heterokaryosis was further reinforced by the discovery of somatic incompatibility which restricts hyphal anastomosis to compatible strains and so reduces the chances of widespread hyphal fusions within a species. Moreover, even in natural situations apparently lacking somatic incompatibility, e.g. *Gremmeniella abietina* [AD], there is no evidence for heterokaryosis.

### 8.1.1.4 Heterokaryosis in Basidiomycotina

A dikaryon is, of course, a special and usually stable kind of heterokaryon that normally arises as the result of a sexually compatible hyphal fusion, but dikaryons do not always maintain a 1:1 nuclear ratio. This may reflect an imbalance between resident nuclei of one monokaryon and 'invading' nuclei from another at the time of establishment of the apparent dikaryon. Indeed, in some Basidiomycotina the nuclei of one of the partners may ultimately be totally displaced by 'invading' nuclei, or the resident nuclei may be preponderant. For instance, microsurgical analysis of dikaryotic hyphae of both *Galerina mutabilis* [BA] (Harder 1927) and *Hirschioporus abietinus* [BA] (Fries and Aschan 1952) demonstrated the occurrence of just a single nuclear type in appreciable lengths of the supposedly dikaryotic hypha.

In many other Basidiomycotina compatible hyphal fusions result in a mycelium with multinucleate hyphal compartments either lacking clamp connections or developing more than one at each

septum. The hyphae of rust fungi (Urediniomycetes) are multinucleate and lack clamp connections, but are perpetuated by means of their dikaryotic uredospores. If two different strains anastomose, one nucleus can be derived from each parental strain, i.e. the uredospore would be heterokaryotic. Since the heterokaryon combines characteristics of both parental types, even when meiotic or mitotic recombination is absent or prevented, it can behave as a novel pathotype. One hypothetical situation where an advantage might accrue to such a heterokaryon following hyphal fusion is if the heterokaryon showed improved competitive ability or had an extended host range. Selection for this initial advantage would ensure the continued stability of the heterokaryon. This condition is often described as 'somatic hybridization'. Two notable examples are *Puccinia striiformis* [BR], where sexuality has never been detected (Manners 1988), and *P. graminis* f.sp. *tritici* [BR] in Australia, where sexuality cannot be expressed because the alternative host on which the reproductive structures arise is absent from the flora.

Some 60 pathotypes of *P. striiformis* [BR], the cause of yellow stripe of cereals and grasses, are known in Europe, where many have arisen spontaneously and appear to combine characteristics of existing races. Little and Manners (1969 a,b) tested this hypothesis by inoculating races 2B (40 E72) and 8B (32 E138) on wheat. Three exceptional isolates were initially somewhat unstable but stabilized on further propagation to combine the different pathogenicities of the parental strains. These novel pathotypes had apparently arisen through heterokaryosis, since no diploid nuclei were detected, and were perpetuated by somatically hybrid uredospores. No more than two kinds of recombinant were ever obtained from such mixed inoculations and this supports strongly the notion that whole nuclei, not recombinants, are involved, i.e. heterokaryosis rather than true hybridization. In a subsequent series of experiments, a novel pathotype 105 E137, which combined the virulences of the natural parental races 41 E136 and 104 E137, was obtained from mixed inoculations on wheat seedlings. It is comparable to a *natural* isolate 45 E 140, which combines the virulences of 41 E136 and 37 E132 (Goddard 1976).

The evidence for natural somatic hybridization in *P. graminis* f.sp. *tritici* [BR] is based on the characteristic pathogenicity patterns of various races and isozyme evidence. The differential responses of tester wheat most closely supported the hypothesis that race 34 was a somatic hybrid between races 126 and 21; the former occurred in Australia prior to 1954, and the latter was introduced in 1954 (Luig and Watson 1977). This was confirmed by the enzyme profiles of the eight isozymes studied. The profile of race 34 differed from that of race 21 at only one site, for aspartate aminotransferase, which was typical of the race 126 allele (Burdon *et al.* 1981, 1982).

A number of basidiomycotous heterokaryons show discrepant ratios depending on their origins. A typical example is *Heterobasidion annosum* [BA] which has multinucleate compartments and produces conidia. The conidia of natural heterokaryons are predominately binucleate, i.e. they are dikaryotic and germinate readily. However, the conidia of synthesized heterokaryons behave differently. If derived from *sympatric* homokaryons, whose nuclei originate from sib-related homokaryons, the conidia are predominantly binucleate and the predominant nuclear genotype reflects the *resident* nuclear genotype. In non-sib-related heterokaryons, conidia are predominantly uninucleate and of the *invasive* genome. In synthetic heterokaryons derived exclusively from *allopatric* homokaryons, the conidia are predominantly uninucleate and carry the *resident* genome. It was assumed that the ratio of the nuclear genotypes in the conidia reflected that of the different nuclei in the mycelium, although no evidence is available to support this (Ramsdale and Rayner 1994, 1996). Evidently, some interaction of an unknown nature occurs in unselected (synthetic) heterokaryons, either between the nuclei or between the nuclei and the cytoplasm, which is absent in naturally established heterokaryons. This phenomenon has been described as 'genomic conflict', but neither the nature of the interaction nor the selective effect are understood. If this example reflects the general behaviour of newly established basidiomycotous heterokaryons, it will affect differentially the nature of the asexual propagules produced. In particular, it would tend to favour the *resident* genotypes in any area but, if allopatric heterokaryons are formed, that would enhance the multiplication of *migrant* propagules.

## 8.1.2 Parasexuality

Parasexuality has most plausibly been suggested for cases where apparently recombinant genotypes occur and somatic recombination or heterokaryosis has been suspected. Parasexuality can be demonstrated experimentally using many fungi but very little evidence exists for its occurrence under natural conditions. Detection in nature is difficult, but this is not surprising since the minimal initiating condition could be as ephemeral as the formation of a transient or unstable heterokaryon which would be virtually impossible to detect in nature. Nevertheless, since heterokaryosis is always a potential condition in any mycelium, any selective advantage of the heterokaryotic condition would be enhanced further if it were replaced by a combination of the desirable traits in a single genotype. Such a sequence would be of particular importance to a taxon either lacking sexual reproduction or where sexual reproduction was impaired or very rare.

Therefore it is not entirely surprising that parasexuality has been identified in *Magnaporthe grisea* [AP]. This species often appears to be clonal and is usually sexually impaired, yet it shows considerable genetic diversity and variation (Leung *et al.* 1988; Levy *et al.* 1991, 1993; Zeigler *et al.* 1995; Nitta *et al.* 1997). The course of its discovery, set out below, is a model of how parasexuality could be detected in other fungi under natural conditions.

**1.** Using morphological and biochemical markers in strains lacking any sexual recombination, *laboratory evidence* was obtained for heterokaryosis and segregation of markers through parasexual recombination (Genovesi and Magill 1976; Crawford *et al.* 1986).

**2.** Using *spontaneous DNA markers exclusively*, putative heterokaryons between wild strains were demonstrated *in the laboratory* and evidence obtained for parasexuality in conidial segregants derived from anastomosed hyphae.

**3.** Evidence was obtained for comparable behaviour *in the field* (Zeigler *et al.* 1997). The genotypes of the strains used were highly characterized by DNA fingerprints using the probe MR586. Hyphal anastomoses were observed both within a mycelium and between mycelia, where hyphal tufts were readily formed at the junction. Hyphal tips derived from tufted hyphae were isolated and grown on, and the genotypes of mycelia derived from the uninucleate conidia they bore were characterized. Comparison of DNA fingerprints from parental and derived strains revealed two relatively uncommon situations, namely:

(a) haplotypes with one or more new gel bands that did *not* co-migrate with bands from the other parental type, i.e. mutant types;

(b) similar haplotypes with one or more new bands where at least one of the new bands co-migrated to a site corresponding to a band from the other parental type, i.e. a putative recombinant type.

The latter situation, although rare, was 10 times more common in mycelia derived from *paired* cultures than from *unpaired* ones, supporting the notion that they were the result of recombination. In addition, by using various combinations of probes and restriction enzymes, the putative recombinants were shown to involve only limited elements of the chromosome complement. These features—a low frequency of recombination involving only limited elements of the chromosome complement—are characteristic of mitotic crossing-over.

**4.** Supporting evidence was obtained from *field isolates* from the Philippines and the Himalayas. Putative recombinants were detected in DNA fingerprints in which a band was found in the same location as one which occurred in different but comparable isolates. Moreover, other band patterns were found which were best interpreted as *partial diploids* (merodiploids). In the most thoroughly analysed experimental parasexual situation known, *Aspergillus nidulans* [A] (Pontecorvo and Käfer 1958), diploid nuclei in heterokaryons break down with the successive loss of one or more chromosomes after mitotic recombination to restore the haploid condition. Therefore merodiploidy in *M. grisea* [AP] was considered as further support for parasexuality.

There seems little doubt that the parasexual cycle occurs in this species in both laboratory, i.e. forced, conditions and unforced natural conditions. But how far parasexuality is responsible for the overall genetic diversity found in natural isolates of *M. grisea* [AP] is not known.

Since the essential prerequisite for this analysis, namely a knowledge of the genotype at the nucleotide level comparable with that for *M. grisea* [PA], is becoming increasingly available for other fungi, similar analyses are now possible in other putative cases of parasexuality. These include, for example, apparently spontaneous recombinants of some human pathogens which appear to lack any known sexual stage, e.g. *Candida albicans* [FA] (Gräser *et al.* 1996) or *Coccidioides immitis* [FA] (Burt *et al.* 1996).

*8.1.2.1 Parasexuality and the 'Buller phenomenon'*
Parasexuality is known experimentally from a few Basidiomycotina but the situation where the best evidence for somatic recombination in them is available is the so-called 'Buller phenomenon' (Quintanilha 1937). It can be manifested:

• when a dikaryon is confronted by a monokaryon (a di-mon mating), e.g. *Schizophyllum commune* [BA] (Crowe 1960)

• if the fungus is not dikaryotic, when a heterokaryotic is confronted by a homokaryotic strain (a 'he-ho' confrontation), e.g. *Stereum hirsutum* [BA] (Coates and Rayner 1985b)

• between a diploid and haploid strains of the same species, e.g. *Armillaria gallica* [BA] (Carvalho *et al.* 1995).

Laboratory investigations show that in such cases transient heterokaryons can form at the site of hyphal fusion. One or more of the following events can then take place:

• one nucleus may be replaced by another to give rise to a different viable dikaryon that can reproduce sexually

• the newly formed heterokaryotic mycelium may continue to grow as such

• rather infrequently, nuclear fusion, mitotic recombination and haploidization may occur in a new transient heterokaryon.

These processes have *never* been observed in the wild. Nevertheless, such juxtapositions probably occur quite frequently in nature. For example, basidiospores are often deposited close together and uneven germination is likely to establish one or more putative dikaryons adjacent to monokaryotic germ mycelia, conditions which promote the Buller phenomenon. Even the occurrence immediately adjacent to each other of a number of different newly formed dikaryons having a small number of MTFs in common, e.g. *Coriolus versicolor* [BA] (Burnett and Partington 1957), might promote such fusions despite the SI reaction between larger and more mature dikaryotic mycelia. If the Buller phenomenon were to occur in natural situations, parasexuality could be more common in Basidiomycotina than is generally supposed.

### 8.1.3 The significance of heterokaryosis and parasexuality for fungi

Heterokaryosis is probably appreciably more common in nature in the Eumycota than in the Straminipila where there is little definitive evidence for it. A few cases are known; for example, after inoculating races 1 and 3 carrying different drug-resistant strains of *Phytophthora megasperma* f.sp. *glycinea* [O] in the same host, isolates from the root were found to carry both resistances (Layton and Kuhn 1990). Transient sexual heterokaryons can arise also at oospore germination in *Phytophthora infestans* [O] (Pipe *et al.* 2000).

In the presence of different kinds of nuclei within the same mycelium, variation can either be expressed or suppressed as a consequence of complementation or dominance. In general, complementation is effective only when the gene products are diffusable and can interact to restore deficiencies or to enhance expression. At some point, if one nucleus is better adapted, the ratio can shift the mycelium to adapt to the new conditions. Therefore heterokaryons are highly adaptive to environmental conditions, but such behaviour is not immediately obvious in natural situations. This may account, in part, for the view that heterokaryosis is of no great natural significance in most Eumycota. However, if the behaviour shown

by heterokaryons of *H. annosum* [BA] in culture occurs in nature then it will affect population structure. If heterokaryons derived from sympatric components are likely to be propagated more effectively than those derived from allopatric homokaryons, it would enhance inbreeding within local sexually compatible populations. In addition, the effectiveness of gene flow would be affected to some extent, despite the increase in spores carrying an invasive genotype. However, such consequences are hypothetical at present.

Although parasexuality has been demonstrated unambiguously in only one natural situation, there are likely to be many other situations in which it could occur, especially those where the Buller phenomenon is manifest. At present no reliable estimate can be made of its contribution to recombination overall in any population. Nevertheless, it is a plausible supposition that its significance in natural asexual populations, or in those with impaired sexuality, may well be greater than has been recognized, while the Buller phenomenon may promote parasexuality amongst Basidiomycotina.

## 8.2 Hybridization and polyploidy

Hybridization can be defined in various ways. Here, the definition proposed by Arnold (1997) is adopted:

Natural hybridization involves successful matings in nature between individuals from two populations, or groups of populations, which are distinguishable on the basis of one or more heritable characters.

This wide definition avoids the problem of pre-defining the taxonomic status of the participating parental types which, in fungi, is particularly difficult in the present uncertain state of fungal taxonomy.

It is not surprising that there is only limited evidence for hybridization and polyploidy throughout all groups of the fungi. Hybrids intermediate in their characteristics are unlikely to be observed easily if the parental characters overlap, as is often the case. Detecting polyploidy is equally difficult because most fungal chromosomes are small, classical fungal cytology is poorly developed and electrophoretic methods have only recently been introduced (see Section 2.6).

## 8.2.1 Hybridization in Eumycota

Fungal hybrids can often be readily produced in the laboratory where the techniques used are biased to produce hybridization, for example by using an excess of fertilizing agents of one potential parent and selective devices to ensure that even the feeblest hybrid progeny will survive. Such conditions are unlikely to occur in nature! Moreover, pre- and post-zygotic mechanisms often exist to prevent crossing between taxa; indeed, even the hyphal wall prevents crossing between *Penicillium chrysogenum* and *P. cyaneo-fulvum* [both A] although their protoplasts can readily fuse and their nuclei recombine (Peberdy *et al.* 1977).

Very few genuine fungal hybrids have been detected under natural conditions, so that it is difficult to draw any general conclusions concerning predisposing conditions which promote hybridization. Nevertheless, two kinds of situation appear to promote hybridization. One occurs in more primitive fungi when parental taxa exist in a sexual pheromonic milieu common to both, i.e. sexual reproduction is initiated and promoted. The other is when parental taxa come to occupy a common, often novel, environment in which intertaxon crossing is at least as likely as normal intrataxon crossing.

### 8.2.1.1 Hybridization in species possessing a pheromone in common

The first pattern is shown in the primitive Chytridomycotina and possibly Zygomycotina. Hybridization is favoured in *Allomyces* [C] where male gametes of all species are responsive to the pheromone sirenin released by female gametangia. Similarly, the trisporic acids are universal pheromones for the Mucorales. Thus taxa within a pheromone diffusion zone are capable of responding regardless of whether the source is the same or a different species. Of course, this only predisposes the affected gamete or gametangium to sexual fusion since fertilization, or zygotic development, can be blocked at a later stage.

Three taxa were recognized initially in collections of *Allomyces* [C], subgenus Euallomyces, which show a regular alternation of sporophytic and gametophytic generations. They turned out to represent two species, *A. arbuscula*, the more common, and *A. macrogynus* [both C], each occurring as a polyploid series (see section 8.3.2), plus a range of hybrids between them, *A.* × *javanicus* agg. [C] (Emerson 1941; Emerson and Wilson 1954; Table 8.1). In all three, male gametes respond to the same pheromone, sirenin, whether released by the same or a different individual or species (Machlis 1958a,b).

Hybridization was successfully analysed for the following reasons.

• Hybrids were recognizable by the arrangement of the gametangia, which were differently

**Table 8.1** The cytological situation in the *Allomyces arbuscula/A. macrogynus* [C] species complex

| Species | Recorded chromosome numbers (numbers of isolates) | |
|---|---|---|
| *Allomyces arbuscla* (n = 8) | 8 | (2) |
| | 16 | (14) |
| | 24? | (1) |
| | 22–26 | (1) |
| | 32 | (1) |
| *Allomyces* × *javanicus* | 13 | (1) |
| | 14 | (1) |
| | 14 or 15 | (1) |
| | 16 or 17 | (1) |
| | 19 | (1) |
| | 21 | (2) |
| *Allomyces macrogynus* (n = 14) | 14 | (1) |
| | 28 | (3) |
| | 50+ | (1) |
| Artificial hybrids *A. arbuscula* × *A. macrogynus* | 20 | (1) |
| | 22 | (1) |
| | 27 | (2) |
| | 32 | (1) |
| | 36 | (1) |
| | 42 | (2) |
| | 42+ | (1) |
| | 44 | (1) |

A number of the artificial hybrids were between polyploid representatives of the two basic species.
(Reproduced from Emerson and Wilson (1954), with permission from the Mycological Society of America.)

arranged in terminal pairs. In *A. arbuscula* the male gametangium is immediately below the female gametangium (hypogynous), whereas in *A. macrogynus* [C] it is immediately above it (epigynous). A minority of hybrids exhibited the same gametangial arrangement as one of the parents, usually the female, but most were intermediate, with some hyphae having the hypogynous arrangement and others the epigynous.

• Unusually for a fungus, the meiotic chromosomes of *Allomyces* [C] are quite large and stain readily with aceto-orcein, enabling the identity of sporophytes and gametophytes to be determined (Wilson 1952).
• The parental origin of male gametes could be identified as a consequence of the different parental position of their gametangia, so that gametes could be unambiguously isolated once released.
• Hybrids comparable with natural ones could be produced experimentally (Emerson and Wilson 1954).

Although crosses sometimes failed completely, viable resistant sporangia usually developed on the F1 hybrid sporophytes. Their viability ranged from less than 1 per cent to 45 per cent, depending upon the hybrid combination, compared with

68–90 per cent in normal fertilizations. The gametophytes used for the crosses came from widely based locations but there was no obvious correlation between hybrid success and geographical origin of parents, although some strains appeared to be more effective than others.

Experimental hybrids showed two kinds of chromosomal behaviour related, in part, to the ploidy of the parents involved in the cross (Figure 8.2):

• in type A, little chromosome pairing took place, one to five bivalents at most, with the remaining chromosomes forming univalents
• type B exhibited far more bivalents and correspondingly fewer univalents.

It was believed that some chromosome doubling had taken place in the sporophytes of the latter *before* meiosis, since spontaneous chromosome doubling took place in this way in some strains of *A. arbuscula* [C]. Type B sporophytes may well be favoured in nature since their pairing is more effective.

Despite the reduced viability shown by F1 hybrids, it proved possible to produce selfed F2 individuals and later generations. The average parental viability was 63 per cent; that of

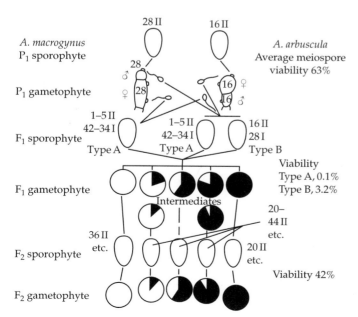

**Figure 8.2** Hybridization in *Allomyces* [C]. In the F1 and F2 gametophyte generations, 'intermediate' indicates the morphological development of the mycelia and antheridia. White sectors denote hypogynous and black sectors denote epigynous, while the numbers of bivalents are indicated in the sporophyte generations.

F1 hybrids was 0.11 per cent, but that of F2 hybrids was 41.7 per cent, with some in the range 50–80 per cent. Thus appreciable fertility was restored by the F2 generation. Common aberrations in some parental strains, especially in *A. arbuscula* and a few hybrids, or hybrid derivatives, are the parthenogenetic development of female gametes and that some sporophytes were amictic. How these condition arose has not been investigated.

It is not known with certainty whether parents and hybrids frequently occur together in nature. It may be that species of different ploidies are distributed worldwide and that parental types and hybrids occur together, but the structure of natural populations in subgenus Euallomyces is not clear, particularly on the following points:

• whether single species populations of mixed ploidy occur naturally;
• whether the different species, or polyploids, occupy different ecological niches or frequently grow together;
• whether hybridization is common or sporadic.

A more detailed examination of localized populations of *Allomyces* spp. [C] is needed to resolve these issues.

Another group of fungi in which hybridization may play a significant role is the Mucorales or 'pin moulds', although virtually nothing is known about their population structure.

One confirmed hybrid is *Rhizopus microsporus* × *R. rhizopodiformis* [both Z] (Schipper *et al*. 1985), and a natural isolate possessing the same features as this hybrid has been isolated once (Schipper *et al*. 1985). In two attempted crosses only nine zygotes could be germinated, and that with difficulty, and only one of these by means of a germ sporangium; the rest produced a mycelium directly. Most of the progeny were intermediate in diagnostic features between those of the parents, but six were azygosporic, with the azygospores being capable of germinating to form similar mycelia. Sporangiospores from one of these mycelia bred true when tested. Progeny from a cross involving an auxotrophic marker were prototrophic. An azygospore from one such mycelium germinated to give intermediate mycelia, only 12 of which were azygosporic; the remaining four had

lost this ability. Non-azygosporic progeny could be mated successfully with the parental *R. rhizopodiformis* [Z] but progeny from this cross were similar to the other parent, auxotrophic and of the same mating type. Therefore it seems probable that the original hybrid progeny were either diploids or partial diploids, since some were prototrophic and progeny apparently identical with one parent could, ultimately, be recovered.

Schipper's (1973) earlier detailed study of *M. hiemalis* [Z], although primarily taxonomic, involving numerous differences in morphology and sexual behaviour between strains, also bears on the question of hybridization. He opted to describe six previously recognized heterothallic species and forms intermediate between them as *formae* of a single species *M. hiemalis* [Z]. A further homothallic species, *M. genevensis* [Z], was described as otherwise the 'counterpart of *M. hiemalis*' (Schipper 1973). In crosses between the six heterothallic taxa the majority formed zygotes, presumably fertile, although this was not tested. In some cases the interaction was incomplete or, in one case, totally absent (Figure 8.3).

All the taxa were connected directly or indirectly through mating tests, although those forming compatible matings were not necessarily morphologically similar and *vice versa*. The phenotypic variations, presumably reflecting genetic differences, included hyphal size and branching, size variation of both sporangia and sporangiospores, pigmentation and reaction to different natural

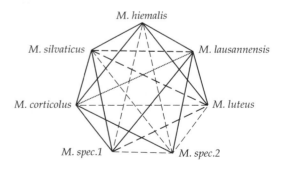

**Figure 8.3** Summarized results of mating tests within the *Mucor hiemalis* [Z] complex: solid line, zygospores formed; broken line, progametangia only formed; dotted line, no interaction. The names are those of the original isolations. (Reproduced from Schipper (1973).

substrate agars. Unfortunately, details of natural substrates and habitats were not available. *M. genevensis* might well represent a natural, potential hybrid, homothallic derivative. It is not clear whether these taxa represent species or ecotypes, but it is not in question that forms intermediate between them occur in nature.

The reproductive behaviour of Mucorales had been explored originally in some depth by Blakeslee and his collaborators (Blakeslee 1904a,b; Blakeslee and Cartledge 1927; Blakeslee *et al.* 1927; Satina and Blakeslee 1930). 'Imperfect' sexual interactions occurred between different heterothallic species, and between either mating type of a heterothallic species and a homothallic species. In most cases of inter-specific crosses azygospores, supposedly non-sexual structures resembling zygospores, developed, usually on one of the conjugating strains only. These early experiments on crossing different species provide no evidence for or against true hybridization as neither zygospores nor azygospores were germinated or examined for segregation. Since both kinds of structure have proved intractable material for both cytological investigations and germination, e.g. Gauger (1965) was unable to germinate less than 1 per cent of the zygospores of *Mucor hiemalis* [Z], considerable uncertainty still exists as to the location or occurrence of meiosis and segregation in Mucorales. However, now that it is known that they react to identical pheromones (Austin *et al.* 1969), the situation resembles that in *Allomyces* [C] in this respect. Therefore, by analogy, it does not seem entirely implausible that genuine hybridization could occur from time to time. Indeed, the observed, low germinability of the zygospores of many 'normal' strains when tested could conceivably be due to their unrecognized hybrid nature.

### 8.2.1.2 Hybrids associated with neutral situations

The second pattern occurs with both Asco- and Basidiomycotina. The characteristic features are that the taxa involved come to occupy a common ecologically neutral or open site, in the sense that it is not biased against inter-taxon crossing.

The smut fungi (Ustilaginomycetes) were the earliest fungi in which experimental hybridizations were successful, and natural hybrids on cereals have been detected in the genera *Ustilago* [BS] and *Tilletia* [BS]. The taxa involved originally infected wild grasses but these were replaced by extensive areas of genetically uniform susceptible cereals. Thus the conditions favouring natural *Tilletia* [BS] or *Ustilago* [BS] hybrids, namely an ability to infect a new ecologically neutral common host and extensive stands of mixed host crops, increased the possibility of cross-infection and hybridization.

For example, intermediates between the two apparently closely related species which cause bunt of wheat, *Tilletia tritici* ( = *T. caries*) [BS] and *T. laevis* ( = *T. foetida*) [BS], have often been isolated from infected areas (Fischer and Holton 1957). They are highly comparable with segregants produced from experimental crosses (Holton 1944, 1951). A third species which causes dwarf bunt, *T. controversa* [BS] also hybridizes both experimentally and naturally with *T. caries* [BS] (Holton 1954). In addition, *T. bromi* [BS], a morphologically similar species to *T. controversa* [BS], often occurs in the wheat crops on weedy *Bromus* spp. It has been implicated as hybridizing with *T. controversa* [BS] although it cannot infect wheat nor can wheat bunt species infect *Bromus* (Meiners 1958). Using both RAPDs and ribosomal markers it became evident that *T. laevis* and *T. tritici* are the most similar and that *T. controversa* [BS] is closer to them than to *T. bromi* [BS] (Shi *et al.* 1996). *T. controversa* × *T. bromi* [BS] hybrids have been made experimentally but showed greatly reduced fitness (Pimental *et al.* 2000). It was considered improbable that these hybrids would be viable in the field.

The situation is a little different in *Ustilago* [BS] where there have been many investigations of experimental hybridization, especially between *U. segetum* var.*segetum*( = *U. kolleri*) [BS], the cause of covered smut of barley (*Hordeum*) which is also found on oats (*Avena*), and its other variety, var.*avenae*( = *U. avenae*) [BS], which is confined to oats. Viable hybrids have been obtained experimentally. Since both varieties carry identical virulence/avirulence genes (Fullerton and Nielsen 1974), it has been suggested that this is due to introgression following hybridization, but the evidence for natural hybrids is not strong. If hybrids do occur they are likely to be associated with the common susceptible host, i.e. oats.

Poplar rust hybrids between *Melampsora medusae* and *M. occidentalis* [both BR] occur naturally in mixed stands of poplar species and thrive on hybrid poplars between *Populus deltoides* and *P. trichocarpa*, the respective hosts of the two rust species (Newcombe *et al.* 2000). Two causes have brought these poplars together. The natural distribution of *M. occidentalis* [BR] parallels that of *P. trichocarpa* but *M. medusae* [BR] has a wider host range. Historically, *M. medusae* [BR] spread from the southeastern USA northwards and westward into the main region occupied by *M. occidentalis* [BR], where both host species now occur in mixed stands. Secondly, man-made hybrid poplars, which can be infected by both *M. medusae* [BR] and the hybrid, have increasingly been planted. Experimentally, at least, the hybrid rust *M.* × *columbiana* [BR] has a wider host range than either of its parents and, to judge by samples examined, it dominates on hybrid poplars in nature. It is now so well established that evidence was obtained for F2 and back-cross generations so obviously introgression could now occur. Thus hybridization has been favoured and become established in situations comparable with those described in general as 'hybridization of the habitat' (Anderson 1948), a situation promoted in part by human activities.

A less fully documented rust hybrid, *M.* × *medusae-populina* [BR], has been found in New Zealand. It is derived from *M. medusae* × *M. larici-populina* [both BR]; the latter is an established imported species in New Zealand, and the former is well established in Australia. After an earlier *M. medusae* [BR] infection in New Zealand had died out, a further introduction, probably on trans-Tasmanian winds, resulted in hybridization (Spiers and Hopcroft 1994). Regardless of its precise mode of origin, this hybrid also conforms in the circumstances of its establishment with the two previous examples. Mixed stands of host poplars provided the opportunity for cross-fertilization and the introduction of hybrid poplars particularly susceptible to the new hybrid rust fungus produced ecological conditions favouring its establishment.

A not dissimilar situation in California has resulted in natural hybridization between different host-specialized types of *Heterobasidion annosum* [BA].

In Europe, the P-type is mainly confined to Pinus and the S-type to *Abies* and *Picea*, and in southern and southwestern Europe, an F-type is confined to *Abies alba*. Host specificity is appreciable but not absolute, a finding borne out by the inability of separating the S- and F-types on the basis of either ITS or IGS sequence data (Harrington *et al.* 1997), although in Italy the F-type is appreciably more distinct and can be separated on the basis of RFLPs and isozymes (Goggioli *et al.* 1998). They are predominantly intersterile in nature but some 4–10 per cent interfertility was recorded experimentally between P- and S-forms and much higher values between S- and F-forms (Korhonen 1978a; Stenlid and Karlsson 1991; Capretti *et al.* 1990).

In North America, the S-type originally had a somewhat wider host range in California than in Europe, but climatic change to more xeric conditions favoured growth of hosts of the P-type and, finally, twentieth-century commercial planting of firs (S-type hosts) has resulted in a mosaic of mixed and pure tree stands. Eleven natural F1 hybrid isolates, confirmed by isozyme analysis, were detected in such a stand at Modoc, California, using taxon-specific competitive PCR primers (TSCP-PCR) for the ITS region of rDNA. In all, 54 per cent of the trees at this site were infected by the P-type, 35 per cent were infected by the S-type and 13 per cent were hybrids on a pine stump, adjacent pines and junipers (Garbelotto *et al.* 1996). Stump infection by hybrids is not surprising, since in Europe it had already been noted that stumps on which *H. annosum* [BA] grows saprophytically provide a site for co-infection and potential hybridization (Otrosina *et al.* 1992; Harrington *et al.* 1989), and this was borne out by the Californian work (Stenlid and Karlsson 1991; Garbelotto *et al.*, unpublished). In addition, elsewhere in California, 7 per cent of the apparently S-type isolates were found using a P-type mitochondrial fragment as a marker, to carry the P-type fragment, indicating introgression (Garbelotto *et al.* 1996). Therefore natural hybridization and introgression appear to be favoured in mixed-host areas subject to felling. Both in Europe and the USA it has been shown, experimentally, that the hybrid does not readily infect intact trees (Garbelotto *et al.* 1996). Mixed-host stands enhanced by plantation forestry and felling are therefore likely to increase the probability of hybrids arising naturally. Here too, human activities have provided the appropriate habitat for promoting hybridization.

A less well authenticated example from the State of Washington, USA, in areas now cultivated for wheat, is hybridization between *Typhula ishikariensis* [BA], a typical soil fungus of forest areas (Bruehl and Cunfer 1975), and *T. idahoensis* [BA], primarily from grasslands. These taxa, hitherto ecologically isolated, have been brought together with a novel host, wheat, in a new environment. Experimental hybridizations were successful although most F1s were not very vigorous but, compared with natural isolations, some were both as vigorous on wheat and comparable in their infectivity to wheat. Variation of natural *Typhula* [BA] isolates from wheat fields is virtually continuous between each parent and includes the range shown by experimental hybrids. Therefore it does not seem unreasonable to assumee that some of these isolates are, in fact, putative hybrids (Bruehl *et al.* 1975; Bruehl and Machtmes 1980).

Natural hybridization in Ascomycotina is less fully documented. Natural hybrids have been detected between *O. ulmi* [AP], the original cause of Dutch elm disease, and both the Eurasian (EAN), ssp. *novo-ulmi*, and the North American (NAN), ssp. *americana*, subspecies of the new species *O. novo-ulmi* [AP]. Only nine out of more than 11 000 samples proved to be unique hybrids (Brasier *et al.* 1998), one from Poland and the other eight from the *same* bark sample in Portugal. However, each isolate proved to be a unique hybrid, showing various combinations of the characters of their putative parents but having in common a faster radial growth rate in culture coupled with a 'waxy' appearance due to the absence of aerial hyphae.

Genealogical evidence suggests a closer relation between these two species and other Ophiostomas pathogenic to hardwood trees, including *O. himalulmi* [AP], another elm pathogen endemic in the Himalayas (Brasier *et al.* 1995; Pipe *et al.* 1997). They differ appreciably in their pathogenicity to elms and, in particular, in their production of the toxin cerato-ulmin, but comparisons of the nucleotide sequences of the toxins indicate an ancient divergence of the species (Brasier 1990, 2000).

Nevertheless, *O. novo-ulmi* [AP] first appeared as two distinct clones, presumably as a result of inadvertent human introduction, in the Great Lakes region of the USA in the 1940s and at much the same time in Eastern Europe. In Europe, subspecies *novo-ulmi* has hybridized with the resident population of *O. ulmi* [AP] and introgression is occurring, although the latter species is rapidly being brought to extinction. A pathogenic gene present in *O. ulmi* [AP] has been transferred in this way to *O. novo-ulmi* [AP] (Abdelali *et al.* 1999). In addition, segments of *O. ulmi* DNA have been found flanking novel DNA *vic* genes in populations where hybridization is occurring, suggesting the transfer of *O. ulmi* [AP] *vic* genes to *O. novo-ulmi* [AP] where they replace the resident *vic* genes. This only occurs in the super-clonal *O. novo-ulmi* [AP] which carries only the single *A* MTF which also acquires the opposite MTF, conceivably from *O. ulmi* [AP] also (Brasier 2001). A second type of hybridization is also occurring between ssp. *americana* and ssp. *novo-ulmi* in those parts of Western Europe where ssp.*americana* has been introduced and spread. Presumably such hybrids will involve other crosses with the existing *O. ulmi* × *O. novo-ulmi* hybrid, and with further backcrossing and subsequent crosses an entirely new range of genotype is becoming available for selection (Abdelali *et al.* 1999).

It is not known how the hybrids arose, but there are two possible routes to hybridization—somatic hybridization or a sexual cross. When *O. novo-ulmi* [AP] invades an already infected elm it will be surrounded by the mycelia derived from numerous genets of the resident *O. ulmi* [AP], providing numerous opportunities for hyphal contact and fusion. That this can occur is supported by two North American isolates of *O. novo-ulmi* [AP] which exhibit typical cDNA profiles but *O. ulmi* [AP] mtDNA profiles (Bates *et al.* 1993), a condition more probable after somatic contact than through fertilization. Alternatively, if a heterokaryon chanced to form between sexually compatible mycelia, hybrid perithecia could develop. A hybrid origin through normal cross-fertilization seems more probable. If mites, the natural carriers of spores, are employed to transfer conidia from *O. novo-ulmi* [AP] to *O. ulmi* [AP] colonies, abundant perithecia develop but not vice versa (Brasier 1978; Kile and Brasier 1990). Both mating types are likely to be present in the *O. ulmi* [AP] genets, although advancing fronts of *O. novo-ulmi* [P] usually carry only a single *B* mating type (Brasier 1988). Therefore a possible origin of the natural hybrids could have been an *O. ulmi Mat A × O. novo-ulmi* ssp. *novo-ulmi Mat B* cross, since both sites were within the advancing edge of Eurasian subspecies epidemics.

What is more extraordinary is that any hybrids had persisted long enough to be detected, since synthetic hybrids are appreciably less fit than their parents in every respect tested, especially in pathogenicity, and many are female sterile (Kile and Brasier 1990). It has been suggested that such natural hybrids probably have a transient existence although they could interbreed with the invasive parent to give rise to introgressive progeny. Indeed, *Pvu II* digests of mtDNA demonstrated the occurrence of introgression of *O. ulmi* [AP] mitochondrial material into two apparently *O. novo-ulmi* ssp. *americana* [AP] isolates (Bates *et al.* 1993). The observation that two of the hybrids were experimentally capable of fertilization by *O. novo-ulmi* [AP] supports this hypothesis.

Incidentally, there is also evidence for appreciable crossing between the subspecies, and vigorous natural hybrids have been detected in Western Europe where the subspecies distribution overlaps. Experimentally, when ssp. *novo-ulmi* is the donor, the cross is fully fertile with ssp. *americana*, but the reciprocal cross produces fewer perithecia some of which only develop partially (Brasier 1986a,c).

### 8.2.1.3 *Hybridization involving anamorphic forms*
An important and biologically significant class of hybrids are strictly anamorphic. Many are widespread and also of appreciable economic as well as biological importance. Their existence strengthens the view that hybridization is not necessarily uncommon in fungi, provided that the circumstances are appropriate.

*Verticillium longisporum*(= *V. dahliae* var. *longisporum*) [FA], a pathogen of oilseed rape and other crucifers, combines some of the characteristics of *V. dahliae* and *V. albo-atrum* [both FA] although its pathogenicity is quite novel. Experimental

heterokaryons made between the two putative parents provided evidence for parasexuality (Fordyce and Green 1964), but when Hastie (1973) repeated the cross he obtained relatively stable hybrids. Although recognizing three species but not the hybrid nature of *V. longisporum* [FA], Morton *et al.* (1995a,b) showed nucleotide differences between them as well as differences in the sub-repeat sequences in the rRNA intergenic region. Finally, total DNA measurements showed that *V. longisporum* [FA] is a near diploid compared with *V. dahliae* [FA], possessing 0.044–0.53 pg per DNA nucleus compared with 0.25–0.030 pg per DNA nucleus in the latter. Clear distinctions could be made between all three in the gel patterns of their RFLPs (Messner *et al.* 1996;. Karapapa *et al.* 1997). Therefore *V. longisporum* [FA] is almost certainly a near-diploid natural hybrid. It is also capable of being back-crossed to *V. dahliae* [FA] to give rise to recombinants resembling *V. dahliae* [FA] in their RFLP patterns and in showing reduced or no pathogenicity to oilseed rape. Similar natural isolates have also been found (Karapapa *et al.* 1997). This is one of a few known cases of a natural spontaneous fungal hybrid with entirely novel pathogenicity compared with its parents (see also section 8.2.2).

The most fully authenticated examples of anamorphic hybrids are grass endophytes involving *Epichloë* [AP] and *Neotyphodium* ( = *Acremonium*) [AP]. Many temperate zone grasses carry symptomless systemic infections due to anamorphic fungi classified as *Neotyphodium* [FA]. Their aconidial hyphae grow in the intercellular spaces, closely applied to the cell walls, but can penetrate the ovules and so the fungus can be transmitted by seed. Similar endophytes, but possessing a teleomorphic *Epichloë* [AP] phase, develop characteristic infections of 'choke' diseases. In the spring, a cylindrical mycelial stroma develops around the upper parts of some tillers, external to and enclosing the leaf that sheathes the inflorescence, thus inhibiting its normal development—hence the descriptive name. Conidia which are attractive to the fly *Phorbia phrenione* form on the stromata. These flies eat stromatic hyphae and deposit both eggs and faeces on it. Cross-fertilization is effected by the transfer of conidia between stromata on the mouth parts, bodies or faeces of the flies. *Epichloë* spp. [AP] are dimictic, with the flies effecting fertilization. Once that is achieved, the stromatata develop abundant immersed perithecia which discharge their ascospores when mature (Kohlmeyer and Kohlmeyer 1974; Bultman *et al.* 1995).

All the teleomorphs are morphologically very similar, but crossing experiments demonstrate that they comprise nine reproductively isolated 'mating populations' (MP I–MP IX), most of which are confined to a single, or a few related, grass species. Exceptionally, however, *E. typhina* MP I [PA] has a broad host range on *Anthoxanthum, Dactylis, Holcus, Phleum, Poa* and *Lolium*. There are also several distinct *Neotyphodium* [FA] anamorphs but in many cases these are hybrids.

The most fully authenticated case is an endophyte, designated *Neotyphodium* LpTG-2, which occurs on rye grass (*Lolium perenne*). It is a hybrid derivative of *E. typhina* MPI × *N. lolli* [AP/FA], both of which also occur on rye grass. The evidence comes from a comparison of the molecular attributes of the three taxa, using DNA extracted from mycelia derived initially from single uninucleate conidia (Schardl *et al.* 1994). LpTG-2 possesses:

- the rDNA typical of *E. typhina* [PA]
- the mtDNA profile of *N. lolli* [FA]
- two copies of each gene, one derived from each of the putative parents except for one isozyme which was totally lacking, at nine other loci (*tub2*, *pyr4* and seven allozymes)
- a chromosome number $n = 13$ (those of *E. typhina* [AP] and *N. lolli* [FA] are $n = 5$ and $n = 8$ respectively) with genome sizes to match (Kuldau *et al.* 1999).

Therefore *Neotyphodium* LpTG-2 is almost certainly a hybrid derivative which has lost part of a chromosome carrying the missing enzyme locus. The status of the parental anamorphic form *N. lolli* [FA] is not known, but it could have been derived from *E. festucae* [AP] MP II since its single *tub* allele is identical with that in MPII isolates. Using this gene as a marker, several other hybrids have been identified (Table 8.2).

There can be little doubt that hybridizations, possibly ancient, have occurred in these fungi but it is not clear how they arose. *Neotyphodium* spp. [FA] are not known to produce stromata or conidia, although external conidia are formed by *E. typhina* [A] on a sparse external mycelium which tends to run parallel to the long axis of the leaf, as well as on the stromata. If such an epiphyllous conidium-bearing mycelium occurred on any strictly anamorphic *Neotyphodium* [FA], the conidia might be transported to stromata of sexually reproductive forms and fertilize them since some anamorphic forms are known to carry a MTF, e.g. *Neotyphodium* LpTg-2 carries *mat2*.

Alternatively, if a grass already infected by an anamorphic form was secondarily infected by ascospores or conidia from a teleomorphic form, the mycelia might meet and anastomose. Subsequently, nuclei from each parent could then fuse to form a hybrid nucleus. This is a possible route since, in culture at least, there appear to be no barriers to heterokaryon formation within or between sexual *Epichloë* spp. [AP] (Chung and Schardl 1997). Unfortunately, heterokaryon formation between strictly asexual forms, or between asexual and sexual forms, has not been investigated. That a heterokaryotic origin is less likely was shown by experimentally infecting both tall fescue and perennial rye grass with pairs of the teleomorphs, or with an anamorphic and a teleomorphic form. Infections were successful and both forms were found in the same tillers, but one always predominated. There was no evidence for heterokaryosis *in planta* (Christensen *et al.* 2000) and after a few months dually infected tillers died out.

It is not known why Neotyphodiums cannot form stromata, but nothing is known about the genetics of stroma formation. Its absence could be due to some kind of loss mutation which might even have occurred before hybridization. It is possible that when a species infects an unusual host stroma formation could be inhibited although the mycelium could still be propagated via seed. It might then loose the genetic ability to develop a stroma. In line with such a hypothesis, Schardl *et al.* (1994) suggested that *N. lolli* [FA], which is probably related to MPII, could have lost its ability to form a stroma when it changed from its normal host fescue (*Festuca*) to rye grass (*Lolium*).

**Table 8.2** The known or probable relationships between anamorphic *Neotyphidium* (=*Acremonium*) grass endophytes and species of *Epichloë* (published and unpublished data provided by C. L. Shardl, *in litt.*, 2002).

| Host grass[a] | Geographic origin[b] | Anamorphic species[c] | Closest non-hybrid groups[d] |
|---|---|---|---|
| Poeae | | | |
| *Lolium arundinaceum* | Eu, NAf | *N. coenophialum* | E. *festucae* |
| | | | E. *typhina* complex[e] |
| | | | *Lolium*-associated clade[f] |
| *L. perenne* | Eu | *N. lolii* | E. *typhina* complex |
| *L. perenne* | Eu | LpTG-2 | E. *festucae* or *N. lolii* |
| | | | E. *typhina* complex |
| *L. pratense* | Eu | *N. siegelii* | E. *bromicola* |
| | | | E. *festucae* |
| *L. pratense* | Eu | *N. uncinatum* | E. *bromicola* |
| | | | E. *typhina* complex |
| *Lolium* sp. | Eu, NAf | FaTG-2 | E. *festucae* |
| | | | *Lolium*-associated clade |
| *Lolium* sp. | Eu, NAf | FaTG-3 | E. *typhina* complex |
| | | | *Lolium*-associated clade |
| *Lolium* annual spp. | Eu | *N. occultans* | E. *bromicola* |
| | | | *Lolium*-associated *clade* |
| *Poa huecu* | SA | *N. tembladerae* | E. *festucae* |
| | | | E. *typhina* complex |
| Aveneae | | | |
| *Echinopogon ovatus* | Au | *N. aotearoae* | *N. aotearoae* |
| *E. ovatus* | Au | *N. australiense* | E. *typhina* complex |
| | | | E. *festucae* |
| Meliceae | | | |
| *Melica decumbens* | SAf | *N. melicicola* | E. *festucae* |
| | | | *N. aotearoae* |

[a] In this table *Lolium* is used *sensu* Darbyshire (1993, *Novon*, **3**, 239–243) to include the broad leaved fescues. Annual species include *L. canariense*, *L. multiflorum*, *L. persicum*, *L. remotum*, *L. rigidum* vars. *rigidum* and *rottboeliodes* and *L. temulentum* all of which are sometimes considered to be subspecies of *L. perenne*.

[b] Regions indicated: Au, Auastralasia; Eu, Eurasia; NA, North America; NAf, North Africa; SA, South America; SAf, South Africa.

[c] As anamorphs are isolated they are initially given temporary prefixes indicating the host from which they have been derived, e.g. LpTH-2 = *Lolium perenne* Taxonomic Group 2; Fa = once thought to be *Festuca arundinacea* = *L. arundinaceum* but phylogentic analysis of chloroplast material demonstrated that this was not the case, moreover, taxonomically, the North African tall fescue is now thought to be different from the European species.

[d] 'Closest groups' based on comparisons of sequence relationships of introns of *β*-tubulin gene, translation elongation factor I-alpha and actin (except for *N. occultans*).

[e] 'The '*E. typhina* complex' includes *E. typhina*, *E. clarkii*, and *E. sylvatica*, whose species limits are not clear.

[f] The '*Lolium*-associated clade' is a 'clade of sequences found in several hybrid endophytes of *Lolium* (*sensu* Darbyshire) spp. and includes *E. amrillans*, *E. baconii*, *E. festucae* and others.

### 8.2.1.4 *Other possible hybrids*

The cases cited above do not cover all the claims which have been made for hybridization in the Eumycota where the evidence is either indirect or less substantial. The evidence comes from two sources. One source is evidence of hyphal or gametangial intercompatibility, but either lacks evidence of further development, or the production of only partially developed carpophores or meio-spores. The other source is from molecular markers, some of which may be common to two or more taxa, or evidence of anomalies, particularly reticulate connections, in phylogenetic trees based on molecular data.

In many groups of fungi, **sibling species**, i.e. taxa that are outwardly very similar or iden-tical, are common and successful pairings between them have been used as an indicator of identity or close relationships. For instance, extensive crossing for many years between some 4600 strains of conidia-bearing *Neurospora* [AP] species have been used as the basis for determining species limits. Crosses result in various degrees of development of perithecia and ascospores, but although crosses between the most closely related species, *N. crassa* and *N. intermedia* [both AP], usually result in perithecium formation, less than 19 per cent of the ascosopores turn black and show some viability. Despite this suggestive evidence, only about 10 cases in all have been judged to provide evidence for true hybridization despite the many thousands of crosses made (Turner *et al.* 2001).

In a great many other Ascomycotina some degree of fertility, comparable with that just described for *Neurospora* [AP], is found when crosses are made in the laboratory. Although this indicates close relationships (see Chapter 11, section 11.3), there is not much reason to believe that similar crosses occur in nature or, if they do, that the progeny will be viable. Nevertheless, an occasional cross between particularly compatible isolates could give rise to more viable offspring which might back-cross to one of the parental species, potentially giving rise to introgression. Such instances, if they occurred, would be difficult to detect in the field. However, the fact that many closely allied Ascomycotina can potentially cross suggests that interspecific hybridization should at least be regarded as possible.

After somatic fusion between monokaryons of various sibling species of Basidiomycotina, a dikaryon develops which either produces clamps irregularly or produces incomplete clamp connections although it continues to grow. This has been taken by some as providing evidence for putative hybridization but by others as a compatibility test without necessarily implying a hybrid condition (Boidin 1986). Because of the complexity of finding the right conditions for basidiocarp development, it is usually not known whether such mycelia represent genuine hybrids. However, the phenomenon is so widespread that it suggests that hybridization could occur between closely related Basidiomycotina.

Molecular evidence for hybridization or, in the case of phylogenetic anomalies, even ancient hybridization events is far more convincing. For incontrovertible confirmation came from comparisons of both RAPDs and mtDNA RFLPs of the

example, using RFLPS of the ITS region, three distinct haplotypes of *Flammulina velutipes* [BA], each with a distinct biogeographical range in North America, Europe and Asia respectively, have been recorded. Gel patterns interpreted as due to hybrids between the European and North American haplotypes have been recorded in Europe and also in Argentina where both are probably immigrant material (Methven *et al.* 2000).

## 8.2.2 Hybridization in Straminipila

In 1991, Brasier commented that:

The occurrence of species hybrids has scarcely been addressed in *Phytophthora* systematics. Their potential as a complicating factor needs to be recognized.

However, as yet, the evidence is indicative rather than definitive, most coming from morphologically unusual, and often sterile, forms in populations of mixed species. For example, mixed populations of *P. nicotianae*, *P. citrophora* and *P. citricola* [all O]occur on citrus plants in Corsica. They can be separated on the basis of their enzyme electrophoretic patterns and there are also two types of *P. citricola* [O]. However, three isolates having the morphological characteristics of *P. nicotianae* [O] were found to carry the minority *P. citricola* [O] enzyme pattern. Secondly, when isoelectric gels of total protein, which are normally species-specific, were used, two *P. citricola* [O] isolates differed from other isolates of the species both morphologically and in three enzyme gel systems which resembled *P. citrophora* [O] (Erselius and de Vallavieille 1984; de Vallavieille and Erselius 1984). These anomalous isolates are possible putative hybrids between the species. A similar but less well-documented situation was detected in mixed populations of *P. gonapodyides*, *P. cryptogea* and *P. dreschleri* [all O] (Brasier *et al.* 1989).

Two unambiguous natural hybrids have recently been identified. One was detected in Dutch commercial hydroponic systems, and the other, *Phytophthora* sp-'h'[O], was shown to be a *P. nicotianae* × *P. cactorum* [bothO] cross.

In the former, a heterodimeric hybrid band was detected using malic enzyme isozymes. Further putative parents and hybrid (Man in 't Veld *et al.* 1998). This ecologically novel situation is one with a high probability that resident fungi will be juxtaposed and hybridization could occur. Other hybrids might be expected in this unusual closed system.

The second hybrid is responsible for a newly detected disease of alders (*Alnus*). Alders were not known to be susceptible to infection by any species of *Phytophthora* [O] until the 'alder' *Phytophthora* [O] was first detected in Britain in 1993 (Brasier *et al.* 1999). It has killed over 10 000 riparian alders within 3 years in Britain and is now known from other western European countries. Comparison of the ITS region of rDNA, AFLPs, chromosome cytology and experimentally produced hybrids indicate that the hybrid is, in fact, a range of natural heteroploid hybrids between *P. cambivora* [O], not known to infect alder, and a fungus similar to *P. fragariae* var. *rubi* [O] found on *Rubus* species. Both species can infect Rubi and it is possible that this is where hybridization first occurred. How pathogenicity to alder has been acquired is not known but must presumably reflect in some way the genetic consequences of hybridization. Thus hybridization can produce genuine novelties possessing new features not shown by the parents.

A further case that is not wholly resolved is that of *P. meadii* [O], a parasite of rubber in Sri Lanka. It occurs as a sterile form with $2n = 9–12$, but larger fertile oospores can develop with $2n = 18–24$. Free-living forms which are similar to the latter also occur. The diploid *P. meadii* [O] is thought to be a sterile hybrid whose parentage is not known, with the fertile form being a tetraploid derivative. Since rubber was imported into Sri Lanka, one parent could have been introduced at the same time and hybridized with a resident species (Sansome *et al.* 1991). If this hypothesis is correct, it would resemble the conditions which favour hybridization in some other Eumycota, namely mixed populations in an ecologically novel environment.

Although spontaneous hybridization in nature evidently does occur in the higher Eumycota, it will be equally evident that, in many of the reported cases, human activities have provided circumstances which favour it. The number and range of these is now so great that it would not be surprising if other cases of fungal hybrids, especially of pathogenic forms, did not come to light in the foreseeable future.

### 8.2.3 Polyploidy

The evidence for widespread polyploidy in the Eumycota is not strong but it appears to be not uncommon in the Straminipila, at least in *Phytophthora* [O].

#### 8.2.3.1 Polyploidy in Eumycota

The only wholly unambiguous example of polyploidy in the Eumycota occurs in *Allomyces* [C]. Each species appears to exist as a polyploid series; *A. arbuscula* [C] has $n = 8$, 16, 24, 32 and *A. macrogynus* [C] has $n = 14$, 28, 50+. Their origin is obscure but parthenogenetic development of unfertilized female gametes is known from *A. arbuscula* [C] and parthenogenesis could be a contributory factor in these species. However, since meiotic pairing in each member of the polyploid series within a species is normal and *no* multivalents have been detected, autopolyploidy in *Allomyces* [C] may well be of ancient origin. No allopolyploids of hybrid origin have been detected as yet, but their existence is not inconceivable since hybridization appears to be a contemporary phenomenon (see section 8.2.1.1).

An example of a possible ancient polyploid is baker's yeast *Saccharomyces cerevisiae* [A]. Molecular evidence for duplication at many loci has led to the suggestion that the entire genome was duplicated at some time, although meiosis now appears to be normal (Wolfe and Shiels 1997). In addition, a contemporary tetraploid form (=$8n$) of *S. cerevisiae* [A] is also known (Byers and Goetsch 1975). Other occasional diploid forms are known, especially amongst the dikaryotic Basidiomycotina, e.g. *Armillaria mellea* [BA] (Franklin *et al.* 1983) and some other *Armillaria* [BA] species (Kim *et al.* 2000).

The diploid condition appears to be unstable in the hyphae of most normally haploid or dikaryotic Eumycota, and when diploid nuclei arise they eventually appear to revert to the haploid condition. Presumably there is positive selection against diploidy in the Ascomycotina and Basidiomycotina as a whole, although the situation is unclear in the Chytridiomycotina and Zygomycotina, apart from *Allomyces* [C].

In other Eumycota evidence for polyploidy is largely circumstantial or indicative. For example, in

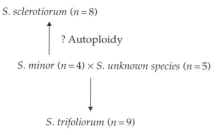

**Figure 8.4** Possible origins of polyploidy in *Sclerotinia*.

the genus *Sclerotinia* [AD] the haploid chromosome numbers recorded have suggested the relationships shown in Figure 8.4 (Willets and Wong 1980). As more electrophoretically or mitotically determined chromosome numbers become available, other such suggestive cases may come to light.

An even more clear-cut case using both chromosomal and molecular evidence is now available, namely *Botrytis aclada* ($=B.\ allii$) [FA] subgroup AII (Shirane *et al*.1989). Mitotic chromosome counts of *B. byssoidea*, *B. cinerea*, *B. squamosa* and *B. tulipae* [all FA] gave counts of $n=16$, but *B. aclada* [FA] (as *B. allii*) showed two clear-cut values, $n=16$ or $n=32$, and these were associated with an approximate doubling of the size of the conidia, namely $88.0–99.1\,\mu m^3$ or $103.8–200.3\,\mu m^3$ respectively. Later, the 16 chromosome types were designated as sub-group AI and the 32-chromosome types as sub-group AII which, it was suggested, was an autopolyploid of AI. Subsequently, using ITS rDNA RFLPs, Nielsen and colleagues (Nielsen and Yohalem 2001; Nielsen *et al*. 2001) demonstrated a clear difference between *B. aclada* AI and *B. bysssoidea* [both FA] and the occurrence in the sequence in AII of 23 ambiguous sites where the L45–550 DNA region was duplicated. By comparing sequences, three duplicated sites were found to match regions in AI of *B. aclada* and *B. byssoidea* [both AF]. This, together with the chromosomal evidence, strongly suggests that *B. aclada* AII [FA] is an allopolyploid derivative of the cross *B. aclada* AI × *B. byssoidea* [both AF]. Further suggestive evidence is provided by the intermediate size of the conidia of *B. aclada* AII [FA] between those of sub-group AI and *B. byssoidea* [FA]. It is not clear how the allopolyploid arose. Both these species cause neck rot of onions,

so that they can occur juxtaposed enabling hyphal fusions to take place. Therefore somatic nuclear fusion could initiate polyploidization (Nielsen and Yohalem 2001). However, *B. byssoidea* [FA] can develop a teleomorphic phase, *Botryotinia allii* [AD], so that fertilization of this species by conidia of the strictly anamorphic *B. aclada* [FA] is not inconceivable.

Apart from these examples, the only other suggestive evidence available is:

• the occurrence of multivalent formation at meiosis in *Cyathus stercoreus* [BA] (Lu 1964) and *Xylaria curta* [AP] (Rogers 1968), suggesting autoploidy
• different depths of nuclear staining in different species of the same genus under controlled conditions by DNA-specific stains such as DAPI (see Chapter 2, section 2.6.1), e.g. *Pleurotus* [BA] (Bresinsky *et al*. 1987)
• quite high haploid chromosome numbers within a genus, e.g. $n=18$ or 22 (Boehm *et al*. 1992; Sansome 1959) in *Puccinia* spp [BR] suggestive of possible polyploid series.

Although the impression given is that polyploidy is not a common occurrence amongst the Eumycota, its scattered nature in all the major taxonomic groups suggests that it may be more common than has been supposed. The difficulty of identifying the condition could also account for the few cases recorded. On the other hand, the fact that so few Eumycota are even diploid suggests that there is either strong selection for the haploid condition, or equally strong selection against the diploid condition or higher ploidy.

### 8.2.3.2 Polyploidy in Straminipila

In contrast with the Eumycota, polyploidy seems to be much more common in the Straminipila. Shortly after the discovery of the diploid condition in *Phytophthora infestans* [O], Sansome (1977) detected a tetraploid in Britain and a subsequent review demonstrated that $2n$, $3n$ and $4n$ isolates and a probable aneuploid isolate occur widely in the USA and Europe, although Mexican populations appeared to be strictly diploid (Tooley and Therrien 1991). The observation of chromosome rings, chains and multivalents in putatively polyploid isolates suggested autoploidy (Sansome 1977), and this seems even more likely since both diploid and tetraploid nuclei have been detected in the same mycelium in Dutch isolates (Tooley and Therrien 1977). In addition, *P. megasperma* vars. *sojae* and *megasperma* [both O] are diploid and tetraploid respectively (Sansome and Brasier 1974).

In view of the ability of several *Phytophthora* [O] species to hybridize, allopolyploidy could well occur. One probable fertile allopolyploid, *P. meadii* [O], has already been mentioned (see section 8.2.2) but other examples are not yet known.

### 8.2.4 The significance of hybridization and polyploidy for fungi

Hybridization can occur in Eumycota and appears to be more prevalent in some groups, such as *Achlya* [C] or the smut fungi (Ustilaginomycetes), than in others. However, it is not easy to detect and may be more widespread than is presently apparent. Much the same applies to polyploidy, but the available data are so slight that any conclusion must be highly conjectural. In contrast, to judge by *Phytophthora* [O] species, hybridization and polyploidy appear to be much more common in the Straminipila.

The impression given by the very limited data is that there is often positive selective pressure against hybridization in Eumycota, and even more so against polyploidy. Strong intraspecific antagonism is widespread in Ascomycotina and Basidiomycotina, and there are also strong and widespread mechanisms that regulate hyphal fusion between species. The only uncertainty is that occasional hybrids between sibling species could lead to limited introgression but, if so, its scale is unknown. As has already been noted, diploid nuclei break down sooner or later in normally haploid fungi and there is virtually no evidence for aneuploid or merodiploid forms persisting in nature, although they can be maintained in the laboratory in some cases (see section 8.1.2). Therefore it is not surprising that the best examples of hybridization occur in the more primitive Chytridiomycotina where there is no evidence for barriers to hyphal fusion and the occurrence of a normal diploid phase appears to be associated with both hybridization and a potential for polyploidy.

Lastly, as has already been noted, human influence in juxtaposing formerly separated host species and in creating new habitats can provide appropriate conditions in which hybridization and hybrid progeny are favoured (Hansen 1987; Brasier 2001).

## Further reading

*Heterokaryosis*

Caten CE and Jinks JL (1966). Heterokaryosis: its significance in wild homothallic ascomycetes and fungi imperfecti. *Transactions of the British Mycological Society*, **49**, 81–93.

Jinks JL (1952). Heterokaryosis: a system of adaptation in wild fungi. *Proceedings of the Royal Society of London*, **B140**, 83–99.

Pontecorvo G (1946). Genetic systems based on heterocaryosis. *Cold Spring Harbour Symposia on Quantitative Biology*, **11**, 193–201.

*Parasexuality*

Pontecorvo G (1954). Mitotic recombination in the genetic system of filamentous fungi. *Caryologia (Supplement)*, **6**, 192–200.

Zeigler RS, Scott RP, Leung H, Bordeos AA, Kumar J, and Nelson RJ (1997). Evidence of parasexual exchange of DNA in the rice blast fungus challenges its exclusive clonality. *Phytopathology*, **87**, 284–94.

*Hybridization*

Brasier CM (2001). Rapid evolution of introduced plant pathogens via interspecific hybridization. *BioScience*, **51**, 13–133.

Emerson R and Wilson CM (1954). Interspecific hybrids and cytogenetics and cytotaxonomy of *Euallomyces*. *Mycologi*, **46**, 393–434.

Holton CS (1951). Methods and results of studies on heterothallism and hybridization in *Tilletia caries* and *T. foetida*. *Phytopathology*, **41**, 511–21.

Schipper MAA, Gauger W and van den Ende H (1985). Hybridization of *Rhizopus* species. *Journal of General Microbiology*, **131**, 2359–65.

*Polyploidy*

Rogers JD (1973). Polyploidy in fungi. *Evolution*, **27**, 153–60.

Tooley PW and Therrien CD (1991). Variation in ploidy in *Phytophthora infestans*. In *Phytophthora* (ed. JA Lucas, RC Shattock, DS Shaw and LR Cooke), pp. 205–17. Cambridge University Press.

Wolfe KH and Shiels DC (1997). Molecular evidence for an ancient duplication of the entire yeast genome. *Nature*, **387**, 708–13.

# CHAPTER 9

# Change in gene frequency in fungal populations—I. General considerations: selection for oligogenically determined traits

No organisms, fungi included, are exempt from the process of natural selection (Darwin 1859). The ways in which selection operates and affects fungal populations will be considered in this and the next chapter.

Direct evidence of selection operating in *natural* situations is sparse, although some information is available from agricultural situations and experimental investigations of economically important characters of pathogens of crop plants in growth rooms, glasshouses or, less frequently, the field.

A wide range of characters has been studied, including selection for particular pathotypes in samples taken from natural populations, selection between different pathotypes of the same taxon for increased aggressiveness, for resistance to fungicides or for a variety of secondary metabolites in various industrial fermentations, most notably penicillins. Selection for quantitative (metric) characters, usually studied in laboratory strains, include differences in growth rate, time to sexual reproduction, ascospore size and a variety of correlated responses to selection.

Despite the dearth of direct evidence from natural situations, most experimental approaches have had some relevance to actual situations under natural conditions in two ways. First, they have demonstrated some of the potential mechanisms that could bring about selective response. Second, they have usually involved features of potential importance under natural or quasi-natural conditions, such as occur in agriculture, and have shown

that they are actually affected by selection. In addition, because correlated responses have been demonstrated in some experiments, albeit in far from natural conditions, observers have been alerted to this possibility in natural situations. Therefore experiments provide models of how selection can act on fungi, but extrapolation to natural conditions, where environmental heterogeneity is far greater, needs to be undertaken cautiously.

In all, the number of experimental studies on selection using fungi, compared with those done with diploid Eukaryotes, has not been great, and there is even greater ignorance of the operation, magnitude or scale of selection in natural fungal populations. Nevertheless, even these limited experimental investigations are helpful when considering the data available from fungi for, as Endler (1986) has written in a wider context:

Laboratory populations serve as models of nature, and help to test specific predictions or conjectures about the way nature works; but without extensive knowledge of natural selection in the wild, we have no idea how relevant experiments or theory are to the evolution of natural population.

## 9.1 General considerations

Except for genetically homogenous clones, some individuals in any fungal population will exhibit intrinsic differences that will be reflected in some way in their phenotypes. Of course, phenotypic traits can vary either continuously (quantitative or metric traits) or discontinuously (polymorphic traits). However, in any given environment selection acts on the organism as a whole, i.e. on the individual phenotype. Nevertheless, allowing for the fact that every gene contributes to an organism's gene complex as a whole, so that its contribution must be pleiotropic to some extent, selection can conveniently be thought to act in one of two ways:

• on a variety of alternative polymorphic traits, each determined by a single pair of alleles, individually having clearly distinguishable effects (**oligogenic** action)
• on quantitative traits determined by several genes, each contributing a small amount to the trait, but whose individual contributions are not readily distinguishable because the effect of environmental variation exceeds the contribution of each individual gene, so that variation in the population appears to be continuous (**multigenic** or **polygenic** action).

Oligogenic selection results in change in gene frequency (Figure 9.1(a)). However, the individual phenotypic contributions of polygenes to a continuous character are usually much smaller than the total phenotypic variation, so that the observed variation of most quantitative traits arises from both genetic and environmental causes. The range of variation in metric characters in a population, appropriately scaled if necessary, often approximates to a normal curve that is best described by its mean and variance (Figure 9.1(a)).

Although, most usually, several genes of individually small effects are involved, this is not necessarily the case. Indeed, segregation of as few as three genes, one having three times and another twice the effect on the third, is sufficient to produce a range of expression approximating to a normal distribution, although 90 per cent of the variance is due to segregation at only two of the loci (Figure 9.1b) (Thompson 1975).

Regardless of whether polymorphic or quantitative characters are involved, the effects of selection on a population can be described in general terms as **stabilizing, directional** or **disruptive**. Under relatively constant environmental conditions the

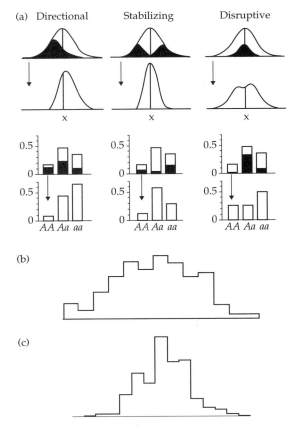

**Figure 9.1** The effects of three modes of selection—directional, stabilizing and disruptive—on segregations due to multigenic or oligogenic segregation. (a) This diagram illustrates selection for both quantitative traits (upper curves) and oligogenically determined traits (lower histograms). In the upper diagram of each set, the area under the curves/histograms represents the total number of individuals and the shaded areas are those individuals at a *relative disadvantage* under the particular mode of selection. The lower diagrams illustrate the consequences of selection. (b) Oligogenic segregations can approach multigenic segregations in appearance if their individual contributions differ in value. The histogram represents a random sample of 150 individuals from a theoretical F2 population in which three such loci are segregating but one gene has three times the effect and another twice the effect of the third on the character. The first two loci account for 90 per cent of the variance. The effects of environmentally induced variation have been excluded but in (c) segregation is only occurring at one locus but environmental variance is included. ((a) Reproduced from Endler (1986), with permission from Princeton University Press; (b) reproduced from Thompson (1975) with permission from Nature Publishing Group.)

most common outcome is that the optimum phenotypes for any particular environment are favoured. In the case of a quantitative trait, optimum phenotypes may well be genotypically different because of phenotypic equivalence of different combinations of polygenes and their differential environmental interactions (Figure 9.1(a)) (Mather 1973).

Even under variable environmental conditions the phenotype selected is likely to be that which performs best under the range of conditions, but if a sudden change occurs in some environmental variable a new optimum phenotype may be rapidly selected. Brasier (1987) has described the latter situation as **episodic selection** as opposed to **routine selection**. Whatever form the selective process takes it involves directional, disruptive or stabilizing selection to different degrees.

Stabilizing selection tends to reinforce the mean phenotypic expression and reduce variability. It is less well documented for fungi than is directional selection.*

Directional selection occurs when it results in the shift of a trait, or a population's traits, away from the existing mean value in one direction or another. Selection of characters determined oligogenically is almost entirely of this type. Most of the available information on selection in fungi is concerned with directional selection of such characters.

There is little direct evidence for disruptive selection in fungi, i.e. selection away from the mean to two or more optima. Nevertheless, it presumably occurs in situations such as selection for different hosts (see Chapter 10, sections 10.3.2 and 10.3.3). Apart from this kind of selection in pathogenic fungi and selection for different environments, disruptive selection has received little attention, even experimentally.

A fourth kind of situation is where two or more forms occur in the same environment in proportions such that the rarest cannot be maintained by

mutation alone (Ford 1965). In fact, they are maintained by some kind of **balancing selection**. This situation, a **balanced polymorphism**, can arise from a variety of causes. In diploids and dikaryons, a balanced situation can arise if the heterozygous condition is at an advantage when measured against the homozygous condition. Inevitably, homozygous forms will be produced constantly by sexual reproduction but maintained at some equilibrium value in the population through selection. This mechanism cannot apply to haploid fungi, but if selection is **frequency dependent** then a balanced situation can arise. For example, if the relative fitness of an individual determined by one of a pair of alleles changes adversely with increased frequency of one of the alleles, such individuals would increase only to the point where they became disadvantaged compared with individuals whose fitness was determined by the alternative allele. Individuals carrying the first of these alleles would then be selected against. Eventually a stable but dynamic balance between the two forms would be achieved. The most common genetical situation is when a rare allele is at a selective advantage but becomes less so, or even disadvantaged, when it becomes common. There is not much evidence at present for balancing selection in fungi by any of these mechanisms, although cases are now beginning to be found (see section 9.4.3.2).

**Density-dependent selection** also occurs, often as a result of intra-taxon competition between different forms whose competitive abilities vary with the overall population density (see section 9.4.3.1).

Some traits, however determined and depending on their total environment, will affect the survival and fertility of each individual. Therefore some individuals survive better and increase at the expense of the others. Relatively, they are fitter. The **relative fitness** of an individual is defined as a measure of its ability to transmit its genes to the next and subsequent generations relative to the abilities of other individuals in the population. Therefore the effects of selection on populations are consequential. In other words, the fittest contribute relatively the greatest proportion of progeny to the next generation, and so differences in the relative fitness of different individuals result in changes

---

* An important caveat on usage needs to be given when consulting the plant pathological literature. Here, 'stabilizing selection' has, unfortunately, been widely *misused* by van der Plank and others (Van der Plank 1968) to describe supposed selection against 'unnecessary' virulence genes in crops whose hosts lacked the complementary resistance genes (see section 9.4.1.3).

in gene frequency. Finally, since selection results in an increase in some members of a population and a decrease in others, it is a process which affects population dynamics as well as population genetics. The former is, of course, profoundly affected by the ecological situation in which a fungus exists. Therefore to understand fully how selection operates, both aspects—the ecological and the genetical—need to be considered together. In addition, fitness needs to be quantified and the features which determine its nature analysed if the behaviour of fungal individuals, or the effects of selection on the distribution of genes in their populations, is to be understood.

## 9.1.1 Quantifying selection and fitness in fungi

Selection is inevitably a complex dynamic process which often differs in different parts of the range of a species or population. Consequently, only its average effects over time at a particular locality can be quantified. In the majority of cases selection in fungi has been considered in a very simple context, and the analysis and measurement of selection in the field is a somewhat intractable matter. It has become increasingly clear that this is inadequate for many purposes. More sophisticated theoretical approaches are being developed which are either generally applicable (Clarke and Beaumont 1992) or applicable in relation to specialized situations such as co-adaptation or pathogen–host situations (Leonard 1977; Leonard and Czochor 1980; Frank 1991, 1993; Brown 1995) (see also Chapter 10, Appendix). To date, the available models remain controversial. However, a very simple measure which gives some indication of the average direction and magnitude of selection is the selection coefficient.

### 9.1.1.1 Oligogenic selection: selection coefficients

Consider first a simple example of a mixed population of two strains, A1 and A2, of a haploid fungus such as fission yeast which grows by simple division. Assume that the strains differ in their rates of growth which are determined by a pair of alleles $A$ and its mutant $a$, whose relative fitnesses are $w_1$ and $w_2$, or 1 and $(1-s)$ respectively. The

selection coefficient $s$ is a measure of the lower rate of growth and multiplication of strain A2 compared with that of strain A1, i.e. the consequential relative disadvantage of the $a$ allele compared with $A$ which is set arbitrarily at unity. Assume also that growth is exponential, that the cell generations are non-overlapping and that the initial frequencies of the two strains are $p_0$ and $q_0$ respectively, where $p_0 + q_0 = 1$. In the next generation the new frequencies $p_1$ and $q_1$ will be

$$p_1 = \frac{p_0}{1 - sq_0}$$

and

$$q_1 = \frac{q_0(1-s)}{1 - sq_0}$$

or

$$p_1 = p_0/w$$

and

$$q_1 = (q_0 - sq_0)/w$$

where $w = (1 - sq_0)$ is the mean fitness. The selection coefficient $s$ can then be estimated from the known frequencies in the two generations from the rearrangement

$$\frac{q_1}{p_1} = \frac{q_0(1-s)}{p_0}.$$

In general, for $n$ generations, on the same assumptions, provided that selection remains constant over the generations and is not affected by genotype frequencies or population density,

$$\frac{q_n}{p_n} = \frac{q_0(1-s)^n}{p_0}$$

or

$$\ln(q_n/p_n) = \ln(q_0/p_0) + n\ln(1 - s).$$

This relationship enables an alternative approximate estimate to be made of the **average selection coefficient** $s_m$ over the generations, since a plot of $\ln(q_n/p_n)$ against $n$ will have a slope $\ln(1 - s_m)$ from which $s_m$ can be calculated. The estimate can only be approximate since in natural situations none of the assumptions set out above are likely to be valid over the period studied.

Within all the constraints just mentioned, the change in $q$ per generations is given by

$$\frac{q - s_m q}{1 - s_m q} - q \quad \text{or} \quad -\frac{s_m q(1 - q)}{1 - q}$$

or, in terms of relative fitnesses,

$$[q w_2/(p w_1 + q w_2)] - q \quad \text{or} \quad (w_2 - w_1) p q / w.$$

Therefore the new value of $q$ after selection, $q$, is

$$\Delta q = -\frac{s_m q(1 - q)}{1 - s_m q}$$

or

$$\Delta q = \frac{-(w_1 - w_2) q(1 - q)}{w}.$$

Because fungal hyphae are coenocytic, or include quasi-coenocytic regions, selection can occur not only between individuals or genets but also intra-hyphally, between nuclei. Therefore the simple treatments of selection coefficients just described can be applied equally effectively to inter-nuclear selection within a heterokaryon. There is clear evidence for such selection (see section 9.3.2 and 9.3.3).

The situations just described are applicable when the growth of the population is exponential and generations are discrete. If growth can be considered as continuous, an estimate can be made of the number $n$ of generations required to bring about a change in gene frequency from the initial value $q_0$ to the desired value $q_n$. This is possible because if $sq$ is very small, as will often be the case, the rate of change can be considered as a continuous function and

$$dq/dt = -sq(1 - q).$$

Therefore integrating,

$$\int_{q_0}^{q_t} dq/q(1 - q) = -s \int_0^t dt$$

and, taking logarithms,

$$t = \frac{\ln[q_0(1 - q_t)/q_t(1 - q_0)]}{s}.$$

Models for selection under particular circumstances, such as estimating the coefficient of selection for haploid fungicide-resistant pathogens, a widespread problem in agriculture, have been developed (Milgroom *et al.* 1986). Similar models have been used by MacKenzie (1978) and by Groth and Barrett (1980), and a slightly different approach has been adopted by Chin (1987). All give similar results.

The theoretical treatment of selection in straminipilous fungi is in every way comparable with that in other diploid organisms, except that many of them can persist as asexual clones for many generations, e.g. *Phytophthora infestans* [O]. The situation in a diploid straminipilous population analogous to that just described for a haploid in a sexually breeding population follows a similar argument. Consider a situation where the homozygous dominant and heterozygous individuals $AA$ and $Aa$ are equally fit ($w_1 = 1 = w_2$) but the homozygous recessive $aa$ is less fit ($w_3 = (1 - s) < w_1$ or $w_2$), where $s$ is the selection coefficient, and the genotypic frequencies of $AA$, $Aa$ and $aa$ individuals in the parental population are $p_0^2$, $2p_0q_0$ and $q_0^2$ respectively. Then in the next generation

$$p_1 = \frac{p_0^2 w_1 + p_0 q_0 w_2}{w}$$

and

$$q_1 = \frac{p_0 q_0 w_2 + q_0^2 w_3}{w}$$

where

$$w = p_0^2 w_1 + 2p_0 q_0 w_2 + q_0^2 w_3.$$

Hence the rate of change after one generation is

$$\Delta p = \frac{p_0 q_0 [p_0(w_1 - w_2) + q_0(w_2 - w_3)]}{w}.$$

Most of the selection coefficients obtained in experimental studies with fungi are based on the various simple calculations set out above. In practice, most published data have been obtained from reproductively isolated asexually reproducing populations, or over a time period of less than one sexual generation, e.g. from infection to production of a further generation of spores or conidia.

Published fungal estimates of selection coefficients are almost entirely derived from various pathogenic fungi under experimental situations and are minimally indicative. However, the majority do give an indication of sufficient value to demonstrate the existence of selection and some idea of its order of magnitude over these short periods. Collected values of $s$ range from 0.025 to 0.804, with the majority ranging from 0.3 to 0.5 (Ennos and McConnell 1995). The average is 0.122 for situations involving a single marker locus, and 0.322 when the groups compared differed at several loci. Despite the necessary reservations in comparing these results with those found for other Eukaryota, it is quite clear that the values of selective coefficients found for fungi are quite high but broadly comparable with many listed for diploid animals and plants. It would be useful to remedy existing deficiencies in the kind of fungal data already reported through appropriate design of experiments, but a greater need is to determine the order of magnitude of selection in natural situations, for which adequate information is lacking.

Endler (1986), discussing data from a wide range of diploid Eukaryota, has drawn attention to the numerous sources of errors involved in many estimates and the difficulties of comparing estimates derived from quite dissimilar situations, constraints which apply to most published fungal estimates. Provided that sufficient data are available, or can be collected, it is possible to make appreciably more reliable and precise estimates of selection coefficients. No fungal data obtained to date are sufficiently sophisticated to achieve this.

### 9.1.1.2 Multigenic selection: selection differentials

Quantitative variation can usually be represented by a normal curve (Figure 9.1) and, since both genetic and environmental factors contribute to the phenotypic variance $\sigma_P^2$,

$$\sigma_P^2 = \sigma_G^2 + \sigma_E^2$$

where $\sigma_G^2$ represents the genetic contribution and $\sigma_E^2$ the environmental contribution. However, $\sigma_G^2$ is complex as it is made up of a number of contributions:

- contributions from each of the genes involved—the additive component $\sigma_A^2$
- the contribution due to interactions between the genes—the epistatic component $\sigma_I^2$
- in a diploid or dikaryotic fungus, the contribution due to dominant alleles $\sigma_D^2$.

Therefore, for a haploid fungus

$$\sigma_G^2 = \sigma_A^2 + \sigma_I^2$$

and for a diploid/dikaryotic fungus

$$\sigma_G^2 = \sigma_A^2 + \sigma_I^2 + \sigma_D^2.$$

The quantity $\sigma_A^2/\sigma_P^2$ is described as the **narrow sense heritability** and is denoted by $h^2$.

If, as a result of selection, the mean value of the population is shifted to a new value $\mu_N$, then the **response to selection** $R$ is obviously $\mu_N - \mu$. It can be shown that

$$R = h^2 S$$

where $S$ is the **selection differential** which is the product of the **intensity of selection** $i$ and the phenotypic variation, i.e. $S = i\sigma_P^2$. Therefore

$$R = ih^2\sigma_P^2.$$

The change in fitness of a multigenic character under selection was initially established by Fisher (1930) in his 'fundamental theorem of natural selection':

The rate of increase of fitness of any organism at any time is equal to its genetic variance in fitness at that time.

However, it should be noted that, in Fisher's context, 'genetic variance' refers strictly to the additive variance of fitness and also that the theorem is strictly true only when successive generations in the population overlap.

Fitness is best measured, ultimately, by the numbers of viable offspring. Thus, if an individual leaves $n$ offspring and the population as a whole in that generation leaves a mean number of offspring $\tilde{n}$, the relative fitness $W$ of that individual is $n/\tilde{n}$. If the mean relative fitness of the population $\hat{W}$ is set at 1 and there are $N$ individuals in the parental generation, the **selection differential for relative fitness** is

$$S_W = \sum \frac{W(W - \hat{W})}{N}$$

and it can be shown that $S_W = V_{P(W)}$ (Falconer and Mackay 1996). Hence, the **response to fitness**

$R_W$ is equal to the additive genetic variance of fitness, i.e.

$$R_W = h_W{}^2 V_P(W) = V_A(W).$$

It should always be remembered that the phenotypic variance shown by characters determined by many genes is in part also determined by the environment, so that environmental changes can play a significant part in determining the course of selection.

It has to be said that these quantities have rarely been calculated for fungal situations but this reflects more the dearth of investigations rather than their utility.

## 9.2 Special aspects of fungal selection

The unique nature and behaviour of the fungal mycelium should be taken into account when considering the action of selection on fungi. Three aspects are of particular importance for selection.

First, because of the structure of the fungal mycelium, both intra-mycelial and inter-mycelial selection must be considered. Intra-mycelial selection can bring about genetic change even in homokaryotic haploid strains or in heterokaryons. If conditions alter, the nuclear selection involved may be reversible and so provide a sensitive response to local conditions. Intra-nuclear selection can affect any cellular component. All such responses need to be distinguished from the metabolic adaptation discussed above. Although selective effects on different genets are the most common forms of selection studied in fungi, any heterokaryotic genet may, potentially, be exposed simultaneously to *both* intra- and inter-mycelial selection.

Second, since many Ascomycotina and Fungi Anamorphici—the vast majority of Eumycota—are haploid and clonal, or behave as such, it is of particular interest to see whether the processes involved and their consequences are notably different from those of the majority of eukaryotes in which selection has been demonstrated, since they are neither haploid nor clonal. Deleterious mutants cannot be 'protected' by heterozygosity in haploids and therefore are more liable to be lost, but if nuclei which carry them are constituents of heterokaryons or dikaryons they may be

'protected'. Epistatic interactions are not, of course, inhibited in either haploids or diploids. The Straminipila, being diploid, are genetically comparable with other Eukaryota.

Lastly, the widespread phenotypic plasticity of a fungal mycelium (see Chapter 1, section 1.6.2) and its ability to respond to its immediate environment, either in a particular region or as a whole, results in situations that need to be distinguished from genetic selection. Such adaptive but generally reversible responses are best described as acclimation to distinguish them from true genetic adaptation. Examples are adaptive metabolic behaviour such as the utilization of different substrates in mixtures or successively, as in the decomposition of complex substrates. Acclimation is brought about by the interplay of intracellular inducer and repressor systems but is readily reversible by altering the external conditions. Analogous morphological responses, such as diffuse versus dense localized branching in relation to the nutritional situation in the vicinity of growing hyphal apices, are similarly adaptive and reversible. Of course, fungi do differ in their genetic ability to respond in these ways but virtually no information is available on how selection operates in such situations.

## 9.3 Intra-mycelial selection

The evidence for intra-mycelial selection is almost entirely derived from experimental situations, although presumably it can and does occur in nature together with inter-mycelial oligogenic and multigenic selection.

Intra-mycelial selection can act on cell organelles such as mitochondria, on extra-nuclear components in the cytoplasm, on alien elements such as plasmids and dsRNA viruses, on nuclei or on more than one of these. Since all components within mycelia can undergo mutation, a heteroplasmic or heterokaryotic condition can arise, albeit in a small region in any mycelium, and, depending on the selective advantage of the mutant and environmental conditions, may persist or be lost. Two important cases of nuclear selection in heterokaryons are the fate of a nucleus carrying a new mutant in an otherwise homokaryotic mycelium and selection between nuclei in two or more component heterokaryons. Little is known about any of these situations except from experimental investigations because so few cases of natural heteroplasmons or heterokaryons have been identified.

## 9.3.1 Non-nuclear intra-mycelial selection

Theoretically, at least, intra-mycelial selection in strictly homokaryotic haploid strains must be due to selection for extra-chromosomal factors but no unequivocal examples are known. A possible case is that of the *kalilo* transposable element in *N. intermedia* [AP]. Mitochondria carrying this element gradually increase in number and normal mtDNA-carrying mitochondria decline during hyphal growth until, eventually, the new senescent *kalilo* phenotype develops (Bertrand *et al.* 1985). Presumably, therefore, *kal* mtDNA has a selective advantage over normal mtDNA.

Other examples are known where the causal agent has not been identified. For example, selection of conidia of *Aspergillus glaucus* [A] was effective for increased or decreased linear growth rate in successive subcultures of an apparently homokaryotic clone isolated from nature (Jinks 1957) and similarly for perithecial density in *A. nidulans* [A] (Jinks 1966). Normal subculturing resulted in no change in the expression of the character concerned, but if selection was exercised for its increase or decrease, the outcomes were significantly different (Figure 9.2).

### 9.3.1.1 Mycoviruses
Mycoviruses can be involved in intra-mycelial selection. The majority of known fungal viruses appear to have little phenotypic effect on their host fungi, but there are exceptions. The hypoviruses of the chestnut blight fungus *C. parasitica* [AP] have notable effects on mycelia carrying them, including reduced growth rate and conidial production as well as inhibition of female fertility, so that isolates are effectively female-sterile (Nuss and Koltin 1990; Nuss 1992). This clearly reduces their competitive and reproductive ability compared with normal strains. Infectivity is not apparently reduced, but their pathogenicity is because the fungus only produces superficial cankers on trees.

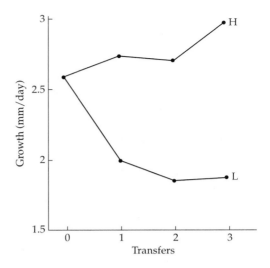

**Figure 9.2** Selection in *Aspergillus glaucus* [A] for high and low growth rates starting from a single clone: high growth rate (curve H) and low growth rate (curve H) over three successive transfers. (After Jinks (1957), with permission from the Royal Society of London).

The mycoviruses are transmitted horizontally through hyphal fusions and these are controlled by *vic* genes. A recipient mycelium assumes the phenotype of the virus-carrying donor mycelium in most cases, but some *vic* genes are known to inhibit virus transmission (Huber and Fulbright 1994). Thus selection operates at the time of hyphal fusion and reflects a selective conflict between the nuclear genomes and virus elements in both mycelia.

D-factors in *O. novo-ulmi* [AP] are supposedly dsRNA units. They have similar effects to the hypoviruses of *C. parasitica* [AP], and are transmitted in the same way. Hyphal growth and perithecium production are reduced but when the fungus adopts a yeast phase, as when invading xylem tissue, the phenotypic effects appear to be lost. Evidently internal selection against these dsRNa elements is related to the internal metabolic condition of the cell, but the situation is not entirely clear (Brasier 1986b).

## 9.3.2 Experimental inter-nuclear intra-mycelial selection in heterokaryons

The origin of a heterokaryon can also involve selective processes. Two-component heterokaryons can be established by allowing compatible uninucleate spores to germinate in close proximity so that their germ tubes anastomose. Those with heterokaryotic tips in a spatially favourable position to grow out have an enormous advantage over any others. Therefore the nuclear ratios of these spatially favoured tips would determine, initially at least, the growth rate of a heterokaryotic mycelium. Further inter-nuclear selection during subsequent growth is not thereby precluded. Such experiments provide a model for the genesis of natural heterokaryons and for selection amongst conidia and spores which have been dispersed in, or accumulated as, masses, which frequently occurs in nature. Therefore the chance advantage conferred by position may be important under natural conditions, especially if inter-nuclear selection is weak.

Another way in which chance, selection or both can operate is when multinucleated spores derived from a heterokaryon germinate. Even if the heterokaryon has a 1 : 1 ratio of nuclei in its mycelium, this is not necessarily the case in its spores, and even when the 1 : 1 ratio is maintained within the spores, a particular nuclear type may gain an advantage at germination. For instance, in many fungi with multinucleate spores only a single nucleus enters the germ tube and, by its subsequent divisions, gives rise to the nuclear population of the mycelium. This nucleus may enter the germ tube either by chance because of its spatial position, or because it multiplies more rapidly as its genome endows it with some selective bias.

Environmental selection for a particular nuclear component of a heterokaryon can modify any intrinsic advantages of that nucleus and the effect can be reversible or irreversible. For example, a natural heterokaryotic isolate of *Penicillium cyclopium* [A/FA], which was unstable when grown on Czapek–Dox agar, sectored spontaneously to give two distinct homokaryons. Each carried a different but genetically uncharacterized kind of nucleus *4A* or *4B*; homokaryons of the former grew much faster. When different amounts of 10 per cent apple pulp were added to the synthetic medium on which the heterokaryon was grown, the relative proportions of the two kinds of nuclei and growth rate could be varied. The proportion of *4A* nuclei reduced progressively with increased amounts of pulp and *vice versa*, resulting eventually in mycelial instability (Jinks 1952 and Table 9.1). However, in an appropriate constant nutritional environment, stabilizing selection maintained a constant nuclear ratio within the heterokaryon. In this example, there was virtually a one-to-one relationship between the nuclear ratio and the phenotypic response in those environments where selection was reversible, but this is not always the case.

Davis (1959) investigated the rate of loss of a nuclear component from a heterokaryon when

**Table 9.1** Estimated nuclear ratios in an dual heterokaryon of *Penicillium cyclopium* [FA] carrying 4A and 4B nuclei grown on different amounts of minimal media and 10 per cent apple pulp

| Medium composition | | No. of colonies scored | Percentage of 4A nuclei estimated in heterokaryon |
|---|---|---|---|
| 10% apple pulp in medium | Minimal medium | | |
| 100 | 0 | 713 | 13.89 |
| 75 | 25 | 463 | 14.49 |
| 50 | 50 | 550 | 17.24 |
| 30 | 70 | 1056 | 23.25 |
| 15 | 85 | 155 | 46.67 |
| 5 | 95 | 463 | 33.33 |
| 0 | 100 | — | — |

The homokaryons differed in morphology and growth rate.
(Based on Jinks (1952), with permission from the Royal Society of London.)

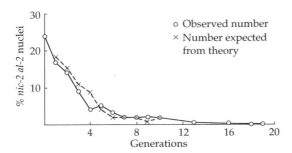

**Figure 9.3** Loss of nuclei selected against from a heterokaryon of *Neurospora crassa* [AP] over successive conidial transfers. The constitution of the heterokaryon was *pan-1 al-1/nic-2 al-2* and it was grown on pantothenate agar so that *nic-2 al-2* nuclei were selected against. Note that, even though selected against, *nic-2 al-2* nuclei still persisted after 20 transfers. (After Davis (1959).)

selected against by the external medium. Successive transfers of 100–600 conidia from a *nic-2 al-2/pan-1 al-1* heterokaryon were made to media supplemented with either pantothenate (*pan*) or nicotinic acid (*nic*). On the pantothenate-supplemented medium the proportion of *nic-1/al-2* nuclei fell rapidly at first, but then more and more slowly, and were only eliminated after 20–30 transfers or more (Figure 9.3).

The rate of loss follows formulae developed by Davis (1959) from Atwood and Mukai's (1955) original formulae describing the proportions of heterokaryotic and homokaryotic nuclei which would enter the multinucleate conidia developed on a heterokaryon. If $r$ is the proportion of heterokaryotic nuclei, $a$ and $b$ are the proportions of the two kinds of nucleus in a two-member hetrokaryon and $\bar{n}$ is the mean number of nuclei per conidium, the proportion $p$ of nucleus $a$ will be

$$p \cong \frac{r(1-r) + a(\bar{n} - 2r)}{\bar{n}(1-r)}.$$

Substituting $b$ for $a$ gives the proportion of $b$ nuclei. Under selection, the nuclei selected ($a$) will contribute a reducing proportion $x$ at each subsequent transfer according to (Davis 1959)

$$p_{t+1} = \frac{r(1-r) + a(\bar{n}r - 2r) + \bar{n}a(1-r)x}{\bar{n}(1-r)(r+ax)}.$$

Thus, coupled with chance fluctuations, it is a common finding that, even with selection,

complete elimination of one component of a heterokaryon is often difficult to achieve as in this experimental transfer.

Cultures are often found to be 'carrying' mutant nuclei. For example, cultures auxotrophic for lysine (*lys*⁻) are often inhibited by arginine, but some appear less sensitive to such inhibition. The latter were found to be heterokaryotic for both *lys*⁻ and spontaneous mutant genes which reduced sensitivity to arginine inhibition (MB Mitchell, quoted by Emerson (1952)).

An issue of some practical importance in nature is the inter-mycelial fate of nuclei carrying spontaneous lethal mutations. Model experiments show that they can persist. Twenty-four out of 26 spontaneous mutants which arose in a synthetic heterokaryon of *N. crassa* [AP] capable of normal growth on minimal medium proved to be lethals in the sense that, when isolated in homokaryons, they would not grow on minimal medium (Atwood and Mukai 1953) (see Chapter 6, section 6.1.1). Clearly, such heterokaryons would persist in nature.

Even when no phenotypic change is obvious and overall mycelial growth appears to be insensitive to the nuclear ratio, internal selection for genetically distinct nuclei can occur. For instance, in synthesized two-component heterokaryons of *Neurospora* [AP] whose nuclei carried different complementary auxotrophic mutants, growth on minimal medium was normal over short periods and the growth rate was unaffected although the initial and final nuclear ratios varied widely, e.g. in *N. crassa* [AP] heterokaryons between *pan/al-1* carrying nuclei together with *nic-2/al-2*, *arg-6*, or *lys-3* nuclei (where *pan*, *nic*, *arg* and *lys* are mutants requiring supplementation with nicotinic acid, pantothenate, arginine and lysine respectively, and *al* is an albino culture) (Pittenger *et al.* 1955). After linear growth along a growth tube for 1000 mm, the nuclear ratio towards the apical region often settled down to an unpredictable different one from that determined initially by mixing uninucleate conidia in specific ratios. The nature of the internal selective process involved is not clear, but the heterokaryon appeared to show rather poorly controlled stabilizing selection.

The causes of selection are not always obvious. For instance, an unusual kind of nuclear selection occurs in dikaryotic diaphoromictic fungi such as *Schizophyllum commune* [BA], when the condition arises from a dikaryotic–monokaryotic (di-mon) mating. In general, a dikaryon $AxBx + AyBy$ will mate with a monokaryon $AzBz$ to give rise to two new dikaryons $AxBx + AzBz$ and $AyBy + AzBz$. In some matings equal numbers of the two new dikaryons are produced, but in others the numbers are unequal. This bias depends upon the genetic substructure of the MTFs, which were originally thought to comprise two very closely linked multi-allelic subunits, separable by crossing over, e.g. $Ax = Ax_{(1-1)}$ where (1–1) represent the two subunit alleles (Raper *et al.* 1958).

The usual criterion for a compatible mating, i.e. alleles should be different, applies to subunit alleles as much as to the complex alleles. Hence the possible dikaryons are as follows:

• doubly heterozygous in respect of $A$ and $B$ alleles (1/2): $(Ax_{(1-1)}Bx_{(1-1)} + Ay_{(2-2)}By_{(2-2)})$
• doubly heterozygous in respect of $A$ alleles but with one $B$ allele in common (e.g. 1/2 and 1/1): $(Ax_{(1-1)}Bx_{(1-1)} + Ay_{(2-2)}By_{(1-2)})$
• doubly heterozygous in respect of $B$ alleles but with one $A$ allele in common (e.g. 1/2 and 1/1): $(Ax_{(1-1)}Bx_{(1-1)} + Ay_{(1-2)}By_{(2-2)})$
• one $A$ allele and one $B$ allele in common (e.g. 1/1 and 1/1) $(Ax_{(1-1)}Bx_{(1-1)} + Ay_{(1-2)}By_{(1-2)})$.

Thus, in a di-mon mating, the monokaryon has a choice between two possible partners. When the choice lies between mating with a dikaryon with a common $A$ or a common $B$, there is a heavy bias in favour of new dikaryons doubly heterozygous for $B$ subunit alleles. Such a cross was as follows (Raper and Ellingboe 1962):

$$A41_{(3-5)}B41_{(2-2)} \text{ (monokaryon)} \times (A47_{(1-1)}B47_{(3-2)}$$
$$+ A42_{(3-4)}B42_{4-4)} \text{ (dikaryon)}$$

gave the dikaryotic progeny

$32(A47_{(1-1)})B47_{(3-2)}) + A41_{(3-5)})B41_{(2-2)})$,    i.e. doubly heterozygous for $A$ alleles ($A1/3$ and $1/5$)

$274(A42_{(3-4)})B42_{(4-4)}) + A41_{(3-5)})B41_{(2-2)})$,    i.e. doubly heterozygous for $B$ alleles ($B4/2$ and $4/2$).

Conversely, when $B$ mating type factors are selectively neutral, there is a bias in favour of new

dikaryons heterozygous for $A$ subunits. Presumably this is due to some form of inter-nuclear competition and, since homo- and heterozygosity are concerned, the MTF composition of a population where di-mon matings are taking place is probably the result of balancing selection between the subunits. This situation will need further investigation in the light of the more complex structure now assigned to diaphoromictic MTFs (see section 9.4.3.2) (May *et al.* 1991; Kües and Casselton 1992).

### 9.3.3 Inter-nuclear intra-mycelial selection under industrial conditions

In many industrial situations the objectives are:

• to select new mutants with enhanced productivity of some kind
• to ensure that equilibrium conditions obtain
• to prevent strain degeneration.

Experimental studies on inter-nuclear selective processes inform and complement important situations which often occur in industrial processes.

The famous mutation–selection 'Wisconsin' series of strains of *Penicillium chrysogenum* [FA], which resulted in increased yields of penicillin, illustrate some of these features (Backus and Stauffer 1955). The Wisconsin laboratory was extremely successful in increasing the yield of penicillin in fermenters through a combined programme of selection and mutagenic treatment. However, as yield increased both growth rate and conidiating ability of the strains decreased and eventually tended to level out. This correlated change appeared to be associated with inter-nuclear selection. On plating out uninucleate conidia from the starting strain *Q176*, derived by limited selection from the original penicillin-producing strain *NRRL* 1951, five main colony types were detected. Thus *Q176* was evidently a multiple heterokaryon or a mixed culture in respect of genetic factors determining colony size, shape and growth rate. Selection was associated with an approach to homokaryosis strain *51–20* producing colony type C predominantly (94 per cent). However, sooner or later, most of the colonies isolated sectored, presumably due to spontaneous mutation and intra-mycelial nuclear selection for increased growth rate. By selecting from such sectors it proved possible to isolate the spontaneously produced strain *F3–64* which retained the high yield of penicillin but had recovered the original ability to grow and conidiate. The precise nature of the genetic changes involved in the 'Wisconsin' series has never been analysed. It is not clear whether penicillin yield was determined polygenically or oligogenically although, if the latter, more than one gene was involved, as demonstrated subsequently (Ball 1971). It is also possible that the selective processes represented, in part, inter-mycelial selection between genetically different strains producing the different colony types, rather than selective segregation within a complex heterokaryon. It seems probable that both forms of selection could have acted, singly or together, at different stages of the selective sequence.

Wiebe *et al.* (1991, 1993, 1996) have clarified the selective processes more fully using continuous-flow cultures of *Fusarium graminearum* [FA] A3/5 (the source of the edible industrial mycoprotein Quorn) in dilute glucose medium. Highly branched recessive mutants with restricted growth ('colonial' mutants), detectable after 200–600 h, have a selective advantage and, ultimately, replace the parental strain (Wiebe *et al.* 1991). In another series they also tested for dominant cycloheximide resistance by periodically plating out the multi-nucleate, initially uninucleate, macroconidia that are formed in continuous culture. Periodic peaks of cycloheximide-resistant macroconidia occur at intervals of approximately 34 generations (124 h) at dilute glucose dilution rates (Figure 9.4) (Wiebe *et al.* 1993), but colonial mutants are only recovered much later (648 h).

'Periodic selection' of mutants is a feature common to continuous cultures of both bacteria and fungi, e.g. 5-fluorouracil resistance in *Saccharomyces cerevisiae* (Paquin and Adams 1983a,b). Some selectively neutral mutants probably exist undetected in the original culture, but others arise periodically. If any one is positively selected, it increases linearly, but if another arises that is more advantageous than a pre-existing mutant it gradually replaces it—hence the apparent peak followed by a decline and replacement of the

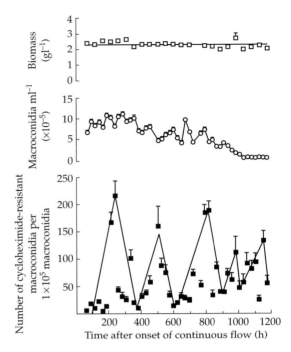

**Figure 9.4** Periodic selection of *Fusarium graminearum* [FA] strain A3/5 in glucose-limited chemostat culture. The lowest curve shows the number of spontaneous cycloheximide-resistant macroconidia that occur naturally over the period of the culture. These mutants were *neither* selected for by the investigators *nor* morphologically distinct, although they were selectively neutral. (Reproduced from Trinci (1994) with permission of the Society for General Microbiology.)

original mutant. Because this sequence occurs more than once, it gives rise to the apparent periodicity. One such cycle of selection is illustrated in Figure 9.5.

There is no reason to assume that this situation, a continuous production of nuclear mutants, some of which become selectable although successively being replaced by better-fitted mutants, is confined to the particular cultural conditions in which periodic selection becomes evident. Within a mycelium this sequence of changes must occur continuously, although the rate of mutant production and replacement is probably appreciably lower than in the particular situation described above. However, inter-nuclear selection is a dynamic process which occurs within any and every mycelium, although it may not be detectable

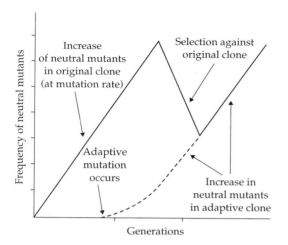

**Figure 9.5** Diagram to illustrate how adaptive mutants are successively selected in an asexual population. Once a slightly better adaptive mutant arises in a population, the original population is selected against and the new mutant increases (dotted curve), eventually replacing the original population until it is similarly replaced in its turn. (Reproduced from Pacquin and Adams (1983a), with permission of Nature Publishing Group.)

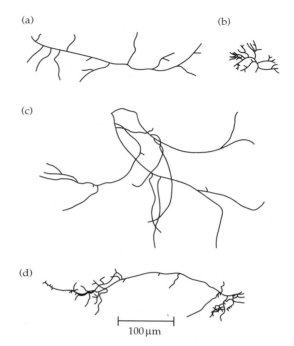

**Figure 9.6** Spontaneous colonial mutants that arise in *Fusarium graminearum* [FA] strain A3/5: (a) normal hyphae; (b) colonial mutant; (c) apparently normal sheared hypha with minority of mutant nuclei; (d) sheared hyphae with accumulation of mutant nuclei at tips and colonial phenotype beginning to be expressed. The colonial mutants are not expressed when they arise, but persist in heterokaryotic hyphae. In culture hyphae are sheared, but if such a length contains predominantly mutant nuclei, the phenotype is expressed. (Reproduced from Trinci (1994) with permission of the Society for General Microbiology.)

unless appropriate conditions for its detection arise.

In the experiments performed by Wiebe and colleagues, colonial mutants behaved somewhat differently. Being recessive, they were not detectable until they occurred in homokaryotic hyphae. This can come about by chance in the growing mycelia. However, under the conditions of culture used the hyphae tend to shear, resulting in fragments carrying unbalanced ratios of mutant to wild-type nuclei. This can aid the selection process if, by chance, a terminal region of a fragment becomes homokaryotic for mutant nuclei, or predominantly so. The terminal growth habit then assumes typical colonial-type growth (Figure 9.6).

Further selection and hyphal shearing ultimately result in selection for wholly mutant mycelia (Wiebe *et al.* 1996). This sequence of events accounts for the much later detection of the colonial mutants in continuous culture. Morphologically differentiated mutant types, comparable with those in *P. notatum* [FA], are not uncommon in the highly controlled conditions of industrial cultures and fermentations (Trinci *et al.* 1990). The experimental and industrial situations described in sections 9.3.2

and 9.3.3 demonstrate the following important features.

• Both quantitative and qualitative characters in fungi are amenable to intra- and inter-mycelial selection.

• In interpreting the outcome of selection experiments with fungi, allowance should be made for both cytoplasmic and nuclear elements being affected.

• Serendipitous features, such as spatial position, nuclear distribution within heterokaryons or hyphal fragmentation can result in chance imbalanced nuclear ratios within hyphae or spores, resulting in spontaneous sectoring. These features should not be neglected when attempting to interpret how selection acts.

Although based on artificial selection, these industrial examples demonstrate how complex selective processes can be in fungi. It is not known how far cytoplasmic or nuclear characters affect intra-mycelial selective processes in nature.

## 9.4 Inter-mycelial selection

Most observations and experiments on inter-mycelial selection in fungi have been made with oligogenically determined characters such as fungicide resistance or specific virulence genes which result in shifts in population means. Differences in oligogenic characters are equally commonplace between different natural races, ecotypes or species, but far fewer observations and experiments have been made on quantitative traits which will be considered in Chapter 10.

### 9.4.1 Selection for oligogenically determined traits relevant to agricultural situations

Three situations of great practical importance for agriculture are:

- selection for fungicide resistance in pathogens
- selection for increased, or specific, pathogenicity to the host crop
- selection by the host crop.

Many, but not all, the experimental and field observations concerning pathogenicity relate to gene-for-gene situations (see Chapter 2, section 2.4.1.1). They can mostly be related to real problems in agriculture, for instance host selection exercised on pathogen populations when new cultivars are introduced.

#### 9.4.1.1 Selection for fungicide resistance
Fungicide resistance of various kinds has developed amongst pathogens, especially over the last 50 years during which the number of site-specific fungicides has increased to more than 100! However, multigenic fungal resistance to the older general fungicides, such as copper- or sulphur-based preparations, has hardly developed at all.

Specific fungicide resistance is generally oligogenic and, initially, episodic since the application of

a fungicide is a profound and novel change in the environment. Claims have been made that it can be polygenic in nature, e.g. resistance to dodine in *Venturia inaequalis* [AP] (McKay and MacNeill 1979), but no unequivocal example has yet been identified. Single-gene resistance is characterized in the field by a rapid exponential rate of increase in the frequency of resistant strains from first detection, whereas with polygenically determined resistance the rate of increase is expected to be slower (Bent *et al.* 1990). Laboratory investigations of field isolates have often demonstrated that oligogenically inherited resistance results from either mutation in one or more of a multiple allelic series, or mutation at a number of different loci. Thus, benzimidazole resistance in *V. inaequalis* [AP] has been found at three levels—moderate, high and very high—determined by three distinct multiple alleles (Shabi *et al.* 1983).

Response to selection under agricultural conditions can be exceedingly rapid and persistent. For instance, over the two years 1972 and 1973 benzimidazole used initially to control *Botrytis cinerea* [AD] in Swiss vineyards failed to control the pathogen as a result of *de novo* mutation to, and selection for, single-gene fungicide resistance in those vineyards where it was used heavily. Even after it had been withdrawn from use, it was found that 60 per cent of the *Botrytis* [AD] population was still resistant 4 years later (Schüepp and Küng 1981; Staub 1991). In this case the fitness of the resistant mutant was evidently comparable to, although somewhat less than, that of the natural unmutated strains, but this is not always the case.

In the Champagne region of France, dicarboximides were used to control *B. cinerea* [AD] between 1980 and 1989. Fungicide-resistant strains increased rapidly to over 90 per cent in 1981 in heavily treated vineyards. Even in untreated vineyards, 54 per cent of the strains in these populations had become resistant by 1983, presumably as the result of gene flow through spore dispersal during 1981–1982. The use of this fungicide was discontinued at about the same time, and by 1986 resistant strains only accounted for 20 per cent at untreated sites. Thus, once selection had been relaxed, fungicide-resistant strains, being less fit, declined rapidly. However, the dynamics of the acquisition of fungicide resistance in the field is an even more complex process than this simple account suggests. This was demonstrated, for example, by the much slower development of resistance in the pathogen population when the fungicide was used at lower concentrations, or where the incidence of disease was only slight.

### 9.4.1.2 Selection for pathogenicity under agricultural conditions

The most striking examples of selection for pathogenicity are often found when epidemics occur. The southern corn blight epidemic of 1970–1971 is an example and demonstrates clearly the fortuitous nature of episodic selection.

Prior to the epidemic, the predominant pathogenic strain of *Cochliobolus heterostrophus* [AP] ( = *Bipolaris* (*Helminthosporium*) *maydis*) was race O. This shows non-specific virulence to maize and a number of wild grasses. It occurs virtually universally on maize in the southeast USA without causing any extensive damage there and elsewhere where maize is grown. In essence, race O was universal and apparently in parasitic equilibrium with its maize host (Nelson *et al.* 1970b). During the epidemic, race O was largely replaced by race T. The characteristics of these races are compared in Table 9.2. Note particularly that race T sporulates more vigorously on susceptible plants than race O and also produces a pathotoxin.

Prior to the epidemic, race T occurred sporadically. From isolates maintained in culture it had evidently occurred on maize in North Carolina as early as 1955, in Nigeria in the early 1960s and on various native grasses in England, Scotland, Switzerland, Nigeria, Guinea, Argentina, Brazil, Mexico and the USA during the same period. However, in 1961 race T caused a serious outbreak on maize carrying a male-sterile gene in the Philippines. The significance of this passed unnoticed for, by 1970, 85 per cent of the USA maize crop had been bred to carry the Texas male-sterile gene (*cms-T*) to facilitate maize breeding.

In 1969–1970 race T, exclusively mating type *A*, rapidly increased to the west of the main area where race O predominated. An exceptionally large airborne inoculum was generated, spread into and became superimposed on the area dominated by race O. This spread was accompanied by appreciable rainfall and persistent high humidity, favouring heavy infection with devastating effects on the crop. It rapidly became clear that the susceptibility of the crop to race T was also related to the presence of the *cms-T* gene in maize. Race O is apparently unaffected by the nature of the host cytoplasm, whether normal or *cms-T*. The course of the epidemic was rapid (Figure 9.7) (Moore 1970). More than 90 per cent of the isolates were race T, *Mat A* in North Carolina by 1970 and 1971, but the original equilibrium was more or less restored when maize with normal cytoplasm was planted in 1971–1972. Race T isolates were then reduced to about 1 per cent and only occurred sporadically wherever the few plants carrying *cms-T* occurred (Leonard 1987).

**Table 9.2** Difference between races O and T of *Cochliobolus heterostrophus* [AP], the cause of the southern leaf blight of maize

| Race O | Race T |
| --- | --- |
| Little or no specificity to host cytoplasms | Highly specific to host cytoplasms, especially if carrying T-type male sterility |
| Infects a wide range of hosts effectively | Only a weak parasite except to hosts with specific cytoplasms |
| Little pathotoxin produced *in planta* or *in vitro* | Produces high amounts of pathotoxin *in vitro* and *in planta* when infectious |
| Primarily infects leaves with small lesions | Leaves, leaf sheaths, husks and stalk tissue all attacked when infectious |
| Reproduces less rapidly than T on susceptible plants | Reproduces rapidly on susceptible plants |
| Limited by temperature and climate (relative humidity) to warmer parts of USA: spreads best in hottest part of growng season | Probably has a lower temperature optimum than O: spreads rapidly throughout growing season in USA |

Data summarized from Hooker *et al.* (1970) and others.

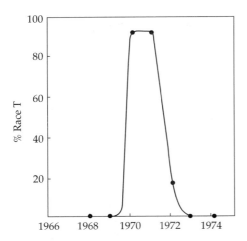

**Figure 9.7** Distribution of race T of *Cochliobolus heterostrophus* [AP] in North Carolina between 1966 and 1974 over the period of the southern leaf blight epidemic. (Reproduced from Leonard (1987), with permission from Blackwell Science.)

The relative fitness of race T has been variously calculated from field data on normal maize plants as 0.61 and from experimental data as 0.7–0.9, but on *cms-T* maize its relative fitness is enhanced to the order of 3.0–5.0. However, such estimates only give a partial indication of the nature of its selective advantage. Leonard (1987) also attributes it to the much greater sporulation ability of race T on susceptible plants rather than to effective competition with race O, i.e. to more efficient reproduction favoured by climatic conditions rather than competitive ability.

The situation in host–pathogen combinations when *specific* virulence is involved is rather different. Once the specific pathotypes of various rusts showing specific virulence to particular host cultivars (see Chapter 2, section 2.4.1) were inferred from observations of rust populations in the field, the changes in them over successive years could be followed. There was an obvious parallel between specific pathotype frequencies and the specific resistance genes carried by different cultivars. This suggested that directional selection favoured pathotypes carrying virulence genes matching the resistance genes carried by the hosts. More detailed examinations of such situations have shown that the selective process is more complex, involving more than the simple gene-for-gene relationship between a virulence gene and a host resistance gene. For example, the replacement in North America and Canada of race 56 of *P. graminis* f.sp. *tritici* [BR] by race 15B followed by the further domination in Canada of race 56 could not be explained in this way since the latter carried fewer genes for virulence. Recombination could not account for these changes because both races are propagated in the field in North America and Canada by asexually produced uredospores only.

Race 56 was first detected in North American wheat fields in 1931 and predominated from 1934 to 1949 despite the use of cultivars resistant to it from 1935 onwards. In 1950 it was suddenly displaced by race 15B until *cv Selkirk*, resistant to both races, was introduced in 1953, when race 56 increasingly dominated again, even though race 15B was potentially favoured by its broader host range.

A partial analysis of the situation was achieved through carefully designed experiments with these pathotypes in growth chambers, held constant except for temperature. Differences were found between them when grown singly or as a mixture on three different susceptible wheat cultivars (Katsuya and Green 1967). With equal amounts of inoculum, race 56 predominated at temperatures of 20 and 25 °C, but at 18 °C this was reversed and race 15B predominated. Infection density also affected the outcome. Light infections favoured race 56 whereas heavy infections favoured race 15B, while, for both, uredosorus (= sporangium) size declined with increasing infection density. There are other differences:

- the incubation period to uredosorus formation is shorter in race 56
- uredosoral development is more rapid in race 56, but the uredosori of race 15B eventually become larger than those of race 56
- uredospore output per uredosorus is greater in race 56
- the uredospores of race 56 are more infective than those of race 15B.

Despite the artificial conditions under which these experiments were performed, Katsuya and Green (1987) were able to relate some of their experimental findings to the field observations.

They noted that low ambient temperatures prevailed over the period when race 15B had arisen and dominated in the wheat fields, but moderate or warm temperatures predominated after 1951. The former conditions would be expected to favour race 15B. It would be expected to respond by heavier infections which would favour it over race 56. After 1951 the reverse was true and, despite the wider potential virulence of race 15B, race 56 would be expected to dominate. Therefore Katsuya and Green suggested that the selective differential was conditioned by ambient temperature as well as by the genetic constitution of the two races involved. Other investigations have sought to investigate the effects of intrinsic factors which affect virulence, such as frequency- and density-dependent selection and inter-racial competition (see section 9.4.3).

Nevertheless, it remains the case that, where gene-for-gene relationships occur under agricultural conditions, a broad correlation is maintained between the virulence genotypes of the predominant fungi and the contemporary resistance genotypes of the hosts. However, once resistance genes are rendered ineffective, breeders replace the non-resistant cultivars with ones carrying new resistance genes. Thus there is a tendency for a succession of correlated cyclical changes to occur as new resistance genes replace ineffective ones, and new virulence genes arise and are selected (Person 1966) (Figure 9.8).

Under such a regime, the number of virulence genes in the rust pathogen population as a whole will, potentially, tend to increase. Some strains, at least, will carry increasing numbers of virulence genes, some of which will no longer be matched by complementary resistance genes in the cultivar population. In this sense they become 'unnecessary' since their virulence attributes are no longer necessary for the strain's survival. Van der Plank (1968, 1978, 1982) repeatedly claimed that a pathotype carrying 'unnecessary virulence genes' would be selected against if it infected a susceptible host. These genes, he claimed, would eventually be lost, since: 'Because of stabilizing selection the races fittest to survive are those with no unnecessary virulence' (van der Plank 1968). This claim became a highly controversial issue.

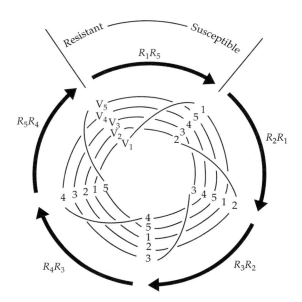

**Figure 9.8** Diagram to illustrate successive interactions between pathogen virulence genes and host resistance genes when a gene-for-gene relationship exists. As novel resistance genes R are successively introduced by breeders, the corresponding virulence gene V is selected, thus establishing a cyclical polymorphism. (After Person 1966.)

### 9.4.1.3 Selection for 'unnecessary' virulence genes: a controversy

Originally, using collected epidemiological data over several years for *Melampsora lini* [BR], a rust pathogen of flax, Flor (1953) found that those pathotypes with the *least* number of virulence genes necessary for survival predominated. As he put it (Flor 1956), 'The tendency has been towards the loss of unnecessary genes for virulence'.

This situation was then generalized by van der Plank (1968) and has led to numerous investigations of the fate of 'unnecessary virulence genes'. Only a few long-term studies (10 years or more) of actual field populations have been made, some with a sexual phase and others reproducing entirely asexually. Sexual recombination occurs regularly and annually in the combinations flax rust–flax (Flor 1953) and barley mildew–barley (Wolfe and McDermott 1994), but reproduction was entirely asexual in stem rust–wheat in Australia (Watson and Luig 1963) and in potato blight–potato (Van der Plank 1968) prior to about 1990. In three of these combinations there have been undoubted *decreases* in the number of virulence genes over time with the successive introduction of new resistance genes and the discarding of ineffective ones. However, in barley mildew (*B. graminis* f.sp. *hordei* [A]), the average number of virulence genes per haplotype has risen steadily throughout most of Europe from 4.8 in 1984 to 5.2 in 1990 and 5.5 in 1991 (Wolfe and McDermott 1994). This result is not universal, since in some regions of France the numbers stayed about the same over the 5-year period from 1986 to 1990 (Andrivon and de Vallavieille-Pope 1993).

A careful re-examination of the available published data on population trends in both the biotrophic pathogenic Puccinia rusts and powdery mildews of cereals worldwide has demonstrated that generalizations such as that of Van der Plank are invalid (Andrivon and de Vallavieille-Pope 1995). One methodologically significant finding was that to keep pace with increasing knowledge and use of resistance genes, the numbers of available differential hosts and changes in them over time resulted in frequently unrealized bias when investigating apparent increases or decreases in virulence genes in agricultural populations. Thus like was often not being compared with like, and when allowance is made for this some of the reported changes are less striking. Nevertheless, it does appear that the numbers and complexity of virulence genes do indeed increase over time in rusts, whereas in powdery mildews both numbers and complexity remain stable or only rise very slightly. The loss and replacement of virulence genes runs parallel with discarding ineffective resistance genes and the introduction of new ones. The differences between these patterns are attributed more to the fact that most of the rust populations studied are effectively propagated asexually, whereas powdery mildews regularly undergo sexual reproduction and a population bottleneck on an annual basis. It is also evident that local conditions can play a significant role in determining the selective process.

This re-assessment of the population genetics of virulence genes in agricultural situations is in accord with a theoretical model developed by Brown (1995). He showed that the occurrence or absence of recombination was of paramount importance to the persistence of virulence genes. Unnecessary virulence genes can be maintained in a population, especially an asexual one, through hitch-hiking with another positively selected virulence gene. Indeed, this mechanism can be operative despite recombination, although less effectively, and even if there is selection against unnecessary virulence genes. The presence of virulence genes in populations of *B. graminis* f.sp. *tritici* [A] unrelated to the resistance genes currently deployed in barley have been attributed to genetic hitch-hiking. For example, throughout the 1980s the complementary virulence genes for resistance genes *Mla2* or *Mla10* were present in mildew populations, although neither of the resistance genes were employed in British barley cultivars over the period (Brown *et al.* 1991). In general, Brown *et al.* concluded that:

...the direction in which the frequency of an unnecessary gene for virulence will change cannot be predicted if the pathogen mainly reproduces asexually and unnecessary virulences have little or no disadvantage in terms of fitness.

If there is no selection against unnecessary virulence genes:

...the long term dynamics of genotype frequencies in a largely asexual population may be unpredictable.

There is some observational support for this last generalization. For instance, the majority of asexually reproducing pathotypes of *P. graminis* f.sp. *tritici* [BR] in the USA between 1973 and 1978 were complex, possessing six or more virulence genes, whereas the minority were simple, carrying no more than one virulence gene. The proportions remained more or less constant over this period, but between 1978 and 1980 the former declined and the latter increased. Yet, throughout the whole period 1973–1980 there was no appreciable change in the resistance genes carried by the wheat cultivars (Nagarajan 1983). During the years of change, the climatic conditions were highly unfavourable for disease spread and it seems likely that the ability of any genotype to survive under extreme conditions was a more important factor than the numbers of virulence genes carried. Environmental tolerance could well have been the dominating selective attribute of all the pathotypes.

In conclusion, therefore, it seems as if undue attention has been given to the notion that the fitness of a pathogen is largely determined by its virulence genes alone. In one respect, virulence genes might be regarded as representing a loss in fitness. They are now known to be alleles which have lost their initial function (see section 6.2.1) and so stop production of the molecules recognized by the products of the host resistance genes. The question is whether this loss of function reduces the fitness of the pathogen *in the absence* of the corresponding resistance gene? This needs to be investigated, but determinants of fitness can certainly lie elsewhere in the genome. Conceivably it is a matter of balance—the loss of the initial function against the gain thereby of access to the host's metabolites and subsequently improved reproductive capacity of the pathogen. The situation demonstrates clearly the relative nature of genetic fitness.

Nevertheless, it is reasonable to regard Van der Plank's generalization on the fate of unnecessary virulence genes, in the form in which he proposed it, as inadequate. Each case is best considered on its particular merits.

### 9.4.2 Specific virulence genes in wild populations

Gene-for-gene relationships were first detected and characterized in a wide and diverse range of crop plants and their pathogens. Therefore it was suggested that it might represent an artefact of the agricultural situation resulting from resistance genes being transferred from their normal genetic backgrounds into cultivated crops (Day 1974). However, this seems to be unlikely, since gene-for-gene relationships have now been detected in at least 15 wild host species and their pathogens (Burdon *et al.* 1996). Only two cases have been fully studied: the powdery mildew *Erysiphe fischeri* [A] on groundsel (*Senecio vulgaris*) in Britain and the rust *Melampsora lini* [BR] on isolated populations of the woody native Australian *Linum marginale.* These cases provide interesting contrasts with similar situations in crop plants.

Although *M. lini* [BR] occurs naturally on *L. marginale* in Australia, very recently it has also been inadvertently re-introduced with commercial flax (*L. usitassimum*) on which it has been known as a pathogen for many decades. The pathogen populations studied in the Kosciusko National Park, New South Wales, do not coincide spatially with commercial flax populations and, as far as is known, no recombination has occurred since the rust on commercial crops is apparently propagated almost entirely asexually (Burdon and Roberts 1995). Nevertheless, the *M. lini* isolates appear to be conspecific since, experimentally, an Australian isolate from *L. marginale* has been successfully crossed with a flax isolate (Lawrence 1989). Races of *M. lini* [BR] derived from *L. marginale* can be distinguished by means of a series of 12 differential hosts carrying different resistance loci and matching specific virulence genes in the fungus (Burdon and Jarosz 1991, 1992). These *M. lini* [BR] isolates carry an entirely different spectrum of virulence genes from those carried by *M. lini* [BR] derived from cultivated flax (Lawrence 1989; Lawrence and Burdon 1989). Thus the two populations of *M. lini* [BR] pathogenic to different hosts are adapted to and, presumably, have been selected for entirely different spectra of host resistance genes in the two *Linum* species. In addition, *M. lini* [BR] populations on *L. marginale* have a very different structure from those of the pathogen on cultivated flax. The Kosciusko population exists as a series of sub-populations of varying sizes and degrees of isolation reflecting the distribution pattern of its hosts, whereas the flax populations exist over continuous areas of, genetically, relatively homogenous flax cultivars. The two environments as well as selection for pathogenicity are also different.

The *E. fischeri* [A]–*S. vulgaris* combination contrasts with both agricultural populations and the *M. lini* [BR]–*L. marginale* situation. *S. vulgaris* is a self-fertile tetraploid weed and both it and *E. fischeri* [A] populations are widespread in Britain, where the latter appears to be asexually propagated although sexual populations occur in continental Europe. The host–pathogen system has only been investigated at a few sites in the UK, near Glasgow and Perth and in the Midlands of England at Wellesbourne (Harry and Clarke 1986, 1987; Bevan *et al.* 1993a,b,c). Two systems appear to be operative: one is a typical gene-for-gene system with well-defined oligogenic resistance/susceptibility and matching avirulence/virulence factors, whereas the other behaves as though determined by multiple genes. The latter system appears to act to modulate pathogenic effects determined by the oligogenic system.

As far as the oligogenic system is concerned, individual mildew strains appear to carry numerous virulence factors. For instance, in a random sample of 24 isolates, 18 different pathotypes were identified, 14 of which possessed virulence factors for all but one of the 14 identified resistance factors in *S. vulgaris*. The remaining 10 isolates showed rather more restricted virulence, but even the most restricted still carried virulence for nine resistance factors and avirulence for the remaining six. Because mildew isolates carry such large numbers of virulence genes, isolates are capable of infecting about 80–90 per cent of the host population. Resistance was dominant to susceptibility in groundsel and the results of crosses suggest that resistance genes were non-allelomorphic although belonging to three gene families within a single linkage group. This resembles the situation in agricultural crops, but the pattern of behaviour of resistant hosts was very different. Host populations were highly heterogeneous in respect of their phenotypic resistance. The most common phenotype was almost always that susceptible to all the mildew races, and one or two of the resistant phenotypes usually occurred at a much higher frequency than the numerous other less frequent resistant phenotypes. A dense population occupying only $1\,m^2$ at Wellesbourne was appreciably more heterogeneous than the Glasgow population and carried one or more factors providing specific resistance to one or more of 10 mildew isolates tested on the plants. In all the populations sampled, only one *S. vulgaris* plant was found to be totally resistant to all the mildew pathotypes tested but it has retained this resistance for over 20 years. These resistance patterns highlight the different kinds of selection to which a wild host had been selected for compared with a highly human-selected agricultural crop.

Not surprisingly, and unlike the rust or mildew situations in crop plants, the pathogen appears to cause little physiological damage to the host despite an extensive mycelium which can bear up to 1000 conidia per $mm^2$ of leaf surface on a susceptible plant whereas, on resistant plants, conidia either just germinate or produce only a microscopically detectable mycelium. In other words, groundsel appears to have developed **tolerance** to the pathogen. This contrasts vividly with the consequences of infection of *S. vulgaris* by novel pathogens such as *Puccinia lagenophorae* [BR] in the 1960s, or a new species of *Albugo* [O] in 1978, which are far more damaging. No gene-for-gene relationship has developed with these pathogens.

Comparison of these population structures suggests that the *E. fischeri* [A]–*S. vulgaris* association is very ancient, typifying a condition resulting from very long-term mutual selection towards a balanced parasitic association of advantage to both host and pathogen alike. The gene-for-gene system is quite unlike the situation found in crop plants where pathogens, successively selected for different virulences, still cause considerable damage to their hosts. It is possible that in the *E. fischeri*–*S. vulgaris* situation a balance has been achieved through the selection of non-allelomorphic genes which modify the physiological consequences of the initial gene-for-gene interaction. The genes for virulence are presumably either selectively neutral or nearly so and have accumulated, or are retained, through genetic hitch-hiking. It is not yet clear whether the polygenic or the oligogenic system developed first, whether they developed simultaneously, or how they interact genetically or physiologically.

It is probable that this situation has had far longer to evolve in nature than systems in agricultural situations. Although the location and genetic architecture of the naturally occurring host resistance genes appear to be similar to those in crops, the impression given is that this pathogen carries appreciable numbers of virulence genes, possibly more than would be found in the pathogens of crops. The interaction between host and parasite over far longer time in the *E. fischeri* [A]–*S. vulgaris* case suggests that section has resulted in the co-evolution of other genetic factors that have ameliorated pathological expression and max-imized both growth and reproduction of host and pathogen alike. An alternative view is that the agricultural situations represent transient poly-morphisms but that in a natural situation, a stable balanced polymorphism eventually develops.

### 9.4.3 Density-dependent and frequency-dependent selection

There is considerable evidence for density-dependent selection in fungi but less for frequency-dependent selection. This probably represents an unbalanced view of the actual situation, reflecting the limited information avail-able on selection of all kinds in fungi. However, recent general treatments of fitness and considera-tion of a number of cases of balancing selection suggest that frequency-dependent selection is probably not uncommon in fungal situations.

*9.4.3.1 Density-dependent selection*
Many aspects of fungal growth are density dependent, such as spore germination, infection, mycelial size and sporulation, representing as they do intraspecific competition resulting from the different densities of adjacent individuals of the same species. One of the most obvious effects is the limit imposed on infection as spore density increases because the available sites for infection are limited. Therefore it is not surprising that density-dependent selection plays an important role in fungal population genetics. Most of the available data relate to experimental situations relevant to agriculture, but they have obvious applications to natural populations as well. A clear experimental demonstration is afforded by com-parisons of the numbers of uredospores produced over one cycle of spore production by different competing pairs of *Puccinia hordei* [BR] on barley. The species is propagated solely by asexual uredospores (Falahati-Rastegar *et al.* 1981). Five naturally occurring pathotypes, A, D, F, J and K, were used in addition to an induced mutant $A^m$, identical in its pathogenic behaviour to pathotype *A* but possessing easily recognizable orange-coloured pustules. This enabled the proportion of $A^m$ spores to be quantified in mixtures. Seedling

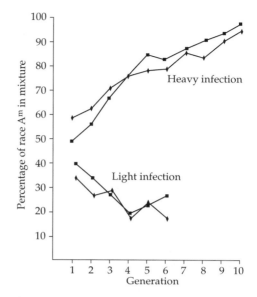

**Figure 9.9** An example of density-dependent selection in *Puccinia hordei* [BR] on barley. The curves show the changes in the proportion of race $A^m$ in duplicate experiments over successive generations using in each case heavy and light levels of infection obtained from the previous generation. Initially, barley seedling leaves, *cv* Proctor, were inoculated with a heavy and a light 50 : 50 mixture of uredospores of mutant race $A^m$ and race F and grown at 14 °C for nine successive generations. (Reproduced from Falahati-Rastegar *et al.* (1981), with permission from the British Mycological Society.)

leaves of barley cv *Proctor* were infected with equal numbers of uredospores at two densities, one sufficient to induce about 250 pustules per cm (high density) and the other 20 pustules per leaf (low density). Figure 9.9 demonstrates the sub-sequent spore production of races $A^m$ and F at high and low initial densities. At high densities race $A^m$ was preferentially selected, but at low densities it was selected against.

When the starting proportions of races $A^m$ and F were 2 : 1 and 1 : 2 respectively, race $A^m$ was still favoured to virtually the same extent, allowing for the different initial frequencies. Therefore in this case selection was density dependent but *not* fre-quency dependent. Incidentally, these experimental results were no guide to the outcome in a quasi-agricultural situation for, when equal inoculum mixtures were applied under field conditions, race F was strongly selected. This result emphasizes the difficulty of using experiments to attempt to analyse what is actually happening in a field situation but does not invalidate the potential importance of density-dependent selection.

Comparable evidence for density-dependent selection has been shown experimentally in *B. graminis* f.sp. *tritici* [A] (O'Hara and Brown 1996) and *Puccini graminis* f.sp. *tritici* [BR] (Newton *et al.* 1997).

### 9.4.3.2 *Frequency-dependent selection*

Wright (1960) and others have shown that in plants the numbers of incompatibility alleles in a population are maintained by frequency-dependent selection. By analogy, this seem to be a likely mechanism for determining the numbers of MTFs in populations of uni- and bifactorial diaphoromictic fungi and there is some supporting evidence. Not surprisingly, it also applies to the numbers of *vic* or *het* alleles in populations of dimictic Ascomycotina.

If a number of populations are sampled and frequency-dependent selection is effective, the frequency of MTFs in each population and in pooled data, allowing for population size, should be equal. If selection is effectively neutral, equal numbers would not be expected in the populations. Only one, rather unsatisfactory, test of this expectation has been made, with *Ustilago maydis* [BS]. This species has an unusual diaphoromictic mating system with two alleles at the *mat A* locus and at least 24 at the *mat B* locus; the necessary condition for a successful mating is, as usual, that the alleles at each locus should be different. Therefore the mycelium is heterozygous at both loci and several alleles are found in any population at the *mat B* locus. Clearly, a new allele in a population, whether derived by mutation or migration, will be at an advantage since it will be able to mate with the rest of the population. Therefore it will increase in frequency. On the assumption that the *mat B* alleles are maintained by frequency-dependent selection, equal numbers of MTFs will occur in the population at equilibrium. The situation was investigated at four sites in Minnesota, USA (Zambino *et al.* 1997). Diploid teliospores were collected and allowed to undergo meiosis, releasing haploid sporidia. The four populations comprised eight, eight, nine and fifteen in all, and 18 MTFs were identified in the pooled sample of only 27 sporidia that were isolated. A Mantel test was employed to check for equal frequencies of MTFs where

$$\chi^2 = (n-1)\frac{\sum_{j=1}^{n}(C_j^2 - 4r^2)/n}{2r - 4r/n},$$

*n* is the number of MTFs recovered, *r* is the number of diploids tested and $C_j$ is the number of times allele *j* appears in the sample. Unfortunately, the numbers isolated in each population were too small to determine whether the MTFs occurred at equal frequencies, but a Mantel test of the pooled data was significant for no departure from equality. Thus, while it seems probable that the *mat B* MTFs in *U. maydis* [BS] are maintained by frequency-dependent selection, a neutral hypothesis cannot be entirely ruled out.

A rather different kind of frequency-dependent selective situation occurs with the three known alleles at the *het-C* locus in *Neurospora crassa* [AP], one of the 11 loci that determine somatic incompatibility in this fungus (Wu *et al.* 1998). The three alleles, *het-C*[OR], *het-C*[PA] and *het-C*[GR], originally isolated at Oakridge, Panama and Graveland respectively, possess characteristic sequences whose SI specificity resides in a variable region 34–48 amino acids long that includes a high proportion of non-synonymous substitutions. In several nonclonal *N. crassa* [AP] populations which were sampled, the three alleles were found to occur in equal numbers, indicating that they had reached an equilibrium, with frequency-dependent selection being responsible for maintaining the balanced polymorphism. Intriguingly, the same three *het-C* sequences are found in three other species of *Neurospora* [AP], three of *Sordaria* [AP] and one of *Gelasinospora* [AP], all phylogenetically closely related species and genera. These three sequences, which show over 66 per cent identity, were the *only* ones found in a collection of some 40 related species and genera. Therefore this polymorphism must be very ancient, preceding the divergence of the genera *Sordaria* [AP] and *Neurospora* [AP]. Using rDNA sequence comparisons and assuming a 2 per cent substitution rate every $100 \times 10^6$ years, the generic divergence is estimated to have occurred 36 million years ago. Therefore it is inferred that these three sequences are not only very ancient but have remained virtually unaltered over this period as the result of balancing selection.

Using a similar argument, the polymorphic variation in the nucleotide sequence of one of the subunits $b_1$ of the *mat A* MTF of *Coprinus cinereus* [BA] was also shown to reflect the outcome of very ancient balancing selection (May *et al.* 1999). This work provides a possible molecular basis for the observed stability of the MTFs of the diaphoromictic fairy ring fungus *Marasmius oreades* [BA] over a relatively short term, estimated as some hundreds of years (Burnett and Evans 1966), or of the MTFs of clone I of *Armillaria gallica* [BA] (Smith *et al.* 1992), postulated to be more than 1000 years old.

These two innovative examples have opened a wide range of potential situations for investigation at the molecular level as well as demonstrating the profound, long-lasting and stabilizing effect of balancing selection despite the inevitable diversity of environmental changes which the fungi concerned must have experienced over these very long time periods.

## Further reading

*General*

Endler JA (1986). *Natural selection in the wild*. Princeton University Press.

Mather K (1973). *Genetical structure of populations*. Chapman & Hall, London.

Wolfe MS and Caten CE (ed.) (1987). *Populations of plant pathogens: their dynamics and genetics*. Blackwell Scientific, Oxford.


*Specific*

Clarke DD (1997). The genetic structure of natural pathosystems. In *The gene-for-gene relationship in plant–parasite reactions* (ed. IR Crute, EB Holub and JJ Burdon), pp. 231–43. CAB International, Wallingford.

Ennos RA and McConnell KC (1995). Using genetic markers to investigate natural selection in fungal populations. *Canadian Journal of Botany,* **73** (Supplement 1), S302–10.

Jinks JL (1966). Mechanisms of inheritance. 4. Extranuclear inheritance. In *The fungi*, Vol. 2 (ed. GC Ainsworth and AS Sussman), pp. 619–660. Academic Press, London.

Ramsdale M (1999). Genomic conflict in fungal mycelia. In *Structure and dynamics of fungal populations* (ed. JJ Worrall), pp. 139–74. Kluwer Academic, Dordrecht.

# CHAPTER 10

# Change in gene frequency in fungal populations—II Natural selection for multigenically determined traits: competition and fitness

Few experiments and observations have been carried out on multigenically determined traits in fungi but they are sufficient to give some idea of how selection operates on such traits in natural populations. They have demonstrated that both directional and stabilizing selection occur in experimental and natural situations, and that the consequences of selection differ on the same fungus in different environments. Disruptive selection in fungi has not been specifically investigated experimentally but some natural situations such as host adaptation are highly suggestive of its effects. An important consequence of selection is that, once different genotypes exist in the same habitat or environment, **intraspecific competition** becomes important. An indication of how such competition can come about is given, but it is a topic that has not yet received the attention it deserves in fungi. Finally, in the light of the diverse ways that selection operates on fungi, a somewhat more sophisticated consideration of the relative fitness of a genotype than that suggested in Chapter 9 is outlined in the appendix to this chapter.

## 10.1 Experimental selection for multigenically determined traits in fungi

The most extensive experiments described to date include selection for growth rate of both dikaryons and monokaryons of *Schizophyllum commune* [BA], for growth rate in *Neurospora crassa* [AP] and for

increased ascospore size. The first is of particular interest since it raises the question of how an identical multigenic trait exhibited both by monokaryotic and dikaryotic phases of the same species is controlled and selected.

### 10.1.1 Experimental selection for growth rate in fungi

Artificial selection has been practised for growth rates both in haploid *Neurospora* [AP] (Papa *et al.* 1966, 1967) and *Aspergillus nidulans* [A] (Caten and Jinks 1979) and in dikaryotic and monokaryotic *Schizophyllum* [BA] (dikaryon by Simchen and Jinks (1964); monokaryon by Simchen (1966) and Connolly and Simchen (1968)). Results were comparable in both genera (Figure 10.1).

The following features are common to all these experiments.

• Selection, both plus or minus, is usually effective for several generations; the result is achieved gradually but reaches a plateau. When the starting population is exposed to both kinds of selection, the results are usually asymmetrical.

• During selection, whatever the character selected, some lines become infertile, but this can usually be overcome if selection is relaxed in some way. Nevertheless, there is always at least one line where such problems are not encountered.

• Genetic variance usually decreases with selection but its degree is variable. When lack of response

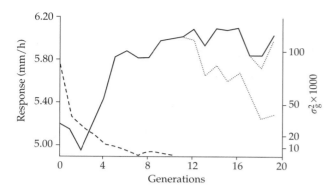

**Figure 10.1** Selection for fast growth rate and reverse selection in progeny initiated by a cross between sexually compatible strains of *Neurospora crassa* [AP]: positive selection is shown by the solid curve as the mean of two replications to generation 10, by which time the variance (dotted line) had fallen to zero; reverse selection is shown by the broken curve. (After Papa *et al.* (1967) with permission of Nature Publishing Group.)

occurs it is usually, but not necessarily, correlated with low genetic variance. Environmental variance usually remains virtually unaltered.

• Reverse selection from lines where the response has reached a plateau can usually be made to show some further response indicating that potential genetic variability is still available. For example, progeny of crosses between lines of different origins but comparable in the level of selection obtained usually restore the ability to respond further to selection.

• The response to selection appears to be mainly polygenic. The number of **effective factors** (blocks of closely linked polygenes), or **quantitative trait loci** (QTLs), is usually quite small ( < 10) and in some cases they can be located on specific chromosomes.

To date, QTLs, in the sense of specific chromosome segments carrying linked groups of polygenes which affect a particular trait, have only been identified in *Pleurotus ostreatus* [BA] (Larraya *et al.* 2002), although the approximate location of some polygenes has been recorded in both *Schizophyllum* [BA] (Simchen 1966) for growth rate and *Neurospora* [AP] for ascospore size (Lee and Pateman 1959) (see section 10.1.3). In *P. ostreatus* [BA] some of chromosomes carry more QTLs than others, so that almost 39 per cent of the effect on dikaryotic growth rate is determined by QTLs on linkage groups IV and VIII and 42 per cent of the effect on monokaryotic growth by QTLs on linkage groups IV, VII and XI.

The results of selection broadly parallel those obtained in selection experiments with other (diploid) eukaryotes and are amenable to similar explanations (Mather 1943, 1973; Falconer and Mackay 1996). One difference is that the decline in genetic variance with selection is usually more rapid in haploids/monokaryons than in dikaryons/diploids because of the absence of any masking effect due to dominance.

In those fungi, such as *S. commune* [BA], which involve both a monokaryotic haploid and a dikaryotic phase, the question arises as to whether the same genes that determine such characters as growth rate in the haploid phase also determine it in the dikaryotic phase. Partitioning the variance for growth rate in a dikaryon showed that it could be accounted for entirely by heritable additive and dominance effects. Moreover, dominance appeared to be ambidirectional, suggesting that growth rate had been subject to stabilizing selection. By a regression analysis of growth rate of the dikaryons on the mid-values of their monokaryotic parents, it appeared that most of the variation in the dikaryon was *not* correlated with that in the monokaryons. Therefore genes determining growth rate either act differently in the two phases or, more probably, different genes are responsible in the two phases. For any fungal trait which behaves in a similar manner to growth rate in *Schizophyllum* [BA] in the monokaryotic and a dikaryotic growth phase, the expectation is that selection will act *differently* on the two phases (Simchen and Jinks 1964) but this is not necessarily the same for all traits or all fungi.

The location of the genes or polygenic complexes determining growth rate in *S. commune* [BA] is not known, but in *Pleurotus ostreatus* [BA] it has been demonstrated that 12 presumptive polygenic sites which determine the growth rate of dikaryons are located throughout the chromosome complement and that five of these regions are also implicated in determining growth rate in monokaryons. Whether or not the sites are identical was not determined (Larraya *et al.* 2002).

The most detailed analysis of selection for growth rate over nine generations of *monokaryotic* strains has been made with *S. commune* (Simchen 1966). Variation was partitioned into genetic and environmental contributions and, in addition, an estimate was made of the number of effective factors contributing to the variation. In the inevitable absence in a haploid of dominance effects, the markedly asymmetrical result of high and low selection was attributed to unidirectional non-additive interactions, i.e. epistasis.

A surprising feature was that the genetic variance $V_G$ remained high even after selection. A possible reason for this is that in the intervening dikaryotic phases, linked alleles affecting growth which were initially in repulsion were brought into coupling as a consequence of crossing over:

$$\left(\frac{+-}{-+}\right) \Rightarrow \left(\frac{++}{--}\right).$$

In the absence of linkage, the recombination frequency $p$ would be $1/2$ and the variance simply $kd^2$. However, under repulsion the variance is $(2p)^{k/2}kd^2$ and under coupling it is $2(1-p)^{k/2}kd^2$.

In the absence of any other information, it was assumed that each effective factor contributed an amount $d$ to the growth rate. Therefore, assuming $k$ factors, the range of the progeny in each generation is $2k\bar{d}$ where $\bar{d}$ is the overall effect of all the $d$ factors. However, the genetic variance $V_G$ is $k\bar{d}^2$, and hence an estimate of the number $\bar{k}$ of effective factors is

$$\bar{k} = \frac{0.25 \, (\text{range of progeny})^2}{V_G}.$$

Using this formula, the probable numbers of effective factors was estimated as ranging from 6 to 11, with an average just under 9, although of course the actual numbers of individual polygenes contributing might have been much larger.

Interactive effects between genes can occur when crosses are made between haploid, diploid or

dikaryotic individuals from populations isolated in some way, spatially or genetically, and as a consequence of intensive selection. Since haploid fungi are homozygous, one important but significant difference from dikaryotic or diploid forms is, of course, that gene interactions can only be between non-allelomorphic genes, i.e. they can only be due to epistasis.

To investigate this aspect of the polygenic control of growth rate, crosses were made between isolates from different VCGs in haploid *A. nidulans* [A] (Jinks *et al.* 1966), or between dikaryotic populations of *S. commune* [BA] isolated by distance (Simchen 1967). Heterokaryon incompatibility in *A. nidulans* [A] imposes some degree of somatic isolation resulting in a relatively low incidence of outcrossing in this taxon. Nevertheless, crosses in both taxa were notable for demonstrating interactive, i.e. epistatic, effects.

In the absence of interactive effects, the progeny of a haploid cross are expected to be distributed symmetrically about the parental mean. However, in the *A. nidulans* [A] crosses, the mean progeny growth rate differed considerably from the parental mean and progeny growth rates showed highly skewed distributions. This is a clear indication of epistatic effects. The three parental isolates of *S. commune* [BA] were specially chosen because their growth rate was determined exclusively by dominance and additive effects and because at most heterozygous loci the gene frequencies did not differ significantly from 1:1. However, crosses between them showed appreciable interactive components $\sigma_I^2$. Thus, although in each initial environment selection for dominance and additive components had resulted in an adapted, balanced and stabilized outcome, when brought together in a common cytoplasm in a novel environment the polygenic systems were unbalanced, as demonstrated by the interactive effects. Adaptation had been lost. The effects were even larger in *A. nidulans* [A] than in *S. commune* [BA], but this difference in magnitude was expected since the former is normally inbreeding and the latter is outbreeding. The result was in accordance with theoretical expectations (Crow and Kimura 1966). They showed that co-adaptation within a population and interaction between genotypes from different populations would be more marked in haploid than in diploid (or presumably dikaryotic) sexually reproducing species.

## 10.1.2 Experimental selection for ascospore size

Ascospore size has been shown in *N. crassa* [AP] to be a multigenically determined character (Pateman 1955) which exhibits polygenic segregation and recombination in ordered tetrads (Pateman and Lee 1960) and linkage with the albino *al-2* locus on chromosome 1, presumably the site of a QTL (Lee and Pateman 1959). The polygenes contributed both additive and interactive (epistatic) components to the genetic variance (Pateman and Lee 1960). Ascus shape and, to some extent, size also is affected by both a major gene and polygenic action.

Correlated responses in ascus size, shape and ascospore behaviour accompanied selection for ascospore size. Over about eight generations of selection, three kinds of change took place.

• Asci became broader and shorter. In eight-spored asci, mean ascospore size increased from about 27 μm × 15 μm to 37 μm × 18 μm when size reached a plateau (Figure 10.2).
• The numbers of four-spored asci containing fertile spores heterokaryotic for mating type increased and the dimensions of their ascospores were comparable with those of the naturally occurring *N. tetrasperma* [AP].
• The proportion of ascospores of different lengths and degrees of sterility, as well as the numbers of aborted asci, increased in parallel.

Although suggestive, these experiments throw no light on how four-spored asci arose historically in *Neurospora*. However, the findings do suggest that if any mutations had arisen in eight-spored ancestors which affected ascus dimensions, correlated response could have played some part in their further development and stabilization.

Further selection was attempted over 11 generations for increased, or decreased, length, starting with the five longest and five shortest ascospores from a sample of about 2000 normal ascospores derived from eight-spored asci of *Neurospora crassa* [AP] (Lee and Pateman 1961, 1962). In each case

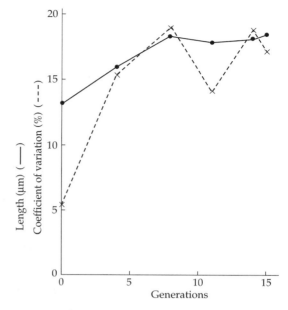

**Figure 10.2** Selection for increased ascospore length in *Neurospora crassa* [AP]: solid line, ascospore length; broken line, coefficient of variation. (Reproduced from Pateman (1959) with permission of Nature Publishing Group.)

there was a tendency for the coefficient of variation to decrease, but not significantly, although fertility decreased as a result of a decline in numbers of both perithecia and viable asci. Selection for increased size in the high selection line achieved only a small increase, about 2 μm compared with the 7–10 μm achieved in the initial selection. In contrast, selection for decreased ascospore length was ineffective and fertility declined within three generations, with the numbers of aborted asci and sterile ascospores increasing rapidly (Lee and Pateman 1962).

These findings reinforce the similar consequences between haploid selection for quantitative characters and those achieved in dikaryotic fungi or other diploid Eukaryota.

## 10.2 Multigenic selection in agricultural situations

Selection in respect of specific virulence in relation to specific host resistance genes has been described earlier (see section 9.4.1.2), but situations are known where non-specific aggressiveness increases over time, presumably due to selection for this quantitative trait. This can be reflected in increased relative fitness. For example, isolates of race T of *Cochliobolus heterostrophus* [AP], which caused the southern leaf blight epidemic in 1970, had materially increased their fitness compared with 1955 isolates. For example, compared with 1955 isolates, the 1970 population required shorter exposures to dew at 12–14 °C, an essential condition for effective leaf infection. In addition, leaf colonization was more rapid and spore production greater. Indeed, the improvement was so great that it was surmised that race T would have been incapable in 1955 of initiating an epidemic even in maize with the increased susceptibility due to its male-sterile gene (Nelson *et al.* 1970)!

Evidence for the multigenic determination of aggressiveness comes from several examples. Pathogenicity to barley by the covered smut fungus *Ustilago hordei* [BS] is determined by a recessive gene *uh-v1* (Sidhu and Person 1971). However, the percentage of genetically uniform barley plants infected, i.e. a measure of its aggressiveness, ranged from 17 to 83 per cent when different fungal isolates, all homozygous for *uh-v1*, were tested. Aggressiveness, assessed by the percentage of infected spikes tested on the same barley (cv Vantage) was determined for the complete progeny from a cross between two such isolates in all possible combinations. Although all were, inevitably, homozygous for *uh-vl*, aggressiveness ranged from 20.1 to 68.9 per cent. Partitioning of the variance showed that although 34.8 per cent could be attributed to environmental variation ($V_E$), 65 per cent was due to genetic causes, specifically 43.9 per cent to an additive component ($V_A$) and 21.3 per cent to a non-additive genetic component ($V_I$) (Emara and Sidhu 1974). Thus, although the pathogenicity of the fungus is due to a single recessive gene, its magnitude (aggressiveness) is determined polygenically. Presumably, therefore, by analogy with the cases described in the previous section, directional selection for polygenes could increase or decrease aggressiveness.

A similar situation occurs with maize smut *Ustilago maydis* [BS] where aggressiveness is determined polygenically, a substantial component of $V_G$ being accounted for by non-allelic interaction.

Galls sampled from a population of maize that had probably been exposed to unregulated infection for 15–20 years on a trial ground showed an excess of highly aggressive isolates. Earlier experiments had shown that the aggressiveness of *U. maydis* [BS] could be increased rapidly if passaged through resistant maize cultivars (Sidenko 1965). In view of the long association between successive maize crops and the same smut population on the trial ground, natural selection for increased aggressiveness might, therefore, be expected to have occurred (Bassi and Burnett 1980).

Another feature of the *U. maydis* [BS]–*Zea mais* relationship is that the host's resistance to smut is also determined polygenically (Bojanowski 1969). This may be a common feature whenever aggressiveness is significant, as has been shown to be the case in partial resistance of barley (*Hordeum vulgare*) to leaf rust due to *Puccinia hordei* [BR]. Partial resistance was shown to be determined polygenically when the latent period from infection to development of uredospores and the infection frequency, i.e. uredosori units per leaf area, was determined for all combinations in the field using three fairly resistant cultivars and five different rust isolates (Parlevliet 1977). An analysis of variance showed highly significant interactions between host resistance levels, as measured in the field, and the experimentally determined latent period of the fungal isolates. These interactions were those expected by a gene-for-gene polygenic model for aggressiveness such as that developed by Parlevliet and Zadoks (1977). Therefore it seems probable that polygenic complexes in a host and its pathogen can develop a gene-for-gene relationship analogous to that described earlier for oligogenes (see Chapter 2, section 2.4.1.1 and Chapter 9, section 9.4.1.2).

## 10.3 Multigenic selection in natural populations

Most of the evidence for selection in natural populations is circumstantial and related to selection by environmental factors such as host or substrate, i.e. ecological adaptation, but the genetic basis for selection has been investigated only infrequently. Quantitative traits such as growth rate (see section 10.1.1), shown to be subject to selection under experimental conditions, are also selected under natural conditions. The relative fitness of different genotypes or taxa is another area relevant to selection and some evidence for this is also available.

## 10.3.1 Selection for quantitative traits in natural populations

There is clear but limited evidence from samples of various fungi taken from nature for stabilizing and directional selection affecting different traits in the same mycelium or for the same trait under different circumstances.

The linear growth rates of 77 dikaryons obtained from a single relatively isolated wild population of *S. commune* [BA] were measured in culture together with those of 72 synthesized dikaryons derived from matings between compatible basidiospores (monokaryons) produced by two of the original dikaryons (Figure 10.3) (Brasier 1987).

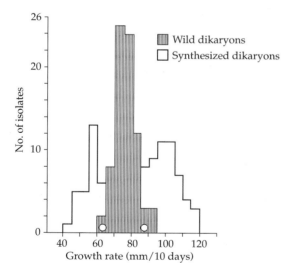

**Figure 10.3** The effect of stabilizing selection. The shaded area represent the growth rates of a sample of 77 dikaryons of *Schizophyllum commune* [BA] isolated from nature compared with the growth rates of two sets of synthesized F1 dikaryons. The latter were obtained by inbreeding two of the wild dikaryons indicated by the open circles. Note that the range of growth rates of the synthesized F1 dikaryons far exceeds that of dikaryons from the wild which had been exposed to natural selection. (Reproduced from Brasier (1987), with permission from the British Mycological Society.)

The potential range of linear growth rates in the derived crosses far exceeded that of the parental and other dikaryons actually found in nature, suggesting that stabilizing selection had operated in favour of dikaryons of intermediate growth rates. In contrast with the approximately normal distribution of linear growth rate of the original dikaryon sample, the distributions of time to sporocarp production and numbers of sporocarps per dikaryon were highly skewed (Figure 10.4). Such distributions strongly suggest the occurrence of effective directional selection towards small numbers of sporocarps which develop rapidly, together with some degree of stabilizing selection for those traits.

The control of growth rate and the time of sporocarp formation have been analysed in both *Collybia velutipes* [BA] and *S. commune* [BA] (Simchen 1965, 1966). Growth rate is largely determined by polygenes, showing additive and dominance effects, but the timing of basidiocarp production and basidiocarp weight was more complex, involving gene interactions. The genes involved in basidiocarp characters also showed unidirectional dominance which is strongly suggestive of directional selection, although in the same isolates growth rate was subject to stabilizing selection. Thus particular characteristics within one individual may, simultaneously, be subjected to selection of entirely different kinds.

Six isolates of *S. commune* [BA] were investigated, two from the England and the remainder from Massachusetts, USA—regions with very different climatic conditions. The American isolates were found to develop basidiocarps more rapidly and also showed higher $V_E$ components for dikaryotic growth rate. The sample is very small so that it is uncertain whether or not these differences were correlated with geographical or environmental differences, but their consistency is at least suggestive evidence for adaptation to different environments (Simchen 1966).

## 10.3.2 Host adaptation

Adaptation to different hosts is widespread in many fungi and this form of ecological adaptation presumably involves disruptive selection.

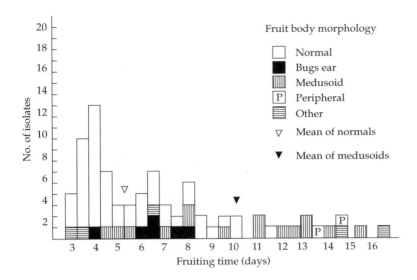

**Figure 10.4** The distribution of days to development of normal and mutant basidiocarps of natural isolates of *Schizophyllum commune* [BA]. Note the bias to early development of normal basidiocarps (open histograms); medusoids and most other mutants developed later. Basidiocarp development is under polygenic control. (Reproduced from Brasier (1970) with permission of Chicago University Press.)

The apparently simplest situation is when a pathogenic fungus in a gene-for-gene relationship with its host acquires a new mutant specific virulence gene which can overcome a novel resistance gene carried by its host. This adaptation can certainly happen rapidly, but whether such a simple change is all that is involved seems doubtful. For instance, the acquisition of a novel virulence gene by *P. graminis* f.sp. *tritici* [BR] to wheat carrying a new resistance gene in the case of races 15B and 56 is only one of the various adaptive changes that have been demonstrated (see Chapter 9, section 9.4.1.2).

The same uncertainty surrounds a change to a novel host. The rice blast fungus *Magnaporthe grisea* [AP] occurs widely on different rice cultivars, but it also occurs on a number of other generically distinct graminaceous crops as host-restricted races, e.g. *Setaria* (foxtail millet), *Panicum* (common millet), *Eleusine* (finger millet), *Digitaria* (crab grass), *Andropogon* (weeping lovegrass) and recently, in Brazil, wheat (*Triticum*). These races can be interbred to give offspring of differing viabilities. Further analysis has shown that a specific gene pair determines specificity between the *Setaria* and *Andropogon* races, another gene pair differentiates between races on *Setaria* and *Triticum*, and three major loci differentiate between the *Oryza* and *Setaria* races (Valent and Chumley 1991; Murakami *et al.* 2000). All appear to have the characteristics of specific avirulence/virulence genes such as occur in typical gene-for-gene situations. Therefore it seems possible that selection for different hosts has come about in *M. grisea* [AP] through selection at these specific loci. However, effective infection probably requires further adaptation such as increased aggressiveness on the new host, leading to a situation such as that described for *Ustilago hordei* [BS]–barley (see section 10.2). After the switch to the new host, induced by a single major gene which determines the new specificity, its effect is reinforced by polygenes that initially enhance aggressiveness. Ultimately, however, in a successful pathogen, stabilizing selection in both host and pathogen results in a balanced host–pathogen relationship such as that described earlier for *Erysiphe fischeri* [A]–*Senecio vulgaris* (see Chapter 9, section 9.4.2).

The complexity may be even greater. For example, *Cochliobolus carbonum* [AP] is found on at least 222 other grasses in addition to cultivated maize. Pathogenicity to each grass is determined by one, or sometimes two, complementary genes; in all, some 35 different genes for pathogenicity were proposed (Nelson *et al.* 1970; Nelson and Kline 1963, 1969, 1971).

Many pathogens are confined to a host in a single family or a group of closely related families, or to dicotyledonous but not monocotyledonous plants. Pathogenicity to peas in *Nectria haematococca* [AP] is determined by the presence of pisatin demethylase carried on a minute B chromosome. However, when the southern leaf blight fungus *Cochliobolus heterostrophus* [AP], normally a pathogen restricted to maize and grasses, was transformed by the gene determining this enzyme, it became capable of infecting and causing limited disease on stems and leaves of pea although it could not infect pea roots as can *N. haematococca* [AP]. In contrast, when *Aspergillus nidulans* [A], a saprophyte, was similarly transformed, it was incapable of inducing disease in peas at all (Schäfer *et al.* 1989; Schäfer and Yoder 1994). It appears that if a fungus already possesses the necessary genetic make-up to ensure stem and leaf penetration, e.g. *C. heterostrophus* [AP], the presence of a novel host-specific virulence gene—in this case pisatin demethylase—enables it to infect a new host. This experiment supports the view that selection for considerable genetic change other than to a single virulence gene is necessary to adapt a fungus to become pathogenic to a different host.

A more complex example is that of the rust *Melampsora lini* [BR] in Australia whose virulence genes have been differentially selected for infecting either the herbaceous *Linum* spp., including commercial flax (*Linum usitatissimum*), or the arborescent *Linum marginale* (Lawrence 1989; Lawrence and Burdon 1989). That the rusts belong to the same species was shown by their morphological identity and the complete fertility between a Western Australian *L. marginale* isolate and an introduced flax isolate. There is also little doubt that the *M. lini* [BR]–*L. marginale* rust races are indigenous since they were found in Australia in 1885, while flax had only been grown at the time of testing for 15 years with no evidence of rust infection over that period. However, when tested on the standard flax differentials a large number of indigenous *M. lini* [BR] were incapable of infecting them except when they carried the *N* resistance gene. Similarly, a wide range of *M. lini* [BR] derived from flax were incapable of infecting the 26 differential *L. marginale* which discriminated between the indigenous rust races found on them, except those capable of overcoming the *N* resistance gene. Evidently this gene is the only complementary virulence gene common to the two host-specific rust populations; otherwise they differ in possessing about 10–15 different virulence genes and no doubt other reinforcing genes.

Rare hybridizations between S- and P-types of *Heterobasidion annosum* [BA] in California have been described already (see Chapter 8, section 8.2.1.2). The limited degree of interfertility between Finnish and North American isolates has enabled more detailed genetic differences between the two forms to be identified. They turn out to be quite complex despite the forms being otherwise remarkably similar. Inter-sterility/fertility is determined by five allelic pairs, and the inter-relationships of the Finnish and North American S- and P-types are shown in Figure 10.5 (Chase and Ullrich 1990a,b).

The *S* and *P* loci determine host specificity through regulating dikaryon formation independently of the MTFs. Thus all P-types carry $P^+$ and all S-types carry $S^+$. For a successful mating isolates must be homoallelic at either the *S* or *P* loci and heteroallelic for MTFs. Interfertility is also affected by three other loci, *V1*, *V2* and *V3*, and a further necessary condition for a fertile cross between an S and P form is that at least one of these loci should be homoallelic$^+$. Therefore $S^+P^+$ forms can be made experimentally between North American S- and P-forms because they have *V3* in common, but not between Finnish isolates because they would be heteroallelic at all five loci (Figure 10.5). Nevertheless, $S^+P^+$ forms have never been identified in nature although, presumably, the recently discovered hybrids must have this constitution. Presumably, the condition is selected against normally. In addition to these genetic differences,

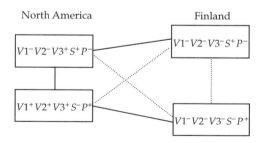

North America           Finland

$V1^- V2^- V3^+ S^+ P^-$    $V1^- V2^- V3^- S^+ P^-$

$V1^+ V2^+ V3^+ S^- P^+$    $V1^- V2^- V3^- S^- P^+$

**Figure 10.5** The five pairs of allelomorphic genes in *Heterobasidion annosum* [BA] which determine interfertility between S- and P-group isolates from Finland and North America. Other possible combinations of the five genes are known (see text). Genotypes are represented in boxes: upper left, S-group; lower left, P-group; upper right, S group; lower right, P-group; solid lines, interfertile crosses; dotted lines, intersterile crosses. (Reproduced from Chase and Ullrich (1988), with permission from Academic Press.)

experiments on infectivity demonstrated that the two forms differ in their aggressiveness. S-types have an advantage over P-types in attacking spruce and vice versa, although P-types are generally more aggressive on both hosts (Stenlid and Swedjemark 1988). This supports field observations that pines are seldom infected by the S-type but other conifers, including spruce, as well as pines can be infected by the P-type (Stenlid 1987). Thus, even if the $S^+$ and $P^+$ genes act as the primary 'switch genes', determining potential host adaptation, further selection of reinforcing factors, both oligogenic and polygenic, are necessary to maintain host specificity in the natural situation

There is suggestive evidence that selection for host specialization is widespread in many other Basidiomycotina but genetic information is lacking. In many cases, two or more morphologically very similar taxa differ principally in that one taxon is restricted to one or a few host species while the other(s) can grow on many host species, often including those on which the more restricted taxon is found. However, in most of these cases little or no genetical information is available. One example is the intriguing situation that occurs in three similar but inter-sterile species of *Peniophora*, namely *P. mutata*, *P. populnea* and *P. heterocystidia* [all BA].

*Peniophora mutata* [BA] includes two morphologically extremely similar types; one is confined to poplars but the other grows on several broadleaved trees including poplars (McKeen 1952). Crosses between monokaryons of the two types were almost always ineffective, but in a very few cases dikaryons were formed. Unfortunately, these were not grown on so that neither their stability nor reproductive fitness is known, but the occurrence of even rare sexually competent fusions suggests a limited degree of genetic affinity. No other genetic information apart from the fusion data is available, but the fusion data are highly suggestive either for host selection from taxa with a wider host range for poplar or vice versa. The existence of the other two species suggests that a parallel kind of disruptive selection had already gone further and led to speciation. Thus *P. populnea* [BA] is confined to poplars but *P. heterocystidia* [BA] is capable of infecting both poplars and several other broad-leaved trees. Many other cases like this occur in the Basidiomycotina

and would repay further investigation using molecular methods.

When pathogens occur on more than one host, or on both a cultivated host and wild plants, an intriguing aspect, as yet little investigated, is the role if any played by isolates growing on the other host plants. Do they exist as predominantly asexual populations or can they cross with the main race? Do they act as reservoirs of races found on the main host? Does selection have the same outcome on other hosts as on the main host, or is hybridization possible on one or more of these alternative hosts?

Nothing about the relationship between isolates from wild plants and from cultivated crops is known for *M. grisea* [AP], *C. carbonum* [AP] or *M. lini* [BR], but some information is available in other situations, e.g. *Ustilago avenae* [BS]. This loose smut fungus infects the inflorescences of both cultivated oats and the wild oats *Avena fatua* and *Avena sterilis*. Collections from cultivated oats tested against a range of seven oat cultivars could be grouped into 10 virulence patterns, usually showing high virulence to four or five of the tester cultivars. In contrast, only 10 of 13 isolates from wild oats either possessed avirulent genes or were weakly virulent to not more than two of the tester cultivars. The three remaining wild oats isolates were far more virulent and showed a wider range of virulence to the tester cultivars. Significantly, these three isolates had been collected close to plots used for testing resistance of oats to *U. avenae*. It seems plausible that these isolates owed their pathogenicity to gene flow from the domesticated to the wild oat populations (Holton 1967). Therefore, in principle, some populations of a pathogen on a wild species could become potential reservoirs of infective strains on related cultivated crops.

Lastly, some suggestive examples of historical selection for changes in host specificity can be inferred from some downy mildews and rust fungi. *Blumeria* (*Erysiphe*) *graminis* f.sp. *hordei* and *tritici* [A], from the main commercial barley- and wheat-growing areas of the world, are inter-sterile although morphologically and behaviourally they are virtually identical. However, in Israel, which represents the probable centre of origin of both these crops and their pathogens, the two *formae speciales* are not so readily distinguishable in their natural situations. Moreover, the anamorphic phase, *E. graminis* [FA], can infect a far wider range of grasses including a number of species in common with those infected by the *formae speciales* (Eshed and Wahl 1970, 1975; Wahl *et al.* 1978). Three stages in a plausible historical scenario can be suggested for this situation.

• In the original open habitat of sparsely scattered grasses, some hybridization and transgressive segregation occurred in the fungal population, resulting in the contemporary highly heterogeneous mildew population in respect of pathogenicity and ability to infect different grass hosts.
• As cereal cultivation developed, larger hectarages of single increasingly uniform cereal host species were grown and the increasing abundance of specialized cultivated host species exerted increased selection on the heterogeneous *E. graminis* [A] population.
• Eventually, this has resulted in the highly adapted *formae speciales*, which now dominate commercial wheat or barley fields. Their distinction is now so great that although a viable cross can be made experimentally between the forms with difficulty, the F1 is only very weakly pathogenic (Hiura 1978).

A similar selective pattern probably accounts for the development of host-specialized rusts, e.g. *Puccinia coronata* f.sp. *avenae* [BR], where, from the numerous wild hosts in Israel, an isolate was recovered in which 'one spore culture covers a wider range of hosts than the range covered by all other isolates studied elsewhere in the world combined!' (Eshed and Wahl 1975).

The host adaptations shown by the *formae speciales* of *P. graminis* [BR] are less well marked. Although they are largely confined to agricultural situations on particular hosts (wheat, barley or oats), they are also capable of infecting other grasses less effectively, some of them in common.

Many of the *formae speciales* can be crossed, although only limited progeny, often with greatly reduced vigour, are obtained. In general they are more weakly pathogenic than their parents but at least some of the progeny possess a wider host range than either of the parental types. For instance, the cross between f.sp. *tritici* and *secalis*, which is

the most successful, produced progeny some of which resembled one or the other parent, some intermediate but capable of weakly infecting barley and, rarely, others capable of strongly attacking wheat, rye and barley (Johnson 1949; Levine and Cotter 1931.) It has been suggested that this indicates that, originally, the host range of *P. graminis* was much wider than that of any of the present *formae speciales* and that it was then a weak parasite on a wide range of grasses (Green 1971). Selection associated with increased virulence to different hosts took place subsequently but it has either been less effective than in the powdery mildews or has not been so severe, since the *formae speciales* of *P. graminis* [BR] are not as completely isolated as those of *Blumeria/Erysiphe* [A]. Plausible though this hypothesis may be, it provides no insight into how the pycnidial and aecidial stages came to occur on an alternate host, nor how selection was, or is, exercised on such alternating stages in the life-cycle.

### 10.3.3 Ecological adaptation

Fungi adapt in a far wider variety of ecologically significant ways than host specialization to their environment. There is little direct evidence for selection in such cases but it can be inferred from the close, often highly significant, correlations between fungal traits and environmental variables, their favoured substrates or some combination of these.

*Neurospora intermedia* [AP] provides one of the earliest examples known of classical ecotypic adaptation from fungi. Distinct ecotypes are correlated with substrate and the geographical area in which they occur (Turner 1987). Worldwide, samples collected from burnt substrates develop orange mycelia bearing macroconidia 5–7.5 µm in diameter with relatively few (two or more) nuclei. Unusually, this form, which is the most common, can also occur on unburnt organic wastes such as vegetable refuse, bagasse and the like in the USA, Brazil, Haiti and Japan. However, in Australia, China, Taiwan, Indonesia, the Philippines, Truk, Congo and the Ivory Coast, a different and less abundant form, with a lemon-yellow mycelium bearing macroconidia 9–11 µm in diameter, each with more than

12+ nuclei, occurs most commonly on this latter type of substrate.

Each type breeds true, although yellow strains show low fertility. When allowance is made for this, laboratory crosses between yellow and orange forms are as fertile as crossed yellow forms. Hybrids segregate for mycelial colour, which is determined by two pairs of alleles, but macroconidia are immensely variable in both size and nuclear number, presumably owing to the segregation of polygenes. Therefore the two types are capable of crossing, but this is not common in nature because intermediate types have never been found. That the two types are ecologically selected by substrate is strongly suggested by each form being found on its characteristic substrate in very close proximity in Australia. This suggests that the primary selective agent might have been the nature of the substrate, but the origin and selective processes resulting in the other phenotypic characters is not clear.

Selection can act on the same taxon across its ecological or geographical range with different results. The different pattern of RFLPs, indicating genetical differences between populations of *Fusarium oxysporum* [FA] from cultivated and uncultivated soils, were described in Chapter 4 (see section 4.3). An even more striking example is provided by the ectomycorrhizic fungus *Cenococcum geophilum* [BA] from roots of *Pinus virginiana* growing on serpentine and adjacent non-serpentine soils in Maryland, USA (Panaccione *et al.* 2001). Using RFLPs derived from the ITS region of repeated rDNA units and PCR-amplified β-tubulin gene fragments, the serpentine isolates, with one possible exception, resembled each other more than they did non-serpentine isolates. Moreover, all serpentine isolates lacked a group I intron usually found in the 18S region in this taxon. When the isolates were re-examined using AFLPs, to give greater discrimination, a clear distinction was obtained when they were subjected to UPGMA analysis.

An excellent example of genetically determined traits correlated with geographical distribution is provided by *Cochliobolus carbonum* [AP] in the USA. This fungus infects and causes leaf lesions on maize and several other grasses; it is presumed to overwinter on litter. Five different races are recognized on maize.

**Race 0** is only known in its anamorphic state (*Bipolaris zeicola* [FA]) from two fields in North Carolina where it accounted for over 50 per cent of all isolates before declining rapidly to less than 1 per cent . It is avirulent on all maize cultivars which are susceptible to the other races, causing only flecks or minute lesions on the leaves. It appears to be somewhat distinct genetically from races 1, 2 and 3 (Welz *et al.* 1994).

**Race 1** used to be the most common form, demonstrating host-specific selection by producing a toxin, specific to maize homozygous for the *hm1* gene, and developing large necrotic lesions on the leaves. The relevant toxin-producing locus is absent in the other races. Race 1 has now fallen to less than 1 per cent over its North American range as a consequence of negative selection due to the introduction of resistant hybrid cultivars carrying *Hm1*. Indeed, isolates from North and South Carolina were found to constitute only a single clone (Welz *et al.* 1994).

**Race 2** is now the most common and most widely distributed in the USA. It develops as small rounded lesions on grasses but causes little significant damage and produces relatively small lesions on maize leaves.

**Race 3** is the most prevalent race on maize throughout the Appalachians from Georgia to Pennsylvania (Leonard and Leath 1990). It causes long linear lesions, but lesion size in races 2 and 3 is controlled polygenically (Dalmacio *et al.* 1979).

**Race 4** forms characteristic large round lesions on inbred maize cultivar *W64A* and is known only from the Midwest of the USA and southern France (Dodd and Hooker 1990).

In North Carolina, up to about 1974, race 3 was confined to the Blue Ridge Mountains. It behaved as a montane ecotype representing virtually 100 per cent of the population above about 450 m where the mean summer temperature does not exceed 21.5–22 °C. However, in the early 1970s it began to spread eastwards and downwards through a steep transitional zone, 7–20 km wide, to the foothills below. This zone marked an abrupt discontinuity in the mean summer temperature which increased eastwards through the main maize growing area (Figures 10.6 and 10.7).

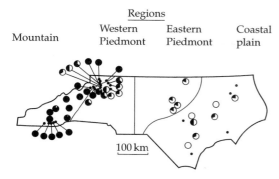

**Figure 10.6** The proportions of races 2 and 3 of *Cochliobolus carbonum* [AP] at sampling sites in North Carolina in 1976–1977: white circles or areas, race 2; black circles or areas, race 3. Pie diagrams are for samples where $N = 2$. Note that the Western Piedmont is separated from the Eastern Piedmont by a steep scarp. (Reproduced from Lodge and Leonard (1984), with permission from the National Research Council of Canada.)

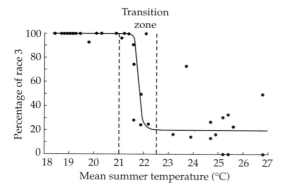

**Figure 10.7** Frequency of race 3 of *Cochliobolus carbonum* [AP] in relation to the mean summer temperature in North Carolina. The mean temperature is for the months July to September inclusive and the frequency of race 3 is its percentage presence in samples where $N = 4$. (Reproduced from Lodge and Leonard (1984), with permission from the National Research Council of Canada.)

In this zone the two races form a cline. Race 3 coexisted with race 2, reaching almost 20 per cent in the foothills by 1985 although varying from 20 to 46 per cent at different sites. Since then, it appears to have declined somewhat overall. Race 3 has not advanced further eastwards and race 2 prevails in the eastern plains to the seaboard. Phenotypically, in the clinal region, race 2 is more variable than race 3 with some phenotypes in

common. In mixed populations, the Shannon index of diversity was:

race 2: 2.35 in 1985 to 2.37 in 1987
race 3: only 1.67–1.34 in the same two years.

However, race 3 in the mountains is phenotypically as variable as race 2 in the lowlands (Lodge and Leonard 1984; Leonard and Leath 1990; Welz and Leonard 1993). Each race maintains its identity in the clinal zone, although they differ in fitness, with race 2 being fitter than race 3 ($w = 1.0 : 0.87$).

Although sexual reproduction within and between each race is readily achieved in laboratory conditions, the races do *not* interbreed in the field. Both races behave predominantly as asexual populations and a gene that prevents the formation of the ascocarp is common in this species (Nelson 1964; Lodge and Leonard 1984). Gene frequencies in samples from nature have remained constant over 15 years and are *not* those expected from random segregation. However, both mating types occur in both species at relatively stable frequencies (30–70 per cent). Although departures from a 1:1 ratio of MTFs were more common in race 2, the variation in frequency of MTFs varied in parallel in the two races, presumably in response to some common selective factor (Leonard and Leath 1990). Detailed experimental analysis of isolates of races 2 and 3 failed to demonstrate a significant $I_A$, suggesting that sexual reproduction does occur, however infrequently, within each race. Therefore it is somewhat puzzling that no evidence for past or current crossing was detected, but it is possible that parental differences in fitness resulted in crossed progeny which were even less fit. Significant selection for particular haplotypes probably takes place during the predominant successive asexual phases (Welz and Leonard 1995).

As there is no introgression in the clinal region, presumably only a limited number of isolates of race 3 are capable of growing in and extending into this region of climatic transition. Thus there is presumptive evidence for a further clear selective effect correlated with montane versus lowland climate. It appears to prevent either race growing successfully in the alternative climatic region, although permitting their coexistence in the narrow transitional zone.

Another situation, which is not uncommon, is when correlated selection for a variety of traits occurs as in *Gremmeniella abietina* [AD]. This asco-mycotous fungus causes blight and cankers on shoots, branches and even the main trunk of various conifers. In Europe its range extends from northern Scandinavia southwards to southern Europe, where a morphologically defined variety, var. *cembrae*, has been recognized from the Swiss Alps.

Two distinct growth patterns occur on infected trees, and isolates from the two types show a con-sistent correlated number of differences in infection pattern, morphology and physiology, and in their genotypes. Comparisons of banding patterns of 28 stable DNA fragments derived by either RAPD or PCR techniques and used to characterize isolates derived from northern Europe divided the isolates into two unique groups (Hellgren and Högeberg 1995). These were correlated with small trees in Northern Sweden and Finland, or large trees in Norway. Moreover, the small tree types (STTs) and large tree types (LTTs) coincide with two distinctive growth habits, previously recognized. The LTTs coincided with the 'tree-crown type', where repro-ductive structures developed in the uncovered crowns of the trees, and the STTs coincided with the 'winter-buried' type, where the structures were in the lower part and became buried under snow. Surprisingly, there was more variation in the band-ing patterns of the LTTs than of the STTs, although the latter produced more apothecia despite being under snow. No intermediate banding patterns were detected, suggesting an absence of interbreeding between the types and, indeed, there appears to be some degree of genetic isolation. Experimental crosses between Finnish LTTs and STTs propagated on young pine stems gave rise to apothecia with reduced numbers of ascospores which were less infective than those of either parent (Uotila 1992).

The most probable selective agency affecting the origin of the two types would appear to be the environment. There seems little doubt that in Scandinavia, at least, the two forms represent dis-tinct ecotypes, each adapted to the different sites they occupy and showing no apparent genetic exchange. It is not clear how far this distinction holds outside Scandinavia. Infection by *G. abietina* var. *cembrae* [AD] in the Swiss Alps results in a similar infected growth type to the STT ecotype. This suggests selection by comparable climatic conditions. Although isolates from the Alps differ in a number of characteristics, the somewhat less definitive technique of soluble protein electrophor-esis revealed protein similarities between the STT type and var. *cembrae* from the Alps. Moreover, these profiles did differentiate var. *cembrae* from other lowland European populations and the Scandinavian LTT (Petrini *et al.* 1990). Comparison of RAPD fragment patterns derived from the 5.8S rDNA and ITS regions of *G. abietina* also suggests some affinity between var. *cembrae* and the STT (Hamelin *et al.* 1996; Hamelin and Rail 1997). Therefore there seems little doubt that the growth habit of the STT ecotype and var. *cembrae* represent a genetic response to identical environmental con-ditions. Whether the STT and the variety represent disjunct remnants of a once common population, or two independent but parallel consequences of cli-matic selection, has yet be determined. Elsewhere in this group of fungi climatic factors appear to have played a less significant selective role, since other apparently related fungi are confined to dif-ferent host conifers, e.g. *G. laricina* [BA] on *Larix* (Hamelin and Rail 1997).

### 10.3.4 Experimental transplantation as a tool to detect selection

Transplants, either reciprocal transplants between sites or transplantation to a common environment, provide a powerful experimental tool to investigate habitat adaptation. Transplant experiments are possible with fungi. In an artificial sense, every isolation from nature grown in plate culture is equi-valent to transplanting to a common environment. Such cultures usually only provide information concerning mycelial behaviour and asexual spore production but, potentially, can be informative.

The common experience of 'degeneration' of cul-tures isolated from nature is typical of a transplant experiment as is the comparable phenomenon of the dual phenomenon which has already referred to (see Chapter 8, section 8.1.1.3). Soil Fusaria isolated to nutrient agar in plate culture rapidly develop so-called 'patch mutants'—flattened areas with fluffy aerial hyphae, compacted hyphal masses or sclerotia

(Miller 1946). In contrast, isolates kept in soil tubes remain unchanged. These observations suggest that such 'mutants' are selected against in the soil environment, but selection is relaxed and the mutant forms are favoured in a physically different high-nutrient environment.

Transplantation from one natural site to another has been achieved successfully with the ascomycotous *Crumenulopsis sororia* [AD] (Ennos and McConnell 1995). In Scotland, *C. sororia* invades the stem and branches of stressed Scots pine over a wide range of sites ranging in annual precipitation and height above sea level. Single ascospore isolates from three sites (Loch Maree, 2000 mm annual precipitation, 50 m elevation; Glen Affric, 1700 mm anual precipitation, 250 m elevation; Abernethy, 770 mm annual precipitation, 300 m elevation) were isolated, grown on, characterized by means of alternate alleles at an isozyme locus (*Got*) and used to prepare mixed inocula (Table 10.1).

Their relative fitness in each environment was investigated. Mixed conidial inocula of known composition, derived from the isolated ascospores, were used at the same three sites to inoculate two points on branches of each of 80 trees within a 30-m$^2$ quadrat. One was on first-year wood, and the other on second-year wood. Inoculations were 59–95 per cent successful and altogether 49–84 per cent produced pycnidia whose conidia were sampled. Since the composition of each mixture had been defined, the original strain frequencies ($p_0q_0$) and those derived from samples of conidia from the inoculated trees in each quadrat ($p_1q_1$) were determined. This enabled the relative fitness of

each genotype to be obtained. At Loch Maree and Glen Affric, statistically significant differences in performance were found between samples from first- and second-year branch inoculations which were therefore scored separately. Overall relative fitness for the combined results at all sites is shown in Figure 10.8.

Significant differences between the original populations were detected in half the cases and the magnitude of some of the differences was surprisingly large. Isolates derived from Loch Maree populations did best at their own site and better than those from Glen Affric or Abernethy, although the differences were not always significant. Overall, Loch Maree isolates showed greater relative fitness in competition than did isolates from the other two sites. Thus, in an approach to natural conditions, there is evidence for differences in relative fitness between differently located populations and some suggestion of localized adaptation as

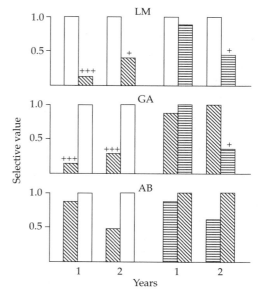

**Figure 10.8** Estimated relative fitness of three geographically separated populations of *Crumenulopsis sororia* [AD] grown as mixtures at each site: LM, Loch Maree, open histograms; GA, Glen Affric, oblique hatching; AB, Abernethy, horizontal hatching. Inocula comprised five ascospores from each of two populations in all possible combinations and were used to infect first- and second-year wood of Scots pine at each site. The data represent the overall estimates of relative fitness: $+++$, $p < 0.001$; $++$, $p < 0.01$; $+$, $p < 0.05$. (Reproduced from Ennos and McConnell 1995, with permission from the National Research Council of Canada.)

**Table 10.1** Composition of mixed inocula of *Crumenulopsis sororia* [AD] used in transplant experiments at different sites in Scotland

| Mixture | Initial frequency of alleles | | Inoculation sites |
|---|---|---|---|
| | *Got* f | *Got* S | |
| LM *Got f* + AB *Got s* | 76 | 75 | LM and AB |
| GA *Got f* + LM *Got s* | 80 | 77 | GA and LM |
| AB *Got f* + GA *Got s* | 36 | 50 | AB and GA |

AB, Abernethy; GA, Glen Affric; LM, Loch Maree.
*Got* is the locus for various alleles for glutamate oxaloacetate transaminase (EC 2.6.1.1).
Reproduced from Ennos and McConnell (1995), with permission from the National Research Council of Canada.

measured by conidial production after a year's growth at the original sites. Although the outcome of this particular experiment was not wholly clear cut it was both indicative and, most importantly, demonstrated that fungal transplant experiments can be made under a close approach to natural conditions. This, like the experiments with *Laccaria bicolor* to be described in section 10.4.2, has opened up a new approach for detecting, measuring and investigating natural selection in fungi in the field.

## 10.4 Intraspecific competition and relative fitness

In Chapter III of *The origin of species* (Darwin 1859), Charles Darwin not only stressed the importance of competition for natural selection but also commented succinctly: 'struggle for life most severe between individuals and varieties of the same species'. Therefore it is disappointing that there is surprisingly little information concerning intraspecific competition in fungi compared with the many studies of interspecific competition.

Competition has been variously defined by mycologists (Cooke and Rayner 1984; Lockwood 1992; Wicklow 1992), but it is convenient to recognize the following:

• **exploitation competition** defined as depletion of resources by one individual, organism or population without reducing the access of another individual organism or population
• **interference competition** which is a term applied to behavioural or chemical mechanisms by which access to a resource is influenced by a competitor.

Of course, it is possible for a fungus to employ both modes of competition at different sites or times. Although couched in terms of simultaneous demand on a resource, the consequences may be manifested in a variety of ways—reduced growth, reduced reproduction or, ultimately, extinction.

Interference competition is very widespread in fungi, although most studies have been concerned with interspecific rather than intraspecific competition. An example of the latter is the existence of 'killer' yeasts. Killer and sensitive strains exist

in species of both *Saccharomyces* [A] and *Pichia* [A]. One of the causes of spoilage of brewery fermentations is the presence of killers amongst the normal sensitive *S. cerevisiae* [A] in the inoculum. Provided that the proportion of killer is high enough, they lyse the sensitives by the release of specific toxins during the logarithmic growth phase, resulting in failure of the fermentation. However, at low cell concentrations in continuous culture, the sensitives can utilize the available sugar more effectively and so suppress killer growth, i.e. through **exploitation** competition (Young 1987). Evidently, in nature, some kind of selective balance is struck since killers represent, on average, only about 17 per cent of the yeast populations investigated.

Although less dramatic, SI interactions, especially in Basidiomycota, often involve some kind of antagonistic reaction including lysis between interacting mycelia. Such reactions occur between mycelia of *Stereum gausapatum* [BA] and conceivably this is one reason why branches are often only infected by a single isolate (Boddy and Rayner 1982).

A few cases of intraspecific competition involving interference or exploitation competition in natural or quasi-natural situations have been recorded. Although not analysed in this way, variation in the volume of decay columns due to genetically different dikaryons of wood-destroying fungi, e.g. in heart rot caused by *Fomes cajanderi* [BA] in young Douglas fir (Adams and Roth 1969) or by *Coriolus versicolor* [BA] in various hardwoods (Rayner and Todd 1979), reflect their competitive ability (see Chapter 2, Figure 2.1). Unfortunately, in the natural situation it is not possible to control other variables, e.g. whether different dikaryons had all infected at the same time, which is a necessary requirement if the volume of decay columns is to be compared and used as a measure of competitive ability. Experimental inoculations of the horizontal cut surface of vertically buried beech logs, or even the surface of timber slices 2 cm thick, with different genets can provide an approach to the natural situation. The depth or area of infected wood within zone lines due to antagonistic SI reactions between different genets can then give some measure of competitive ability (Rayner and Todd 1979). Using this technique, they showed clearly how different isolates of *C. versicolor* exploit very different areas, reflecting differences in the ability of different dikaryons to exploit the substrate.

## 10.4.1 Experimental studies on intraspecific competition

Most of the available evidence for intraspecific competition from fungi comes from experimental situations. A very simple example is the growth of dikaryons of *Microbotryum* (*Ustilago*) *violacea* [BS] in species of red or white campion (*Silene* spp.), resulting in invasion of the inflorescence, flowers and anthers, sexual reproduction, and the liberation of meiospores within the anthers. Stems of *S. alba* inoculated with six genetically marked monokaryotic strains of *M. violaceum* [BS] enabled nine potential dikaryons to form one or more systemic mycelia. These are thought to grow to the meristems where they are maintained as a single apical hyphal segment whose growth is maintained in step with the meristematic cells, while the rest of the mycelium dies off (Fischer and Holton 1957). When the shoot finally differentiates into branching inflorescences, mycelial growth resumes and invades the anthers. Spores can then be isolated and the dikaryons identified. Competition could be expressed at various growth stages:

• at dikaryon formation by competition between the monokaryons through differing growth rates of the dikaryons towards the apical meristem, and thereafter within the meristem tissue
• at infection of and growth into flower primordia
• by competition between monokaryons or dikaryons for available nutrients at any or all these locations.

A typical example of the results obtained is shown in Figure 10.9. Five flowering shoots were

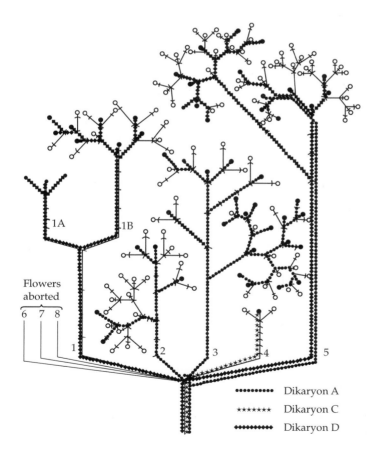

**Figure 10.9** Competition between dikaryons of *Microbotryum* violaceum [BS] infecting white campion (*Silene alba*). Three dikaryons (A, C and D) infect the main stem of the host, but competition results in different branches becoming infected. (Reproducd from Day (1980), with permission of Chicago University Press.)

•••••••• Dikaryon A
★★★★★★★ Dikaryon C
✦✦✦✦✦✦✦ Dikaryon D

infected, four by single different dikaryons (A, C or D) and the fifth by two dikaryons (A and D). Dikaryon D infected a complete branch of this last shoot, but in the other branch A and D continued growth together and in one flower both actually infected the same anther. Evidently, six of the possible dikaryons either failed to develop or were unable to compete with the three successful ones. In the flowering shoot as a whole D was marginally the most successful but clearly A and C were comparable. Interference competition could have been involved, but the occupation of a particular branch could simply have been a matter of chance, for example the spatial chance that a dikaryon occupied a particular meristematic cell at the time when the floral primordia were differentiated. By analogy with *Heterobasidion annosum* [BA], where multiple persistent monokaryotic infections occur on the surface of stumps (Garbelotto *et al.* 1997), they compete and, with time, the number of genetically distinct monokaryons is reduced, but usually only one monokaryotic or dikaryotic clone fully infests the root system (Stenlid 1985; Garbelotto *et al.* 1997). Interference competition would seem most probable in both situations.

In a similar experiment carried out by infecting stems of rape with different strains of *Sclerotinia sclerotiorum* [AD] (Maltby and Mihail 1997), the mass of the sclerotia produced on the stem surface was used as a measure of their competitive ability. Infection with pairs of known naturally occurring genotypes usually reduced total sclerotial production, but in some combinations a mosaic of genotypically different sclerotia developed, while in others sclerotia were of a single genotype only. The first two cases were regarded as evidence of interference competition, and the last as the result of exploitive competition. Results were also affected by entirely serendipitous factors which ultimately favoured exploitative competition. For instance, an advantage lay with the first inoculated genotype, provided that there was an interval between inoculations, or with a genotype inoculated at a lower level of the stem. The former presumably gave the first inoculated genotype time to occupy an appreciable volume of stem tissue; the latter enabled the first genotype to block the upward growth of the second inoculum. The experiment

also demonstrated that selective advantage may reflect different attributes, e.g. growth rate and speed of sclerotium development.

The interaction of growth and reproductive capacity are widespread determinants of fitness in fungi. For example, in *Schizophyllum commune* [BA] in nature, slow-growing dikaryons often produce basidiocarps earlier and vice versa (Brasier 1970). Which growth and reproductive pattern will prove to be most advantageous is likely to be determined by quite different extraneous factors and to differ at different sites at other times.

Numerous experiments have been made with pathogenic fungi in the glasshouse, growth chambers or small field plots to investigate competition between genetically different rust races on the same or a different host. These results frequently bear little relationship to agricultural situations. For instance, although races 56 and 17 of *P. graminis* f.sp. *tritici* [BR] were most common on a range of wheat cultivars, including cv *Mindum* grown in the USA between 1938 and 1943, attempts to replicate these results by growing mixtures of races on these cultivars in the greenhouse failed to replicate the field situation (Loegering 1951). For example, various proportions of mixtures of two races (17 and 19) of *P. graminis* f.sp. *tritici* [Br], were grown on three cultivars, all susceptible to both races, for six asexual generations. Regardless of the starting proportions, race 17 survived better than race 19. However, in the field on *cv Mindum*, one of the hosts used in the experiment, race 19 predominated. Similarly, as mentioned in Chapter 9 (see section 9.3.2.1), the results of competition experiments using races of *P. hordei* [BR] on barley bore no relation to those obtained in field trials or in agricultural conditions (Falahati-Rastegar *et al.* 1981).

Much the same applies to straminipilous fungi. When mixtures of various races of *Phytophthora infestans* [O] were grown together under greenhouse conditions, race 0 was both more infectious and apparently antagonistic to other races, as judged by the inhibitory effects of water in which its sporangia had germinated on the germination of sporangia of other races. However, similar experiments carried out in the field resulted in races 2 and 4 predominating under such conditions, with race 0 being the least effective of all those tested (Thurston 1961)!

These experimental results, although useless for predicting behaviour in the field, do demonstrate how very susceptible the outcome of selection is to environmental conditions.

### 10.4.2 Field experiments on intraspecific competition

Ideally, investigations of competitive ability or relative fitness of different races or genets should be carried out in the field.

The transplant experiments using *Crumenulopsis soraria* [AD] represent one approach (see section 10.3.4). Others using *Laccaria* [BA] involve intercontinental transplants. The latter experimental approach, more comparable to conditions found in nature, involved the introduction of strains of the toadstool *Laccaria bicolor* [BA], an ectomycorrhizal species, into plantations where indigenous strains of both that species and the similar *L. laccata* [BA] occurred. All the strains, indigenous and introduced, were typified genetically and could be identified and distinguished unambiguously. Genetic assay was achieved by using electrophoretic patterns of RAPD-amplified DNA fragments derived from the IGS2 region of rDNA and from the large ribosomal subunit of mitochondrial DNA (Lr mtDNA).

In the most informative series, seedlings of Douglas fir (*Pseudotsuga menziesii*) were inoculated in 1985 with known genets of *L. bicolor* [BA] in fumigated nursery soil and grown on there. Three treatments were compared:

- firs inoculated with a North American strain *S238N*
- firs inoculated with a French strain *81306*
- uninoculated fir plants.

Two years later the seedlings were transplanted to defined plots in a forest plantation at Saint-Brisson (Morvan, France). Trees inoculated with strain *S238N* were planted in the middle of one plot and surrounding plots were planted with either uninoculated firs or firs inoculated with *S238N*. It was known that, *in culture*, the strains used as inocula could cross-breed with some of the native strains already resident in the plantation (Selosse *et al.* 1998a,b, 1999).

In the *S238N* plots, a random sample of 100 basidiocarps was collected in 1994 and 1996; in 1995, their positions were mapped and each was assayed genetically. Since no direct underground sampling was attempted, this procedure gave information on the distribution of the fungus only when and where it produced basidiocarps. On the other hand, the samples could, potentially, give information on mycorrhizal persistence, reproductive capacity, genetic introgression and any growth away from the inoculated root systems.

Clear differences between each genet could be identified. The American strain (*S238N*) persisted for 10 years and appeared to have spread throughout the plot as a single extensive genet but it neither colonized areas occupied by the eight resident *Laccaria* (i.e. French) strains, which were identified as either *L. bicolor* or *L. laccata*, nor introgressed with them. The restriction of the characteristic Lr mtDNA fragment to *S238N* genets only confirmed this. All the *S238N* isolates exhibited identical patterns of RAPD fragments and possessed the characteristic LrmtDNA fragment. Apparently, therefore, no new mycelia developed to a reproductive stage, despite a heavy annual rain of basidiospores on the plot from the transplanted American mycelia. In contrast, judged by the spatial size and distribution of the basidiocarps of the other native *Laccaria* spp., their mycelia remained fairly small and spatially constrained.

In other plots sampled annually between 1994 and 1997, where the French strain *81306* had been used as the test inoculum, a very different result was obtained. Strain *81306* colonized the plot effectively and also colonized an adjacent plot of uninoculated firs either by mycelial growth, since the RAPD patterns were identical, or by rapid growth of an ectomycorrhizal root infected with *81306*.

The rate of expansion (1.1 m per year) was appreciably faster than had hitherto been recorded for *L. bicolor*. By chance, one of these plots (different from that described previously) was also invaded by the *S238N* American genets from an adjacent plot. But its estimated growth rate was only 0.47 m per year, which is more in keeping with growth rates recorded for this species. This fortuitous observation demonstrated that the American strain *is* capable of colonizing under appropriate conditions, even though that had not occurred in the main experiment. However, the French strain appeared to be relatively more vigorous. In addition, a number of indigenous native genets were isolated which carried appreciable numbers of the rDNA fragments of strain *81306*. This, although not entirely definitive, suggests that introgression into the resident population had occurred as had been found in laboratory crosses.

The results raise a number of questions. Did the presence of a North American *S238N* genet in the American–French transplant plots result in the inhibition of the growth of the indigenous established French *L. bicolor* [BA] and *L. laccata* [BA] mycelia? Was mating of intercontinental genets inhibited in some way? Why did growth inhibition not occur between a French *81306* genet and indigenous (French) genets, and what enabled mating, almost certainly, to occur?

Insufficient information is available to answer these questions but data from other species may be indicative. The SI phenomenon in some Basidiomycotina a can involve growth exclusion between genetically distinct mycelia, e.g. *Rhizoctonia solani* [BA] (see Chapter 2, section 2.3.2.3). Mating between mycelia can also be inhibited even though they are carrying compatible MTFs, and a dikaryon, so formed, can show weak or irregular growth or be sterile. This might have occurred in the American–French transplant plots. On the other hand, it is not uncommon for isolates separated for a long period, as the American and French strains had been, to develop incompatibility or even antagonistic reactions (see Chapter 11). Further investigation of the *Laccaria* [BA] situation, especially the mycelial situation, is obviously desirable in addition to the available basidiocarp information.

A second series of experiments provides some possible answers. These involved inoculating spruce (*Pices abies*) with *S238N* under similar conditions to the previous experiment and, after 2 years in the nursery, planting the saplings out either at a former spruce–birch forest site or in a former chestnut–birch–oak forest area at different locations in France (Martin *et al.* 1998). Samples of the root tips of the infected trees were examined for mycorrhizal infections 2, 3 or 4 years after

transplanting. Unfortunately, a comparable series involving strain *81306* was not attempted.

- In the **nursery**, 80 per cent of the short root tips were infected with *S238N*. No other fungi were detected although non-inoculated seedlings were colonized, as much as 99 per cent, by other indigenous mycorrhizic fungi.
- After 2 and 4 years transplantation in the former **spruce–birch site**, only 18 per cent and 3 per cent of the tips respectively carried the American strain *S238N*.
- In the **chestnut–birch–oak site**, the American strain accounted for only 15 per cent of infected tips 1 year after planting.
- In both transplant sites short root tips, originally inoculated with *S238N*, were replaced by a variety of indigenous mycorrhizal species.

The clear implication of these results is that strain *S238N* is unable to compete under these conditions. These findings cannot be compared directly with those from the previous experiments with fir as host because an entirely different method of assessment was used. Whether, in the first experiment, the small percentage of infected short roots of spruce would have been capable of producing numbers of basidiocarps comparable with those on fir within 10 years is not known, nor has the percentage of short-root infection on the latter been measured. At first sight, however, it does appear as if the American strain of *L. bicolor* [BA] mycorrhiza on spruce in French habitats is not as competitive on fir.

These experimental approaches under quasi-natural conditions have not provided definitive information on whether, or how, potentially mycorrhizal genets compete, but it has shown the great analytical power of the molecular methods that are now available. It seems probable that a longer experimental timescale will be needed with fungi such as these and that, ideally, subterranean mycelia should be sampled in addition to aboveground sporophores.

## 10.5 Complex intraspecific competition in nature

One well-documented case of significant and complex intraspecific selection in nature has been recorded. In addition to the clear-cut interspecific competition in Britain between the resident species of the Dutch elm disease fungus *Ophiostoma ulmi* [AP] and the invasive Eurasian ssp. *novo-ulmi* and North American ssp. *americana* of *O. novo-ulmi* [AP], a clear case of natural *intraspecific* competition occurs between the latter two subspecies.

When the patterns of VCGs of *O. novo-ulmi* [AP] were first analysed from the advancing edge of aggressive populations derived from the host xylem, one or two different VCGs were found to be appreciably more frequent. In contrast, other populations behind the advancing front were highly heterogeneous for VCGs, each occurring at low frequency (Brasier 1988). The one or two preponderant VCGs, termed 'supergroups' by Brasier, were originally thought to owe their dominance to chance founder effects at the time that the fungus first invaded Britain. However, when the epidemic populations throughout Europe were analysed in more detail, this seemed less likely. 'Supergroups', usually associated with *mat B* alone or, rarely, with very small numbers of strains carrying *mat A*, are normally the sole component of every aggressive advancing epidemic front (Table 10.2), although exceptions do occur. Since *mat A*-carrying strains are *less* aggressive this was not surprising. Therefore, without exception, populations contributing to an aggressive front were genetically highly homogenous asexually reproducing clones.

Similarly, all older established populations behind the advancing front were highly heterogeneous for VCGs with greatly increased frequencies of *mat A*. Indeed, the frequency of the 'supergroups' was reduced even in populations of the saprophytic phase which occur in the bark and consolidate the initial pathogen invasion. *Mat A* was present and was also present at a high frequency in the associated highly heterogeneous VCGs. Increased VCG heterogeneity is not surprising since *mat A* is highly fertile so that sexual recombination is inevitably promoted and the resultant recombination generates increased numbers of VCGs.

In addition, d-factor (virus) infection was quite high in the frontal aggressive populations but reduced in the saprophytic populations. However, d-factors are transmitted via hyphal anastomoses but not through ascospores; hence aggressive strains are prone to become infected and their vigour thereby reduced. In contrast, in the saprophytic phase, where sexuality is rife, cross infection will be reduced because of the high frequency of d-factor-free ascospore infections.

Therefore, as an epidemic develops, asexually propagating clonal *mat B* 'supergroups' have a selective advantage over other strains carrying *mat A* because of their greater pathogenicity. However, this advantage is lost in the saprophytic consolidation phase because the vigour of the 'supergroups' is reduced by d-factor infection, to which they are prone, and their frequency is further reduced by the generation through recombination of additional VCGs, which are less prone to infection because they originate from ascospores. Therefore, within the epidemic populations, there is a selective balance between exploitation and consolidation, i.e. between the clonal *mat B* 'supergroups' and other VCGs, based on their relative pathogenicity. The epidemic advances by the preponderance of potentially aggressive 'supergroup' strains, some of which are, by chance, less heavily infected by d-factors and have retained their infective vigour. If this does not occur, the epidemic front becomes stabilized. Conceivably this is what happened to *O. ulmi* [AP] over the 20-year period of the original European epidemic infection, by which time it had become reduced to less pathogenic sporadic infections. Rendered less aggressive in this way, it has now been virtually replaced by the more aggressive *O. novo-ulmi* [A].

## 10.6 General comments

It will have become evident that the information reviewed in this chapter and in Chapter 9 only begins to scratch the surface of the somewhat intractable problem of how selection operates on fungi in nature.

A much broader descriptive treatment of fungal selection than the specific examples described here has been attempted by classifying them according to the kind of strategy for survival imposed by selection, namely *r*- and *K*-selected types. These parameters are derived from the logistic equation used to describe growth, namely the intrinsic

**Table 10.2** Occurrence in Portugese populations of largely clonal strains of *Ophiostoma novo-ulmi* [AP] at an advancing pathogenic front of epidemic infections of Dutch elm disease compared with a saprophytic area where the epidemic had passed on

| Phase | Ssp. *americana* sample | Uniform or supergroup component | | | | Heterogeneous component | | | |
| --- | --- | --- | --- | --- | --- | --- | --- | --- | --- |
| | | Sample (%) | A type (%) | d-infected (%) | % sample | vc group frequency | | | |
| | | | | | | No. groups/no. isolates | Isolates/groups | | |
| *Tomar site—epidemic front* | | | | | | | | | |
| Pathogenic | 59[a] | 93 | 4 | 13 | 7 | (3/4) | (1.33) | (75) | (0) |
| Saprophytic | 150[a] | 64 | 5 | 44 | 36 | 9/45[b] | 5.0 | 91 | 38 |
| *Mafra site—recent front* | | | | | | | | | |
| Pathogenic | 57 | 37 | 0 | 0 | 67 | 21/29 | 1.38 | 19 | 6 |
| Saprophytic | 102 | 9 | 33 | 67 | 91 | 32/83 | 1.6 | 46 | 21 |

Note the large differences in clonal populations in samples from the Tomar and Mafra sites. The pathogenic phase is isolated from twigs, and the saprophytic phase from bark.

[a] The non-aggressive *O. ulmi* still comprised 21% of the overall twig sample and 41% of the overall bark sample but was absent from Mafra.

[b] One vc group of 24 isolates comprised 20 A types and four B types plus two vc groups of five isolates, two of three isolates, one of two isolates and three of one isolate, all A types.

(Reproduced from Brasier (1988) with permission of Nature Publishing Group.)

growth rate $r$ and the carrying capacity $K$ of the environment (see Appendix).

The concept was derived from a consideration of the operation of selection at different phases in the colonization of an island habitat (MacArthur and Wilson 1967) and is related to the interaction between competition- and density-dependent selection. Obviously, in uncrowded conditions those individuals with a high specific rate of increase $r$ will be at an advantage since they will exploit the environment most effectively, whereas in subsequent more crowded conditions those individuals which best exploit the carrying capacity $K$ of the environment will be at a selective advantage.

The concept has limitations. The most serious is that density is only one of the conditions affecting selection (see Chapter 9, section 9.4.3.1), and secondly the concept is not applicable to organisms that show modular growth, such as fungi. Nevertheless, it has been applied, although usually in an ecological rather than a genetical context (Pugh 1980; Andrews 1984; Cooke and Rayner 1984).

Those fungi that have a short somatic life expectancy during which an appreciable proportion of the available resources are directed into reproductive effort are described as $r$-selected fungi. In contrast, $K$-selected fungi have a long life expectancy and, although they may transform a good deal of the available resources into reproductive effort over their lifetime, at any one period the proportion devoted to reproduction is small. Thus, for example, a *Mucor* [Z] or an *Aspergillus* [A] can be described as an $r$-selected fungus, whereas a perennial *Armillaria* [BA] is clearly a $K$-selected fungus. Therefore the terms have some utility as descriptive ecological terms.

Nevertheless, there are further problems when the concept is applied to organisms with modular growth and the consequences are different. If mycelial growth is of the 'phalanx' type, i.e. compact with short hyphal internodes, frequent branching and hence considerable overlap of resource depletion zones (RDZs), $r$-selection will act to favour fungi with the same characteristics as have just been described. However, with 'guerilla' growth, i.e. where hyphal internodes are long, branching is infrequent and RDZs do not overlap, both $K$- and

$r$-selection will favour perennial non-reproductive clones (Sackville-Hamilton *et al.* 1987).

Useful as such considerations may be in surveying broad trends within the Fungi, these general approaches do not provide much information on how selection operates on fungi in any particular instance. Moreover, they give little idea of the complexity of selective processes involved.

Selection is manifested at various levels—within an individual mycelium either between genetically different nuclei or cell organelles, or between individual mycelia. Since the majority of mutants arise within a multinucleated mycelium, intra-mycelial selection within the heterokaryon or heteroplasmon so formed is an essential first step in *all* mutant selection. Despite most of the available information on selection in fungi being concerned with intermycelial situations, the necessity for prior intramycelial selection should not be neglected.

There is no doubt that fungi provide many opportunities for simple or more complex patterns of selection to occur. There is still an appreciable dearth of information about the processes which contribute to selection in any one fungus. Indeed, there is a great dearth of knowledge on this topic in fungi as a whole. The genetic bases underlying apparently simple responses, such as the effect of a single-gene mutation conferring fungicide resistance, or the factors which enabled race T of *Bipolaris maydis* (FA) to exploit its notable selective advantage over race O once its host carried the gene for Texas male-sterile cytoplasm, are still only partially understood. Similarly, while there can be little doubt that the phenomenon of host specificity represents a consequence of initially disruptive selective processes, little is known of their nature or how they act. Nevertheless, many of the known examples of selection in fungi reinforce the notion that natural selection results in sequences of constantly shifting balances due to the interaction of new intrinsic genetic combinations and constantly varying environmental conditions. Ways in which this can be investigated more deeply are being developed.

Of course, ignorance of how selection operates is widespread in most other organisms, but there are particular attractions to the fungi for studies on selection. Their relatively simple metabolic and physiological behaviour, coupled with many possessing predominantly haploid status, provide opportunities to explore such issues in simple eukaryotes which could provide a model for higher Eukaryota. This possibility is reinforced by the fact that the fundamental processes—change in gene frequency, its various determinants and its consequences—appear to be virtually identical with those in higher Eukaryota.

## Appendix: More complex approaches to estimating fitness in fungi

An estimate of the value of the fitness of a population or individual can, of course, be obtained from the equations developed in Chapter 9, section 9.1.1. But fitness estimated by this simple selection model has two drawbacks.

- It only gives approximate results when dealing with selection over several sexual or asexual generations.
- It does not give much insight into the nature of fitness or the features which can affect it.

On the other hand, depending on how the estimates are made, they can be informative on some aspects, notably the change in fitness over a growing season. For example, the relative proportions of the three races of *Cochliobolus carbonum* [AP] on cultivated maize were shown to change over the season in North Carolina. Assuming various generation times and using the relationship (see Chapter 9, section 9.1.1.1)

$$\ln \frac{q_n}{p_n} = \ln \frac{q_0}{p_0} + n(\ln W_q),$$

the relative fitnesses of races 0 and 3 were shown to decline relative to that of the predominant race 2 in the plains region at the foot of the montane region of North Carolina (Table 10A.1) (Welz and Leonard 1993). In general, however, a more sophisticated approach is desirable.

Two kinds of approaches have been used. The first is to attempt to determine which features contribute to fitness and to develop some kind of index which can be used to compare isolates. The other is to develop a more realistic approach to selection, making allowance for various complicating factors such as frequency- and density-dependent selection and interaction between competing isolates.

### 10A.1 Fitness indices

The first approach is typified by attempts to partition and evaluate the different biological components of fitness either over a period of the life-cycle or over a number of asexual generations. In essence this covers three elements:

- spore production and dispersal
- establishment and growth
- reproduction, both asexual and sexual.

The partitioning can be very simple, but an attempt is usually made to include parameters of biological significance which reflect the capacity of the fungus to survive, grow and reproduce. For example, in an experimental comparison of the fitness of asexual with sexual populations of

**Table 10A.1** Relative fitness of races 0 and 3 relative to race 2 of *Cochliobolus carbonum* [AP] at two sites at different elevations

| Race | Fitness in Wilkes County site (285 m) | | Fitness in Yadkin County site (350 m) | |
|------|-------------------|--------------------|-------------------|--------------------|
| | 7-day/generation | 10-day/generation | 7-day/generation | 10-day/generation |
| 0 | 0.42*** | 0.29*** | —[a] | |
| 2 | 1.0 | 1.0 | 1.0 | 1.0 |
| 3 | 0.82** | 0.76** | 0.84* | 0.78* |

Fitness values significantly less than 1.0 are indicated as follows: * $P < 0.05$; **$P < 0.01$; *** $P < 0.001$.
[a] Race 0 was recovered only once from Yadkin County.
(Reproduced from Welz and Leonard (1993) with permission of American Phytopathological Society.)

*Phytophthora infestans* [O], leaflets of young potato plants were exposed to suspensions of each strain containing identical number of sporangia and left for a fixed period in a growth chamber. Three features considered to be of biological significance were measured and combined into a 'fitness index':

- infection frequency (IF)—the proportion of leaflets which became infected
- lesion area (LA)—the area ($cm^2$) of lesions per leaflet
- sporulation capacity (SC)—the number of new sporangia produced per $cm^2$ of leaflet.

The composite 'fitness index' used was simply

$$IF \times LA \times SC$$

This is a very simple index of limited usefulness but it does, at least, indicate the complex contributions of different components, both somatic and reproductive, and implies that comparable fitness could be achieved in different ways. In other kinds of situation, such as a competitive situation, a different kind of index would have to be devised.

## 10A.2 More complex analyses of fitness components and selection

An alternative approach, namely to attempt to describe the components contributing to selection in a more sophisticated way, has been developed with other organisms. A general approach applicable to both haploids and diploids has been developed by Clarke and Beaumont (1992). Their approach has two main features. First, they recognize that population growth is best described as logistic, for, although growth often commences exponentially, it slows down in proportion to the population size, resulting in the characteristic rising S-shaped growth curve represented by

$$N_t = N_0 - rN(K - N_0/K)$$

where $N_t$ is the size at time $t$, $N_0$ represent the initial population size, $r$ is the intrinsic rate of growth and $K$ is the carrying capacity of the environment. Second, they also recognize that selection can be density dependent, frequency dependent or both, and that competing organisms usually interact. All these features are included in their model.

First, consider two haploid genotypes *A1* and *A2* with initial frequencies of $1-q$ and $q$ respectively. The model then allows for density-independent rates of increase $w_1$ and $w_2$, an overall population density $N$, carrying capacities $k_1$ and $k_2$, competition coefficients $\alpha$ and $\beta$, which represent the effects of each genotype on the other, and constants $b_1$ and $b_2$ which measure the strengths of feedback between density and selective value. The selective values of the two haploid genotypes can then be represented as follows:

$$A1 \quad \frac{w_1}{1 + \{N[w_1(1-q) + \alpha w_2 q]\}/k_1^{b_1}}$$

$$B2 \quad \frac{w_2}{1 + \{N[w_2 q + \beta w_1(1-q)]\}/k_2^{b_2}}.$$

Second, for diploid (or dikaryotic) genotypes where *A* is completely dominant to *a* and the frequency of genotypes *AA* and *Aa* is $(1-q^2)$ and that of genotype *aa* is $q$, the selective values are

$$AA \quad Aa \quad \frac{w_1}{1 + (\{N[w_1(1-q^2) + \alpha w_2 q^2]\}/k_1)^{b_1}}$$

$$aa \quad \frac{w_2}{1 + (\{N[w_2 q^2 + \beta w_1(1-q^2)]\}/k_2)^{b_2}}.$$

One advantage of this theoretical treatment is that it can be simulated over many more generations than can be achieved in an experiment and so can be used for prediction and compared with what occurs over long or short periods under natural conditions. Computer simulations for various values for some of these quantities over 200 generations of both a haploid and a diploid (or dikaryotic) organism are shown in Figure 10A.1. The former would represent a eumycotous haploid fungus such as *Aspergillus* [A], and the latter either a dikaryotic fungus, such as an agaric or polypore, or a diploid straminipilous fungus.

The most important difference between the results of this model and the simpler formulation considered in Chapter 9, section 9.1.1.1, is that an average value of $w$ over the generations can be replaced by successive values which do not remain constant but fluctuate to a greater or lesser extent. Moreover, the course of selection is affected by interaction between the phenotypes, e.g. competition when $\alpha > \beta$ or $\alpha < \beta$. Even when there is *no*

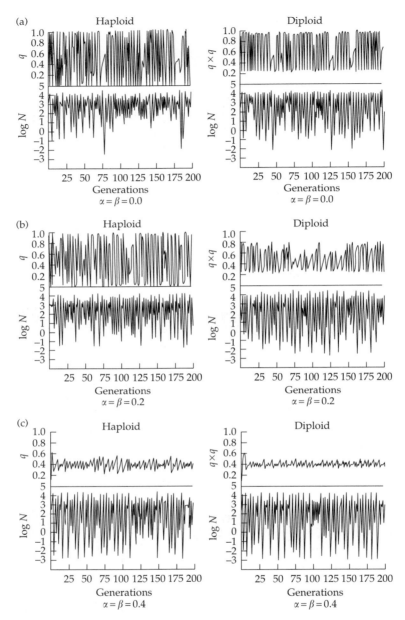

**Figure 10A.1** Simulated changes in genotype frequencies and population densities with progressively higher levels of competition in haploid and diploid/dikaryotic organisms respectively. The simulations are based on the formulae in the text. The haploid curves show competition between two clones. The diploid/dikaryon curve illustrates competition between the dominant $Aa$ and $AA$ phenotypes. In the last set of curves (f) the polymorphism has become unstable and allele $A$ moves to fixation. To illustrate this $q$ is increased to 0.9 to enhance the effect. In *all* cases $w_1 = 100$, $w_2 = 120$, $k_1 = 24\,000$, $k_2 = 20\,000$ and $b_1 = b_2 = 5$. The starting value of $N$ is 10 and that of $q$ is 0.1 (except in (f) where it is 0.9). Note that the oscillations in genotype frequency become progressively smaller as the level of competition rises and that, in general, $q^2$ in diploids/dikaryons shows less variation than $q$ does in haploids. (Reproduced from Clarke and Beaumont (1992), with permission from Blackwell Science).

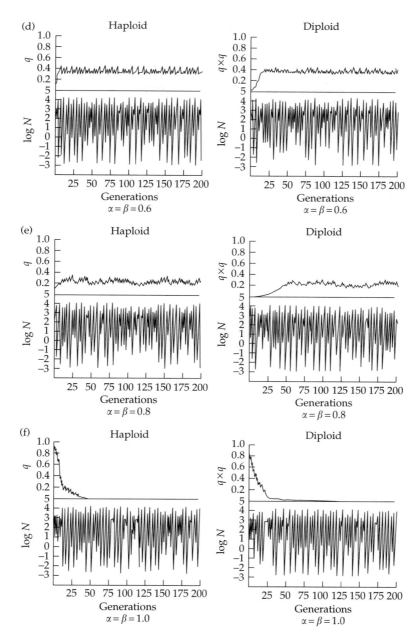

**Figure 10A.1** (*Contd*).

competition, i.e. $\alpha = \beta$, because, for example, they occupy different ecological niches, the gene frequencies still fluctuate widely. However, the fluctuations become progressively less as competition rises and eventually one genotype is eliminated. It is worth noticing that population density fluctuates widely. This is a not uncommon feature of natural populations, where it is often attributed to environmental interactions rather than to these purely stochastic effects.

## 10A.2.1 An experiment and a complex fungal model

The approach just described has not been applied in this form to fungi but a number of its features, such as the effects of density- and frequency-dependent selection and an assessment of components of fitness, have been incorporated into a fungal model which has been tested both theoretically and experimentally by Newton *et al.* (1996, 1997, 1998).

### 10A.2.1.1 The experiment

The model was developed in connection with an experiment on competition between two strains of *Puccinia graminis* f.sp. *tritici* on wheat. The strains could be distinguished by the colour of their uredosori (uredospore-bearing pustules), which were grey in race SR41(*Str1* in the model) and orange in race SR22 (*Str2* in the model). Equivalent inocula of a 1:1 mixture of spores (strictly, dikaryotic uredospores) of each strain were applied under identical conditions to leaflets of comparable young plants of the universally susceptible wheat (cv *McNair 701*). Ten days after inoculation, newly formed spores were collected every two days for a further fortnight, by which time spore production had ceased and sufficient measurements of the various characters recorded had been obtained.

The model included two complex components of fitness: the infection component $U$, i.e. a measure of competitive ability, and the sporulation component $Q$, i.e. a measure of the reproductive capacity. In addition, when estimating the relative fitness of each strain it allowed for continuous *intra*-strain competition in single-strain infections, and for both *intra*- and *inter*-strain competition in mixed infections by comparing each infecting spore population with that subsequently produced by the new pustules.

Separate estimates were made for each strain of the various parameters believed to be biologically significant and to contribute to the components of fitness. The six biologically significant parameters recognized and used in the model over the sequence from infection to the production of the next spore generation were as follows: the inoculum dose $S$, the maximum sporulation capacity

$M$, the infection efficiencies $E$, the carrying capacity $K$ of the leaf for pustules, the sporulation efficiency $D$ and the number $U$ of pustules per leaf. Allowance was made also for any competitive interactions, namely the competitive effect $\alpha$ of *Str1* on spore production by *Str2* and the competitive effect $\beta$ of *Str2* on spore production by *Str1*, as well as the effects of both density- and frequency-dependent selection. All the components were finally combined to obtain a single numerical estimate of the relative fitness of the two strains over the period. Both the experiment and the model developed to analyse it only applied to a limited period of the lifespan of the fungus, namely the period between the initial infection of the leaves and subsequent mycelial growth culminating in a measure of spore production by the new spore-bearing pustules after infection. Nevertheless, it provided an insight into the kinds of interactive processes that can be involved in selection between two coexisting foliar pathogens.

The experimental results are summarized in Figure 10A.2 and Table 10A.2. As inoculum density increased the production of new sori clearly became limited and reached saturation. Strain SR22 developed more sori than SR41 at *all* densities, and similarly it produced more spores per sorus. Hence both the infective efficiency $E$ and carrying capacity $K$ of strain SR22 was superior to that of SR41 (Table 10A.2). Moreover, the values of both $\alpha$ and $\beta$ (the competitive effects) on sori and sorus production for SR22 were greater than the reciprocal values for SR41. Thus, when competing in a mixture, the relative fitness of strain SR22 was higher than that of SR41 because of its greater infection efficiency, its greater carrying capacity and its maximum sporulation capacity $M$. Nevertheless, in competition, SR41 did appreciably reduce the capacity of SR22 to produce both sori and spores per sorus. That relative fitness is density dependent is clearly demonstrated by the decline in spore production at increasing densities of mixed inocula. The relative fitness of SR22 was also appreciably greater at lower than at higher densities. Evidently, with this pair of strains, high reproductive potential compensated for inferior competitive ability under the experimental conditions and so resulted in greater reproductive

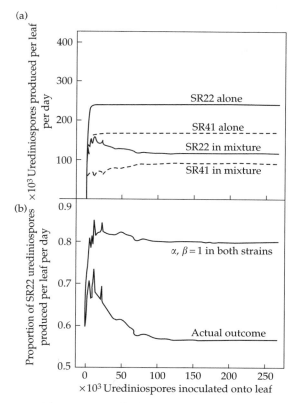

(a)

(b)

**Figure A10.2** Competition between races of *Puccinia graminis* f.sp. *tritici* when inoculated on to wheat leaves. (a) Relationships between the number of uredospores applied per leaf and the subsequent uredospore production per day by races SR22 and SR41 both alone and as a 1 : 1 mixture. Because the ratio of the mixture cannot be absolutely controlled, the curves for the mixed inocula are not a smooth as those for single inoculations. (b) Relationship between the proportion of uredospores of SR22 produced per day and the number of uredospores applied per leaf in mixed strain inoculations. The upper curve is a theoretical expectation if the competitive interaction between the strains was equal and $\alpha = \beta = 1$; the lower curve is the experimental result. There is clearly competitive interaction. (See also Table 10A.1.) (Reproduced from Newton *et al.* (1997), with permission of the Ecological Society of America.)

**Table 10A.2** Estimates of the fitness parameters for the two competing strains of *Puccinia graminis* f.sp. *tritici* [BR] derived from the experiment based on the model for density-dependent reproduction and interstrain competition

| Parameter | Strain | Value |
|---|---|---|
| Carrying capacity $K$ | SR22 | 432.7 |
| | SR41 | 384.3 |
| Infection efficiency $E$ | SR22 | 0.080 |
| | SR41 | 0.033 |
| Maximum sporulation capacity $M$ | SR22 | $2.39 \times 10^5$ |
| | SR41 | $1.67 \times 10^5$ |
| Sporulation efficiency $D$ | SR22 | 0.024 |
| | SR41 | 0.028 |
| Competitive effect on uredinial formation $\alpha$ | SR22 | 0.52 * |
| | SR41 | 1.99 * |
| Competitive effect on urediniospore production $\beta$ | SR22 | 0.62 * |
| | SR41 | 1.52 * |

All the comparisons between strains were significantly different at the $P < 0.05$ level except $D$; in the cases of $\alpha$ and $\beta$, an asterisk indicates that they also differed from 1.0 at the same level for $P$.
(Reproduced from Newton *et al.* (1997) with permission of the Ecological Society of America.)

output by SR22 in mixtures. This experiment and the implications of the model demonstrate some of the complexity underlying selective processes which is likely to be even greater under more variable natural conditions.

Thus relative fitness is a complex property arising from the balance between different reproductive capacities—infection efficiency, carrying capacity and maximum sporulation capacity—and competitive ability. In this case, the reproductive ability of SR22 is higher but its competitive ability is lower than that of SR41 so that, under the conditions of this experiment, high reproductive potential has compensated for less effective competitive ability and resulted, ultimately, in a greater reproductive output. It is evident that a complex combination of these factors determined, in part, by genotypic differences between coexisting strains will determine the success of any one strain, and since these will differ under different environmental circumstances and at different phases of the growth, exploitation and reproductive phases of a fungus, the outcome will not always be easily predictable. Moreover, it is evident that the relative fitness of a fungal isolate will vary in both time and space according to the balance of these interacting components.

One important factor that affects the fitness of pathogenic fungi which could not be measured in this experiment nor allowed for in the model was the effectiveness of the pathogen in transmitting offspring, however produced, to a new infection site. Of course, this determines the spread of a pathogen and is obviously an important factor in the overall success of any pathogen.

### 10A.2.1.2 The model

Although the approach adopted in this model was developed around a particular experiment, it is especially valuable for three reasons:

- it is a biologically realistic treatment
- it has wide applicability since, in principle, the strains could have been taxa differing at any level (genotypes, strains, races, sub-species, species or even different genera)
- it is potentially applicable to any fungus, pathogenic or non-pathogenic, where a substrate is utilized in common and the other assumptions described apply.

Inevitably, in developing the model, a number of assumptions had to be made. Since there could be many infections, it was assumed that:

- there was a random distribution of infecting spores of each strain
- all sites on the leaf surface were equally susceptible to infection
- every spore had an equal chance of infection.

An estimate of the relative numbers of spores per leaf required to produce the number of pustules detected was made using Gregory's multiple infection transformation (Gregory 1944). This provided an estimate of the efficiency of infection. It was assumed that:

- spores competed for infection at the *local* level only
- each pustule developed from a *single* effective infective spore.

In contrast, it was assumed that competition for pustule production, which affects both the numbers of pustules produced per leaf and the total number of spores which they produce, occurs across the whole leaf, since it is known that this fungus acts as a metabolic sink for the leaf as a whole.

The **infection component** $U$ of a single strain was estimated from $E$, $K$, and $S$. Since the probability of a leaf becoming infected is determined by its carrying capacity, the probability of one strain achieving a single infection is

$$1 - [1 - (E_1/K_1)]^{S_1}$$

It follows that the numbers of pustules that will develop in the absence of competition is

$$U = K_1[1 - (1 - E_1/K_1)]^{S_1}.$$

When both strains are infecting a leaf there will be competition and three possibilities will exist at *every* potentially infectable site on the leaf:

- infection by *Str1* only
- infection by *Str2* only
- infection by both *Str1* and *Str2*, although only a *single* pustule can develop at that site from only *one* of the competing strains.

However, although the initial ratio of strains in the inoculum used is $1:1$, their competitive abilities are *unequal*. It follows that, although the unbiased probability is that both strains will infect the same site, the probability that the pustule that develops will be derived from a *Str1* spore, for example, is

$$\left(\frac{E_1}{E_1 + \alpha_2 E_2}\right) \times \left[1 - \left(1 - \frac{E_1}{K_1}\right)^{S_1}\right]$$
$$\times \left[1 - \left(1 - \frac{E_2}{K_1}\right)^{S_2}\right].$$

(NB: If the ratio of strains in the original inoculum is not $1:1$, $E_1/(E_1 + \alpha_2 E_2)$ will have to be replaced by the appropriate value of *Str1* infections per site.) Therefore, however it has originated, the probability $U_1$ that a pustule will be of *Str1* type is

$$U_1 = K_1 \left\{ \left(1 - \frac{E_1}{K_1}\right)^{S_1} \left(1 - \frac{E_2}{K_1}\right)^{S_2} + \left(\frac{E_1}{E_1 + \alpha_2 E_2}\right) \right.$$
$$\left. \times \left(1 - \frac{E_1}{K_1}\right)^{S_1} \left[1 - \left(1 - \frac{E_2}{K_1}\right)^{S_2}\right] \right\}$$

where the subscripts indicate the strain number involved. An exactly comparable formulation for $U_2$ can be made in respect of *Str2*.

The **sporulation component** $Q$, i.e. the number of spores produced per leaf, was similarly estimated from the components $U$, $M$ and $D$ for both strains.

Since the resources of a leaf are finite, the numbers of pustules which can be developed in successive time intervals *declines* as their numbers increase, and their number will be limited. This slowing down of pustule development will be proportional to the difference between the maximum and actual numbers of spores produced. Hence

$$dQ/dU = D(M - Q)$$

or

$$Q = M - Ce^{-DU}$$

But as $Q = 0$ when $M = 0$,

$$Q = M(1 - e^{-DU}).$$

$C$ is a constant and $D$ is a rate parameter related to the approach to the maximum sporulation capacity $M$. It can be expressed in terms of $A$, the number of spores produced by a single pustule in the absence of any other competing pustules. However, when $U = 1$, $A = Q$ and, solving for $D$,

$$A = M(1 - e^{D}).$$

Substituting this value for $D$, $Q$ becomes

$$Q = M\{1 - [1 - (A/M)]^{U}\}.$$

This relationship is clearly analogous to that derived above for the infection efficiency of a single strain (recall $1 - [1 - (E/K)]$) and $A$ is evidently analogous to $E$. A similar argument concerning competition can be developed. Therefore the number $Q_1$ of pustules per leaf of *Str1* in relation to pustule number is

$$Q_1 = \left( \frac{U_1}{U_1 + \beta_2 U_2} M_1 \right)$$
$$\times \left[ 1 - \left( 1 - \frac{A_1}{M_1} \right)^{(U_1 + \beta_2 U_2)} \right].$$

The two components of fitness, $U$ and $Q$, can now be combined to give an estimate of the relative fitness of *Str1* relative to *Str2*, since they have pustule number in common, as pustule production is defined in terms of the inoculum dose $S$ and sporulation as a function of pustules produced. therefore the relative fitness $w_1$ is

$$\frac{[Q_1/(Q_1 + Q_2)]/[S_1/(S_1 + S_2)]}{[Q_2/(Q_1 + Q_2)]/[S_2/(S_1 + S_2)]} = \frac{S_2 Q_1}{S_1 Q_2}.$$

This is, in effect, the eventual spore production per spore inoculated of *Str1* relative to *Str2*. If required, an analogous measure of the relative fitness of *Str2* can be obtained in the same way.

This model has been used for a number of model calculations relating the relative fitness of *Str1* to the frequency of *Str2* and to the population density (spores $\times 10^4$) for different values of the competitive effect of one strain on the other $\alpha$ and $\beta$ (Newton *et al.* 1998). They demonstrate clearly how relative fitness can vary with different values of the parameters recognized in this model.

The effects of population density and genotype frequency on the contributions of specific traits to relative fitness given in some of their results are shown in Table 10A.3. They frequently result in selection being dependent on density or frequency for both components of fitness—infection and sporulation—even when all parameters are identical except one, e.g. inequality between any *one* of $A$, $K$, $M$, $\alpha$ and $\beta$. In the majority, the effect increases with an increasingly greater proportions of the fitter *Str1* in the initial inoculum.

More unexpectedly in this model, since it has not often been detected in experiments, a number of parameters promote frequency-dependent effects on selection, notably when either interactive effects or the $\alpha$ or $\beta$ values of the strains differ, all other parameters being identical. Less surprisingly, when there are large differences in the inoculum doses of the two strains, although every other parameter is equal including the coefficients of competition, frequency-dependent effects occur. A significant finding with this model is that whenever frequency-dependent selection occurs it is *always* associated with density-dependent selection. Relative fitness is far from being a static entity.

Experimental data such as those just described together with the model developed to estimate fitness components demonstrate how the intensity of selection exerted on a fungus and its fitness are dynamic quantities, varying from day to day throughout the life of the fungus and, indeed, from site to site in different parts of a growing mycelium. Therefore it is highly questionable whether an adequate description of how selection acts upon a

**Table 10A.3** The effects of population density and genotype frequency on the contributions of specific traits of *Puccinia graminis* f.sp. *tritici* on relative fitness of strains 1 and 2 according to the model

| Parameter | Element of model | Parameter combination | Density dependent[a] | Frequency dependent[b] |
|---|---|---|---|---|
| | | Equal baseline values for both strains | No | No |
| Reproductive efficiency | Infection | $E_1/E_2 = 1.5$ | No | No |
| | Sporulation | $A_1/A_2 = 1.5$ | Yes ($-$) | No |
| Carrying capacity | Infection | $K_1/K_2 = 1.5$ | Yes ($+$) | No |
| | Sporulation | $M_1/M_2 = 1.5$ | Yes ($+$) | No |
| Competitive ability | Infection | $\alpha_1 = 1.5; \ \alpha_2 = 1.0$ | | |
| | | $\alpha_1 = 1.0; \ \alpha_2 = 0.667$ | Yes ($+$) | Yes ($+$) |
| | Sporulation | $\beta_1 = 1.5; \ \beta_2 = 1.0$ | Yes ($+$) | Yes ($+$) |
| | | $\beta_1 = 1.0; \ \beta_2 = 0.667$ | Yes ($+$) | Yes ($+$) |
| | | $\beta_1 = \beta_2 = 1.5$ | | |
| | | $S_1 < S_2$ | Yes ($-$) | Yes ($+$) |
| | | $S_1 > S_2$ | Yes ($+$) | Yes ($+$) |
| | | $\beta_1 = \beta_2 = 0.667$ | | |
| | | $S_1 < S_2$ | Yes ($+$) | Yes ($-$) |
| | | $S_1 > S_2$ | Yes ($-$) | Yes ($-$) |

The symbols $A$, $E$, $K$, $M$, $S$, $\alpha$ and $\beta$ are those defined for the model.

[a] The symbols $+$ and $-$ denote an increase or decrease respectively in the relative fitness of strain 1 as total inoculum density increases. When $\beta_1 = \beta_2 \neq 0$, the direction in which the relative fitness of strain 1 changes with increasing density depends on whether its proportion in the inoculum is less than ($S_1 < S_2$) or greater than ($S_1 > S_2$) that of strain 2.

[b] The symbols $+$ and $-$ denote an increase or decrease respectively in the relative fitness of strain 1 as its proportion in the inoculum increases. (Reproduced from Newton *et al.* (1998) with permission of the American Phytopathological Society.)

fungus can ever be provided and it is only possible to deal with average effects or trends.

## Further reading

Caten CE (1979). Quantitative genetic variation in fungi. In *Quantitative genetic variation* (ed. JN Thompson and JM Thoday), pp. 35–59. Academic Press, London.

Clarke BC and Beaumont MA (1992). Density and frequency dependence: a genetical view. In *Genes in ecology* (ed. RJ Berry, TJ Crawford and MG Hewitt), pp. 353–64. Blackwell Science, Oxford.

Ennos RA and McConnell KC (1995). Using genetic markers to investigate natural selection in fungal populations. *Canadian Journal of Botany*, **73** (Supplement 1), S302–10.

Lockwood JL (1981). Exploitation competition. In *The fungal community: its organization and role in the ecosystem* (2nd edn) (ed. GC Carroll and DT Wicklow), pp. 243–63. Marcel Dekker, New York.

Newton MR, Kinkel LL and Leonard KJ (1998). Determinants of density- and frequency-dependent fitness in competing plant pathogens. *Phytopathology*, **88**, 45–51.

Trinci APJ, Robson GD, Wiebe MG, Cunliffe B and Naylor TW (1990). Growth and morphology of *Fusarium graminearum* and other fungi in batch and continuous culture. In *Microbial growth dynamics* (ed. RK Pole, MJ Bazin and CW Keevil), pp. 17–38. IRL Press, Oxford.

Wicklow TD (1992). Interference competition. In *The fungal community: its organization and role in the ecosystem* (2nd edn) (ed. GC Carroll and DT Wicklow), pp. 265–74. Marcel Dekker, New York.

# PART IV

# Species and speciation

While dealing with the population genetics of the fungi, the problem of the taxonomic definition of an organism, in particular the definition of a species, has not so far been addressed nor have the processes which result in the formation of species been considered.

The definition of a species, whether as an operational unit or as a theoretical concept, has been endlessly and continuously discussed by successive generations of biologists with varying degrees of agreement. One handicap to a population approach is that the concept is essentially taxonomic, although the processes that result in such units being recognized are essentially population phenomena. Another source of confusion is the approach adopted, whether practical and operational, or theoretical. There is still no universal agreement on what constitutes a species, whether different kinds of species exist, or on the process or processes involved in speciation. However, these uncertainties have not inhibited a variety of proposals being made for possible processes involved in species formation. The uncertainty lies in knowing how many of these are effective and actually operate in nature. Therefore, even on these functional matters, there is far from universal agreement on processes or their relative importance.

Until relatively recently speciation issues have been virtually ignored in the Fungi, and species have been almost entirely described in taxonomic terms based on descriptions of their phenotypes—so called 'morphospecies'. Increasingly, however, problems have arisen with morphospecies, not least the occurrence of biologically recognizable groups of practical significance for field mycologists. In particular, many kinds of infraspecific entities such as races or *formae speciales* have been described which are often of great practical significance to agriculturalists and others. Moreover, the fact that a high proportion of fungi are known only from the anamorphic state has precluded them from being recognized as species by some taxonomists and biologists, and their relation to obviously similar but holomorphic species has been left unclear. The relationships of all these entities to recognized species and their origins in populations need to be considered.

Therefore, in this section, the constraints which fungal features impose on taxonomic species descriptions and the biological situations which have been found to underlie taxonomically described fungal species will be described first. Then the process or processes that may have given rise to these conditions, namely speciation, will be considered, although the available information on speciation in fungi is still very limited.

# CHAPTER 11

# Fungal species

Mycology as a recognizable activity only really got under way in the eighteenth century and then as a branch of botany rather than as a science in its own right. Therefore fungal taxonomists inevitably adopted the procedures and concepts developed by botanical taxonomists and did their best to fit the fungi into a framework largely developed for flowering plants. Species, as originally conceived, were recognizable as distinct entities which enabled living organisms to be classified, i.e. they were strictly taxonomic units, and the category species could be defined as a group of organisms having more features in common between its members than between any one of its members and that of any other species. Species were based on the recognition of distinct morphological patterns to be observed in nature, and the limits of a species were determined by the degree of discrimination adopted by the classifier.

Thus, although simple, this definition was not always easy to put into practice with fungi where the available morphological characters were very different from those studied hitherto and their variability appeared to be much greater. Even so, much of the variability described could be circumscribed within the available standard taxonomic framework of species and varieties. This was not without its difficulties for the macrofungi and more obvious pathogens such as rusts and mildews, but discrimination was almost impossible for microfungi until the advent of precise isolation procedures and pure culture methods towards the end of the nineteenth century. This was, in part, a reflection of the extreme phenotypic plasticity of many fungi. However, the application of cultural techniques to microfungi permitted a greater degree of standardization to be achieved, thereby promoting more effective comparisons and, incidentally, greatly increasing the available numbers

of characters to be recorded or measured. Once in culture and separated from other species, additional attributes—biological, physiological or biochemical—were discovered and were exploited for taxonomic purposes. When sexuality was recognized at the start of the twentieth century, new taxonomic horizons were opened up, although one of the first consequences was the growing realization that large numbers of fungi were apparently entirely lacking in sexuality. This did not pose a major taxonomic problem as long as species were separated on morphological grounds alone. As knowledge accumulated, increasing realization of the very specific roles that fungi can and do play in natural communities and in plant disease led to more attention being paid to their behaviour and their precise habitat conditions in the field. Plant pathologists found that they needed to discriminate between a variety of biologically recognizable but taxonomically undefined groups. These entities were of significance in the field but could not be adequately defined within the adopted contemporary infraspecific taxonomic framework. One of the earliest problems was the discovery of 'biological races', followed by the realization that many species comprised groups of sibling or cryptic species which were virtually indistinguishable morphologically. Some of these could be distinguished using more sophisticated morphological distinctions so that standard taxonomic practice could still be employed, and the characteristic divisions between 'splitters' and 'lumpers' of taxa increased. In an increasing number of cases, however, this kind of solution was either impossible or ineffective. Therefore there was a growing problem of how to deal with such situations and an alternative taxonomy was developed to cover population units which did not fit the classical taxonomic categories of species, subspecies, varieties and *forma*. Even so, the over 75 000 contemporary described fungal species are still defined in largely morphological terms together with some biochemical, physiological and habitat information.

Meanwhile, changing ideas on the nature of species developed elsewhere in biology—in particular, the notion that species were the result of processes which also gave rise to evolutionary change. This resulted in species coming to be regarded increasingly as dynamic entities which arise, persist for a longer or shorter period, change, decline and then become extinct and are replaced by other species. As a consequence, a variety of species concepts have been proposed to reflect this dynamic aspect in some way. They differ mainly in stressing some particular aspect which is regarded as of particular evolutionary or biological significance and raised the question of how species arose, i.e. the twin problems of the mechanism(s) of speciation and the true course of evolutionary descent or phylogeny. Evolutionary and molecular biology, largely neglecting the fungi, expanded understanding of these issues immeasurably in the last half of the twentieth century. Now it is necessary to reconcile and expand fungal taxonomic concepts to the new knowledge and to begin to understand fungal speciation. This chapter describes and applies some of the ideas developed for other organisms to the fungi while drawing attention to some of the novel issues which the recognition and definition of fungal species pose.

In a recent general survey of speciation, some 22 different definitions of species were listed and compared (Mayden 1997). One of these, the **evolutionary species**, differs from all the others in that it is universally applicable and embodies a dynamic population approach. It is a primary concept defined as follows (Wiley and Mayden 1997):

An entity composed of organisms which maintains its identity from other such entities through time and space, and which has its own evolutionary fate and historical tendencies.

This definition is adopted here as a *strictly conceptual* definition for fungal species although it can hardly be applied in any *practical* sense. However, four other concepts are, to a greater or lesser extent, applicable as practical working concepts both for taxonomic purposes and also for a consideration of how species arise:

- the **morphospecies** concept—a recognizable group of individuals distinguished predominantly on morphological characters
- the **agamospecies** concept—recognizable coherent groups of individuals which lack normal sexual reproduction but are capable of asexual propagation of some kind
- the **biological species** concept—a group or groups of interbreeding individuals, reproductively isolated from other such groups
- the **phylogenetic species** concept—the smallest recognizable group of individuals sharing a common monophyletic lineage.

The application of these definitions to fungi will be considered in this chapter together with an account of the increased complexity that has accumulated and enriched the original taxonomic species concept in fungi. A description of some of the population processes involved in fungal speciation follows in the final chapter.

## 11.1 Fungal species as morphospecies

The original simple concept of the morphospecies, adopted from plant taxonomic practice, suffers from the same kind of discriminatory problem with fungal species as it does with plants and other organisms. In essence, this is the problem of judging when a group of organisms is sufficiently distinct to regard it as a species or when a better course would be to regard very similar but not constantly different individuals to be members of the same highly variable polymorphic species. The latter solution is most usually adopted by fungal taxonomists. In fungi, taxonomic difficulties are further exacerbated by:

- the problem of defining a fungal individual
- the phenotypic plasticity of many fungi and their variability in response to environmental conditions of all kinds
- the relatively limited range of macroscopic form and structures available for description, although this is balanced by a wealth of microscopic features

- the many potentially taxonomically useful features not evident in the field which necessitate making collections at different times and thereby increasing the liability of wrongly associating them with a particular isolate
- the widespread occurrence of cryptic/sibling species and, in pathogenic forms, biological races which require additional culture studies.

The problems associated with defining the nature of a fungus individual (see Chapter 4) make it evident that it is not always easy to be certain of the limits of an individual from field observations alone.

Other peculiarly fungal problems arise because of their great phenotypic plasticity and response to environmental conditions. Differentiation can be suppressed or alterations induced in structures or their dimensions. For example, basidiocarp features and spore dimensions in many agarics can be affected by the relative humidity during development. The most extreme situation is seen in dimorphic fungi when the growth habit is entirely altered, e.g. mycelial and yeast-like forms of species of *Mucor* [Z]. Another widespread feature, especially common in Ascomycotina, is the variety of conidia and conidiophores which develop under different environmental conditions. When first recognized in the eighteenth century, discrimination was largely confounded by the inability to decide which spore form was associated with which mycelium. Consequently, the same species was sometimes isolated as two or more apparently distinct species, or spore forms from different species were incorrectly associated, thereby adding to the taxonomic confusion! Such confusions multiplied until it was possible, over a century later, to isolate spores and grow fungi in pure culture. Isolates could then be compared under conditions *in vitro* where, with appropriate manipulation, it was often possible to produce the alternative spore forms from the same inoculum.

Features of the teleomorphic phase, if known, tend to be more constant and provide many reliable taxonomically useful characters. Unfortunately, in many fungi only the anamorphic state is common and the teleomorph is not known, has not been associated with all of its anamorphic phases or

does not exist. The first two situations are especially true of pleomorphic fungi, especially if the teleomorph occurs at a different time and at a different site from the asexual phase; for example, *Mycosphaerella graminicola* [AD] occurs on cereal stubble and its anamorph *Septoria nodorum* [FA] is a widespread pathogen on growing wheat leaves. In other cases, the sexual reproductive phase is completely unknown in nature, having only been obtained in culture, e.g. *Nectria haematococca* [AP], the teleomorph of *Fusarium solani* f.sp. *cucurbitae*. More commonly, sexual reproductive structures are rare in nature or difficult to obtain in culture. Most gilled and pore-forming Basidiomycotina either lack an anamorphic phase or it is not readily visible in the field, e.g. the oidial stages of Coprini. What is now evident is that if both phases occur then an adequate taxonomic description requires knowledge of *both* phases, i.e. the whole fungus or holomorph (Kendrick 1979; Reynolds and Taylor 1993). Implicit in these considerations is that potentially three kinds of morphospecies could occur, i.e. anamorphic species, teleomorphic species and holomorphic species. Fungi only exhibiting the anamorphic phase can be described as agamospecies (see section 11.2).

The problem of the dearth of morphological characters is obviously greatest in unicellular or few-celled fungi such as yeasts or Laboulbeniales, but it is also the case with mycelial forms. For example, the hyphal features of most fungi in the anamorphic phase contribute only a few discriminatory characters such as pigmentation, type of cross-wall and pore, presence of clamp connections and wall ornamentation. Therefore the bulk of the description depends on the features of spore-producing structures. These can include sporophore or conidiophore development and structure, mode of spore formation, and mature spore features such as wall structure and ornamentation, compartmentation and pigmentation. The numbers of permutations and combinations of such a limited number of structural features is quite high and provides sufficient information to enable several hundreds of different form genera of the Fungi Anamorphici to be distinguished, but they are still relatively few at the specific level. Quantitative measurements, of necessity, provide important discriminatory characteristics but are often confounded by environmentally imposed variation resulting in uncertainty or overlapping measurements between different species. Consequently, many species are either unsatisfactorily, or only notionally, distinguished by differences in *mean* measurements. This can lead to real problems of identification, especially if only small numbers of specimens are available. Culture under standardized conditions often enables comparable measurements to be taken of conidial dimension, but even ascertaining appropriate cultural conditions is often laborious or even impossible. Such requirements add appreciably to the time taken to identify a fungus satisfactorily.

As a consequence of these various problems, many fungal species are described in very broad terms and regarded as highly variable. In some cases appreciable variation occurs over the spatial or host range of a species, resulting in different names being applied to entities which later come to be regarded as members of a single highly variable species. In other cases attempts were made to deal with the variability by defining classical infraspecific taxa such as varieties or subspecies. With pathogenic fungi especially, strains restricted to a particular kind or family of host plants, or to a specific habitat, were simply described as biological races or *formae speciales* and regarded as infraspecific entities.

The rust fungi provide classic examples of a variable taxonomic species; for example, *Puccinia graminis* [BA] comprises several 'biological races' recognized in the field by pathologists as biologically significant units and defined by their host ranges. Many years after the *formae speciales* of powdery mildews and rusts had been described, experimental attempts to cross them showed that varying degrees of infertility existed between them (see Chapter 8, section 8.2.1.2, and Chapter 12). Interfertility or its lack as an indicator of specific differences was not considered when they were first recognized and *formae speciales* continued to be regarded as an expression of infraspecific variation outwith the classical taxonomic categories. How far 'biological races' or *formae speciales* were truly taxonomically or biologically comparable with other infraspecific entities was not clear. Some further clarification has been obtained with the introduction and interpretation of intercompatibility tests (see section 11.3).

Numerical taxonomic techniques (Sneath and Sokal 1973) enjoyed a brief phase of popularity as an apparently objective means of delimiting species. They were used to identify so-called operational taxonomic units (OTUs), and the technique has sometimes been applied to fungi in an attempt to resolve groups of very similar but variable organisms such as isolates of *Aspergillus* [FA] and other Fungi Anamorphici which have defied more usual taxonomic treatments. Insofar as such techniques use all available characters and are based essentially on comparison and the detection of similarities they have some usefulness, but it is very unclear how far OTUs found in this way can be compared with species as recognized by more conventional methods. In particular, it is not possible to predict how they will behave in nature. In general numerical taxonomic methods have neither received much support from fungal taxonomists nor many applications in mycology. On the other hand, many of the techniques used have found application in the practical development of cladistic methods now widely used in determining phylogenetic relationships (see section 11.4).

## 11.2 Fungal species as agamospecies

Because anamorphic species are so widespread in fungi it is convenient to consider their nature before that of the more common, sexually reproducing species. They probably occur in all major groups of the Eumycota, but few anamorphic Basidiomycotina have been recognized.

Many species of Chytridiomycotina are asexual, although sexual reproduction also occurs. It is generally presumed that the asexual forms are primarily asexual and that sexuality may have evolved in this group. At present, so little is known about the group as a whole that the relationships of sexual and asexual species can only be conjectural.

The fact that a fifth to a quarter of all other described species of Eumycota are anamorphic, i.e. the Fungi Anamorphici, raises a taxonomic issue which it has been suggested can be dealt with by regarding them as different in some way from 'true' species and recognizing them as agamo-species. Despite the convenience implied by such a treatment in the past, this attitude is becoming increasingly untenable. Morphological similarities between entirely anamorphic species of *Aspergillus* [FA], *Penicillium* [FA] and the relatively few other comparable species which possess a teleomorphic stage have often been recognized in taxonomic arrangements of these abundant and widespread genera. These morphological similarities have been confirmed by the use of molecular analysis. Close genetic links between strictly anamorphic species and sexually reproducing species have been increasingly demonstrated. For example, some species of *Penicillium* [FA] are now known to differ from species of *Talaromyces* [A] by the loss of the sexual stage from the latter (LoBuglio and Taylor 1993; LoBuglio *et al.* 1993) (see Chapter 12, Figure 12.4). Indeed, this change has been shown to have occurred more than once with different species pairs. The likelihood is that asexuality in these fungi is probably always secondary, although the existence of primary asexual species cannot be ruled out entirely.

The demonstrations of secondary asexuality in ascomycotous Fungi Anamorphici has led to the proposal that they should all be reclassified in the Ascomycotina under the appropriate teleomorphic names. Progress in determining genetic associations has been so rapid that in the latest edition of the *Dictionary of the Fungi* (Kirk *et al.* 2001), about 2800 (15 per cent) of those previously classified as Fungi Anamorphici have been so transferred. A complete transference of all Fungi Anamorphici may be possible eventually, but the available data are insufficient to support a complete change as yet. However, these developments make the necessity of recognizing agamospecies increasingly redundant.

Nevertheless, three other kinds of problem still exist in recognizing species in the Fungi Anamorphici:

- the association of very similar anamorphic forms with several different teleomorphs
- the occurrence of different but morphologically distinct mating groups and their anamorphic associates

- the occurrence of very similar forms on an immensely wide range of hosts with, in some cases at least, evidence for specific host specialization, i.e. *formae speciales*.

These situations exemplified by the Fusaria are illustrated in the following examples.

First, very similar species of *Fusarium* [A] can be associated with teleomorphs of the genera *Nectria*, *Gibberella*, *Calonectria* and *Micronectriella* [all AP]. Moreover, since Fusaria are mostly found and isolated from their substrate as anamorphs, they may only be known in this form. Therefore identification involves appreciable laboratory investigations to sort out which anamorph is associated with which teleomorph. Morphological criteria alone are not always sufficient to achieve this, but the use of experimental crosses or molecular characters has now resolved many such problems.

Secondly and conversely, a large number of similar Fusaria, based on their anamorphic characters, can be associated with a single teleomorphic form, e.g. *Gibberella fujikuroi* [AP]. It was therefore treated as a single species with a number of varieties. However, on the basis of their mating compatibility, many of these anamorphic forms proved to be separable into at least seven different mating populations (MPs)—MP *A* to MP *G*—between which there is apparently complete incompatibility and sterility. In some cases the anamorphic states of different MPs are not morphologically separable, e.g. mating groups MP *A* and MP *F* both exhibit '*F. moniliforme*' [FA] anamorphic states (Leslie 1995). Moreover, with the exception of some isolates of MP *F* which have been isolated from pine or rice, both MP *A* and MP *F* forms have been isolated from the same hosts, i.e. bananas, figs, maize, sorghum and the soil. Therefore without a mating test to separate the two MPs, confusion is possible. This is even more likely since other unrelated differences can occur between isolates in the dimensions of their conidia and other characters such as different host specificities. It is practically important that they should be separable in some way but, in the case of '*G. fujikuroi*', it has not proved possible to distinguish all seven MPs on morphological grounds so that their recognition could only be achieved by a mating test with standard MP isolates. Subsequently, separation has proved possible by comparing nucleotide sequences derived from three different gene regions and the demonstration that different MPs also occupy different geographical ranges. For example, MP *A*, MP *F* and MP *M* are basically African in their distribution and are also related to a variety of other anamorphic *Fusaria* [FA] in that region (O'Donnell *et al.* 1998). This has reduced the potential confusion considerably. Similar situations may well exist in other Fungi Anamorphici.

Lastly, in *F. oxysporum* [FA] over 100 *formae speciales* have been recognized, each supposedly confined to one or more different host species or to a particular kind of host; for example f.sp. *conglutinans* is found on *Brassica, Raphanus* and other Cruciferae (Armstrong and Armstrong 1981). In some cases, the *formae speciales* themselves are subdivided into different races on grounds of morphology or host preferences. Thus three races are recognized within *F. oxysporum* f.sp. *conglutinans* [FA]: race 1 on Brassicas, race 2 on radishes and race 3 on stocks. One outcome of such treatments has been considerable confusion; for example, mycologists working in different areas have recognized different taxa as a result of using different criteria. This kind of confusion was particularly prevalent with large widespread Fungi Anamorphici such as *Fusarium* [FA] before generally agreed criteria were adopted. In some cases the supposed specificity was eventually demonstrated to be either spurious or only manifest by some isolates or by isolates which differed in their degree of specificity to the host species. The majority of *formae speciales* appear to consist of one or a few genetically closely related clones, but there is no evidence of sexuality. Similar pathogenic and non-pathogenic forms occur, sometimes together or in other cases as different and very distinct clones. A large number of clones are known from soil and it has been speculated that pathogenic forms may be selected from this reservoir. How the different clones have arisen is not understood, and at present taxonomic discrimination is based largely on their morphological features and host specificity (Gordon and Martyn 1997). Until the evolutionary relationships of *F. oxysporum* [FA]

have been further elucidated, it is a species which can genuinely and most usefully be described as an agamic species.

## 11.3 Fungal species as biological species

Sexually reproducing fungi are more comparable to the bulk of other species in other groups which have been recognized as various kinds of species. For example, Mayr (1940) defined a **biological species** as follows:

... groups of actually or potentially interbreeding natural populations which are reproductively isolated from other such groups.

Although originally devised to describe the situation in birds, the biological species as defined above is probably the most widely adopted species concept for all organisms. Subsequently, Mayr amended this definition in various ways and, most significantly, recognized the importance of the relationship between a species and its habitat and ecology. For instance in 1982 he wrote:

A species is a reproductive community of populations (reproductively isolated from others) that *occupies a specific niche*. [emphasis added]

This is an important amendment whose significance has not always been acknowledged, although it makes the biological species concept more attractive in many ways. For instance, Levin (2000) has developed the concept of the 'ecogenetic species' which, in effect, adopts the later views of Mayr (1982), by defining it, somewhat more explicitly, as follows:

A species has a unique way of living in and relating to the environment and has a unique genetic system ... that governs the intercrossability and interfertility of individuals and populations.

This usage has even wider applicability than Mayr's original formulation.

In the Eumycota, individual mating populations and many cryptic species meet both Mayr's original and amended definitions as do many biological races and *formae speciales* which, historically, were treated as infraspecific entities. However, there is still considerable reluctance amongst mycological taxonomists to adopt the concept as a day-to-day working definition of a species, although the modified definition is more applicable in many ways.

The two essential elements in defining a biological species are:

• whether isolates are members of an interbreeding population
• whether they are reproductively isolated from other populations of other species.

These are not always easy to determine. In fungi it is often difficult to determine whether a population is sexually reproductive or strictly clonal, a matter addressed earlier (see Chapter 7, Appendix). Then there is the inherent difficulty, by no means confined to fungi, of knowing whether or not individuals in geographically isolated populations are capable of crossing in nature, especially when this appears to be unlikely, for example populations in different continents or separated by some physical barrier such as a mountain range.

Interfertility can, of course, be tested by making experimental crosses but successful experimental crosses between adjacent populations do not necessarily occur under natural conditions. Therefore the experimental approach is only an indication of the potential ability of two populations to interbreed. Experimental test crosses can also provide ambiguous results. Although interfertility can usually be readily determined, not every combination may be be equally fertile and some crosses may even be infertile. Perkins (1994) has drawn attention to this difficulty in determining and interpreting infertility in ascomycotous crosses. The problem is greater with Basidiomycotina because conditions for the development of basidiomata in culture are often unknown. Test matings are rarely taken beyond the stage of identifying hyphal fusions and the initiation of dikaryotic/heterokaryotic hyphae, rather than to the development of a fully differentiated functional basidioma. Therefore such a test is no guarantee of fertility, so that Boidin (1986) described such test matings as **intercompatibility tests**. However, these tests been

widely used to indicate potential cross fertility although this is frequently unjustified.

Intercompatibility tests are now frequently employed and are equally applicable to all fungal groups, although in Fungi Anamorphici only somatic fusions can be tested. However, since many isolates of Fungi Anamorphici turn out to be unrecognized single mating-type strains of dimictic Ascomycotina, the mating tests not only enable anamorphs to be related to a teleomorph but also provide evidence for reproductively distinct MPs as described in section 11.2. Intercompatibility tests have also revealed that many *collective* species, i.e. morphologically identical or very similar species (cryptic or sibling species), which cannot be reliably separated on morphological grounds, can be distinguished by an intercompatibility test. Even with this test, results are not always clear cut and some collective species show total compatibility between some isolates but only partial compatibility with others. The taxonomic interpretation of such partial compatibility has often been a matter for debate. Nevertheless, crossing as a routine tool for defining species was first used by Shear and Dodge ( 1927) with Ascomycotina for *Neurospora* [BA] and, in the same year, by Vandendries (1927) with *Coprinus* [BA] for Basidiomycotina. The technique was applied from the 1930s onwards by Mounce and her colleagues (Mounce and Macrae 1938) to many pore-forming Basidiomycotina which had proved difficult to separate, and by Boidin and Hallenberg in many papers to the even more taxonomically difficult 'crust' fungi (Corticiaceae).

Results in Ascomycotina differ somewhat from those in Basidiomycotina.

## 11.3.1 Biological species in Ascomycotina

Since the pioneering work of Shear and Dodge (1927) in distinguishing between *N. crassa* and *N. sitophila* [both AP] on grounds of both infertility and morphology, the conidial *Neurospora* species, apart from *N. crassa* [AP] and *N. intermedia* [AP], have been fairly clearly distinguished by the crossing test. Phenotypically they are remarkably similar but some species regularly occur intermingled in the same habitats. On the basis of observations and pairings of a worldwide collection of more than 4000 strains, Perkins and Turner (1988) commented:

Although the heterothallic species tend to differ somewhat in size and morphology of ascospore, these differences are only averages. It would be difficult or impossible to assign strains reliably to the heterothallic species either by vegetative morphology or by ascospore size or shape. There is too much variation within species, and there is overlap between species.

and commenting further on the criteria they used for defining species:

In practice, fertility tests have proved to be the only reliable criterion, and much the most convenient.

Fertility or infertility continue to be the definitive indicators of a species amongst the conidial Neurosporas (Turner *et al*. 2001). In many species crosses, meiotic pairing is complete, perithecia and asci develop and in some cases ascospores are delimited, but usually these are almost entirely inviable. For instance, *N. crassa* × *N. tetrasperma* [both AP] develop barren perithecia only, but with *N. intermedia* × *N. tetrasperma* [both AP] white inviable ascospores are formed. Despite this, *N. crassa* [AP] and *N. intermedia* [AP] regularly produce 1–10 per cent viable ascospores when crossed and isolates of the latter from Andhra Pradesh (India), Pakistan, Thailand and Penang (Malaysia) cross unusually well, whereas those from other areas may fail to cross completely with other *N. intermedia* [AP] strains. *Neurospora* demonstrates admirably the advantages and limitations of the intercompatibility test and of the biological species concept.

1. The ability to cross varies with both isolate and, often, location.
2. Reproductive isolation is not necessarily absolute but, when the numbers of exceptions are small, an intercompatibility test, together with morphological characters, is generally sufficient to enable species to be distinguished.

For example, as mentioned earlier (see Chapter 8, section 8.2.1.2), the number of genuine species hybrids found by Perkins and coworkers in Neurosporas (Turner *et al*. 2001) is minuscule when compared with clear-cut infertility reactions.

A less effective pattern of reproductive isolation is the production of poorly viable offspring. This behaviour is found in *Ophiostoma ulmi* and *O. novo-ulmi* [both AP]. Weakly growing hybrid isolates can be produced in culture but are extremely rare in nature, are clearly unable to compete with the more actively growing *O. novo-ulmi* [AP] and do not persist (see Chapter 8, section 8.2.1.2). In contrast, the two subspecies of *O. novo-ulmi* [AP] which have migrated into Europe from North America and Eastern Europe respectively are fully viable and largely interfertile, although when ssp. *novo-ulmi* is the 'female' parent in a cross with ssp. *americana*, perithecial production is reduced by about 90 per cent compared with a cross within ssp. *novo-ulmi* (Brasier 1979).

An even more ambiguous situation was found in *Ceratocystis* [AP] where it was suspected that several species of the highly variable *Ceratocystis coerulescens* agg. [AP] occurred on conifers and hardwoods but could not be clearly distinguished. Their resolution has now been effected by a combination of molecular investigations and the application of a fertility test. Ten enzymes were investigated in a sample of almost 100 isolates of the complex obtained from conifer and hardwood hosts and their isozyme profiles were determined. Five distinct patterns were found in those considered to be *C. coerulescens* [AP] plus two other species similar to isolates of the *C. coerulescens* complex, namely *C. polonica* and *C. laricicola* [both AP], which were morphologically indistinguishable and regarded by some as conspecific (Harrington *et al*. 1996). The close relationship but individual distinctness of these seven putative species was confirmed by an investigation of DNA sequences of the isolates (Witthuhn *et al*. 1998) and, finally, by a fertility test. This was complicated by the ability of some strains carrying the *MAT-2* mating type (the species are dimictic) to self-fertilize *in vitro* (Harrington and McNew 1998). However, by using self-sterile *MAT-2* strains or those that produced only a few selfed perithecia, it proved possible to make crosses between the putative species. Twenty-four out of 74 interspecific crosses produced perithecia with a few misshapen ascospores which germinated poorly and up to 2 per cent were fertile in about half the remaining 50 crosses. Genetic markers indicated that mycelia from those ascospores which germinated were genuinely hybrid, but their vigour was so low that it seemed unlikely they would persist in nature. No natural hybrids have been detected.

Once the seven putative species had been defined by this combination of molecular and mating tests, it proved possible to distinguish and describe them using newly recognized reliable morphological and habitat characteristics.

Similar behaviour has been found in several other Ascomycotina so that a comparison of the vigorous fertility of intraspecific crosses with very low but variable infertility of interspecific crosses is usually sufficiently clear cut for specific distinctions to be accepted. Moreover, it often proves to be the case, once biological species have been recognized, that they are then seen to differ in their habitat preferences and in hitherto unrecognized morphological differences. But whether or not morphological specific differences are found, the reality of such biological species is not in doubt. Biological species in Mayr's sense appear to occur commonly in the Ascomycotina. Although positive compatibility can often be demonstrated *in vitro*, crossing appears to be largely ineffective in nature.

Incidentally, post-zygotic failure after crossing sibling or cryptic species, such as shown by *Neurospora* [AP] or *Ceratocystis* [AP], is common in the Ascomycotina but is uncommon in other Eumycota. However, it does occur in the Blastocladiales (see section 11.3.3).

## 11.3.2 Biological species in Basidiomycotina

Intercompatibility tests are not always as informative in the Basidiomycotina as in the Ascomycotina but they have led to the recognition of unsuspected cryptic species. In the first case studied, Vandendries (1927) investigated isolates of *Coprinus micaceus* [BA], then regarded as a single variable species. Although his results are somewhat confused, matings between monokaryons derived from different isolates entirely failed to anastomose, confirming his view that there were, indeed, three morphologically very similar but reproductively isolated species. He also suggested that strains from closely located sites were intercompatible but those

several hundreds of kilometres apart were not. This suggestion has not received much support from subsequent work.

Specific distinction within some sections, at least, of the genus *Coprinus* [BA] are particularly difficult to make using morphological characters alone. When Lange (1952) used intercompatibility tests to distinguish or confirm species in the Sect *Nudi*, however, he found a correlation between distinguishing morphological characters and incompatibility in most cases. In a few species he found incompatible groups within the limits of a *single* morphologically defined species but did not suggest that they should be recognized taxonomically. This latter pattern occurs more frequently in other Coprini, where Kemp (1983, 1985) demonstrated the occurrence of numerous cryptic species in Sect *Lanatuli*. All isolates possess tetrapolar diaphoromictic mating sytems and two isolates regularly produce two-spored basidia only. The 162 isolates studied fell into 25 clearly defined groups on the basis of a total absence of anastomoses between them, i.e. absolute incompatibility. Fourteen of these groups were represented by a single isolate and only eight of the groups, including the two-spored forms, had been assigned names, but these were not consistently applied and many were regularly misidentified. In culture, only seven of these biological species were induced to develop basidiocarps. Although some groups had distinct morphological or other traits such as smell, all differed in their basidiospore dimensions, and it was claimed that these differences were sufficiently diagnostic to enable a separation to be made even without an intercompatibility test. Thus, in many cases, basidiomycotous biological species recognized by an intercompatibility test can eventually be distinguished on morphological grounds.

Widespread species are often found to include incompatible groups which may or may not be distinguishable morphologically, but this is not always the case. *Panellus stypticus* [BA], a 'good' morphospecies, is widely distributed. Isolates from Europe, the USA, Canada, Russia, Japan and New Zealand are all fully intercompatible. Nevertheless, the presence of a gene for bioluminescence, present in North American isolates, is totally absent from isolates from the rest of the world; a similar distinction is evident in the taste of the basidiome, which is mildly acrid in the USA but tasteless elsewhere (Macrae 1937; Petersen and Bermudes 1992).

Molecular methods have enabled further intraspecific differentiation to be identified (Jin *et al.* 2001). Biogeographically significant variation was detected in the variable length of the ITS DNA region as well as by different characteristic RFLP patterns over the same region based on digestion with the restriction enzymes *BstZ* I, *Dra* I and *Hinf* (Figure 11.1). First, there is a clear distinction in ITS length between Australasian isolates and those from the northern hemisphere, with the former being shorter than the latter because of several deletions. Second, the RFLP data enable distinctions to be made between some of the northern hemisphere isolates. In particular, European isolates share a unique haplotype and Asian and Northwestern American isolates share another. Surprisingly, isolates from Eastern North America were distinct from both. However, ITS sequence data as well as the presence of a group I intron in both the northern hemisphere and New South Wales isolates led to the conclusion that there was, or had once been, a probable connection between them via Southeast Asia. Thus today, *P. stypticus* [BA], although a single potentially interbreeding biological species, is becoming differentiated in

different geographical regions into distinct genetically recognizable populations. Nevertheless, these are intercompatible and together with their morphological similarities plus the widespread distribution of the species suggest that it is a single relatively ancient species.

A superficially similar case is the highly variable species originally known as the honey or bootlace fungus, *Armillaria mellea* [BA], which is a major worldwide root pathogen of ornamental, fruit and forest trees and of herbaceous plants. It has even been known to form an endotrophic mycorrhizal association with an orchid! *A. mellea* [BA] had long been suspected as being a highly polymorphic species complex but it was never satisfactorily resolved into component species until the extensive application of the intercompatibility test. European isolates then fell into five incompatible groups (Korhonen 1978b) and similar investigations revealed nine North American groups (Anderson and Ullrich 1979), three of which corresponded to European groups (Guillaumin *et al.* 1991), i.e. 11 incompatible groups in all. These were given species recognition. In some cases limited compatibility occurs between some strains with very low frequency. For example, some isolates of North American *A. sinapina* [BA] paired with the unnamed North American biological species X or with the

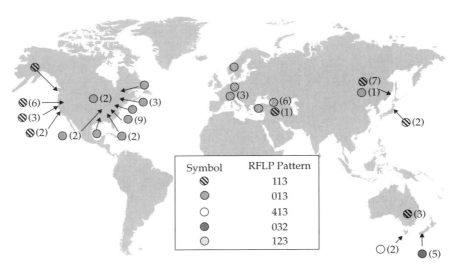

**Figure 11.1** World distribution of different haplotypes of *Panellus stypticus* [BA]. The haplotypes represent different restriction fragment patterns of the rDNA ITS region. (Reproduced from Jin *et al.* (2001), with permission from the Mycological Society of America.)

European *A. cepistipes* [BA], sometimes showed a positive pairing reaction of up to 6.5 per cent (Anderson *et al.* 1980; Banik and Burdsall 1998). This was taken to indicate some degree of genetic similarity but was not sufficient to affect their recognition as distinct biological species. Nuclear flow cytometry has since indicated that one isolate of *A. sinapina* [BA] is probably triploid, but the origin of this ploidy is unknown; the majority are diploids as is normal in Armillarias (Kim *et al.* 2000). Other incompatible groups, often previously recognized taxonomically by some, occur in Africa (three), Japan (two) and Australia (four). All but two of the North American isolates and all of the European isolates differ in small but detectable morphological characters and either coincide with previously named but doubtful species or have now been named. The African, Japanese and Australian groups have also been named.

In addition to reproductive isolation and morphological differences, a number of Armillarias have now been recognized to vary in habit and pathogenicity. Some isolates are strongly pathogenic, e.g. *A. mellea* s.str.(*sensu stricto*) [BA] and *A. ostoyae* [BA], but others, such as *A. gallica* [BA], a circumboreal species, and *A. calvescens* [BA], limited to Eastern North America as a component of hardwood forest mycofloras, are mainly saprophytes although capable of colonizing trees under stress or when in decline.

The biological species now redefined in a strict sense (*A. mellea* s.str. [BA]) occurs in all the continents except Australia and is morphologically very distinct, setting it somewhat apart from the other biological species, a situation confirmed from DNA–DNA hybridization data between North American species (Miller *et al.* 1994). All isolates of *A. mellea* s.str. [BA], like those of *P. stypticus* [BA], are compatible *in vitro* regardless of their geographical origins, although pairing is difficult or reduced in some cases, e.g. between Japanese and Eastern North American isolates (Ota *et al.* 1998b).

Just as with *P. stypticus* [BA], cytological, genetical and molecular data indicate that the species as a whole is differentiated into recognizable biogeographical groups. European and North American *A. mellea* s.str. [BA] are heterothallic and possess diploid basidia, African isolates, with one

exception, are said to be homothallic, and Japanese forms show anomalous nuclear behaviour in the basidium, suggesting behaviour intermediate between primary and secondary heterothallism (homo-heterothallism) (Ota *et al.* 1998a). These differences in behaviour largely account for some of the difficulties encountered in crossing experiments. Nevertheless, sequence analysis of the both the ITS and IGS regions of the rDNA locus have further differentiated the species into European, Eastern North American, Western North American and Asian groups. Although shown by the ITS sequences but not by the IGS sequences, Californian isolates most resembled those from Japan and Korea. Overall, the four biogeographical groups provide little or no evidence of inter-crossing in nature, despite retaining their complete or partial compatibility. The characteristics of the non-heterothallic Armillarias from Africa and Japan have been compared with some European and a few North American heterothallic isolates using SI reactions, isozymes and RAPD polymorphisms. Four Japanese SI groups were detected, one of which (SIG A) was the most common in Japan and also identical with the African isolates tested. Both these populations were clearly genetically distinct on all grounds from the European heterothallic isolates which, in turn, were distinct from the North American isolates, despite their partial compatibilities (Ota *et al.* 2000). Therefore the fourth biogeographical group has been expanded to become an Afro-Asian group. However, these groups have not been recognized taxonomically.

The one exceptional African group consists of heterothallic isolates of *A. mellea* s. str. [BA] from the garden of the former Dutch East Indies Company in Capetown, South Africa. Molecular methods show that these are identical with European isolates and were presumably introduced by the Dutch when they established and later planted up the garden. The estimated age of these isolates, based on the area they cover, is 108–575 (mean 340) years, depending on the growth rate, which accords reasonably well with the known establishment of the garden in 1653 (Coetzee *et al.* 2001). They have evidently not crossed with the indigenous African species of *Armillaria* [BA], although retaining compatibility with their European progenitor populations.

These two species demonstrate that appreciable genetic differentiation can occur in widely separated populations where the probability of intercrossing is virtually negligible and over very long periods of geological time *without* either the loss of *in vitro* compatibility or the development of effective incompatibility. Nevertheless, they apparently remain distinct in nature. Experimental support for this view comes from the quasi-natural experiments using North American and French isolates of *Laccaria bicolor* [BA] (see Chapter 10, section 10.4.2).

In contrast, in other species, incompatibility occurs in populations occupying broadly the same region even though they are compatible, or at least partially compatible, with distant populations from which they may be physically isolated. The polyporaceous *Hirschioporus* (*Polyporus*) *abietinus* [BA] species complex illustrates this as well as some other aspects (Macrae 1967). The intercompatibility test distinguished five groups, all with bifactorial diaphoromictic mating systems, although in some cases partial compatibility was shown in some matings (Figure 11.2).

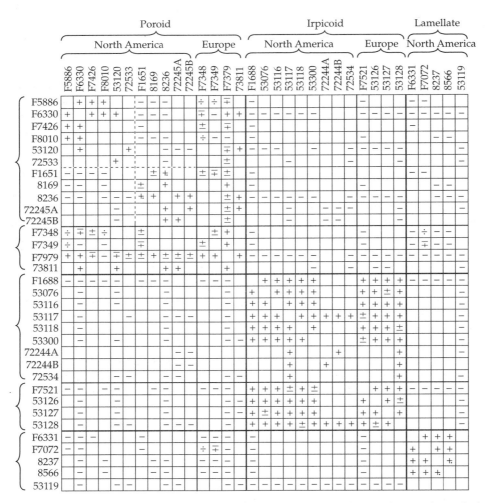

**Figure 11.2** Interfertility relationships between morphologically different isolates of the *Hirschioporus abietinus* [BA] complex in Europe and North America: +, 100 per cent formed clamped dikaryons; +̇, clamped dikaryons in every pairing but one; ±, more than 50 per cent of pairings formed clamped dikaryons; ∓, less than 50 per cent of pairings formed clamped dikaryons; ÷, only one clamped dikaryon formed; no symbol, untested. (Reproduced from Macrae (1967), with permission from the National Research Council of Canada.)

Originally all the isolates were included under a single specific name, *Polyporus abietinus* [BA], although variation from a typical polypore morphology, namely an irpicoid form (spine-like structures replacing pores) and a lamellate form (elongated folds replacing pores), together with some degree of host specialization, had been noted and sometimes recognized taxonomically in North America and more particularly in Europe (Overholts 1953). Basidiocarps which appeared to grade morphologically from an irpicoid or even poroid form into the lamellate form were found, and so the question arose as to whether they were distinct species. However, the three morphological types appear to be genuinely distinct since there is total incompatibility between them regardless of their location. Therefore Macrae (1967) considered them to be three distinct species.

The pairing behaviour in culture between morphologically similar types raised further questions. Poroid Canadian or northern USA isolates fell into two virtually incompatible groups which could not be separated on morphological or host-specific grounds, although the limited biogeographical information (10 isolates) suggests that one form might be distributed more to the west of Canada and the northern USA than the other, i.e. they can be considered to be two biological species. Unexpectedly, *both* are partially compatible to an almost equal extent with isolates of the poroid form from Europe (Norway and Germany). Macrae evidently decided that since *both* North American forms could be connected by some pairings with European isolates, these three groups should be included in one species *Hirschioporus abietinus* [BA].

In culture, all irpicoid isolates were completely compatible regardless of their geographical origins, although there was a slight reduction in normal clamp-bearing hyphae in a few intercontinental crosses. Therefore they were also recognized as a single species *H. fusco-violaceus* [BA], although the irpiciod form in North America occurs mostly on firs (*Abies*) while that in Europe occurs on pines (*Pinus*).

An intercompatibility test supported the distinctiveness of isolates of the lamellate North American form, now named *H. laricinus* [BA], from all the other groups. Indeed, because pairings in culture

with the other forms developed a strong antagonistic reaction unlike that between any other incompatible groups, it can even be questioned whether the lamellate form belongs to the same genus!

The results of these intercompatibility tests require interpretation not only in taxonomic terms but, more fundamentally, in terms of processes in populations and of phylogeny (see Chapter 12). Using phenotypic characters alone, they are largely unanswerable.

The crossing pattern between the three polyporaceous forms is often described as the A–B–C pattern, i.e. where species A and B are incompatible but both are at least partially compatible with a third species C. Other species behave like this. For example, another polypore, *Fomes pinicola* [BA], includes two intersterile groups in North America each of which is partially compatible with a group of compatible European isolates (Mounce and Macrae 1938). The opposite situation, the occurrence of two incompatible groups in Europe both partially compatible with a third group in northeastern North America, occurs in the corticiaceous *Peniophora cinerea* [BA] complex. Each of the European groups occupies a distinct habitat, one being largely confined to beech (*Fagus*) as its host, and it is these latter isolates which are the least compatible with the North American isolates (Chamuris 1991; Hallenberg and Larsson 1992). Another pattern is shown by *Marasmius rotula* [BA] where European and North American populations are largely intersterile except for partial compatibility between a few isolates from the southeastern USA with some European isolates (Petersen 1995b). These and other variants of the A–B–C pattern are not uncommon and presumably reflect, in various ways, the history of the taxa. In all these cases, a taxonomic judgement has been made as to what constitutes a species even when complete compatibility or incompatibility was not the case. Therefore the criteria for a biological species, *sensu* Mayr, are frequently interpreted by taxonomists rather than being treated as inviolable.

Incompatibility is quite often associated with host specialization, a situation compatible with Mayr's modified definition of the biological species (see section 11.3). The crustose *Hypoderma* (*Peniophora*) *mutata* [BA] species complex in Ontario, Canada, shows this behaviour. It includes two morphologically indistinguishable and partially compatible groups A and B, and a third very similar incompatible but morphologically distinguishable group *H. populnea* [BA]. This last species is confined to poplars as is group A of *H. mutata* [BA]; group B occurs on a range of deciduous trees (McKeen 1952). The jew's ear fungus *Auricularia auricula* [BT] shows even more complex behaviour. European populations are confined to old elder bushes (*Sambucus*) and are morphologically indistinguishable from those in North America where it is normally found on coniferous wood, and rarely on deciduous trees. Intercontinental crosses are almost entirely incompatible, but a few isolates show partial compatibility (Duncan and Macdonald 1967).

The ability to form some somatic hyphal fusions is evidently not easily lost in Basidiomycotina.

The best examples of host-specialized biological species are probably the biological races or *formae speciales* recognized in many powdery mildews and rusts, e.g. *Blumeria* (*Erysiphe*) *graminis* f.spp. [A] and *Puccinia graminis* f.spp. [BR] (see Chapter 10, section 10.3.2, and Chapter 12).

Even more complex patterns are shown by other agaricaceous fungi. At least 20 mainly incompatible groups have been recognized among 110 collections of the ectomycorrhizal-forming *Hebeloma crustuliniforme* [BA] complex from Northwest Europe (65 from The Netherlands alone). Although this fungus is widely distributed in Europe and North America, this is a surprisingly large number of reproductively isolated groups from such a small sample from so small an area. Only one, *H. incarnatulum* [BA], could be unambiguously separated on morphological characters alone, although subsequent analysis enabled four groups to be recognized using several criteria, each including several biological species using incompatibility as the criterion (Aanen 1999; Aanen and Kuyper 1999). Many showed a host preference for Salicaceae, but others were more widely distributed on hardwoods and coniferous trees. Phylogenetic evidence suggested that preference for Salicaceous hosts had arisen more than once and that isolates showing such host preference were not necessarily closely related (Aanen *et al.* 2000a) (see section 11.4). Thus host specificity is not necessarily a guide to close evolutionary or taxonomic relationship.

Biological species in Basidiomycotina appear to differ from ascomycotous ones in that incompatibility is largely expressed at a *pre-zygotic stage*, indeed at the site of plasmogamy, whereas infertility in most ascomycotous biological species is manifested at some *post-zygotic* period. It is possible that this distinction is, in part, an artefact of the intercompatibility test since if apparently healthy basidiomycotous dikaryotic hyphae result from a confrontation, it is assumed that the cross is potentially fertile. Because they are not usually followed through to basidiocarp development, it is not known how many initially apparently viable crosses would actually culminate in meiosis and produce viable offspring. On the other hand, it may well be that an intercompatibility test is a genuine indicator of fertility in Basidiomycotina and that failure at the stage of plasmogamy simply reflects the biology of basidiomycotous sexual reproduction which is initiated through somatic fusion.

### 11.3.3 Biological species in Chytridiomycotina and Zygomycotina

The morphospecies dominates species delimitation in both these eumycotous groups largely because only limited investigations have been made in them. However, some information is available from *Coelomomyces* [C] and the Mucors [Z].

Two similar parasitic species, *Coelomomyces dodgei* [C] and *C. punctatus* [C], occupy identical habitats and show an alteration of a gametophytic phase with a sporophytic phase (Federici 1985). Sporophytes undergo meiosis within *Culex* mosquito larvae and the motile haploid products infect a copepod from which compatible plus and minus motile gametes are released. Compatible gametes fuse in the water and re-infect the mosquito larva. Since the release of the motile gametes of the two species overlaps slightly in time, hybrids can develop between them. However, gametophytes formed from hybrid sporophytes abort at an early stage so that no viable hybrid forms persist. While post-zygotic hybrid inviability ensures genetic isolation of the species, their morphological and ecological similarities and initial inter-compatibility suggest some degree of genetic identity also.

The Zygomycotina behave in a manner intermediate between Ascomycotina and Basidiomycotina. In some cases progametangia develop in incompatible matings but reproduction proceeds no further; in others, plasmogamy occurs but presumably not fertilization, although an azygospore is formed. For example, *Mucor hiemalis* [Z], a highly polymorphic species, was studied extensively for many years and several variants were described as distinct species. Schipper (1973) made an intensive study of the species and related taxa and concluded, after pairing tests (see Chapter 8, section 8.2.1.1 and Figure 8.2), that they constituted a single variable species within which she recognized four varieties and also a sibling species, *M. genevensis* [Z], which differed in being homothallic. The varieties are mostly compatible, but different isolates of the same variety show different responses; in some plasmogamy fails or reproductive differentiation only proceeds to aygospore formation. Apart from the largely quantitative differences in various characters between these varieties, there were also significant differences between them in their response to cultural conditions. Unfortunately, little is known of the natural ecology of these species and so it is unclear whether these cultural differences indicate any correlation between the varieties, different natural habitats and the taxonomic differences.

In contrast, distinct habitat differences exist between another pair of compatible sibling Mucors, *M. mucedo* [Z] and *M. piriformis* [Z] (Schipper 1975). The former grows best at pH 7–8 and is most frequently isolated from rodent dung; the latter is best adapted to pH 4 and occurs most commonly on rotting fruit. Abundant zyygospores are formed *in vitro* on a synthetic intermediate substrate, horse dung–cherry agar, at 15 °C. Despite the ecological distinction and some morphological differences, *M. piriformis* [Z] is incapable of forming zygospores with any species other than *Mucor mucedo* [Z] which strongly suggests a possible genetic affinity between them. Nevertheless, it has proved convenient to name the two forms as separate species because of their ecological differences.

### 11.3.4 Fungi and the biological species

As described in section 11.2, the taxonomy of the Fungi Anamorphici has benefited greatly from the application of the intercompatibility test, resulting in the recognition of biological species exemplified by the seven MPs distinguished in *Gibberella fujikuroi* [AP] and by relating hitherto isolated anamorphic forms to specific teleomorphs.

Biological species of Ascomycotina usually show sufficient morphological and microhabitat differences, albeit very fine ones at times, to be recognized through normal taxonomic procedures as morphospecies. This is sometimes the case in the Basidiomycotina, but many examples exist where the distinctions are too subtle to be detectable. Therefore they have been recognized by taxonomists in two ways although there is no universal recognition of how they should be treated.

First, if constant morphological and other attributes such as habitat preferences can be found, once biological species have been distinguished on the basis of an intercompatibility test, they are often named and recognized taxonomically, e.g. the *Armillaria mellea* [BA] species complex (Guillaumin *et al.* 1985).

An alternative procedure, initially suggested by Watling (1977), is to recognize cryptic biological species as *microspecies* within a taxonomic entity which is essentially the morphologically recognizable *macrospecies*. For example, the (macro)species *Conocybe farinacea* [BA] includes four biological species defined by intercompatibility tests with slight morphological differences between them. These would be defined as microspecies. This approach is essentially one that can be applied at the field collection level, but it may result in very different evolutionary entities being missed or neglected which may, for example, be of phytopathological or commercial significance. To obviate this, Jurand (cited in Kemp 1977) suggested that microspecies (biological species) could be indicated by adding a prefix or suffix to the macrospecific name, e.g. *Psathyrella candolleana* [BA] would have four microspecies, a-*candolleana*, b-*candolleana*, c-*candolleana* and d-*candolleana*. If these suggestions were to be accepted, macrospecies would then become the species of the taxonomists, and

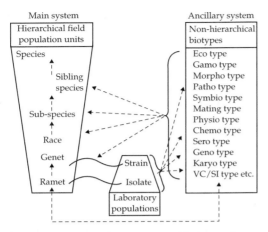

**Figure 11.3** Diagram to illustrate a proposed dual system for infraspecific terminology. The arrows indicate possible progression as information accumulates about population structure and variation. Prefixes to types indicate the nature of the characters in common: Eco, habitat; Gamo, ability to interbreed independent of mating type; Morpho, morphology; Patho, pathogenic ability towards host; Symbio, host specificity; Mating, MTFs; Physio, physiological property relationships; Chem, chemical moiety relationships, i.e. chemical variants; Sero, serological profile; Karyo, chromosome characters; Geno, one or more genes (genets); VC/SI, vegetative or somatic incompatibility types (may or may not belong to the same genet). (Reproduced from Brasier and Rayner (1987), with permission from the British Mycological Society.)

microspecies the tools of those concerned with speciation.

A third, operational, solution, is a proposal for a more elaborate ancillary system of non-hierarchical terms to describe intra-specific situations which can be related to a supplemented list of more conventional taxonomic units, described by the proposers as 'hierarchical field population units' (Brasier and Rayner 1987). The proposals (Figure 11.3) provide a useful form of shorthand for discussing some of the situations which emerge when species are considered as populations, but since it perpetuates the dichotomy between traditional taxonomic approaches and more practical or experimental biological approaches it seems unlikely to be adopted by practising taxonomists.

## 11.4 Fungi as phylogenetic species

An important element of Wiley and Mayden's (1997) definition of the evolutionary concept, namely

A single lineage of ancestor–descendant populations which maintains its identity from other such lineages and

which has its own evolutionary tendencies and historical fate

can be met by adopting the phylogenetic species concept, since the requirement that a species should be descended from a single lineage that maintains its identity can be tested by investigating its phylogenetic tree. This became a more practical proposition following the widespread adoption of cladistic methods, originated by Hennig (1950). Initially the method was applied using only phenotypic characters, but its resolving power has been enormously increased by including molecular characters. A consequence for fungi is that more objective phylogenetic analyses can replace phylogenetic conjectures, the majority of which were based on very few hard facts! In particular, the ability to apply such methods to determining one or more gene genealogies across a group of supposedly related organisms has provided a powerful tool for determining their probable phylogenetic history. With this approach, species are effectively defined in terms of common descent, so that in a phylogenetic tree common descent is represented by the branches and nodes of the tree leading eventually to terminal monophyletic clades. These indicate the units of potential taxonomic significance and can, in general, be equated to phylogenetic species.

Objections have been raised to this procedure—basically that species trees are not necessarily the equivalent of gene trees since the same genes are not necessarily carried by all members of the same species, especially in those with sexual reproduction. Davis (1999) gives a very full discussion of the theoretical and practical issues involved.

A typical example of a phylogenetic species tree based on various characters is that provided by Vilgalys (1991) (Figure 11.4). To investigate the possible evolution of the agaricaceous *Collybia dryophila* [BA] species complex he combined data from morphology, compatibility, isoenzymes and DNA–DNA re-association to produce one of the earliest fungal phylogenetic trees based on whole DNA information. He pointed out that 'different components of divergence (morphology, genetic identity, and intersterility) are not coupled during speciation', and observed that they could best be treated by superimposing the various components

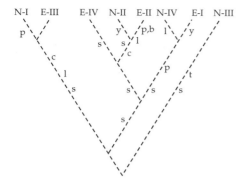

**Figure 11.4** Hypothetical patterns of speciation in the *Collybia dryophila* [BA] complex based on comparisons of whole DNA. Species: N, North American; E, European; I–IV, different species. Characters: p, pileus light to dark, gills white to yellow; b, basidiospores medium to large; c, cheilocystidia (sterile cells amongst basidia) inflated or filamentous; t, stipe base tomentose; l, substrate soil/litter; S, evolution of sterility. (Reproduced from Vilgalys (1991), with permission from the New York Botanical Garden Press.)

on the tree based on DNA–DNA re-association data. When this was done various identical characters such as the colour of the gills and the kinds of cystidia they bore appeared to have evolved repeatedly, providing additional information on the probable course of evolution of the *C. dryophila* [BA] species complex. In particular it drew attention to one of the problems of using morphological characters alone, namely that identical characters are not necessarily homologous in a genetic sense.

Although Vilgalys' approach rationalized the evolution of the entities in the complex, as far as recognition of significant taxonomic entities was concerned, it added little, taxonomically, to what could have been achieved by using intercompatibility tests to determine the reproductively isolated biological species. However, his use of DNA re-association data provided a more fundamental basis for determining phylogenetic relationships.

A possible problem in interpretation can arise when morphological and molecular characters are compared. Most taxonomists regard the identification of homologous characters through comparisons as an essential element of taxonomic reasoning. A possible assumption is that they will have a common genetic basis. However, using gene genealogies alone, it may appear that there is a lack of true genetic homology between similar,

supposedly homologous, characters. It has to be recognized that gene homology can probably be identified more precisely than can homology between similar phenotypic characters so that the former may be more acceptable, but this does not necessarily resolve the issue. In practice, the best way to resolve such issues is to base phylogenetic trees, whenever possible, on combined information from genes that are both conservative and regulate essential fungal functions such as the rDNA or β-tubulin loci. Indeed, this is a widespread procedure for determining phylogenies, e.g. the *Gibberella fujikuroi* [AP] species complex already referred to in section 11.2.

In many cases when gene genealogies are used there is a common identity between species recognized on a basis of terminal clades and those previously recognized in other ways, whether as morphospecies or as biological species. For example, a common identity has been confirmed between the biological species of *Armillaria mellea* [BA] recognized by intercompatibility tests and by a comparison of sequences of part of the rDNA region (Anderson *et al.* 1987; Anderson and Stasovski 1992). It will be recalled that subsequent to the recognition of these biological species it proved possible to identify distinct morphological, habitat and behavioural characters so that they could often be defined by classical taxonomic methods as 'good' morphospecies (see section 11.3). In this case, the results of the different procedures coincided. Molecular analysis provided further information about the degree of relatedness between the species in this complex which was of evolutionary and phylogenetic interest, but this did not appreciably affect the strictly taxonomic basis of species recognition.

A one-to-one identity between previously recognized species and the entities recognized by molecular analysis is not always found. Various situations have been detected.

• In the shiitake mushroom complex *Lentinellus edodes* [BA], a departure from recognized species has been indicated.
• In the human pathogen *Ajellomyces* (*Histoplasma*) *capsulatum* [A], the cause of histoplasmosis and supposedly a single morphospecies with three

varieties, six genetically isolated groups have been demonstrated.

• Known species complexes have been resolved into phylogenetically distinct groups which have illuminated the understanding of the complex, e.g. the *Nectria haematococca–Fusarium solani* [AP/FA] complex but in others, e.g. *Hebeloma crustuliniforme* [BA], the suggested taxonomic treatment has had to be a compromise using a range of characters.

• In *Heterobasidion annosum* [BA] and *Neurospora* [AP] there are either real discrepancies between the existing taxonomic concepts or, in the latter fungus, it has simply not been possible to determine a satisfactory phylogeny to relate to the presently recognized species.

Many cultivars of *Lentinellus edodes* [BA] are known in Asia, but it occurs over a wide area in the wild—from the Himalayas to the north, west to Afghanistan, east to Japan and as far south as Australia and New Zealand (Hibbett *et al.* 1995). Three morphospecies are recognized—*L. edodes*, *L. laterita* and *L. novaezelandieae* [all BA] (Pegler 1983)—but all were found to be mutually compatible (Shimomura *et al.* 1992) so that it had come to be regarded as a single polymorphic species by some. Hibbett *et al.* (1995) compared the ITS sequence of the rDNA of 22 isolates, wild and domesticated, from the Japan–Australasian region and also included data collected by others from the same region as well as RFLP data already available on the mitochondrial small subunit rDNA. Because of the different methods used to obtain the data, he was not able to combine the different data sets, but he was able to carry out both parsimony and distance maximum likelihood analyses and to investigate the congruence of the phylogenetic trees generated from the mtDNA RFLPs and rDNA sequence data. Despite some discrepancies between the mtDNA- and rDNA-based trees, which might reflect different evolutionary histories of the nDNA and mitochondria, support was obtained for four major lineages or **clades**. One monophyletic clade coincided with isolates of *L. novaezelandieae* [BA], i.e. descendants from a common ancestor, and there was similar but much weaker support for *L. edodes* s.str. [BA]. However, it has not proved possible to recognize the two other clades taxonomically, apart

from these analyses, and the position is rendered more uncertain by the mitochondrial data suggesting the presence of a fifth clade! Despite the small sample size, correlations of different clades with distinct biogeographical areas were apparent. The Papua New Guinea group appeared to be the most variable and the New Zealand and Japanese groups the most compact. The Japanese group probably reflects the predominant domestic use of a relatively few quasi-clonal isolates in Japan, but the New Zealand group could well reflect a naturally isolated group of wild isolates.

This investigation does not resolve the taxonomic treatment of the *L. edodes* [BA] complex but it demonstrates the inherent complexity that may lie behind the use of a single morphospecies designation. In this case, the *in vitro* intercompatibility test results probably do not reflect cross-fertility in nature and are best not interpreted as indicating a single species. The taxonomic recognition of two of the morphospecies is supported by their being monophyletic and the information now available indicates lines along which future genetic and taxonomic investigations might usefully proceed to resolve the situation more satisfactorily.

*Ajellomyces* (*Histoplasma*) *capsulatum* [A/FA] is widespread in its anamorphic form. It has been regarded as a single morphospecies with three mutually compatible varieties: var. *duboisii*, which is limited to Africa and has yeast-like morphology, var. *capsulatum*, which is confined to the New World, and var. *farciminosum* which has a worldwide distribution but is confined to contact skin infections of equines. Forty-six isolates were available covering all the varieties. A phylogenetic analysis based on four different but congruent gene genealogies delineated six distinct phylogenetic lineages. One coincided with var. *duboisii*, and the other five fell within the var. *capsulatum* clade which also included isolates of var. *farciminosum*, all of which turned out to be identical. Evidently, var. *farciminosum* is simply a genotype that has become specialized to infect skin and is presumably transmitted clonally from one animal to another. The other four different var. *capsulatum* genotypes coincided in some cases with isolates which had been distinguished earlier on the basis of showing different pathogenicities (Kasuga *et al.* 1999). Thus, while

*A. capsulatum* [A] and its varieties could be regarded as a single polymorphic species on the basis of their compatibility reactions, on phylogenetic grounds it could be regarded either as two phylogenetic species, one of which was highly polymorphic, or as five phylogenetic species—the interpretation preferred by Kasuga *et al*. There was no clear-cut correlation between forms recognized as morphospecies, only with some phylogenetic species; neither of these coincided with the biological species. Indeed, there is no clear-cut determination of which entities could best be regarded, taxonomically, as species. The phylogenetic analysis has revealed clearly the complexity of the situation. Support for recognizing the phylogenetic species lies in the fact that the four different gene genealogies were concordant. Identical tree topologies are best interpreted as indicating that fixation of formerly polymorphic loci has occurred *after* their genetic isolation in a monophyletic clade. If species are regarded as dynamic entities, this analysis gives the most complete picture of the contemporary position, however difficult it might be to recognize them without recourse to genetic analysis. Genetic analysis also enables a more rational approach to be adopted in the practical treatment of this pathogen. The procedure used, namely the concordance of more than one gene genealogy, has been termed genealogical concordance phylogenetic species recognition (GCPSR) and has been promoted by Taylor *et al*. (2000) as '...straightforward in application and interpretation...', and it is claimed that '...GCPSR can be used to diagnose species for any fungi'.

However, when concordance is not achieved between different gene genealogies, the interpretation is more difficult. It may be the result of recombination between individuals within a species, recombination between species or the species being completely clonal. To date, incontrovertible evidence for this last situation has not been found, although only a few of the very large numbers of Fungi Anamorphici have been investigated in this way. It is not uncommon fo find cases where complete gene genealogical concordance is not found, but even then a phylogenetic analysis may be possible as in the case of *L. edodes* [BA] already described.

The *Nectria haematococca–Fusarium solani* [AP/ FA] complex illustrates the resolution that can be achieved by the application of GCPSR to morphologically similar, supposedly related, sexual and asexual entities. O'Donnell (2000) prepared phylogenetic trees from two regions of the rDNA gene, namely the large 28S subunit and the ITS region, and introns and exons within the gene for the translation elongation factor gene (*EF-1a*). The two elements of rDNA data showed significant incongruity when tested, but it proved possible to demonstrate that, nevertheless, the trees could be satisfactorily combined. Each genealogy was investigated separately and also in a combined analysis. The results were remarkably similar. Three main clades showed clear biogeographical relationships— one largely made up of Afro-Asian isolates with a few from South America, a second entirely of South American isolates and a third from New Zealand. It was further inferred that there were, in all, 26 phylogenetically distinct groups within the 36 isolates investigated. Fifteen of these were dimictic, seven were homomictic and four were putatively anamorphic (Figure 11.5).

Five of the seven, biologically distinct, previously recognized MPs I–VII were fully supported as phylogenetically distinct and coincided with entities previously recognized as *formae speciales*. Two additional *formae speciales* were shown to be monophyletically distinct, coinciding with the two entities previously described as distinct on grounds of incompatibility and pathology as races 1 and 2 of f.sp. *cucurbitae*. Although previously regarded as variants of a single species, they turned out not to be closely related. In this case, the historical treatment of *formae speciales* as different variants of a single species was clearly misleading, although the new analysis fully justifies the early recognition given by plant pathologists to these taxa as distinct biological operational units.

It is not always obvious how best to interpret gene genealogies in taxonomic terms. For example, the 20 intercompatibility groups (ICGs) or biological species found in the *Hebeloma crustuliniforme* [BA] complex (see section 11.3.2) were largely supported by a phylogenetic tree based upon the sequence of the ITS rDNA regions (Aanen *et al*. 2000a). The isolates were not a monophyletic group

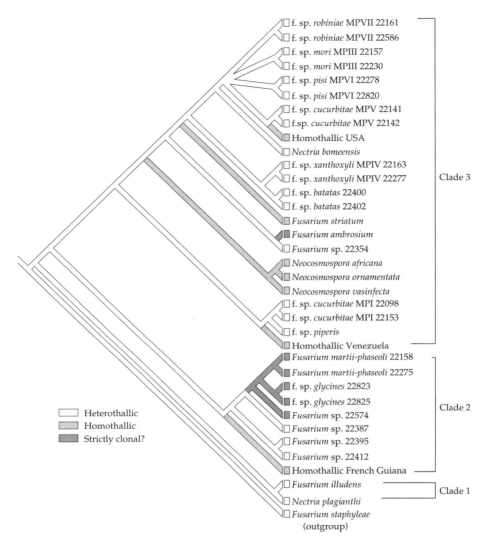

**Figure 11.5** The *Nectria haematococca–Fusarium solani* complex. *Strict* consensus tree from a combined data set to show the distribution of different reproductive modes in the complex. Note that homothallic clades are scattered throughout the tree and that the supposedly strictly clonal clades are few in number. (Reproduced from O'Donnell (2000), with permission from the Mycological Society of America.)

and fell into two clades, of which one (clade I) included ICGs 16, 17 and 18, all of which had been assigned to the morphospecies *H. velutipes* [BA]; the other (clade II) comprised the remaining 17 ICGs. Clade I was itself further divisible into two clades, one of which included ICGs 16 and 17 which had been shown to be partially compatible by intercompatibility tests. All but one of these monophyletic groups could be reliably equated to previously recognized morphospecies.

A further analysis was made in which the association between the ICGs and various morphological features was investigated. As a result of this analysis four taxa were recognized even though this involved paraphyletic groupings, i.e. groups that include a common ancestor and *some, but not all*, of its descendants. This was, in part, justified by the authors on the very close similarity and relationships of the monophyletically defined ICGs. The value of such a taxonomic treatment cannot yet be assessed nor is it evident that if gene genealogies for other gene regions were to be made and combined, they would necessarily support such a treatment. Indeed, it might well be expected that it would be more likely to confirm the original 20 monophyletic groups! This example highlights the kinds of practical problems facing taxonomists if they rely on the phylogenetic species as their species unit. Indeed, although phylogenetic analysis as a guide to defining species in fungi is still in its infancy, there are already several cases where clearly defined clades may, or may not, coincide with entities recognized by other means.

Phylogenetic analysis does not always resolve complex situations. For example, in the *Hetero-basidion annosum* [BA] species complex (Harrington *et al.* 1997) it has provided some support for some groups which had been recognized on morpho-logical and ecological lines, but not for others. The European and North American *P*-forms coincide with two clearly defined clades, supporting the distinction already recognized between these two closely similar groups. However, a third major clade includes *all* the *S*- and *F*-forms with an indi-cation of a distinct North American *S*-group but no such clear separation within Europe, where *S*- and *F*-forms show clear morphological differences as well as differences between their isozyme and RAPD markers. Moreover, although the European *S*-form and North American *F*-form are highly compatible, they are appreciably less so when European *S*- and *F*-forms are tested. In addition, from other data, the occurrence of a group of dis-tinct Japanese *S*-forms within the *S*- and *F*-clade was unexpected.

None of the different forms have hitherto been given formal taxonomic recognition and it is not clear how best *S*- and *F*-forms shoud be treated taxonomically in the light of these further findings. It may be that in both Europe and Japan the *S*- and *F*-forms are beginning to differentiate within the existing *S*- and *F*-clades, with the genetic differ-ences being reflected in their different phenotypic expressions. However, since this monophyletic group has not yet split, the situation could be an indicator of incipient speciation. The phylogenetic data do not justify recognition of the groups as valid taxa and therefore, taxonomically, *H. annosum* [BA] is best regarded as a single polymorphic species.

A further kind of situation arises when a valid phylogeny cannot be determined, as with the conidial species of *Neurospora* [AP]. Apart from support for the separation of *N. sitophila* [AP] and *N. tetraspema* [AP] by an investigating mtDNA and various nDNA nucleotide sequences, no definitive phylogeny is apparent. There are various anomal-ies. One is the lack of concordance between gen-ealogies derived from different regions of nDNA, another is the high variability of the strains within each of the recognized species and the third is the variable mating behaviour between the species despite their having been separated initially on mating compatibility. One worrying alternative is that the *Neurospora* [AP] situation may simply be a reflection of sample size, since a far larger collec-tion of isolates from virtually the complete range of the group have been investigated than for any other species. It would be a matter of some concern if the apparent ease with which phylogenetic spe-cies have been recognized in other species was due to using too small or limited a sample, i.e. the use at most of tens of isolates rather than the hundreds, or potentially, thousands available with *Neurospora* [AP]. At present, a final decision on the desirable sample size and range can only be tested by further investigations.

## 11.5 Conclusions

It will have become evident that different defini-tions of species can often lead to different kinds of entities being recognized as species. Indeed, there is no universally acceptable definition which adequately encapsulates the different viewpoints promoted by those who have defined 'the species'. This is not really surprising because those con-cerned regard different attributes as being import-ant. The taxonomic question is: How much information and what kind of information does a user expect the term 'a species' to convey?

A morphospecies is a perfectly adequate category for purposes of visual recognition, although different limits may be set by different workers on the limits of discrimination to be expected. However, species are not defined solely for pur-poses of categorization but because they are also tools to assist other activities such as discriminating between pathogens. In some cases, e.g. *Armillaria mellea* [BA] or *Ceratocystis coerulescens* [AP], the obvious limitations of the morphospecies arose more from deficiencies of taxonomic perception than because of an inherent property in either organism. Once the biological species had been recognized by intercompatibility tests and confirmed by molecular data, they promoted the subsequent recognition of valid morphospecies based on more discriminating characters than had been used before. The recognition of other attributes (ecological, degree of pathogenicity and distribution) followed naturally.

Unfortunately, it is not clear how far these two examples are typical and, in many cases, recognized biological species continue to be intractable as far as morphological discrimination is concerned. Equally, compatibility or incompatibility are not adequate bases on which to distinguish species reliably in the field since an *in vitro* intercompatibility test is not necessarily a reflection of what happens in nature. Isolates compatible in laboratory tests can be ecologically and biologically distinct in nature, e.g. *Laccaria bicolor* [BA]. The same kinds of uncertainty surround the recognition of phylogenetic species, even in the guise of the GCP species!

However, these uncertainties would seem to reflect the evolutionary history and contemporary state of the organism, matters which may or may not be of value for purposes of classification. It might be simpler if, in practice, it was recognized that, taxonomically, species should be confined to the morphospecies with the exercise of the greatest degree of discrimination readily attainable and that other aspects should be seen as relevant to the study of speciation. Of course, these aspects illuminate the present condition of a species but they cannot, in general, be used easily and effectively in a strictly taxonomic context. There are obvious exceptions such as when there is a practical need to discriminate between morphologically indistinguishable pathogenic fungi. Provided that taxonomists are aware of the nature of the variation masked under morphological identity, particular cases can be dealt with as necessary.

Therefore it is suggested that the most pragmatic way to proceed is to continue to work with the morphospecies as the basic taxonomic unit but to include as supplementary material the additional information supplied by other types of study whether out of practical necessity or as a guide to both phylogenetic trends and the interpretation of evolutionary potentialities. In operational terms this could most easily be achieved by the use of the macrospecies/microspecies concept described in section 11.1. Alternatively, if it is thought to be more useful to incorporate additional information within the species concept, supplementary reference to the entities defined as biological species as defined by Levin (2000), the ancillary terminology suggested by Brasier and Rayner (1987), or the GCP species of Taylor *et al.* (2000) might usefully draw attention to problems inherent in the morphospecies definition. However, even from the few examples cited in this chapter it can be seen that, as always, 'taxonomic judgement' will ultimately have to be exercised in deciding which entites shall be recognized as species.

From the viewpoint of population genetics, however, the information required to define a biological species or the GCP species is of immense interest and relevance to defining and understanding the processes that may be involved in speciation. This topic will be considered in the next and final chapter.

## Further reading

*General*

Claridge MF, Dawah HA and Wilson MR (1997). *Species: the units of biodiversity*. Chapman & Hall, London.

*Fungal species*

Brasier CM (1997). Fungal species in practice: identifying species units in fungi. In *Species: the units of biodiversity* (ed. MF Claridge, HA Dawah and MR Wilson), pp. 135–70. Chapman & Hall, London, UK.

Harrington TC and Rizzo DM (1999). Defining species in the fungi. In *Structure and dynamics of fungal populations* (ed. JJ Worrall), pp. 43–71. Kluwer Academic, Dordrecht.

Natvig DO and May G (1996). Fungal evolution and speciation. *Journal of Genetics*, **75**, 441–52.

Taylor JW, Jacobson DJ, Kroken S *et al.* (2000). Phylogenetic species recognition and species concepts in fungi. *Fungal Genetics and Biology*, **31**, 21–3.

# CHAPTER 12

# Speciation in fungi

Speciation is a population phenomenon, namely the process or sequence of processes that result in the origin of a new species. Characteristically, all species are adapted more or less effectively to particular habitats, and show either very limited or no gene flow between populations of other species. Therefore the basic processes to understand are as follows.

• How do species adapt to different habitats?
• What processes result in a reduction or cessation of gene flow between populations?
• How do these processes interact to result in a unique, discrete new species?

Not much is known in detail about how fungi adapt to their habitats, although there is evidence that they are so adapted and examples of the effects of selection which result in such adaptation have been given in sections 9.4 and 10.3.

Gene flow in fungi is promoted by dispersal which determines, in part, the distance between populations beyond which they are physically isolated and the probability that, once dispersed, genes are incorporated into other genotypes either through hyphal fusion followed by recombination or directly through sexual reproduction. A variety of conditions and processes, singly or collectively, can impair these processes.

What appears to be least essential to speciation is change in morphology as the widespread occurrence of sibling species in fungi demonstrates. An association of morphological change with speciation is not thereby excluded but it does not appear to be essential either for adaptation to a habitat or for reproductive isolation.

How these processes operate in fungi will be the main consideration of this chapter, but first it is useful to set speciation against a historical background based on such limited information as is available.

## 12.1 Historical aspects of speciation

Three kinds of information throw light on historical aspects of speciation.

• In some cases, species distribution can be related to known and dated geological events.
• Phylogenetic data can be related to temporal and other events on the assumption that genes and proteins evolve, on average, at a constant rate, i.e. that there is a 'molecular clock'. Comparisons of homologous sequences of a gene can then indicate substitution rates and, from this, estimates of relative time-scales can be made. Sometimes, phylogenetic sequences can also be related to various other events such as genetic changes and biogeographical information.
• Sometimes the time taken for speciation to occur can be related to events in human history or prehistory especially if the dominant selective agency is human.

The first two aspects will be considered in this section but, since human influence is a significant predisposing factor to speciation for many fungi, it will be considered in section 12.2. Unfortunately, another possible approach, the study of fossil lineages, is not available in the fungi since fossil information, even recent quaternary data, is still so limited that it can only be used for determining gross events in fungal evolution. For example, Glomalean fungi and fungi with septate hyphae, presumably Ascomycotina, occur in the Rhynie Chert, dated at 390 million years ago (Kidston and Lang 1921), and in subsequent deposits. However, fossils of terrestrial green plants, potential hosts for parasitic or mycorrhizic fungi, are known from 460 million years ago and probably arose about 600 million years ago (Gray 1985), so that fungi almost certainly originated more than 400 million years ago. In any event, fossil Glomalean fungi have not been related to contemporary species so that their fossil lineages cannot be traced.

## 12.1.1 Taxonomic data and speciation

Many major geological events, such as the separation of the continents or the development of mountain rages, are now dated with some degree of accuracy. Distribution dates for species can be related to such events; for example, the time of final separation of Europe from North America was some 30 million years ago. Therefore species found on both sides of the Atlantic are likely to be older than this, provided that transport by humans can be eliminated, but related species only found on one side of the Atlantic are likely to be younger. The three species *Hirschioporus abietinus* [BA] (Macrae 1967), *Marasmius androsaceus* [BA] (Gordon and Petersen 1991) and *Flammulina velutipes* [BA] (Hughes *et al.* 1999) illustrate this.

In the case of *H. abietinus* [BA], the American representatives fall into two morphologically inseparable populations which are totally incompatible, although both show limited compatibility with isolates from the European population (see Chapter 11, section 11.3.2). All isolates, regardless of their location, are indistinguishable morphologically. Therefore the parent species must have arisen more than 30 million years ago, but the two groups of North American isolates have become effectively isolated geographically and reproductively within this period and are less than 30 million years old.

A similar situation is known in *M. androsaceus* [BA] where isolates, most commonly found on oak litter (*Quercus*) in the southern Appalachian Mountains, USA, are totally incompatible with both other North American isolates generally found on pine-needle litter (*Pinus*) and European isolates on comparable substrates, which are largely compatible. Therefore the Appalachian population has become adapted to a different substrate and probably became reproductively isolated within the 30 million year period.

Lastly, within the same period, although there are some genotypic differences between European and North American populations of *F. velutipes* [BA], all isolates are completely compatible. Therefore the species is more than 30 million years old.

Thus incipient reproductively isolated species can arise naturally within 30 million years in some cases without a change in habitat and in others with an associated habitat change, while in yet others genetic changes can occur without reproductive isolation. This evidence indicates that fungal species can persist for very long periods (more than 30 million years), but they can also speciate in a shorter period. These periods are comparable with those found for herbs and shrubs (Levin 2000).

## 12.1.2 Assessing species lineages

Relative timing can be determined in a molecular lineage by recording the number of nucleotide substitutions involved, or if an assumption is made about the rate of nucleotide substitution, e.g. 1.26 per million years (Berbee and Taylor 2001), in approximate real time. In principle, all that is required is to estimate the evolutionary distance between the nodes of interest in a phylogenetic tree.

There are only a few examples of the application of the direct 'molecular clock' dating method to fungal species, but it has been used with the parasitic basidiomycotous yeast *Filobasidiella* (*Cryptococcus*) *neoformans* [BT/FA], the cause of cryptococcosis in humans and animals. The fungus is distributed worldwide and is generally isolated in its anamorphic form. It is increasingly found as a secondary infection in patients suffering from HIV infection and, as a result, has probably become even more frequent and widely dispersed recently. Although it is treated as a single biological species three varieties are recognized.

• Var *neoformans* is found in Northern Europe often in association with pigeon droppings, so that the birds probably act as dispersal agents, and is known in serotype A and D forms.

• Var. *gattii* from the tropics and subtropics is often associated with *Eucalyptus* trees, and is known in serotype B and C forms.

• Var. *grubbii*, found worldwide in soil and bird faeces, is known only in serotype A form. It is the most common form in AIDS patients.

Xu *et al.* (2000) sequenced four genes—the ITS, orotodine monophosphate pyrophosphorylase (URA5), laccase (LA) and the large ribosomal subunit of mtDNA—and established multiple gene genealogies. Using the two enzyme markers and assuming a consensus rate of $2 \times 10^{-9}$ nucleotides per year for protein coding genes (Nei 1987), they estimated the divergence times of the varieties and serotypes (Table 12.1).

The divergence time between varieties and serotypes are much greater than within serotypes and varieties. Since these divergence times are greater than or comparable with divergence times between various diploid non-fungal species, the authors suggested that the three varieties should be regarded as potentially phylogenetic species despite the fact that they are compatible *in vitro*. There is also evidence for natural hybridization. Some isolates were inconsistently placed in the individual gene genealogies and their inclusion in the combined genealogy resulted in incongruity which was resolved if they were excluded. The most likely cause for these anomalies is that these isolations were recombinants. Since there was clear evidence from the genealogy for divergence of the different forms and that the anomalies were near the tips of the tree, it was probable that the hybridizations are relatively recent events. The estimated range for such events was 0–2.36 million years, but since there was also evidence for recent dispersal it was suggested that these were associated events. It is of interest that these estimates are not inconsistent with the time-scales suggested from the distribution data already described.

Nested clade analysis and coalescent methods can take such analyses further (see Chapter 5, section 5.3). The only published fungal application to date involving several species, presumed to be closely related, is that of Carbone and Kohn (2001) which was applied to numerous isolates of *Sclerotinia sclerotiorum* [AD] and to the four species *S. sclerotiorum*, *S. minor*, *S. trifoliorum* and an unnamed species *Sclerotinia 1* [all AD]. Six sequences were determined from a small series of 44 isolates of the species plus a much larger one from *S. sclerotiorum* [FA]. In the small sample only the promoter regions of the IGS and ITS1 rDNA region were screened, although in the large sample the complete ITS repeat unit and an enzyme had also been used.

The *S. sclerotiorum* [AD] population sample indicated an initial fragmentation event of the common ancestor. This was followed by dispersal in space and to different hosts resulting in recurrent restrictions on gene flow because of isolation

**Table 12.1** Estimates of the times since divergence among gene lineages in the (*Filobasidiella*) *Cryptococcus neoformans* [BT] complex

| Taxon groups recognized | Time since divergence (million years) |
| --- | --- |
| *C. neoformans* var. *gattii* with *C. neoformans* var. *neoformans* and *C. neoformans* var. *grubbii* | 37 |
| Between *C. neoformans* var. *neoformans* and *C. neoformans* var. *grubbii* | 18.5 |
| Within *C. neoformans* var. *grubbii* | 4.5 |
| Within *C. neoformans* var. *neoformans* | 3.3 |
| Between serotypes B and C | 9.5 |
| Within serotype B | 2.6 |
| Within serotype C | 6.4 |

*C. neoformans* var. *grubbii* includes serotype A, var. *gattii* includes either serotype B or serotype C and var. *neoformans* includes serotype D. The genealogies used for these estimates were based on mtLrRNA, ITS, URA5 and LAC. Times since divergence were calculated on the average divergence for two protein coding genes, Laccaase (LAC) and orotidine monophosphate pyrophosphorylase (URA5), assuming a neutral mutation rate of $2 \times 10^{-9}$ per nucleotide per year for protein coding genes.
(Based on Xu *et al.* (2000), with permission from Blackwell Science.)

by distance and, eventually, to speciation (see Chapter 5, section 5.3). The other *Sclerotinia* [AD] species were clearly differentiated in the species sample and had evidently split from the *S. sclerotiorum* [AD] lineage appreciably earlier than the clades within that species (see Figures 5.6 and 12.1, and Tables 5.6 and 12.2). It was suggested that

the fragmentation event might have been associated with various migrations from Pleistocene refugia, although fragmentation associated with effects due to humans, such as dispersals or even crop domestication, could not be ruled out. Clearly, it will be important to investigate this issue more thoroughly, especially in view of other evidence of

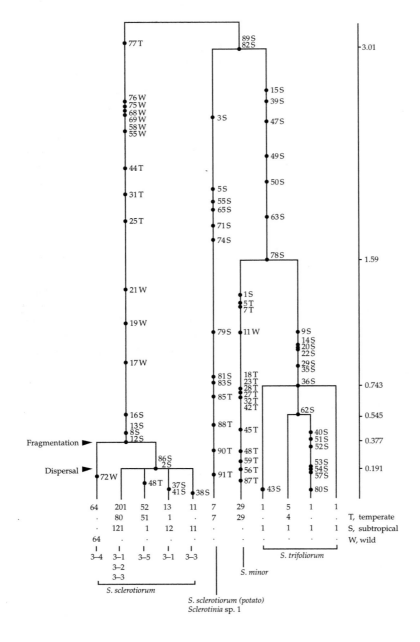

**Figure 12.1** Genealogy of species of *Sclerotinia* [AD] based on nested clade analysis and coalescent inference using multilocus data. Sequence data from seven loci were combined to reconstruct the evolutionary history of four species of *Sclerotinia* [AD]. The time-scale on the right is the estimated time to the MRCA in units of $N_e$ generations (see also Table 12.1 and text). (Reproduced from Carbone and Kohn (2001), with permission from Blackwell Science.)

**Table 12.2** Estimates of the effective numbers of migrants between different geographical regions from a sample of 385 haplotypes of *Sclerotinia sclerotiorum* [FA/AD] and related species using the MIGRATE program

| Sampling areas | Estimates of $2N_em$ | | |
| --- | --- | --- | --- |
| | **1,$x$** | **2,$x$** | **3,$x$** |
| 1. Temperate | — | 2.0 | 1.1 |
| 2. Subtropical | 0.5 | — | 0.4 |
| 3. Wild | 1.0 | 1.3 | — |

The same convention for reading the parameters 1,$x$ etc. applies as in Table 5.6.
(Reproduced from Carbone and Kohn (2001), with permission from Blackwell Science.)

human influences on fungal speciation (see section 12.3). This analysis, the first of its kind for fungi, illustrates how, once multilocus genealogical trees have been established for populations or species, events leading to speciation can be investigated and contemporary variation and distributions interpreted more effectively.

In the case of fungi which are restricted to particular hosts, it is important to know the histories and lineages of *both* the fungi and their hosts. Different host species which now appear to be developing in proximity, apparently infected by sibling fungal species for example, may in fact have arisen historically in quite different circumstances, and their contemporary association may be a relatively recent event. Adaptation of the fungi to each host could have occurred quite separately and at different times. The apparent evolution of sibling species in close proximity would then be quite spurious (see section 12.3.1).

## 12.2 Speciation processes

With other Eukaryota it has become usual to recognize three principal modes of speciation:

• **allopatric** or **geographical** speciation due to the accumulation of genetic differences in populations which are initially isolated by distance (geographical races) but ultimately become reproductively isolated

• **sympatric** speciation due to isolation arising between populations occupying similar or overlapping ranges
• **abrupt** speciation due mainly to polyploidy or chromosomal rearrangements resulting in reproductive isolation.

The first two modes reflect the spatial distribution of populations prior to speciation, but the third describes the event that actually initiates speciation. Superficially, fungi show similar distributions, but although they can be distributed over wide areas, they occupy microhabitats within those areas. Significant spatial separation between species may be as little as their location at different but adjacent sites on a single log. In such circumstances, the traditional distinction between allopatry and sympatry becomes somewhat meaningless. Therefore the application of this approach to fungi does not appear to be very rewarding and it seems more useful to describe speciation in terms of the functional processes leading to it. These can most conveniently be distinguished as those which function as predisposing factors and those which result in reproductive isolation itself.

Notwithstanding the previous remarks, fungal species resemble those of other eukaryotes in tending to recur or persist in much the same habitat over appreciable periods of geological time (Bennett 1997). This reflects the remarkably precise adaptation that they can show to their habitat, although the continuity of the association does not imply that they are not simultaneously undergoing genetic change. Widely dispersed fungal species are usually found to have undergone some genetic differentiation, often associated with different areas, habitats or environments, e.g. *Panellus stypticus* [BA] (genotypic differentiation on a regional scale), ecotypes of *N. intermedia* [AP] (on different substrates) and races 2 and 3 of *Cochliobolus carbonum* [AP] (differential adaptation to summer temperatures) (see Chapter 10, section 10.3.3, and Chapter 11, section 11.3.2) Such genetic changes must often be a slow process if only because of the constantly fluctuating physical and biological environment in which fungi grow and the various opportunities and routes for gene flow which arise over time. Complete fitness, even at the local level, can never be achieved, since no sooner is there selection for a fungus to adapt to the current environment than the latter shifts and re-adaptation inevitably follows. Moreover, no two habitats, however similar, are ever more than approximately identical. Thus, over many aeons, the most that any species can achieve is adaptation to a range of similar habitat and environmental conditions, e.g. polygenic adaptation of different populations of *Schizophyllum commune* [BA] (see Chapter 10, section 10.3.1). Therefore stabilizing selection towards a similar range of norms or to a broad norm is likely to be the predominant selective mode. It must play an important role in maintaining species stabilization in comparable ecological niches and probably ensures an appreciable degree of phenotypic uniformity. Thus multigenic selection is likely to be the most effective kind of selection involved, providing as it does the possibility of fine tuning to the environment and habitat (Mather 1973).

Nevertheless, sooner or later, prolonged species persistence inevitably exposes a fungus to new physical and biological environments and this may result in its being exposed to directional or even disruptive selection of various kinds. Even without environmental change, habitat heterogeneity generally engenders increased genetic variability since heterogeneity provides greater opportunities for selection to act differentially on the genome, e.g. forms of *Marasmius androsaceus* [BA] on pine and oak litter (see section 12.1.1). Habitat heterogeneity on a scale which affects fungal mycelia can equally well exist in a very limited habitat, so that a wide geographical distribution is not a necessary condition for differential selection to occur. It can occur in fungi not only within a single habitat, since there are so many potentially available microsites, but even within a single type of site as has occurred in the evolution of sibling species of *Coprinus* [BA] growing on organically contaminated straw (see Chapter 11, section 11.3.2). One possibility is that disruptive selection may favour extremes of timing of growth and reproductive process in a single environment. Temporal isolation can then be just as effective as spatial isolation as a predisposing factor.

Thus stabilizing, directional or disruptive selection can predominate over different periods or at different locations within the distribution of a species, and intraspecific differentiation may occur over a wide geographical range, within a limited but heterogeneous habitat or even within one site. However, there is no evidence that reproductive isolation is an inevitable consequence. Indeed, because of long species persistence times, it seems probable that it is a comparatively rare event, although in the intervening period genetic differentiation apparently proceeds gradually and continuously. Of course, this may be an artefact since the actual speed of genetic change will depend on many factors and may occur extremely rapidly, as in selection for more pathogenic populations in response to new resistance genes in host populations or periodic selection in industrial batch cultures. If such change is followed by a period of stabilizing selection, change may only appear to be gradual to an intermittent observer. The situation when selection is rapid has been described by Brasier (1987, 1995) as **episodic selection** which he defined as 'the effect of any sudden environmental disturbance likely to lead to a significant alteration in a species' population structure'.

The context in which Brasier appears to have considered such selection to be initiated is the occurrence of external cataclysmic events such as sudden climatic change or the sudden availability of a novel host, vector or competitor resulting in a sudden alteration of the current environmental or habitat balance. Even alterations in the availability or quality of resources on which the species is dependent might be a sufficient trigger to initiate change. However, internal cataclysmic events such as cytological aberrations can result in change of the mating system or in varying degrees of genetic isolation within a population. A further possibility is that envisaged by Carson (1982, 1985) as a consequence of the intervention of chance dispersal events such as drastic reduction in population size or sudden spatial isolation. As a consequence the population would possess only a reduced and possibly unbalanced gene pool which would be exposed to severe novel selection. In such circumstances the likely alternatives would be either extinction or selection for a new effective genetic balance, which could become increasingly adapted to the new circumstances under stabilizing selection. To these must be added inadvertent human intervention. Transfer of host plants to new locations can abruptly expose any pathogenic or mycorrhizal fungus infecting them to novel environments or bring them into proximity with related fungi with which they may hybridize. Hybridization certainly predisposes a population to novel forms of selection and can even result in abrupt speciation through allopolyploidy. There seems little doubt that human activities are probably playing an increasingly important role in promoting species processes.

There is no reason to assume that gradual selection or episodic selection are mutually exclusive. It is just as plausible to suppose that a range of changes could be involved so that at one extreme selection appears to be a gradual process, while at the other genetic change and even speciation are cataclysmically or episodically initiated and appear to be comparatively rapid.

There is little evidence for the process which leads to reproductive isolation, or whether reduction in gene flow is correlated with adaptive change to a different habitat or is a consequence of such a change. In the fungi, at least, it is best considered as a different process to adaptive change. Processes determining adaptation and reproductive isolation will now be considered in more detail.

## 12.2.1 Selection by the habitat

Populations of a species occurring in much the same habitat throughout its range show local adaptation. In *Aspergillus nidulans* [A] in the UK and three populations of *Schizophyllum commune* [BA] (see Chapter 10, see section 10.1.1), differences in growth rate between populations are due to selection for different balanced polygenic complexes, presumably each adapting to its local situation (Jinks *et al.* 1966; Simchen 1967). When representative isolates are crossed, the progeny are less well adapted and there is appreciable epistatic gene interaction, whereas only dominance and additive gene action are evident in the parents.

This is likely to be a widespread kind of adaptive process resulting in optimal adaptation to broadly comparable local conditions.

Physical or temporal isolation, but not necessarily reproductive isolation, can develop as a result of changes in habitat. The vast majority of fungi are saprotrophic and tend to recur in apparently similar ecological habitats. Their presence is maintained by the secretion of extracellular enzymes, known as **exoenzymes**, adapted to particular substrates as long as external factors such as pH enable the batteries of enzymes to function. However, the phenotypic plasticity of many fungi enables them to adjust their exoenzyme patterns to permit growth on the range of different substrates that they may encounter. They may well grow with greatly reduced vigour on a novel substrate outside its normal range, but if they persist for sufficient time, genetic assimilation results in the selection of genotypes better adapted to the new substrate than to the original one. Some phenotypic plasticity might be lost as a consequence, but adaptation to the new substrate, determined by the modified genotype, especially in the absence of intraspecific competition, would become more effective. In this way a micro-habitat change could come about, e.g. utilization of oak litter instead of fallen pine needles by the Appalachian populations of *M. androsaceus* [BA] (see section 12.1.1).

Alternatively, if nutritional conditions are so adverse that growth is in danger of being inhibited, mycelial survival can be ensured by either avoidance or dispersal. Neither of these results in a genotypic change or a change in ecological habitat when growth is resumed. Avoidance is achieved in lower Eumycota or Straminipila by developing dormant, often physically resistant, structures which frequently do not germinate until nutritional conditions are again favourable. These may be spores, chlamydospores or sclerotia which can undergo limited dispersal and germinate preferentially under conditions similar to those in which the original growing mycelia were adapted. Alternatively, the life-style adopted by mycelia of longer-lived or perennial higher fungi, such as Basidiomycotina, often involves establishment in, or on, a 'food base', such as pieces of dead organic matter, from which they grow out and exploit the surrounding substrate. Survival is achieved by remaining embedded in the food base at low metabolic activity until external conditions again become favourable. Of course, mycorrhizal fungi rely on their host's roots as their food base. Thus constancy of adaptation to a particular habitat is ensured.

Dispersal consequent on the induction of asexual or even sexual spores in response to adverse nutritional conditions is widespread in fungi and provides an opportunity for the same kind of habitat to be colonized elsewhere.

Therefore saprotrophic and mycorrhizic fungi are likely to be conservative in the exploitation of new habitats since their survival strategies tend to ensure that a new active growth phase will occur in much the same ecological niche as before. Moreover, during a dormant phase, fungi persist at a metabolically low level inimical to mutation and other processes of genetic change. Selection operates most effectively during growth and at the time of spore dispersal and establishment.

The situation in non-saprotrophic fungi is somewhat different since growth is confined to more or less specific niches or hosts, although most mycorrhizal fungi do not appear to be highly host specific. However, mycorrhizal fungi do share with pathogenic fungi a general ability to infect living plant tissues, but most pathogens differ by often showing a greater degree of host specificity. Little is known of the genetic basis of the general ability to infect plant tissues. There are obviously differences in its genetic determination because pathogens tend to be confined to species within broad groups of higher plants such as dicotyledons or monocotyledons, coniferous or 'hardwood' trees and, even more specifically, particular taxonomic groups of flowering plants, or infect different parts of plants such as shoots or roots. For example, it will be recalled that specific host pathogenicity for peas could only be conferred on a fungus through the acquisition of a mini-chromosome carrying the pisatin enzyme factor, provided that the receptor fungus already possessed a general ability to infect a green plant (see Chapter 10, section 10.3.2). It does not seem unreasonable to assume that selection for these general infective traits was probably a very ancient process, occurring at the time when the progenitors of these groups were first evolving the pathogenic habit.

The subsequent genetic determination of more restricted host specificity within such groups associated with speciation is far better understood. It often appears to be determined by a single or a few pairs of alleles, such as the specific virulence genes which match host resistance genes in gene-for-gene situations (see Chapter 2, section 2.4, Chapter 9, section 9.4.2, and Chapter 10, section 10.3.2). Therefore the basic step in a change of host is often a relatively simple change in one or a few genes. Consequently, the chance deposition of a spore which has undergone a mutation conferring the ability to infect a different species is enough to initiate the process of adaptation to a new host. Thereafter, rapid multiplication through spore formation in the absence of competition from the originating population could enable a sufficiently large inoculum to maintain and spread it on the new host. The occurrence in soil and roots of mixtures of pathogenic and non-pathogenic strains of wilt-infecting Fusaria exemplifies this situation. Both are capable of general root penetration but only the pathogenic forms are capable of penetrating xylem and being carried to the aerial parts of the plant, inducing wilts.

In highly host-specific pathogens, the initial mutation or mutations conferring host specificity can be expected to be reinforced subsequently by polygenic selection, resulting in increased adaptation to the new host. Other kinds of differentiation, such as morphological changes, may follow.

Habitat selection may be confined to site selection within a general habitat as exemplified by *Gremmeniella abietina* [AD]. In Scandinavia and adjacent Russsia, selection for different but adjacent sites on the same species of conifers has resulted in two types developing—the small tree type (STT) which is confined to lower branches, and the large tree type (LTT) which is confined to branches in the crown. Each type is associated with different reproductive patterns and timings (see Chapter 10, section 10.3.3) and therefore the different sites are exposed to different patterns of snow coverage. Their sexual reproductive patterns have become adapted to the conditions under which apothecia can develop and liberate ascospores. These patterns presumably reflect the selective effects of the very different microclimatic environments associated with the different sites to which the two types have become adapted. That climatic factors are likely to be the significant selective factor is shown by the existence of the genetically distinct but similarly adapted alpine variety, var. *cembra*, found some hundreds of miles to the south in the Swiss and Italian Alps. No evidence has been found for hybridization between natural isolates of the two types, although it proved possible to cross them experimentally and produce limited numbers of apothecia which developed some weakly infective ascospores. It seems improbable that such crosses are effective in nature (Hellgren and Högeberg 1995; Uotila *et al.* 2000). A hypothetical course of evolution resulting in the contemporary situation might have been as follows.

**1.** Initially, *G. abietina* [AD] occurs on *all* branches of its coniferous hosts.
**2.** *Either* (a) environmental selection operates differentially on isolates on lower branches from those on upper branches and eventually adaptive differences result in trees of somewhat different habit becoming the preferred habitat of the LTT and the STT, *or* (b) by chance the fungus infects adjacent trees of different habit and strains become adapted to the slightly different habitats (disruptive selection) as the LTT and the STT.
**3.** A consequential correlated effect to site selection arising with either of alternatives 2(a) or 2(b) is that incidental differences in the timing of apothecial production have also developed, i.e. developmental changes.

All these changes could have been be determined initially by a combination of chance and the dispersal of ascospores followed by climatically determined polygenic selection over a very long period until the maintenance of the new reproductive polymorphism had become reinforced through stabilizing selection. In this case compatibility is retained in *in vitro* incompatibility tests, although crossing in nature does not seem to occur.

Thus in *G. abietina* [AD] adaptation to a new spatially and seasonally separated habitat and environment has predisposed the LTT and the STT to the potential development of reproductive isolation.

## 12.2.2 Selection for the timing of development

Differences in the timing of reproductive development are common in fungi and, if these become fixed, can result in populations becoming isolated from each other because their reproductive phases no longer overlap. Ultimately, this could predispose such populations to reproductive isolation.

The timing of the reproductive phase is often variable; for example, different genets of *Pleurotus sapidus* [BA] (Wang and Anderson 1972) enter the reproductive phase at different times which are genetically determined, and this is also true for genets of the ectomycorrhizic agaric *Hebeloma cylindrosporum* [BA] even when growing in adjacent sites (Guidot *et al.* 2001). It is known that the timing of basidioma development is determined polygenically in other basidiomycotous fungi, e.g. *Collybia velutipes* and *Schizophyllum commune* [both BA] (Brasier 1970; Simchen 1965, 1966). Selection for the time of basidioma development could be triggered by factors such as temperature or soil moisture, both of which are known to affect the timing of basidioma development. Therefore environmental selection combined with different genetically determined times of development could result in a situation where the reproductive phases of some individuals in a population would cease to overlap, thus partitioning the potential ability of different genets to cross. This, of itself, need not lead to reproductive isolation, particularly between adjacent genets, since in basidiomycotous fungi somatic fusions amongst compatible monokaryons can result in crossbreeding. However, in *H. cylindrosporum* [BA] there is strong circumstantial evidence that in nature genets are certainly inbreeding and kept apart through an SI reaction. Thus, potentially, two populations could develop in the same area, in effect reproductively isolated.

More generally, it is a common observation that different isolates produce basidomata or ascomata at different characteristic times and, although oligogenes may be involved, polygenic control seems likely to be more common since most fungal growth phenomena are so determined. Thus the kind of selection suggested for *H. cylindrosporum* [BA] is potentially possible in most fungi.

There is no direct evidence for isolation arising in any fungus as a result of selection for developmental timing, but three morphologically extremely similar species of the ascomycotous genus *Lophodermium* [AP] appear to differ primarily as the result of different developmental timing during growth. They may well have arisen through disruptive selection for such characters. They occur over the same geographical and host ranges and are often found on the secondary needles of Scots pine (*Pinus sylvestris*). *L. seditiosum* [AP] is usually the first to infect green needles, *L. conigenum* [AP] occurs predominantly on killed needles and *L. pinastri* [AP] is the predominant form in needle litter, where the two previous forms are virtually unknown. Although their optimal growth rates, conidial development and times of ascomata development differ, they do overlap somewhat but they remain distinct (Minter *et al.* 1978; Minter and Miller 1980a,b; Minter 1981). If these species had a common progenitor, it seems plausible to suggest that their differences arose through disruptive selection for differences in germination time of ascospores, the later timing of ascoma development and, possibly, in the case of *L. pinastri* [AP] with further site or substrate selection. Indeed, of the four predisposing factors for sympatric speciation suggested by Maynard-Smith (1966)—habitat selection, pleiotropism, modifier genes and allochrony (differences in mating period)—all but pleiotropism could well have played some part in their ultimate reproductive isolation.

## 12.2.3 Episodic selection for adaptive response

Apart from mutation to a new host, adaptive change resulting from the two kinds of situation just considered is likely to be gradual and even the complete adaptation to a new host is only initiated by the mutational event. However, adaptive change which predisposes a fungus to reproductive isolation can be initiated in various ways by episodic selection. Three situations will be considered

here—climate change, founder populations and human influences.

### 12.2.3.1 Dramatic climatic change

Dramatic climate change is an example of episodic selection in the sense used by Brasier (1995) (see section 12.2). There is no convincing evidence for the direct causation of speciation processes with dramatic climatic change, i.e. specific events or climate change over, geologically speaking, relatively short time intervals. Species adapted to either high or very low temperatures (e.g. thermotolerant and thermophilic, or psychrotolerant and psychrophilic species) do exist as well as those adapted to limited exposure to novel temperatures such as exposure to forest fires or fresh volcanic debris. There is clear evidence for increased mutation rate with increased temperature (see Chapter 6, section 6.1.2), and this could conceivably result in the origin and selection of novel mutations which would significantly affect subsequent selection. It does not appear to have done so in the case of *Sordaria fimicola* [AP] where the induced changes are within the known natural range of variation of this fungus. Characteristic mycofloras of species adapted to disturbed ecological conditions in areas exposed to fires and volcanic disturbance as well as to arctic-alpine conditions, such as those associated with glaciation, or post-glaciation habitats are known (Watling 1988). Presumably some of these species were selected when such periods predominated locally or more generally during the Earth's history, but selection probably reflected selection to the novel habitats created by the climate rather than to direct climatic effects on the fungi. The association of recent sibling species of *Hebeloma* [BA] with dwarf alpine willows (*Salix*) (Aanen *et al.* 2000a) suggests that they emerged in response to conditions during the Pleistocene glaciations. Also it is suggested that fragmentation and dispersal events which affected both population structure and potential speciation of *Sclerotinia sclerotiorum* [AD] (see Chapter 5, section 5.3, and section 12.2) was correlated, albeit without great precision, to the period of Pleistocene changes in both Europe and North America—a period of changing habitat conditions. As the histories of more species are investigated using similar methods to those used with *S. sclerotiorum* [AD] (Carbone and Kohn 2001), it seems probable that more and closer associations may be found between periods of major climatic changes and local speciation events. However, these are likely to be due to the consequential changes in the habitat rather than to direct effects of climate on the fungi themselves.

### 12.2.3.2 Founder populations, drift and speciation
A situation theoretically capable of predisposing a fungal population to speciation, particularly if it is aerially dispersed, is the establishment of a numerically small and spatially isolated population—the so-called 'founder effect'. If the population is sufficiently small, drift could become significant, especially since the available genotypes would in any case already be limited. An alternative situation where drift might be of limited significance is a dispersed metapopulation whose individual sub-populations are small. However, since limited gene flow is usually still possible in such a situation, the population would be less isolated.

However theoretically plausible, the evidence for genetic change in isolated fungal populations through drift is non-existent. Such evidence as is available concerning population numbers suggests that they rarely fall to a size where drift would become effective, for example in the small populations making up the metapopulation of *Melampsora lini* [BR] on *Linum marginale* (Burdon and Jarosz 1992) in Kosciusko National Park, New South Wales (see Chapter 6, section 6.1). Suggestions that the genotypic structure of the northern Scandinavian-Russian populations of *Fomitopsis rosea* [BA] (Hogberg and Stenlid 1999) or the small scattered populations of *Gaeumannomyces graminis* [AP] in South Australia (Harvey *et al.* 2001) are due to drift are based more on the different and limited heterogeneity of each population and the lack of any evidence of selection. Evidence on population numbers is only available for the Kosciuscko populations of *M. lini* and they are on the limit for drift to be effective (see Chapter 6, section 6.1); otherwise, evidence on population numbers or possible population bottlenecks is lacking. In none of these cases is there any suggestion of even partial reproductive isolation between populations

although, in some cases, their physical isolation is undoubted.

A more likely cause of change than drift in a founder population is either that the balance of in- and outbreeding is altered (e.g. loss of an MTF in a dimictic species resulting in a loss of outbreeding capacity) or there is a change in selection pressure due to the new environment and the reduced range of available genotypes. The former would result in an appreciable change in the ability of the population to respond to selection. Brasier (1995) has proposed a possible scenario relating to clonal selection in which the chance of both physical and reproductive isolation could be increased (Figure 12.2). This change is not implausible but only one example has actually been observed. The leading-edge populations of *Ophiostoma novo-ulmi* [AP] in the epidemic across Europe exhibit a switch to an asexual population but, in this case, sexuality is eventually restored.

Gradual selective changes, as described in section 12.2.1, could result in new kinds of adaptive response because they would now be different from those in the parental population and so might reduce the probability of gene flow into the founder population. But, such an outcome is not predictable.

However theoretically plausible, the lack of evidence for the founder effect in fungi is not surprising. First, by its nature, detection of a founder population is difficult without close monitoring. The most useful situations, which may now become more frequent in Britain as a consequence of global warming, are small invading populations of alien continental fungal species. A programme of monitored genetical analysis involving a comparison with adjacent putative parental populations might provide some evidence for change in the founder populations. Secondly, because most fungi, apart from some Basidiomycotina, are capable of effective asexual reproduction, small population size *per se* is unlikely to persist for long if conditions are favourable for fungal growth and sporulation. If they are unfavourable, extinction rather than speciation would appear a more likely outcome.

### 12.2.3.3 Human influences predisposing to speciation
The consequences of human activities on the natural world and on cultivated plants are likely

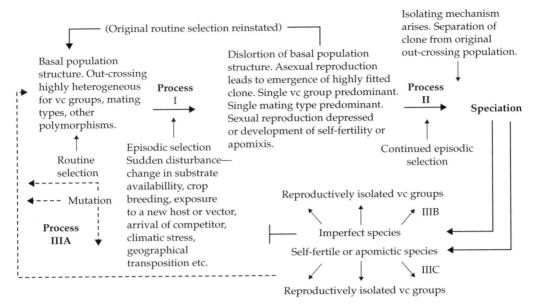

**Figure 12.2** Diagram to illustrate the possible consequence of episodic selection which could result in rapid clonal selection. I, II, IIIA, B, C— possible steps in the process. (Reproduced from Brasier (1995), with permission from the National Research Council of Canada.)

to have profound consequences for population structure and speciation in fungi. Three main types of change, which overlap to some extent, are induced by human activities:

• inadvertent transfer of fungi to new localities and habitats
• drastic changes in selection pressure on pathogenic fungi either from exposure to new habitats or because of new agricultural practices
• creation of novel habitats by human activities.

Inadvertent transfer, by direct transfer on clothes, inanimate objects, road vehicles or aeroplanes, by indirect transfer of substrates, such as soil adherent to plant roots and through infected host plants, or by the deliberate introduction of both new substrates and host plants into existing environments, is perhaps the most common human activity affecting fungi. When the number of trees, shrubs and crop plants moved about the globe by humankind are considered, it is not surprising that pathogenic, endophytic and mycorrhizal fungi must often occur as founder populations in novel situations. Whatever the means of transport, the

consequence is that resident or transferred fungi are exposed to new environments and hence to novel selection. Therefore population change of some sort is predictable, although whether speciation will be initiated is not. To date, most of the known historical examples have only resulted in differentiation at the population level.

One of the clearest examples is that of *Gremmeniella abietina* [AP] in North America (Hamelin *et al.* 1998). A distinct, supposedly indigenous, race occurs and there is evidence for introductions from Europe in the 1970s into the northeastern USA, Quebec and Newfoundland. However, in these latter areas the populations differ from the indigenous race, as expected; in particular, they lack apothecia and infect and kill large trees, but they also differ in the two regions. In Quebec and the northeastern USA, 53–86 per cent of all isolates are a single genotype (#5) plus some 25 rare genotypes not found in Newfoundland. The reasons for the dominance of the genotypic clone #5 are not known. Two possibilities exist. Either it is more aggressive than the other (asexual) clones or it was endophytic in saplings distributed from a single nursery. In contrast, in Newfoundland there are five haplotypes that are rare or absent from Quebec and the northeastern USA. Few of the introduced genotypes resemble those in European populations very closely. Indeed, only one genotype (#15) occurs in Europe, Quebec and Newfoundland, but in quite different proportions, namely 27 per cent, 8 per cent and 58 per cent respectively. Some of the rare genotypes in both populations were probably due to mutation after their introduction, but it is quite evident that two introductions have taken place and that both show the characteristic features of isolated founder populations. The different genotypic structure from that in Europe together with the probable eventual overlap with the indigenous North American populations suggest that evolution in this fungus will be very different from both that of the original North American populations and those of Europe.

Probably the most striking case of human transference of a clonal population is the worldwide spread of *Phytophthora infestans* [O] from South America in the mid-nineteenth century, followed by selection of numerous mutant specific virulence genotypes to resistant potatoes. Most recently, and consequent on the introduction of the second mating type to world populations, sexual recombination has resulted in bursts of new variation in different regions. In the twentieth century, the introduction of two species of Dutch elm disease to Europe and North America has resulted in epidemics followed by new hybridizations with unpredictable consequences (see Chapter 8, section 8.2.1.2). These are typical of the results of the inadvertent transfer of innumerable pathogenic species as a consequence of domestication or export of their host plants to new geographical environments.

Dramatic changes in selective pressure on pathogens of cereals, for example, must have resulted from crop domestication, increased host hectarages and monoculture which resulted in the creation of totally new kinds of habitats through human activities. The evolution of at least three host-specific *formae speciales* of *Blumeria* (*Erysiphe*) *graminis* [A], namely f.spp. *hordei*, *tritici* and *avenae*, from natural *B. graminis* agg. [A] in response to cereal cultivation is a typical example.

These modern *formae speciales* are best regarded as sibling species of *B. graminis* agg. [A] since they are effectively reproductively isolated and show increased host specificity and aggressiveness, a quantitative trait. Contemporary isolates from modern cultivated wheat, barley or oats can be crossed experimentally, although with difficulty, to produce a few weakly viable offspring that are unlikely to persist in nature (Hiura 1978) (see Chapter 8, section 8.2.1.2). Growth on cereals other than their normal hosts is slight and only occurs if the cereal is already infected with its normal pathogen; for example, inocula of f.sp. *tritici* will grow on barley and produce a very few conidia provided the barley is *already* infected with its normal pathogen, f.sp. *hordei*, but f.sp. *tritici* will *not* infect an *uninfected* barley plant. It is just possible that, rarely, insignificant hybridization may occur under these circumstances (Hiura 1978).

Fortunately, at the probable centre of origin of modern wheats and barley in the Israel–Palestine region, *B. graminis* agg. [A] still occurs naturally in open mixed grassland on descendants of the wild progenitors of the cultivated cereals, e.g. *Hordeum spontaneum* and *Triticum dicoccum*. In these natural grasslands, *B. graminis* [A] exists in a genetically highly heterogeneous range of forms, capable of inter-crossing and showing a behavioural range from partial host restriction to a much wider range of hosts than the modern siblings (Wahl *et al.* 1978). Such populations probably resemble the situation before cereal cultivation commenced. However, it seems probable that opportunities for genets of *B. graminis* [A] to become more highly adapted to a particular host species gradually increased as seed gathering was replaced by cultivation and the genetic purity and acreages of host plants increased with modern intensive monocultures. Selection during this era was presumably mainly polygenic and for aggressiveness. Dispersal of the incipient species was no doubt also promoted by human migrations. How or when reproductive isolation arose is not known.

Assuming that the cereal cultivation originated between about 12 000 BC and 10 000 BC, the sibling species of barley mildew evolved over a period of 10 000–15 000 years at most. This is an appreciably shorter time than the orders of magnitude suggested for the natural species, described in section 12.1.1, which were not specifically exposed to human activities.

Natural and human-promoted speciation of other fungi is widespread, for example in the worldwide *Gibberella fujikuroi* [AP/FA] species complex and its associated anamorphic species. The progenitors of this species complex existed some 100 million years ago prior to the break-up of Gondwanaland; its subsequent radiation and wide host range are described by O'Donnell *et al.* (1998). The group also provides a historical example of how a pathogenic fungus can be inadvertently dispersed with its host. *Fusarium ramigenum* and *F. lactis* [both FA] are found in Southern Arabia and nearby Africa where their host, the wild fig (*Ficus carica*), is endemic and also in California where both calimyrna and wild figs have been cultivated for about the last 100 years (O'Donnell *et al.* 1998). The stability of species associations, once established, is clearly demonstrated by this transfer to a very different environment.

Amongst the Straminipila, Hansen and co-workers (Hansen 1987; Hansen and Maxwell 1991) have drawn attention to the *Phytophthora megasperma* [O] species complex with its broad division into legume pathogens and those of various, usually woody, hosts. The legume pathogens occur as sibling species on alfalfa and soybean, both crops which have had an enormous increase in acreage as a result of agricultural practices, although alfalfa was grown as forage for horses at least 2000 years ago. The changes involved are complex, since polyploidy and chromosomal interchange are probably involved as well as host selection. No appropriate genealogical analysis has been made but Hansen suggests that some of these sibling species may have originated, or the speed of their incipient speciation may have been hastened, as a result of selection associated with domestication and more extensive growth of some of the host plants.

There are many other comparable examples, and almost certainly human influence is having an increasing effect on populations of pathogenic fungi by planting schemes, the transfer of host species to new environments and increased selection pressure on the pathogens by measures taken to protect crops. One example of particular interest, since it provides an opportunity of monitoring potential change due to human activities, is the specificity of ectomycorrhizal *Pisolithus* [BA] species, especially those found in Australia in association with species of *Eucalyptus*. Unusually amongst mycorrhizal fungi, species of *Pisolithus* [BA] show high host specificity (Martin *et al.* 2002), but since *Eucalyptus* species are now widely exported to different regions where they are not indigenous, a potential opportunity is provided for change in host specificity and, eventually, speciation is possible.

A different source of potential change is the use of entirely novel synthetic materials including polymers, such as polyurethane and polyvinyl products, or purified natural materials not found free in nature such as petroleum. In most of these cases, existing species have adapted to utilizing these materials (e.g. *Amorphotheca* (*Hormoconis*) *resinae* [A], a soil fungus which is now a serious contaminant of aircraft petroleum tanks), but inevitably selection will now follow a different course from that in the natural environment. Similarly, *Pneumocystis carinii* [A], now a serious secondary infection which causes pneumonia in immunosuppressed AIDS patients, is being subjected to quite different selection regimes from those prior to the worldwide AIDS pandemic. In most of these cases, selection has not been operative for a sufficiently long period for genetical differentiation, let alone speciation, to become manifest but change does seem inevitable.

## 12.3 Modes of reproductive isolation

In section 12.2, a range of processes were described which could predispose some populations to spatial or temporal isolation from others but need not necessarily bring about their reproductive isolation. The poorly understood origins of reproductive isolation are described in this section.

A detailed understanding of how reproductive isolation arises in fungi is not available, but various situations are known that appear to result in some degree of reproductive isolation or even complete isolation.

## Gradual processes

• The gradual development of genetically determined reproductive isolation between populations already isolated in a spatial or temporal way. Gradual spatial and temporal isolation may result in inadvertent reproductive isolation simply because fertilizing agents cannot reach their reproductive targets.

## Episodic processes

• Reproductive isolation due to change to a different functional mating system, or through loss of sexual reproduction altogether.
• Abrupt reproductive isolation due to polyploidy which may be associated with hybridization.
• 'Instant' speciation due to a single event without other appreciable genetic or phenotypic modification.

In most cases, one or other of the predisposing conditions described in section 12.2 may have already been established and persisted for a long time before genetic isolation is achieved, if at all.

## 12.3.1 The gradual development of reproductive isolation

It has been plausibly argued by Mather (1943, 1973) that populations become adapted through selection resulting in the development of balanced polygenic complexes which ensure that individuals are optimally adapted to their local environment. Therefore, when different populations interbreed, there is breakdown of balance with a loss of fitness which continues in subsequent generations. Some experimental evidence for this in fungi was presented in Chapter 10, section 10.1.1. It is assumed that, ultimately, the loss in fitness will become so great that the interbreeding populations will become genetically isolated.

There is experimental evidence that genetic isolation can be achieved rapidly through disruptive selection in Drosophila (Thoday 1972), but similar evidence is not available from fungi. Although several situations described in section 12.2 could result in disruptive selection, its impact on reproductive isolation in fungi is likely to be altogether more gradual for two reasons. First, the frequency of sexual reproduction is probably lower in fungi and, second, the common occurrence of long, interpolated asexual phases is likely to reduce the impact of such selection.

One issue that has only been partially resolved is how genetic isolation is finally achieved. There are a number of ways in which this could occur.

• The least abrupt situation is that hybrid progeny will experience an increasing lack of integration and co-ordination between the normal cellular processes of differentiation, mimicking the gradual changes in the genomes brought about by adaptation to a new environment, and ultimately a hybrid fails to develop at all.
• Alternatively, at some stage in the gradual genomic changes in response to physical isolation, a single essential process which determines effective sexual reproduction fails. An obvious candidate is the physiological failure of a stage in the chain of pheromonic reactions which ensures successful plasmogamy, whether between gametangia, gamete cells or, in Basidiomycotina, compatible hyphae. The investigation of sexual reproduction in the yeasts has demonstrated how a complex series of co-ordinated stages is involved in achieving contact between compatible cells, far less plasmogamy and fertilization (Elliott 1994): loss of one stage could prove fatal.
• Another possibility, especially in the Ascomycotina, is that it occurs inadvertently owing to the ineffective transmission of a fertilizing agent, such as a microconidium, to the receptive trichogyne. It was pointed out earlier (see Chapter 6, section 6.3.1) that virtually nothing is known concerning the dispersal distances over which fertilizing agents such as microconidia can be dispersed while still retaining their viability. Superficially, at least, the likely distances appear to be extremely restricted, so that effective reproduction requires the partners to be in close proximity. Thus isolation by distance, or by host and time, could bring about reproductive isolation simply as a consequence of the physical isolation and it could then be reinforced by assortative mating within the separated populations. No selection for reproductive isolation, as such, would have been involved, but it would

simply arise as a mere byproduct of adaptation to the different habitats or timing of development.

Putative examples for each of these possibilities can be identified but in most cases their demonstration will require further observation and experiment.

Whenever large numbers of widely distributed isolates of almost any fungal species are examined, genetic and phenotypic variation is found in their reproductive behaviour, ranging from complete isolation through partial compatibility to full fertility with other isolates. Typical examples of such a range are shown by several ascomycotous species, e.g. *Magnaporthe grisea* [AP] (Kumar *et al.* 1999; Notteghem and Silué 1992) and *Neurospora crassa* [AP] (Perkins and Turner 1988; Turner *et al.* 2001). Genetic components of the changes which result in reproductive isolation evidently arise fairly frequently and are not an uncommon phenomenon. They could well represent stages in the gradual onset of reproductive isolation.

In Basidiomycotina, isolation is essentially prezygotic. At its most extreme, hyphal fusion is simply not initiated or, in some cases, a lethal or aversion reaction develops, such as a deeply pigmented border to the opposed mycelia between which there is a hypha-free zone of aversion, e.g. *Polyporus ciliatus*, *P. brumalis* and *P. arcularius* (Hoffman and Esser 1978). A striking example of the failure to develop hyphal fusions between sibling species is shown by the Coprini investigated by Kemp (1983, 1985) (see Chapter 11, section 11.3.2). The physiology of hyphal fusion is not well understood, but the fact that anastomoses can be induced at a distance suggests the action of one or more diffusible substances analogous to pheromones when the fusions are sexual. Here, therefore, speciation may well represent a loss or major modification of a diffusible agent essential for initiating hyphal fusions.

Partial or complete incompatibility occurs between monokaryons carrying potentially compatible MTFs in many Basidiomycotina, especially those of closely related species (see Chapter 8, section 8.1.1.4, and Chapter 11, section 11.3.2). It is evident after successful somatic fusion by the development of inadequately differentiated dikaryons, often with either aberrant or unregulated development of clamp connections, or by a weakly growing dikaryon which seems unlikely to develop a mature basidiocarp. Thus reproductive isolation appears to be a gradual process, rather than a simple matter of two alternative conditions. The best evidence for the gradual development of reproductive isolation in the Basidiomycotina is provided by *Heterobasidion annosum* [BA] and *Hebeloma crustuliniforme* [BA].

Different host-specific types of the former species are known (see Chapter 8, section 8.2.1.2) whose intercompatibility is determined by at least five pairs of intersterility alleles. For example, some S- and P-types are infertile when crossed since the S-type carries $(v_1\ v_2\ v_3)^- s^+ p^-$ and the P-type carries $(v_1\ v_2\ v_3)^+ s^- p^+$, and other combinations result in partial compatibility (Chase and Ullrich 1990b). The observation that five unlinked loci are involved in determining reproductive isolation strongly suggests that intersterility must have arisen gradually as the result of successive mutations, probably followed by recombination and selection, since the sudden origin of appropriate combinations of five different alleles would seem highly improbable.

Although the different types of *H. annosum* [BA] have been regarded as incipient species rather than clearly distinct species, the same problem does not arise between the sibling species of the *Hebeloma crustuliniforme* [BA] species complex. Different degrees of compatibility are shown between three ICGs in one clade. One branch in the phylogenetic tree gave rise to two ICGs showing partial incompatibility, with up to 15 per cent of all crosses being compatible. Phylogenetically, however, that branch *preceded* a later branch in the same lineage which gave rise to an ICG which showed even greater incompatibility with the other two ICGs—only 0.2 per cent were compatible. Complete incompatibility was shown between all other sibling species (Aanen *et al.* 2000b). The situation in the three ICGS with partial compatibility suggests that reproductive isolation has developed gradually in a stepwise manner.

Although the strongest case for a gradual development of reproductive isolation is represented by these two cases, many other examples of partial compatibility or complete incompatibility between monokaryons of related basidiomycotous species provide further supporting, if incomplete, evidence.

Other evidence is available from the Ascomycotina, where it was suggested in section 12.2.1 that the adoption of a new host or site appeared to predispose some pathogenic species to speciation. Although each sibling species in the *Ceratocystis coerulescens* [AP] complex on conifers (Harrington and McNew 1998; Harrington and Wingfield 1998) is restricted to one kind of conifer (e.g. *C. pinicola* [AP] to pines (*Pinus*), *C. polonica* [AP] to *Picea abies* and *C. douglassii* [AP] to *Pseudotsuga menziesii*), those that occur on the same species occur in different sites. For example, on spruce (*Picea*), *C. coerulescens* [AP] is largely confined to dead wood or cut timber, *C. resinifera* [AP] is confined to living wood near wounds and *C. polonica* [AP] is associated with the beetle galleries of the bark beetle (*Ips typographus*) and lacks conidia.

The degree of incompatibility differs between the species. Some crosses develop sterile perithecia, e.g. *C. coerulescens* × *C. resinifera*, *C. polonica* or *C. douglassii* [all AP]. In others, some hybrid progeny are produced, of which a very few are exceptionally fertile. For example, *C. laricicola* [AP] forms fully fertile progeny with *C. pinicola*, *C. coerulescens*, *C. resinifera* and *C. douglassii* [all AP]! However, even when apparently mature perithecia are produced in other crosses, viable ascospores are few. The perithecia contain misshapen ascospores and cellular debris suggestive of aborted cells. This range of compatibilities is not unexpected if reproductive isolation is indeed a condition which arises by successive changes.

The *Ceratocystis* [AP] situation just described is the most common situation in interspecific crosses between closely related species of Ascomycotina. Incompatibility in this group is manifest by different degrees of post-zygotic developmental failure, such as incompletely differentiated ascospores, inviable ascospores or seriously impaired growth of germ mycelia, e.g. conidial *Neurospora* species [AP] (Perkins and Turner 1988; Turner *et al.* 2001).

Those cases where closely related species appear to differ in the timing of their development, such as the three species of *Lophodermium* [AP] mentioned

earlier (section 12.2.2), are situations in which reproductive isolation could arise entirely inadvertently. It is unfortunate that details of the reproductive behaviour of these species are not known but, developing reproductively as they do with only a limited period when the processes in the three species overlap, the chances of cross fertilization would be expected to be reduced or even lost. Hence cross fertilization within each species is likely to be more frequent than between species and, eventually, genetic isolation would become complete.

The examples described in this section support the view that reproductive isolation results most commonly from the accumulation in a stepwise manner of alleles, or polygenic complexes, which reduce intercompatibility. It would be expected that the phenotypic effect of each such mutant, or selected complex, would be reinforced by further polygenic selection, so enhancing the cumulative effect until complete isolation became the norm. However, both loss of pheromonic effectiveness or inadvertent failure of fertilization probably occur also.

One common finding in both ascomycotous and basidiomycotous sibling species is that the greater the degree of sympatry shown by any two species, the more effectively they are isolated. For example, the two phylogenetically most closely related species of *Ceratocystis* [AP], *C. polonica* [AP] and *C. laricicola* [AP], have identical ITS sequences, differ in only one isozyme allele, occur sympatrically across Eurasia and are both associated with bark beetles. The former species is pathogenic to spruce (*Picea*) and the latter to larch (*Larix*), yet reproductively they are the most highly isolated of all the species in the complex. Unfortunately, the biology, genotypes and genetics of the situation have not been investigated in the *C. polonica*–*C. laricicola* [both AP] case, or in similar cases. Without additional information an adequate explanation is not possible, but a speculative suggestion is discussed in section 12.3.4.

Two even more striking examples from the Basidiomycotina are the two sibling poroid North American species of *Hirschioporus abietinus* [BA] and 'races' of *Microbotyrum violaceum* [BS]. The former are completely incompatible even though both retain partial compatibility with morphologically identical sibling European isolates. Neither clear habitat differences nor differences in timing of their reproductive phase have been detected between the North American siblings. It is not immediately obvious why incompatibility should be more strongly developed between sympatric sibling species than between non-sympatric siblings although, presumably, competition between them is very strong.

In many cases of apparent sibling species, or incipient species, it is not possible to be certain whether or not isolation has developed while the taxa were juxtaposed. The races of the pathogenic *M. violaceum* [B] are a good example. Although morphologically identical and exhibiting a similar life style, each race is apparently confined to one or a few species of different Caryophyllaceae and were recognized as long ago as 1921 (Zillig 1921). Recently, using SSRs based on five loci, it has proved possible to demonstrate that genetically distinct isolates were confined to *Silene vulgaris*, *S. nutans* and *Dianthus carthusianorum* found growing together in close proximity (Shykoff *et al.* 1999). It was shown experimentally that each isolate was capable of infecting the other host species and that they were capable of conjugating *in vitro*, although some abnormalities were observed. However, there was no evidence for any *natural* crosses at this site. Moreover, various insects which can passively transfer the teleutospore of these fungi between flowers as a result of touching the spore-filled anthers were in fact frequently seen to visit one host plant immediately after visiting another. Therefore there were ample opportunities for the different isolates to infect another host plant, becoming attached when the insects visit the flowers. Thus there are apparently physiological barriers which prevent both cross-infection and cross-breeding between these host-specific sibling isolates.

Two kinds of explanation to account for this situation are possible. It could perhaps reflect an early stage in the development of new sibling species resulting from the adaptation of different populations to different host species. Alternatively, the differentiation of the fungal races and the barriers to crossing may have arisen elsewhere, or at

a different time under different conditions when the ecological niches of the host species were distinct. It is not clear whether or not the present juxtaposition of the three host species is a recent ecological event. Thus, without some knowledge of the history of the host species, discrimination between the possible evolutionary development of the fungal races is not possible.

## 12.3.2 Abrupt reproductive isolation through change of mating system

Many asexual and some sexual species occur in the Chytridiomycotina, the most primitive fungi. It may well be that sexuality evolved in this group from asexual progenitors, but so little is known of these fungi that this hypothesis can only be speculative. Primary asexual fungi may occur in the more advanced fungal groups but the impression given is that in them asexuality is more usually derived. That is the view adopted in this account.

Many genera of fungi include species with different mating systems—heterothallism and homothallism as well as those that have apparently lost all sexuality. In the majority of cases species which differ in their mating systems will not cross with each other, and this is the usual situation between asexual species and apparently related sexual species.

Examples can be found in all the main groups of fungi. In the Zygomycotina, *Mucor genevensis* [Z] is a homothallic homologue of *M. hiemalis* [Z] (Schipper 1973). In the Ascomycotina, heterothallic dimictic, homo-heterothallic and homothallic species occur in the related conidia-forming *Neurospora* [AP] species, with dimictic and homodimictic species evidently being most closely related, e.g. *N. sitophila* and *N. tetrasperma* [both AP] respectively (Skupski *et al.* 1997).

Similar situations occur in Basidiomycotina, e.g. *Marasmius* and *Sistotrema* [both BA]. In some cases, the distribution of mating systems within a genus suggests that the change was a single event giving rise to different clades which adopted the new mating system. For example, all species of the agaric genus *Marasmius* (BA) are diaphoromictic, but four sections are tetrapolar, one section (*Sicci*) is bipolar and a further section (*Globulares*) is mostly bipolar with one tetrapolar species. These last two sections are thought to be closely related, and it has been surmised that the change from a tetrapolar to a bipolar mating system occurred once within the clade that gave rise to these sections (Petersen 1995b) although phylogenetic evidence is not available to confirm this hypothesis. Tetrapolar, bipolar and homomictic species occur in the corticiaceous *Sistotrema* species complex [BA], and there are other species, some evidently related to sexually reproducing forms, some of whose isolates are unable to form dikaryons with clamps or even to undergo hyphal fusions (Hallenberg 1984b; Ullrich 1973). This suggests a stepwise change in mating system rather than its being due to a single event.

Processes involved in the changes from heterothallism to homothallism and thence to asexuality have been documented in some Ascomycotina. Homothallism probably arises most frequently from heterothallism as a consequence of cytological aberrations

One possibility in a dimictic species is a deletion leading to the loss of a MTF. In four of the morphologically very similar homothallic aconidial soil-inhabiting species of *Neurospora* [AP], the *Mat a* locus is absent but the *Mat A* locus, homologous with that in the heterothallic conidial species, is retained. Evidently, the genome still retains the necessary genes to enable fertile perithecia to develop. How the loss has come about is not known. A similar situation occurs in other homothallic members of the Sodariaceae such as *Gelasinospora* [AP] and *Anixiella* [AP] (Glass *et al.* 1990). As mentioned in section 12.2.2, loss of an MTF could occur easily during the establishment of a founder population and this could conceivably have been the genesis of an originating population which gave rise to the soil-inhabiting Neurosporas, especially since their habitat is also so different from that of the conidial species.

The one exceptional homothallic species, *N. terricola* [AP], carries *both* MTFs closely linked in the same mycelium, a situation known also from *Cochliobolus* [FA/AP]. This suggests a possible translocation of one or both MTF loci to an adjacent chromosomal site.

*Cochliobolus* [FA/AP] is another genus where heterothallic, homothallic and apparently asexual species occur, the last being placed in the anamorphic genera *Bipolaris* and *Curvularia* [both FA]. Heterothallic species resemble those of the Sordariaceae in being dimictic, with mating being determined by two alternative idiomorphs *Mat-1* and *Mat-2*. The incorporation of both mating types, tightly linked, in a single strain occurs in homothallic species of *Cochliolobus* [AP], where there is evidence that the initiating event was an unequal cross-over within the MTFs of a heterothallic progenitor (Figure 12.3)

Examination of the nucleotide sequences in the closely related dimictic *C. heterostrophus* [AP] and the homothallic *C. luttrelli* [AP] have identified the site of such a cross-over. This translocation, by itself, may not be sufficient to fully ensure homothallic behaviour because {*Mat-1/Mat-2*} genes from the more distantly related homothallic *C. homomorphus* [AP] transferred to a different (ectopic) site to that normally occupied by the MTF in a strain of *C. heterostrophus* [AP] whose *Mat* gene had been deleted, resulted in selfing but *without* the formation of ascospores. Therefore full fertility may require the presence of some of the flanking genes to the idiomorphic core of the MTF (Turgeon 1998).

Heterothallic

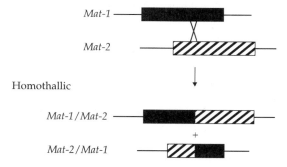

Homothallic

**Figure 12.3** A model based on situations in *Cochliobolus* spp. [AP] for the evolution of homothallism from heterothallic ancestors. The boxes represent the idiomorphs of the two MTFs (*Mat-1* and *Mat-2*) and the consequences of an unequal cross-over between them to generate a fused MTF in one recombinant which would thus acquire homothallic behaviour. (Reproduced from Turgeon (1998), with permission from *Annual Reviews*.)

Species possessing different mating systems occur intermingled in the phylogenetic tree where asexual species appear to be derived from sexual progenitors and are of relatively recent origin (Berbee *et al.* 1999). The asexual *Bipolaris sacchari* [FA], a sister species to *C. heterostrophus* {AP}, carries the same *Mat* genes although differing in the number of nucleotide replacements compared with the sexual species. Transfer of these *Mat* genes from *B. sacchari* [FA] into *Mat*-deleted strains of *C. heterostrophus* [AP] resulted in successful sexual reproduction. However, transfers in the opposite sense were ineffective, indicating that the *B. sacchari* genome, other than the *Mat* genes, was defective or lacking in some way so that sexual reproduction could not be initiated. Further evidence that effective sexual reproduction requires genes other than MTFs is provided by the apparently asexual *Cochliobolus sativus* [AP], where there is clear evidence for the presence of MTFs in populations throughout the world. Experimental crosses achieve varying degrees of success between different compatible isolates, but there is no evidence to support the occurrence of sexual reproduction in the wild (Tinline 1988; Zhong and Steffenson 2001).

Thus, although a cytological abnormality resulting in a homothallic or asexual mating system may initiate change in a mating system, it appears that it may need to be reinforced by subsequent selection to render the new condition functionally more effective and so ensure complete reproductive isolation.

An alternative pattern of change from sexuality to asexuality in Ascomycotina could involve the widespread occurrence of female sterility. Both asexuality and 'female sterility' are widespread in Ascomycotina, and this may indicate that many ascomycotous Fungi Anamorphici have lost sexuality by a gradual process initiated by the loss or mutation of a gene other than an MTF resulting in 'female sterility'. If this is followed by further functional deterioration, it could eventually result in the total loss of sexual reproduction. The diverse patterns shown by fungi such as *Glomerella cingulata* [AP] (see Chapter 7, section 7.2.4) are also compatible with gradual or stepwise processes of change.

The phylogeny of anamorphic Penicillia and related teleomorphic species have been investigated. Asexual and sexual species, whether heterothallic or homothallic, are scattered throughout those phylogenetic trees that have been examined. In many cases a sexual species is more closely related to an asexual species than to any other. For example, *Talaromyces wortmannii* [A] is a sister species of *Penicillium variable* [FA] and both are related to the anamorphic *P. islandicum* [FA] (LoBuglio *et al.* 1993) (Figure 12.4).

There is also clear evidence that loss of sexuality has occurred more than once in *Talaromyces* [A] and, in addition, asexual species are separated from sexual species by short branch lengths in terminal clades which indicates that loss was a relatively recent evolutionary event. No such separations are found earlier in the phylogenetic tree. This can be interpreted in two possible ways. Either loss of sexuality never occurred previously or ancient asexual lines have become extinct (LoBuglio and Taylor 1993). The latter hypothesis seems more probable in view of the evidence that such changes occur quite frequently.

Less is known about change in mating systems in other fungal groups. In Basidiomycotina, asexual isolates (monokaryons which have lost the ability to anastomose with potentially fertile monokaryons) are found in many species when intercompatibility tests are made. It is not known whether homothallic forms can arise from heterothallic ones or vice versa in these fungi.

A similar situation obtains in the Zygomycotina, but since the nature of their MTFs is unknown and little of their genetics is understood, it is not possible to account for the origin of forms with different mating systems or asexual isolates. Homothallic and asexual isolates, morphologically very similar to sexual species, are common in the Mucoraceae but their genetical relationships are unclear.

Similar ignorance characterizes change of mating system in the Straminipila although, from the limited phylogeny available for *Phytophthora* [O] and related species, both mating systems are found together in most clades so that it is not clear whether one mating system is more ancient than the other. Cooke *et al.* (2000) suggest that

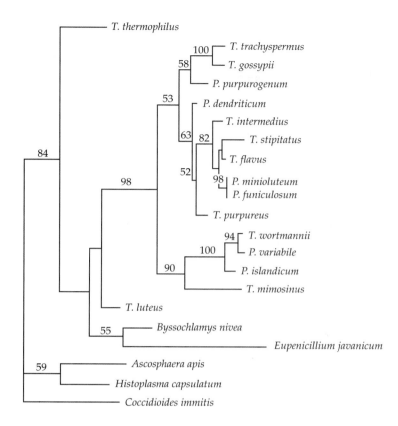

**Figure 12.4** *Talaromyces–Penicillium* species. Phylogenetic diagram based on rDNA, ITS and mitochondrial small rRNA regions to illustrate relationships of sexual to asexual species. Note that asexual species arise sporadically and recently, also that more than one may be a 'sister' species to a sexual form. (Reproduced from LoBuglio *et al.* (1993), with permission from New York Botanical Garden Press.)

heterothallism probably preceded the radiation shown in their phylogenetic tree because of its 'intricacy and universality', which would imply that the homothallic condition was derived, but there is no other support for this view.

### 12.3.3 Abrupt reproductive isolation through hybridization and polyploidy

An obvious way in which reproductive isolation can arise relatively abruptly is through polyploidy, especially allopolyploidy following hybridization. In some Ascomycotina, hybridization alone is known to result in the creation of new species.

Hybridization is an important contributor to speciation in the endophytic *Epichloë–Neotyphodium* [FA/AP] species complex (see Chapter 8, section 8.2.1.3). In these fungi crosses between sexually reproducing species have given rise to

heteroploid anamorphic species which appear to be stable, and in some cases these, in turn, have crossed with other sexual species to produce further stable anamorphic species, often with changed host ranges. For example, *N. coenophialum* [FA] has *N. uncinutum* [FA] and both *E. baconii* and *E. festucae* [both AP] in its ancestry. It is not known for how long the anamorphic species persisted, but even if it was only for a short evolutionary period, presumably these crosses can re-occur periodically, restoring the range of hybrid species. This possibility underpins the immense advantage many fungi have over most other Eukaryota by virtue of their highly effective modes of asexual reproduction. The ability of hybrids to persist by means of asexual propagation provides a potential extended opportunity for polyploidization to occur.

*Botrytis aclada* AII [FA], a well-established species, closely resembles *B. aclada* AI [FA] but is an allopolyploid which may well have been derived in a similar manner to species of *Neotyphidium* [AP], i.e. as the result of a hybridization between anamorphic *B. aclada* AI [FA] and *Botryotinia alii* (*Botrytis byssoidea*) [AP] which is capable of sexual reproduction. Alternatively, it could have arisen from a diploid nucleus which arose in a heterokaryon between the anamorphs. If this was the manner in which the polyploid arose, it would demonstrate yet another advantage that fungi have over other eukaryotes—being able to bring genetically different nuclei together without the intervention of sexual reproduction.

A few cases are known where natural allopolyploids, which have probably arisen some time in the last century, are well on the way to becoming new species (see Chapter 8, section 8.2.3.1). *Verticillium longisporum* [FA] is already established as a pathogen with a unique host range. The new 'alder *Phytophthora*' [O], another polyploid pathogen with a unique host range, is still in a very plastic evolutionary phase, with the chromosome number not yet having stabilized. The consequences of ongoing selection by its host are not predictable but speciation is evidently in progress. In neither of these cases is it known how frequently hybrids have arisen in the past which have either been insufficiently pathogenic to become established or have lacked sufficient, or specific, pathogenicity to novel hosts to be selected. By analogy with allopolyploidy in green plants (Arnold 1997), the 'alder *Phytophthora*' [O] polyploids continue to be created, and polyploid populations across Europe probably represent the outcome of different parental crosses rather than a single clone. Some of the cytological evidence supports such a situation in this case.

In contrast with the previous examples, allopolyploidy in species of *Achlya* [C], appears to be much older. Polyploid series exist together with a range of autopolyploids. They also differ in being morphologically similar to their hybrid progenitors and parental species, and in populations of individuals of different ploidies reproductive isolation is far from complete. The progeny of such crosses are probably inviable or, at least, less viable than their parents in nature, although they can be reared successfully in the laboratory. The allopolyploids may remain distinct in their natural habitat but

information about their ecology and habitats is not sufficient to determine whether they occupy ecological niches distinct from those of their parental species.

Some evidence for autopolyploidy has been provided in Chapter 8, section 8.2, but the cases cited appear to be well-established ancient species in which any undesirable cytological effects of autopolyploidy are not in evidence. The haploid or dikaryotic states from which the autopolyploids have been derived are not known for any of these cases. Presumably they would be morphologically similar, so that the extent and distribution of autoploid species in fungi must await more intensive and widespread cytological examination.

### 12.3.4 'Instant' speciation—a hypothetical situation

A hypothetical possibility to account for abrupt speciation that should be considered is the initiation of reproductive isolation by a single event without other substantial genetic, cytological or phenotypic differentiation necessarily having taken place.

In many Ascomycotina, somatic isolation has developed through somatic incompatibility, although in all cases investigated except one, *Aspergillus amstelodami* [A], it is *not* associated with reproductive isolation. However, in this normally self-fertile species, sexual crosses *within* a VCG resulted in about 40 per cent fertile perithecia, approaching the theoretical maximum, but crosses *between* isolates from different VCGs show greatly reduced fertility (C.E. Caten, unpublished data). The cause of this reduction is not known, but clearly the origin by mutation of any factor comparable to a *vic* locus (or loci) which inhibits *sexual* compatibility with isolates from other VCGs could result in reproductive isolation.

Another simple change which might account for the preceding situation is that reproductive isolation could be brought about by loss or impairment of the molecules responsible for initiating reproduction through pheromonic activity. There is clear evidence for the role of pheromones in the induction of sexual reproduction in the Chytridiomycotina (Machlis 1958a), the

Zygomycotina (Burgeff 1915), the Ascomycotina (Bistis 1983) and the Basidiomycotina (Kemp 1977). Clearly, any genetic change in an individual which resulted in impairment of the production, liberation or specific properties of a pheromonic molecule would result in immediate reproductive isolation of that individual. Asexual propagation would not be expected to be affected, however, so that the individual concerned could continue to be propagated. In some cases, only hyphal fusion might be affected and, if so, the derived population would appear to be another VCG.

A not dissimilar novel situation is known from some Basidiomycotina, especially in the genus *Coprinus* [BA]. Kemp (1980) demonstrated a novel type of incompatibility which overrides the normal MTF compatibility reactions in *Coprinus bisporus* [BA], a two-spored agaric which produces binucleate basidiospores of which about 93 per cent are heterothallic and about 7 per cent are homokaryotic. Although originally interpreted as being due to heterokaryotic incompatibility, it is now considered to be the result of an unusual heterogenic incompatibility reaction (RFO Kemp, unpublished data). Homokaryotic spores fall into three groups (A, B and C) which mate differentially, regardless of their MTF constitution. Group A is fully compatible with both group B and group C, but groups B and C are incompatible. When mycelia of any strain are opposed to oidia derived from either of the other two, their hyphae grow towards the oidia and fuse with them; if mycelia are opposed they also fuse. In a compatible reaction (either A + B or A + C), the hypha–oidium fusion cell develops a viable mycelium, but in an incompatible mating (B + C), fusion hyphae develop followed by a lethal reaction about 3 hours after fusion and the hyphae die. If mycelia have been opposed, the tips of the hyphae between them in a B + C confrontation also die. Therefore the potential exists for isolation to develop between groups B and C, and if by chance group A, which is the least commonly isolated, were to become extinct, groups B and C would be reproductively isolated, could evolve separately and would appear to be a sibling species. To date, this kind of heterogenic incompatibility is only known from two species of *Coprinus* [BA] with two- or three-spored basidia,

although the lethal reaction which is associated with this A-B-C type of heterogenic incompatibility is widespread in the genus, in some cases occurring between known sibling species (Kemp 1977). Both species are also notable for having very few MTFs, although diaphoromictic, suggesting that they are relatively new species themselves. This, if correct, coupled with the scenario just described would appear to result in very rapid speciation indeed.

It will be recalled that in another group of Coprini, 16 totally inter-incompatible sibling species, which were morphologically indistinguishable apart from their basidiospore dimensions, were isolated from the same habitat (see Chapter 11, section 11.3.2). It is not known whether the basis for their incompatibility is similar to that just described, but if a similar type of heterogenic incompatibility to that shown by *C. bisporus* occurs in these species it could be the basis for their reproductive isolation. Another possibility is that, since they occur in an ephemeral habitat (dung-saturated straw), they represent a sequence of populations, each arising from a distinct founder effect. Isolation could arise from time to time in different founder populations but they would be largely replaced by the next population. The sibling species would then represent different reproductively isolated surviving populations. However, there is no evidence for this hypothetical scenario.

It is evident that, in both Ascomycotina and Basidiomycotina, genetic systems exist which could, potentially, result in the division of a species into at least two incompatible populations, possibly by as little as single mutational changes, prior to ecological or morphological selection. However, it has been argued (Hansen and Hamelin 1999) that such an event is unlikely since

an 'instant' species isolated by an intersterility barrier would seem to face a large competitive disadvantage from its numerically and territorially superior progenitor, unless reproductive isolation was preceded or accompanied by some fitness advantage or ability to colonize a new niche.

There is, of course, much strength in this point of view especially since, to date, no examples are known of so-called 'instant' speciation, but it is equally true that if such species did exist, they would be very difficult to detect. Moreover, provided that a newly reproductively isolated species was no less fit than its progenitors, it could surely persist on equal terms and possibly multiply asexually. In this respect, it is obvious that species possessing an effective asexual multiplicative stage would be more capable of avoiding extinction than those lacking such means, so that conidial ascomycotous forms would have some advantage in such a situation compared with most Basidiomycotina which frequently lack an anamorphic phase. Support for 'instant speciation' of this nature must obviously await its demonstration.

## 12.4 Fungal populations and speciation

Because the sample of species studied to date is so small and biased in various respects, there are still many restrictions on fully understanding fungal populations, species and speciation. Of these three, speciation is probably the least understood process or series of processes occurring in fungal populations. There are several reasons for this.

Even less is known about these issues in the fungi than in many other groups of the Eumycota because they have received far less detailed and prolonged study. Investigations of the structure and dynamics of natural or semi-natural fungal populations are still few and far between and all the most detailed, such as those on barley mildew and a few cereal disease fungi are concerned with populations whose structure is largely determined by agricultural activities. Only three detailed investigations, of chestnut blight, Dutch elm disease and *Melampsora lini* [BR] on the wild *Linum marginale*, approach a situation resembling that of most natural fungal populations.

It is difficult to draw general conclusions because most of the available information is biased towards populations of pathogens. This is largely historical and reflects the needs of pathologists to understand the biological units with which they had to work and between which they had to discriminate. Since most fungi are saprotrophic or rely on an extensive saprotrophic phase, as do mycorrhizic fungi for example, this is unfortunate.

A real constraint is imposed by the incomplete understanding of the taxonomy, ecology and biogeography of the fungi compared with that available to those concerned with populations, evolution and speciation in higher green plants, mammals, birds, fishes or *Drosophila*. Inevitably, therefore, fungal situations have been interpreted using concepts developed for groups such as those just mentioned, despite their obvious differences.

Despite these drawbacks, fungi have some intrinsic advantages, notably their haploidy, or the ease with which haploid isolates can be obtained, and also the ability to manipulate them in culture in order to investigate genetical issues and to determine their potential reproductive abilities.

Obvious differences between many fungal populations and those of other eukaryotes are as follows.

• the balance between sexual and asexual phases and the intermittent periods of metabolic activity, often far exceeded by the long dormant phases undergone by so many fungi, from the perennial Basidiomycotina
• the ability of rapid generation of huge clonal populations which far exceed in numbers and frequency the products of the intermittent meiotic products, so that clonal selection is of considerable and continuing importance
• their exposure to much stricter selection because of the prolonged duration of the haploid phase
• the occurrence of large numbers of secondarily asexual species.

Some of these considerations as they apply to speciation in fungi will be discussed in the rest of this chapter. Although there is an increasing body of evidence to indicate that asexual species are often closely related to species with sexuality, it is convenient to consider them separately since they pose different problems from their sexually reproductive relatives.

### 12.4.1 Secondarily asexual species

One novel aspect of speciation which appears to be particularly characteristic of the fungi, especially the Ascomycotina, although less common in other Eumycota, is the frequent occurrence of a secondarily asexual species. How this condition may arise has been discussed in section 12.3.2, but the problem they pose is: How do such species persist?

Phylogenetic studies with some of the genera where asexual species are most common suggest that most are closely related to sexual species and have arisen relatively recently. Because of this connection it has been suggested that they became adapted to their present habitat *prior* to the loss of sexuality. Of course, habitat selection would be more effective and rapid in a sexually reproducing than in an asexual species. Thus asexual species may have been pre-selected and become highly adapted for particular conditions prior to their reproductive isolation, although, in an evolutionary sense, they are probably short lived.

This latter attribute would accord with the view first developed by Muller (1932, 1964) that the accumulation of even slightly deleterious mutations can potentially drive an asexual species to extinction. Muller considered an asexual haploid population in which deleterious mutations occurred which were eliminated by selection. When equilibrium is reached between mutation and selection, even though the numbers of non-mutant individuals may be quite small, there is no way in which the original wholly non-mutant population can be restored. The population has moved to a new optimum with, minimally, one additional deleterious mutation so, like a 'ratchet', the population moves one step nearer to extinction. With succeeding generations there is a further inexorable accumulation of deleterious mutants. Muller's ratchet can only be compensated for in two ways.

First, recombination can reconstitute mutant-free individuals, but this cannot be achieved by an asexual species since restoration can only be brought about by back mutation which occurs with a lower frequency than forward mutations (Haigh 1978). Second, compensatory mutations could restore the situation and there is evidence to suggest that these are not uncommon. They would obviously be positively selected so that, theoretically, they could bring about a slow recovery over time (Poon and Otto 2000). Experimental support from fungi is still needed for this hypothesis.

Haigh (1978) has examined the situation quantitatively. In particular, he has estimated the size of

the optimum class for survival $E(n_0)$:

$$E(n_0) = N \exp(-U/s)$$

where $N$ is the size of the population, $U$ is the mutation rate per genome per generation and $s$ is the selective disadvantage per mutation. In other words, if the population number rises above $n_0$ it is driven down by deleterious mutations. In practice, if $E(n_0)$ exceeds about 1000 the ratchet will not work, but if it is 10 or less the ratchet acts quite rapidly and the population declines. Bidochka and De Koning (2001), using data from various entomopathogenic fungi, have concluded that populations will usually be of a size for extinction to be unlikely, given a plausible mutation rate of $10^{-8}$ per nucleotide. However, they assumed that population size would remain constant and this seems highly unlikely. Clearly, if the population fluctuates, as seems more plausible, the significant size for the ratchet to operate will be its *smallest* value. Therefore eventual extinction may be a real possibility. Extinction could account for the fact that asexual species are not found early in phylogenies but only occur as recent events, i.e. evolutionarily speaking, Fungi Anamorphici really are short-lived species.

How serious are these issues is currently uncertain. Considerable genetic variation is now beginning to be revealed when asexual fungi are subjected to genetic analyses. Whenever particular cases have been examined using genealogical techniques, they have provided supporting evidence for recombination, although whether it occurred recently or in the past has not been determined (Taylor *et al.* 1999). Therefore it must be seriously questioned whether apparently asexual fungi *are* truly asexual and whether or not most undergo recombinant episodes at some time, however infrequent such episodes might be. Only time and further investigation can resolve this issue.

Summarizing the present position, it seems probable that most Fungi Anamorphici are probably derived from similar sexually reproducing species and probably most have originated in, evolutionarily speaking, relatively recent times. Their persistence times are probably shorter than those of sexually reproductive species but their evolutionary span may be extended, in some cases at least, owing to rare reproductive episodes. It is worth noting that non-functional MTFs have been identified as a relict feature in some asexual fungi. It is not inconceivable that they were once functional but suffered progressive loss of function so that recombination could have occurred prior to, or in the early stages of, speciation of the asexual form. A more remote possibility is that in some circumstances the potential but relict MTFs could become functional.

## 12.4.2 Sexual fungal species

The majority of the fungi reproduce sexually and therefore the processes that result in speciation might be expected to be more comparable with those shown by the vast majority of other Eukaryota, even though they are predominantly diploid. This seems to be the case but the basis for such a generalization is very small indeed. There are still many situations and processes which are only understood from one or a few fungi, often in far from natural situations.

Nevertheless, the significant role played by asexual multiplication is evident in the properties of founding populations, whether originating seasonally or through dispersal (see Chapter 7, section 7.3.2). Examples of the former situation are common but only two examples of the latter are known. They appear to demonstrate an important role for clonal populations in the establishment of founder populations of pathogens, at least under quasi-natural conditions.

In North America, two distinct originating clones of *O. novo-ulmi* [AP] were identified (Brasier and Kirk 2000). Only one of the two mating types was present in each, and a similar situation occurred in expanding populations in Europe. These clones were able to generate new genotypes, possibly as a result of hybridization with the former resident *O. ulmi* [AP] populations, thereby integrating the missing MTF through introgression and subsequently reproducing sexually (Brasier *et al.* 1998).

In Switzerland, two newly established populations of *C. parasitica* [AP] north of the Alps were dominated by similar single clones, and a third contained two dominant clones (Hoegger *et al.* 2000). In contrast, a newly established population

south of the Alps within the area already affected by the chestnut blight fungus was much more variable. However, *C. parasitica* [AP] has a secondary mode of sexual reproduction (see Chapter 7, section 7.2.3.2) which could generate new genotypes in due course once mutant loci had accumulated in the genetically homogenous northerly clones. These may be very exceptional situations, but there is every reason, and in the case of *O. novo-ulmi* [AP] clear evidence despite the initially clonal state, to suppose that variation will continue to be generated and that the new populations will flourish.

It is not possible to say whether selection is more significant in one set of circumstances than in another. Selection undoubtedly operates differently at different phases of the life-cycle so that pleomorphism, so common in fungi, will affect selection differentially because the teleomorphic phase generally occurs under quite different ecological and environmental conditions from the anamorphic phase. Dormancy must also affect the speed of selection. Its duration probably exceeds the active growth phase of many Zygomycotina and Ascomycotina, thus reducing, to some extent, the time available for selection although, of course, passive largely random selection between dormant propagules occurs.

Selection does result in genetic change in fungi in much the same way as it does in other organisms, whether it is change in gene frequency or change in quantitative characters. The available evidence demonstrates that polygenic selection results in the establishment of balanced polygenic complexes which adapt individuals to the local environment. Their break-up, consequent on intercrossing, results in the progeny being genetically less well balanced, less fit and selected against. This promotes genetic isolation and is thought to be one sufficient cause for speciation. However, it seems more likely that reproductive isolation is furthered by a greater variety of predisposing factors and develops gradually in more than one way.

Population size is another factor which is not properly understood. For example, it is doubtful whether *natural* populations, even of pathogens, ever resemble populations of pathogens of crops grown in close proximity in vast numbers, often as monocultures. These are the situations which provide most of the evidence for medium- to long-range dispersal.

Host plants of natural pathogen populations are likely to be appreciably more scattered or grow in well-dispersed small clumps and most dispersal is probably over shorter distances. Therefore their fungal pathogen populations are likely to be smaller, possibly more inbred and may well approximate to a metapopulation condition. Populations of the only wild plant pathogens which have been investigated in detail, *Ustilago violacea* [BS] and *Melampsora lini* [BR] in the Kosciusko highlands (but not in the adjacent plains), appear to behave in this way. Chance extinction and re-colonization are more likely to determine a population's genotypic profile under such conditions. Where a more evenly distributed population occurs, which is capable of regular genetic exchange through gene flow, it will tend more towards the hypothetical panmictic condition. Indeed, this appears to be the case with the natural plains populations of *M. lini* [BR], and many pathogens of forest trees also occur as widely distributed populations between which genetic exchange occurs regularly, resulting, effectively, in a single widespread panmictic population, e.g. *Heterobasidion annosum* [BA] or *Gremmeniella abietina* [AP]. All the evidence available in the latter two cases suggests that gradual genetic differentiation is the norm, resulting from host selection in the former species and selection due to different climatic and ecological factors in the latter case.

No detailed information is available concerning the spatial structure of saprotrophic fungal populations. A general consideration of what is known of their biology and ecology (Cooke and Rayner 1984; Hallenberg 1995) suggests that the majority, being broadly adapted to general substrates, will tend towards panmixis since individuals are unlikely to be separated by distances greater than their dispersal range. In general, the most common outcome of dispersal is probably such as to enable compatible matings to occur with adjacent populations, thus maintaining the genetic structure of the more widely dispersed populations roughly constant. Local selection could still result from establishment in a novel environment or habitat

reached through the chance of random spore dispersal. In any of these circumstance, physical isolation, initially by distance, ecologically or through developmental timing, could arise and be followed by reproductive isolation as outlined in sections 12.2 and 12.3.

Amongst those species associated with more exacting substrate requirements, however, a meta-population structure may well be quite common since gene flow between widely separated sites may often be restricted. Far more information is needed before generalizations can be made.

Overall, the limited evidence available suggests that speciation in fungi in their natural habitats is most usually a gradual process in effectively continuous populations, and selection is most probably determined by ecological features, including host selection. However, in metapopulations of individuals with relatively limited propagule dispersal, a mixture of local selection and stochastic processes are more likely to occur and spatial isolation could play a more significant role than in continuously dispersed populations. Consequently, the strength of ecological selection would be enhanced because genetic dilution and panmixia resulting from gene flow would be reduced.

There is no clear evidence as to how reproductive isolation in fungi is actually initiated, although it seems likely that it is a gradual stepwise process. In Eumycota, the physical site of the isolating process appears to be in the cytoplasm during the haploid phase. In Ascomycotina, reproductive incompatibility is nearly always located in the cytoplasm of the immediate post-meiotic products where malfunctioning of the ascospore is by far the most common manifestation. In contrast, in the Basidiomycotina, incompatibility is evident in the fusion cell which initiates the reproductive process or in impaired growth of its cellular progeny.

If reproductive isolation is effectively a quantitative phenomenon, a condition not incompatible with its observed gradual stepwise development, it is likely to be a multigenically determined process. It could involve both oligogenes and polygenic complexes. Thus any or all of a period of physical isolation, growth in a different habitat or growth in a different site would increasingly result in selection for very different balanced polygenic complexes and the fixation of different mutations. In this way, physical or environmental isolation of a population predisposes it to ever-increasing reproductive isolation from any other population as a by-product of developmental malfunction.

One further evolutionary speculation is possible. Conventionally, the Ascomycotina would be described as showing post-zygotic isolation and the Basidiomycotina as showing pre-zygotic isolation. Since incompatibility, as far as it is understood at present, in the Zygomycotina and Chytridiomycotina is also post-zygotic, it appears that this represents the evolutionarily earliest cellular location for reproductive isolation. Post-zygotic isolation also appears to be the norm in the Oomycota and perhaps in the Straminipila as a whole. This would be consistent with their probable earlier origin than that of the Eumycota, with pre-zygotic isolation developing much later and becoming the norm in the Basidiomycotina.

## 12.5 Speciation in fungi: some conclusions

The account in this chapter indicates that the processes of species formation in the fungi are broadly similar to those in predominantly diploid Eukaryota. Nevertheless, there are some differences in the modes of speciation.

Population differentiation is largely the result of selection by the habitat, often leading to some kind of isolation falling short of genetically determined reproductive isolation. Once populations are predisposed in this way, reproductive isolation can build up as a gradual process, mainly determined by polygenic selection. That such selective conditions are very effective is demonstrated by the common and characteristic occurrence of sibling species, frequently evolving in close proximity, even within the same habitat. This might well reflect a greater intensity of selection in predominantly haploid organisms compared with diploid ones.

Human activities are undoubtedly a significant selective force on plant pathogenic fungi both in providing dense stands of highly uniform host species, so increasing the selective effect of the host on the fungus, and by inadvertent transport of spores or infected plant material. This enables fungi to exploit new habitats or, at least, to exploit them more rapidly than normal dispersal would allow. There is evidence that speciation in fungi results from these inadvertent activities quite apart from the more general ecological consequences of pathogen dispersal.

Such evidence as is available seems to demonstrate that hybridity and polyploidy are not common, although whether this simply reflects the limited evidence or is a genuine feature of Eumycota is not clear. Certainly, both hybridization and polyploidy seem to be much more common in the smaller group of the diploid Straminipila which, of course, resemble Eumycota in their life-style.

A striking feature of the Eumycota in contrast to the Straminipila is the widespread occurrence of clonal populations and the very large numbers of secondary asexual species. The former enable habitats to be exploited rapidly but do not fundamentally affect speciation processes. The ability possessed by asexual clones of species capable of sexual reproduction to disperse and potentially to exploit a wide range of habitats may perhaps account, in part, for the very large numbers of fungal species compared with green plants. In many cases asexual species are highly adapted, possibly preadapted, to their habitat or, in the case of pathogens, to their host species, and it seems probable that the majority are derived from sexual species. They represent a particularly effective kind of adaptation, again perhaps reflecting the increased intensity of selection on haploids. Nevertheless, their lack or infrequency of recombination is consistent with evidence to suggest that their evolutionary persistence times are appreciably shorter than those of sexual species. But the probability of extinction is more than balanced by the constant creation of new asexual species, a form of evolutionary compensation for the breakdown of sexuality.

## Further reading

Aanen DK, Kuyper TW, Mes THM and Hoekstra RF (2000). The evolution of reproductive isolation in the ectomycorrhizal *Hebeloma cristuliniforme* aggregate

(Basidiomycetes) in Northwestern Europe: a phylogenetic approach. *Evolution*, **54**, 1119–1206.

Brasier CM (1987). The dynamics of fungal speciation. In *Evolutionary biology of the fungi* (ed. ADM Rayner, CM Brasier and D Moore), pp. 231–60. Cambridge University Press.

Brasier CM (1995). Episodic selection as a force in fungal evolution, with special reference to clonal speciation and hybrid introgression. *Canadian Journal of Botany*, **23** (Supplement 1), S1213–21.

Chase TE and Ullrich RE. (1990). Five genes determining intersterility in *Heterobasidion annosum*. *Mycologia*, **82**, 73–81.

Haigh J (1978). The accumulation of deleterious genes in a population—Muller's ratchet. *Theoretical Population Biology*, **14**, 251–67.

LoBuglio KF, Pitt JI and Taylor JW (1993). Phylogenetic analysis of two ribosomal DNA regions indicate multiple independent losses of a sexual *Talaromyces* state among asexual *Penicillium* species in subgenus *Biverticillium*. *Mycologia*, **85**, 592–604.

Taylor JT, Jacobson DJ and Fisher MC (1999). The evolution of asexual fungi: reproduction, speciation and classification. *Annual Review of Phytopathology*, **37**, 187–246.

Turgeon BG (1998). Application of mating type gene technology to problems in fungal biology. *Annual Review of Phytopathology*, **36**, 115–37.

Wahl I, Eshed N, Segal A and Sobel Z (1978). Significance of wild relatives of small grains and other wild grasses in cereal powdery mildews. In *The powdery mildews* (ed. DM Spencer), pp. 83–100. Academic Press, London.

# References

Aanen DK (1999). Species and speciation in the *Hebeloma crustuliniforme* complex. Wageningen Agricultural University.

Aanen DK and Kuyper TW (1999). Intercompatibility tests in the *Hebeloma crustuliniforme* complex in Northwestern Europe. *Mycologia*, **91**, 783–95.

Aanen DK, Kuyper TW, Boekhout T and Hoekstra RF (2000a). Phylogenetic relationships in the genus *Hebeloma* based on ITS1 and ITS2 sequences, with special emphasis on the *Hebeloma crustuliniforme* complex. *Mycologia*, **92**, 269–81.

Aanen DK, Kuyper TW, Mes THM and Hoekstra RF (2000b). The evolution of reproductive isolation in the ectomycorrhizal *Hebeloma cristuliniforme* aggregate (Basidiomycetes) in Northwestern Europe: a phylogenetic approach. *Evolution*, **54**, 1119–206.

Aanen DK, Kuyper DW and Hoekstra RF (2001). A widely distributed ITS polymorphism within a biological specie of *Hebeloma velutipes*. *Mycological Research*, **105**, 284-90.

Abdelali E, Brasier CM and Bernier L (1999). Localization of a pathogenicity gene in *Ophiostoma novo-ulmi* and evidence that it may be introgressed from *O. ulmi*. *Molecular Plant–Microbe Interactions*, **12**, 6–15.

Adams DH and Roth LF (1969). Intraspecific competition among genotypes of *Fomes cajanderi* decaying young-growth Douglas fir. *Forest Science*, **15**, 327–31.

Ahmad M (1953). The mating system of *Saccharomyces cerevisiae*. *Annals of Botany*, **17**, 329–42.

Aist JR and Morris NR (1999). Mitosis in filamentous fungi: how we got where we are. *Fungal Genetics and Biology*, **27**, 1–25.

Ali-Shtayeh MS, Lim-Ho CL and Dick MW (1986). An improved method and medium for quantitative estimates of populations of *Pythium* species from soil. *Transactions of the British Mycological Society*, **86**, 39–47.

Anagnostakis SL (1982). Genetic analyses of *Endothia parasitica*: linkage data for four single genes and three vegetative compatibility groups. *Genetics*, **102**, 25–8.

Anagnostakis SL (1988). *Cryphonectria parasitica*, cause of chestnut blight. *Advances in Plant Pathology*, **6**, 123–36.

Anagnostakis SL and Kranz J (1987). Population dynamics of *Cryphonectria parasitica* in a mixed-hardwood forest in Connecticut. *Phytopathology*, **77**, 751–4.

Anagnostakis SL, Hou B and Kranz J (1986). Diversity of vegetative compatibility groups of *Cryphonectria parasitica* in Connecticut and Europe. *Plant Disease Reporter*, **70**, 536–8.

Anaya N and Roncero MIG (1995). *Skippy*, a retro-transposon from the fungal plant pathogen *Fusarium oxysporum*. *Molecular and General Genetics*, **249**, 637–47.

Anderson E. (1948). Hybridization of the habitat. *Evolution*, **2**, 1–9.

Anderson JB, Petsche DM and Smith ML (1987). Restriction fragment polymorphisms in biological species of *Armillaria mellea*. *Mycologia*, **79**, 69–76.

Anderson JB and Stasovski E (1992). Molecular phylogeny of northern hemisphere species of *Armillaria*. *Mycologia*, **84**, 505–16.

Anderson JB and Ullrich RC (1979). Biological species of *Armillaria mellea* in North America. *Mycologia*, **71**, 402–14.

Anderson JB, Korhonen K and Ullrich RC (1980). Relationships between European and North American biological species of *Armillaria mellea*. *Experimental Mycology*, **4**, 87–95.

Andrews JH (1984). Relevance of *r*- and *K*-theory to the ecology of plant pathogens. In *Current perspectives in microbial ecology* (ed. MJ Klug and CA Reddy), pp. 1–7. American Society for Microbiology, Washington, DC.

Andrivon D and de Vallavieille-Pope C (1993). Racial diversity and complexity in regional populations of *Erysiphe graminis* f.sp. *hordei* in France over a 5-year period. *Plant Pathology*, **42**, 443–64.

Andrivon D and de Vallavieille-Pope C (1995). Race diversity and complexity in selected populations of fungal biotrophic pathogens of cereals. *Phytopathology*, **85**, 897–905.

Antequera F, Tamame M, Villanueva JR and Santos T (1984). DNA methylation in the fungi. *Journal of Biological Chemistry*, **259**, 8033–6.

Arganoza MT, Min J, Hu Z and Akins RA (1994). Distribution of seven homology groups of mitochondrial plasmids in *Neurospora*: evidence for widespread mobility between species in nature. *Current Genetics*, **26**, 62–73.

Armstrong GM and Armstrong JK (1981). Formae speciales and races of *Fusarium oxysporum* causing wilt diseases. In *Fusarium: diseases, biology, and taxonomy* (ed. PE Nelson, TA Tousson and RJ Cook), pp. 391–9. Pennsylvania State University Press, University Park, PA.

Arnold ML (1997). *Natural hybridization and evolution*. Oxford University Press.

Atwood K and Mukai F (1953). Indispensible gene functions in *Neurospora*. *Proceedings of the National Academy of Science of the United States of America*, **39**, 1027–35.

Atwood K and Mukai F (1955). Nuclear distribution inconidia of *Neurospora* heterokaryons. *Genetics*, **40**, 438–43.

Auerbach C. (1959). Spontaneous mutations in dry *Neurospora* conidia. *Heredity*, **13**, 414 (abstract).

Austin DJ, Bu'Lock JD and Gooday GW (1969). Trisporic acids: sexual hormones for *Mucor mucedo* and *Blakeslea trispora*. *Nature*, **223**, 1178–9.

Avise JC (1994). *Molecular markers, natural history and evolution*. Chapman and Hall, New York.

Aylmore RC and Todd NK (1984). Hyphal fusions in *Coriolus versicolor*. In *The ecology and physiology of the fungal mycelium* (ed. DH Jennings and ADM Rayner), pp. 103–25. Cambridge University Press.

Aylor D (1990). The role of intermittent wind in the dispersal of fungal pathogens. *Annual Review of Phytopathology*, **28**, 73–92.

Bachmann K (1994). Molecular markers in plant ecology. *New Phytologist*, **126**, 403–18.

Backus MP and Stauffer JF (1955). The production and selection of a family of strains in *Penicillium chrysogenum*. *Mycologia*, **47**, 429–63.

Bakalinsky AT and Snow R (1990). The chromosomal constitution of wine strains of *Saccharomyces cerevisiae*. *Yeast*, **6**, 367–82.

Bakkeren G and Kronstad JW (1993). Conservation of the *b* mating-type gene complex among bipolar and tetrapolar smut fungi. *Plant Cell*, **5**, 123–36.

Bakkeren G, Kronstad JW and Lévesque CA (2000). Comparison of AFLP fingerprints and ITS sequences as phylogenetic markers in Ustilaginomycetes. *Mycologia*, **92**, 522–7.

Ball C (1971). Haploidization analysis in *Penicillium chrysogenum*. *Journal of General Microbiology*, **66**, 63–9.

Banik MT and Burdsall HH Jr (1998). Assessment of compatibility *in vitro* among *Armillaria cepistipes*, *A. sinapina*, and North American biological species X and XI, using culture morphology and molecular biology. *Mycologia*, **90**, 798–805.

Barksdale AW (1965). *Achlya ambisexualis* and a new cross-conjugating species of *Achlya*. *Mycologia*, **72**, 493–501.

Barnett HL and Hunter BB (1998). *Illustrated genera of imperfect fungi*. APS Press, St Paul, MN.

Barr DJS (1992). Evolution and kingdoms of organisms from the perspective of a mycologist. *Mycologia*, **84**, 1–11.

Barr DJS, Warwick SI and Désaulniers NL (1996). Isozyme variation, morphology and growth response to temperature in *Pythium ultimum*. *Candian Journal of Botany*, **74**, 753–61.

Barrett DK (1978). An improved selective medium for isolation of *Phaeolus schweinitzii*. *Transactions of the British Mycological Society*, **71**, 507–8.

Barron GL (1977). *The nematode-destroying fungi*. Canadian Biological Publications, Guelph, Ontario.

Barry EG (1966). Cytological techniques for meiotic chromosomes in *Neurospora*. *Neurospora Newsletter*, **10**, 12–13.

Bassi FGA and Burnett JH (1980). The genetic architecture of aggressiveness in *Ustilago maydis*. *Annals of Applied Biology*, **94**, 281.

Bates MR, Buck KW and Brasier CM (1993). Molecular relationships of the mitochondrial DNA of *Ophiostoma ulmi* and the NAN and EAN races of *Ophiostoma novoulmi* determined by restriction fragment length polymorphisms. *Mycological Research*, **97**, 1093–1100.

Bayman P and Cotty PJ (1991). Vegetative compatibility and genetic diversity in the *Aspergillus flavus* population of a single field. *Canadian Journal of Botany*, **69**, 1707–11.

Beerli P and Felstenstein J (1999). A maximum-likelihood estimation of migration rates and effective population numbers in two populations using a coalescent approach. *Genetics*, **152**, 763–73.

Bennett KD (1997). *Evolution and ecology: the pace of life*. Cambridge University Press.

Bent KJ (1982). Case study 4. Powdery mildews of barley and cucumber. In *Fungicide resistance in crop protection*. (ed. J Dekker and SG Georgopoulos), pp. 219-30. Pudoc, Wageningen, The Netherlands.

Bent KJ, Holloman DW and Shaw MW (1990). Predicting the evolution of fungicide resistance. In *Managing resistance to agrochemicals* (ed. MB Green, HM LeBaro and KW Moberg), pp. 303–19. ACS Symposium Series 421, Oxford University Press, New York.

Berbee ML and Taylor JW (2001). Fungal molecular evolution: gene trees and geologic time. In *Systematics and evolution*, Part B, Vol. 7 (ed. DJ McLaughlin, EG McLaughlin and PA Lemke), pp. 229–45. Springer-Verlag, Berlin.

Berbee ML, Pirseyedi M and Hubbard S (1999). *Cochliobolus* phylogenetics and the origin of known, highly virulent pathogens, inferred from ITS and glyceraldehyde-3-phosphate dehydrogenase gene sequences. *Mycologia*, **91**, 946–77.

Bernard U and Küntzel H (1976). Physical mapping of mitochondrial DNA from *Neurospora crassa*. In *The genetic function of mitochondrial DNA* (ed. C Saccone and AM Kroom), p. 105. North-Holland, Amsterdam.

Bertrand H, Chan BSS and Griffiths AJF (1985). Insertion of a foreign nucleotidesequence into mitochondrial DNA causes senescence in *Neuospora intermedia*. *Cell*, **41**, 877–84.

Bevan EA and Makower M (1963). The physiological basis of the killer character in yeast. *Genetics Today*, **1**, 202–3.

Bevan JR, Crute IR and Clarke DD (1993a). Variation for virulence in *Erysiphe fischeri* from *Senecio vulgaris*. *Plant Pathology*, **42**, 622-35.

Bevan JR, Clarke DD and Crute IR. (1993b). Resistance to *Erysiphe fischeri* in two populations of *Senecio vulgaris*. *Plant Pathology*, **42**, 636-46.

Bevan JR, Crute IR and Clarke DD (1993c). Diversity and variation of expression of resistance to *Erysiphe fischeri* in *Senecio vulgaris*. *Plant Pathology*, **42**, 647-53.

Bidochka MJ and De Koning J (2001). Are teleomorphs really necessary? Modelling the potential effects of Muller's ratchet on deuteromycetous entomopathogenic fungi. *Mycological Research*, **105**, 1014–19.

Biggs R (1938). Cultural studies in the Thelephoraceae and related fungi. *Mycologia*, **30**, 64–78.

Birt TP and Baker AJ (2000). Polymerase chain reaction. In *Molecular Methods in ecology* (ed. AJ Baker), pp. 50–64. Blackwell Science, Oxford.

Bissegger M and Milgroom MG (1994). Outcrossing rates of *Cryphonectria parasitica* in European chestnut (*Castanea sativa*) forests. *Phytopathology*, **84**, 1074 (abstract).

Bistis GN (1983). Evidence for diffusable, mating type specific trichogyne attractants in *Neurospora crassa*. *Experimental Mycology*, **7**, 292–5.

Blakeslee AF (1904a). Zygospore formation: a sexual process. *Science*, **19**, 864–6.

Blakeslee AF (1904b). Sexual reproduction in the Mucorineae. *Proceedings of the American Academy of Arts and Sciences*, **40**, 205–319.

Blakeslee AF and Cartledge JL (1927). Sexual dimorphism in Mucorales. II: Interspecific reactions. *Botanical Gazette*, **84**, 51–7.

Blakeslee AF, Cartledge JL, Welch DS and Bergner AD (1927). Sexual dimorphism in Mucorales. I: Intraspecific reactions. *Botanical Gazette*, **84**, 27–50.

Blanch PA, Asher MJC and Burnett JH (1981). Inheritance of pathogenicity and cultural characters in

*Gaeumannomyces graminis* var. *graminis*. *Transactions of the British Mycological Society*, **77**, 391–9.

Boddy L and Rayner ADM (1982). Population structure, inter-mycelial interactions and infection biology of *Stereum gausapatum*. *Transactions of the British Mycological Society*, **78**, 337–51.

Boeger JM, Chen RS and McDonald BA (1993). Gene flow between geographic populations of *Mycosphaerella graminicola* (anamorph *Septoria tritici*) detected with restriction fragment length polymorphism markers. *Phytopathology*, **83**, 1148–54.

Boehm EWA, Wenstrom JC, McLaughlin DJ, *et al.* (1969). Studies of inheritance of reaction to common smut in corn. *Theoretical and Applied Genetics*, **39**, 32–42.

Boehm EWA, Wenstrom JC, McLaughlin DJ, *et al.* (1992). An ultrastructural karyotype for *Puccinic graminis* f.sp. *tritici*. *Canadian Journal of Botany*, **70**, 401–13.

Boehm EWA, Ploetz RC and Kistler HC (1994). Statistical analysis of electrophoretic karyotype variation among vegetative compatibility groups of *Fusarium oxysporum* f.sp. *cubense*. *Molecular Plant–Microbe Interactions*, **7**, 196–207.

Boekhout T and van Belkum A (1997). Variability of karyotypes and RAPD types in genetically related strains of *Cryptococcus neformans*. *Current Genetics*, **32**, 203–8.

Boidin J. (1986). Intercompatibility and the species concept in the saprobic Basidiomycotina. *Mycotaxon*, **26**, 319–36.

Boisselier-Dubayle MC, Perreau-Bertrand J and Lambourdière J (1996). Genetic variation in wild populations of *Mycena rosea*. *Mycological Research*, **100**, 753–8.

Bok J-W and Griffiths AJF (1999). Transfer of *Neurospora* kalilo plasmids among species and strains by introgression. *Current Genetics*, **36**, 275–81.

Bonello P, Bruns TD and Gardes M (1998). Genetic structure of a natural population of the ectomycorrhizal fungus *Suillus pungens*. *New Phytologist*, **138**, 533–42.

Boone DN and Keitt WG (1956). *Venturia inaequalis* (Cke)Wint. VIII: Inheritance of color mutant characters. *American Journal of Botany*, **43**, 226–33

Borchardt DS, Welz HG and Geiger HH (1998). Genetic structure of *Setosphaeria turcica* populations in tropical and temperate climates. *Phytopathology*, **88**, 322–9.

Bosland PW and Williams PH (1987). An evaluation of *Fusarium oxysporum* from crucifers based on pathogenicity, isozyme polymorphism, vegetative compatibility, and geographical origin. *Canadian Journal of Botany*, **65**, 2067–73.

Bourchier RJ (1957). Variation in cultural conditions and its effects on hyphal fusion in *Corticium vellereum*. *Mycologia*, **49**, 20–8.

Brasier CM (1970). Variation in a natural population of *Schizophyllim commune*. *American Naturalist*, **104**, 191–204.

Brasier CM (1978). Mites and reproduction in *Ceratocystis ulmi* and other fungi. *Transactions of the British Mycological Society*, **70**, 81–9.

Brasier CM (1979). Dual origin of recent Dutch elm disease outbreaks in Europe. *Nature*, **281**, 78–9.

Brasier CM (1984). Inter-mycelial recognition systems in *Ceratocystsi ulmi*: their physiological properties and ecological importance. In *The ecology and physiology of the fungal mycelium* (ed. DH Jennings and ADM Rayner), pp. 451–97. Cambridge University Press.

Brasier CM (1986a). The population biology of Dutch elm disease: its principal features and some implications for other host–pathogen systems. In *Advances in Plant Pathology* (ed. DS Ingram and PH Williams), pp. 55–118. Academic Press, London.

Brasier CM (1986b). The d-factor in *Ceratocystis ulmi*—its characteristics and implications for Dutch elm disease. In *Fungal virology* (ed. KW Buck), pp. 177–208. CRC Press, Boca Raton, FL.

Brasier CM (1986c). Dutch elm disease—*Ophiostoma (Ceratocystis) ulmi*. The emergence of EAN and NAN hybrids in Europe. *Report on forestry research 1986*, p. 37. HMSO, London.

Brasier CM (1987). The dynamics of fungal speciation. In *Evolutionary biology of the fungi* (ed. ADM Rayner, CM Brasier and D. Moore), pp. 231–60. Cambridge University Press.

Brasier CM (1988). Rapid changes in genetic structure of epidemic populations of *Ophiostoma ulmi*. *Nature*, **332**, 538–42.

Brasier CM (1990). China and the origins of Dutch elm disease: an appraisal. *Plant Pathology*, **39**, 5–16.

Brasier CM (1991). Current questions in Phytophthora systematics: the role of the population approach. In *Phytophthora* (ed. JA Lucas, RC Shattock, DS Shaw and LR Cooke), pp. 104–28. Cambridge University Press.

Brasier CM (1992). Evolutionary biology of *Phytophthora*. Part I: Genetic system, sexuality and the generation of variation. *Annual Review of Phytopathology*, **30**, 153–71.

Brasier CM (1995). Episodic selection as a force in fungal evolution, with special reference to clonal speciation and hybrid introgression. *Canadian Journal of Botany*, **23** (Supplement 1), S1213–21.

Brasier CM (1997). Fungal species in practice: identifying species units in fungi. In *Species: the units of biodiversity* (ed. MF Claridge, HA Dawah and MR Wilson), pp.135–70. Chapman & Hall, London.

Brasier CM (2000). Evolutionary relationships of the Dutch elm disease fungus. *Ophiostoma novo-ulmi* to other *Ophiostoma* species investigated by restriction fragment length polymorphism analysis. *Journal of Phytopathology*, **148**, 533–9.

Brasier CM (2001). Rapid evolution of introduced plant pathogens via interspecific hybridization. *BioScience*, **51**, 123–33.

Brasier CM and Gibbs JN (1975). Highly fertile form of the aggressive strain of *Ceratocystis ulmi*. *Nature*, **257**, 128–31.

Brasier CM and Rayner ADM (1987). Whither terminology below the species level in the fungi? In *Evolutionary biology of the fungi* (ed. ADM Rayner, CM Brasier and D Moore), pp. 379–88. Cambridge University Press.

Brasier CM, Hamm PH and Hansen EM (1989). Phytophthora diseases: status of *P. gonapodyides*, *P. dreschleri*, *P. cryptogea*. *Report on Forest Research 1989*. HMSO, London.

Brasier CM, Kirk SA and Tegli S (1995). Naturally occurring non cerato-ulmin producing mutants of *Ophiostoma novo-ulmi* are pathogenic but lack aerial mycelium. *Mycological Research*, **99**, 436–40.

Brasier CM, Kirk SA, Pipe ND and Buck KW (1998). Rare interspecific hybrids in natural populations of the Dutch elm disease pathogens *Ophiostoma ulmi* and *Ophiostoma ovo-ulmi*. *Mycological Research*, **102**, 45–57.

Brasier CM, Cooke D, EL and Duncan JM (1999). Origin of a new *Phytophthora* pathogen through interspecific hybridization. *Proceedings of the National Academy of Sciences of the United States of America*, **96**, 5878–83.

Brasier CM and Kirk SA (2000). Survival of clones of NAN *Ophiostoma novo-ulmi* around its probable centre of appearance in North America. *Mycological Research*, **104**, 1322–32.

Brayford D (1990a). Variation in *Phomopsis* isolates from *Ulmus* species in the British Isles and Italy. *Mycological Research*, **94**, 691–7.

Brayford D (1990b). Vegetative incompatibility in *Phomopsis* from elm. *Mycological Research*, **94**, 45–52.

Bresinsky A, Fischer M, Meixner B and Paulus W (1987). Speciation in *Pleurotus*. *Mycologia*, **79**, 234–45.

Bridge P (1998). Numerical analysis of molecular variability: a comparison of hierarchic and non-hierarchic methods. In *Molecular variability of fungal pathogens* (ed. P Bridge, Y Couteaudier and JM Clarkson), pp. 291–308. CAB International, Wallingford.

Bridge PD, Arora DK, Reddy CA and Elander RP (1998). *Applications of PCR in mycology*. CAB International, Wallingford.

Brodie HJ (1936). The occurrence and function of oidia in the hymenomycetes. *American Journal of Botany*, **23**, 309–27.

Bronson CR, Taga M and Yoder OC (1990). Genetic control and distorted segregation of T-toxin production in field isolates of *Cochliobolus heterostrophus*. *Phytopathology*, **80**, 819–23.

Brown AHD (1975). Sample sizes required to detect linkage disequilibrium between two or three loci. *Theoretical and Population Biology*, **8**, 184–201.

Brown AHD and Weir BS (1983). Measuring genetic variability in plant populations. In *Isozymes in plant genetics and plant breeding. Part A* (ed. SD Tanksley and JJ Orton), pp. 291–308. Elsevier, London.

Brown AHD, Feldman MW and Nevo E (1980). Multilocus structure of natural populations of *Hordeum spontaneum*. *Genetics*, **96**, 523–36.

Brown JKM (1995). Recombination and selection in populations of plant pathogens. *Plant Pathology*, **44**, 279–93.

Brown JKM (1996). The choice of molecular marker methods for population genetic studies of plant pathogens. *New Phytologist*, **133**, 183–95.

Brown JKM and Wolfe MS (1990). Structure and evolution of a population of *Erysiphe graminis* f.sp. *hordei*. *Plant Pathology*, **39**, 376–90.

Brown JKM, O'Dell M, Simpson CG and Wolfe MS (1990). The use of DNA polymorphisms to test hypotheses about a population of *Erysiphe graminis* f.sp. *hordei*. *Plant Pathology*, **39**, 391–401.

Brown JKM, Jessop AC and Rezanoor HN (1991). Genetic uniformity in barley and its powdery mildew pathogen. *Proceedings of the Royal Society of London, Series B*, **246**, 83–90.

Brown TA (1987). The mitochondrial genomes of filamentous fungi. In *Gene structure in eukaryotic microbes* (ed. JR Kinghorn), pp. 141–62. IRL Press, Oxford.

Bruehl GW and Cunfer BM (1975). *Typhula* species pathogenic to wheat in the Pacific northwest. *Phytopathology*, **65**, 755–60.

Bruehl GW and Machtmes R (1980). Cultural variation within *Typhula idahoensis* and *T.ishikarensis* and the species concept. *Phytopathology*, **70**, 867–71.

Bruehl GW, Machtmes R and Kiyomoto R (1975). Taxonomic relationships among *Typhula* species as revealed by mating experiments. *Phytopathology*, **65**, 1108–1114.

Bruns TD and Palmer JD (1988). Evolution of mushroom mitochondrial DNA: *Suillus* and related genera. *Journal of Molecular Evolution*, **28**, 349–62.

Bruns TD, Palmer JD, Shumard DS, Grossman GI and Hudspeth MES (1988). Mitochondrial DNAs of *Suillus*: three-fold size changes in molecules that share a common gene order. *Current Genetics*, **13**, 49–56.

Bruns TD, Fogel R and Taylor JW (1990). Amplification and sequencing of DNA from fungal herbarium specimens. *Mycologia*, **82**, 175–84.

Bucheli E, Goutschi B and Shykoff JA (1998). Isolation and characterization of microsatellite loci in *Microbotryum violaceum*. *Molecular Ecology*, **7**, 665–6.

Buller AHR (1931). *Researches on fungi*, Vol. 4. Longmans Green, London.

Bultman TL, White JF Jr, Welch TI and Johnston J (1995). Mutualistic transfer of *Epichloë* spermatia by *Phorbia* flies. *Mycologia*, **87**, 182–9.

Burdon JJ (1987). *Phytopathology*, **87**, 664–9.

Burdon JJ (1993). The structure of pathogen populations in natural plant communities. *Annual Review of Phytopathology*, **31**, 305–23.

Burdon JJ (1997). The evolution of gene-for-gene interactions in natural pathosytems. In *The gene-for-gene relationship in plant–parasite interactions* (ed. IR Crute, EB Holub and J Burdon), pp. 245–62. CAB International, Wallingford.

Burdon JJ and Jarosz AM (1991). Host pathogen interactions in natural populations of *Linum marginale* and *Melampsora lini*. I. Patterns of resistance and racial variation in a large host population. *Evolution*, **45**, 205–17.

Burdon JJ and Jarosz AM (1992). Temporal variation in the racial structure of flax rust (*Melampsora lini*) populations growing on natural stands of wild flax (*Linum marginale*): local versus metapopulation dynamics. *Plant Pathology*, **41**, 165–79.

Burdon JJ and Roberts JK (1995). The population genetic structure of the rust fungus *Melampsora lini* as revealed by pathogenicity, isozyme and RFLP markers. *Plant Pathology*, **44**, 270–8.

Burdon JJ and Roelfs AP (1985a). Isozyme and virulence variation in asexually reproducing populations of *Puccinia graminis* and *Puccinia recondita* on wheat. *Phytopathology*, **75**, 907–13.

Burdon JJ and Roelfs AP (1985b). The effect of sexual and asexual reproduction on the isozyme structure of populations of *Puccinia graminis*. *Phytopathology*, **75**, 1068–73.

Burdon JJ, Marshall DR and Luig NH (1981). Isozyme analysis indicates that a virulent cereal rust pathogen is a somatic hybrid. *Nature*, **293**, 565–6.

Burdon JJ, Marshall DR, Luig NH and Gow DJS (1982). Isozyme studies on the origin and evolution of *Puccinia graminis* f.sp. *graminis* in Australia. *Australian Journal of Biological Science*, **35**, 231–8.

Burdon JJ, Wennström A, Elmqvist T and Kirby GC (1996). The role of race specific resistance in natural plant populations. *Oikos*, **76**, 411–16.

Burgeff H (1915). Untersuchungen über Variabilität, Sexualität und Erblichkeit bei *Phycomyces nitens* Kuntze. II. Flora N.F., 108, 353-448.

Burgeff H (1924). Untersuchungen über Sexualität und Parasitismus bei Mucorineen. I. *Goebel's Botanische Abhandlungen*, 4, 1–135.

Burges NA (1960). Dynamic equilibria in the soil. In *The ecology of soil fungi* (ed. D Parkinson and JS Waid), pp. 185–91. Liverpool University Press.

Burggraaf AJP and Beringer JE (1989). Absence of nuclear DNA synthesis in vesicular arbuscular mycorrhizal fungi during *in vitro* development. *New Phytologist*, 111, 25–33.

Burnett JH (1975). *Mycogenetics*. John Wiley, London.

Burnett JH (1976). *Fundamentals of mycology* (2nd edn). Edward Arnold, London.

Burnett JH (1983). Speciation in fungi. *Transactions of the British Mycological Society*, 81, 1–14.

Burnett JH and Boulter ME (1963). The mating systems of the fungi. II. Mating systems of the gasteromycetes *Mycocalia denudata* and *M. duriaeana*. *New Phytologist*, 62, 217–36.

Burnett JH and Evans EJ (1966). Genetical homogeneity and the stability of the mating-type factors of 'fairy rings' of *Marasmius oreades*. *Nature*, 210, 1368–9.

Burnett JH and Partington ME (1957). Spatial distribution of fungal mating type factors. *Proceedings of the Royal Physical Society of Edinburgh*, 26, 61–8.

Burt A, Carter DA, Koenig GL, White TJ and Taylor JW (1996). Molecular markers reveal cryptic sex in the human pathogen *Coccidioides immitis*. *Proceedings of the National Academy of Sciences of the United States of America*, 93, 770–3.

Burt A, Dechasio M, Koenig GL, Carter DA, White TJ and Taylor JW (1997). Molecular markers reveal differentiation among isolates of *Coccidioides immitis* from California. *Molecular Ecology*, 6, 781–6.

Byers B. and Goetsch L. (1975). Electron-microscopy observations on the meiotic karyotype of diploid and tetraploid *Saccharomyces cerevisiae*. *Proceedings of the National Academy of Sciences of the United States of America*, 72, 5056–60.

Callen EG (1940). The morphology, cytology and sexuality of the homothallic *Rhizopus sexualis* (Smith) Callen. *Annals of Botany (NS)*, 4, 791–818.

Cammack RH (1958). Studies on *Puccinia polysora* Undrw. I. The world distribution of forms of *P. polysora*. *Transactions of the British Mycological Society*, 41, 89–94.

Cammack RH (1959). Studies on *Puccinia polysora* Undrw. II. Consideration of the method of introduction of *P. polysora* into Africa. *Transactions of the British Mycological Society*, 42, 27–32.

Capretti T, Korhonen K, Mugnai L and Romagnoli C (1990). An intersterility group of *Heterobasidion annosum* specialized to *Abies alba*. *European Journal of Forest Pathology*, 20, 231–40.

Carbone J and Kohn LM (1999). A method for designing primer sets for speciation studies in filamentous ascomycetes. *Mycologia*, 91, 553–6.

Carbone I and Kohn LM (2001). A microbial population–species interface: nested cladistic and coalescent inference with multilocus data. *Molecular Ecology*, 10, 947–64.

Carle GF and Olson MV (1984). Separation of chromosomal DNA molecules from yeast by orthogonal-field-alternation gel electrophoresis. *Nucleic Acid Research*, 12, 5647–64.

Carle GF and Olson MV (1985). An electrophoretic karyotype for yeast. *Proceedings of the National Academy of Sciences of the United States of America*, 82, 3756–60.

Carlier J, Lebrun MH, Zapater MF, Dubois C and Mourichon X (1996). Genetic structure of the global structure of banana black streak fungus, *Mycosphaerella fijiensis*. *Molecular Ecology*, 5, 499–510.

Carlile MJ and Machlis L (1965a). The response of male gametes of Allomyces to the sexual hormone sirenin. *American Journal of Botany*, 52, 478–83.

Carlile MJ and Machlis L (1965b). A comparative study of the chemotaxis of the motile phases of Allomyces. *American Journal of Botany*, 52, 484–6.

Carson HL (1982). Speciation as a major reorganization of polygenic balances. In *Mechanics of speciation* (ed. C. Barigozzi), pp. 411–13. Liss, New York.

Carson HL (1985). Unification of speciation theory in plants and animals. *Systematic Botany*, 10, 380–5.

Carvalho DB, Smith ML and Anderson JB (1995). Genetic exchange between diploid and haploid mycelia of *Armillaria gallica*. *Mycological Research*, 99, 641–7.

Casselton LA and Condit A (1972). A mitochondrial mutant of *Coprinus lagopus*. *Journal of General Microbiology*, 72, 521–7.

Caten CE (1972). Heterokaryon incompatibility in imperfect species of *Aspergillus*. *Heredity*, 26, 299–312.

Caten CE and Jinks JL (1966). Heterokaryosis: its significance in wild homothallic ascomycetes and fungi imperfecti. *Transactions of the British Mycological Society*, 49, 81–93.

Caten CE and Jinks JL (1979). Quantitative genetics. In *Second International Symposium on the Genetics of Industrial Organisms* (ed. KD McDonald), pp. 93–111. Academic Press, London.

Cavalier-Smith T (1987). Eukaryote kingdoms: seven or nine? *BioSystems*, 14, 461–81.

Cavalier-Smith T (2001). What are fungi? In *Systematics and evolution*. Part A: *The mycota*, Vol. VIIA (ed. DJ McLaughlin and EG McLaughlin), pp. 3–37 Springer-Verlag, Berlin.

Chakraborty R (1980). Gene diversity analysis in nested subdivided populations. *Genetics*, **96**, 721–3.

Chamuris GP (1991). Speciation in the *Peniophora cinrea* complex. *Mycologia*, **83**, 736–42.

Chase TE and Ullrich RC (1988). *Heterobasidion annosum*, root and butt-rot of trees. In *Genetics of plant pathogenic fungi*, Vol. 6 (ed. GS Sidhu), pp. 501–10. Academic Press, London.

Chase TE and Ullrich RC (1990a). Genetic basis of biological species in *Heterobasidion annosum*: Mendelian determinants. *Mycologia*, **82**, 67–72.

Chase TE and Ullrich RC (1990b). Five genes determining intersterility in *Heterobasidion annosum*. *Mycologia*, **82**, 73–81.

Chen D, Zeigler RS, Leung H and Nelson RJ (1995). Population structure of *Pyricularia grisea* at two screening sites in the Phillipines. *Phytopathology*, **85**, 1011–20.

Chen R-S, Boeger JM and McDonald BA (1994). Genetic stability in a population of a plant pathogenic fungus over time. *Molecular Ecology*, 3, 209–18.

Chin KM (1987). A simple model of selection for fungicide resistance in plant pathogen populations. *Phytopathology*, **77**, 666–9.

Chiu SW (1993). Evidence for a haploid life-cycle in *Volvariella volvacea* from microsphectrophotometric measurements and observations of nuclear behaviour. *Mycological Research*, **97**, 1481–94.

Christ BJ, Person CO and Pope DD (1987). The genetic determination of variation in pathogenicity. In *Populations of plant pathogens: their dynamics and genetics* (ed. MS Wolfe and CE Caten), pp. 7–19. Blackwell Scientific, Oxford.

Christen AA and Bruehl GW (1979). Hybridisation of *Typhula ishikarensis* and *T. idahoensis*. *Phytopathology*, **69**, 263–6.

Christensen MJ, Simpson WR and Al Samarrai T (2000). Infection of tall fescue and perennial rye grass plants by combinations of different *Neotyphodium* endophytes. *Mycological Research*, **104**, 974–8.

Chu G (1989). Pulsed-field electrophoresis in contour-clamped homogenous electric fields for the resolution of DNA by size or topology. *Electrophoresis*, **10**, 290–5.

Chung K-R and Schardl CL (1997). Vegetative compatibility between and within *Epichloë* species. *Mycologia*, **89**, 558–65.

Cisar CR and TeBeest DO (1999). Mating system of the filamentous ascomycete. *Glomerella cingulata*. *Current Genetics*, **35**, 127–33.

Cisar CR, TeBeest DO and Spiegel FW (1994). Sequence similarity of mating type idiomorphs: a method which detects similarity among the Sordariaceae fails to detect similar sequences in other filamentous ascomycetes. *Mycologia*, **86**, 540–6.

Claridge MF, Dawah HA and Wilson MR (1997). *Species: the units of biodiversity*. Chapman & Hall, London.

Clarke BC and Beaumont MA (1992). Density and frequency dependence: a genetical view. In *Genes in ecology* (ed. RJ Berry, TJ Crawford and MG Hewitt), pp. 353–64. Blackwell Science, Oxford.

Clarke DD (1997). The genetic structure of natural pathosystems. In *The gene-for-gene relationship in plant–parasite reactions* (ed. IR Crute, EB Holub and JJ Burdon), pp. 231–43. CAB International, Wallingford.

Clutterbuck AJ and Roper JA (1966). A direct determination of nuclear distribution in heterokaryons of *Aspergillus nidulans*. *Genetical Research*, **7**, 185–94.

Coates D and Rayner ADM (1985a). Genetic control and variation in expression of the 'bow tie' reaction between homokaryons of *Stereum hirsutum*. *Transactions of the British Mycological Society*, **84**, 191–205.

Coates D and Rayner ADM (1985b). Heterokaryon-homokaryon interactions in *Stereum hirsutum*. *Transactions of the British Mycological Society*, **84**, 637–45.

Coates D, Rayner ADM and Todd NK (1981). Mating behaviour, mycelial antagonism and the establishment of individuals in *Stereum hirsutum*. *Transactions of the British Mycological Society*, **76**, 41–51.

Codón AC, Benítez T and Korhola M (1997). Chromosomal reorganization during meiosis of *Saccharomyces cerevisiae*. *Current Genetics*, **32**, 247–59.

Coetzee MPA, Wingfield BD, Harrington TC, Steimel J, Coutinho TT and Wingfield MJ (2001). The root rot fungus *Armillaria mellea* introduced into South Africa by early Dutch settlers. *Molecular Ecology*, **10**, 387–96.

Collins RA and Saville BJ (1990). Independent transfer of mitochondrial chromosomes and plasmids during unstable vegetative fusions in *Neurospora*. *Nature*, **345**, 177–9.

Cong Y-S, Yarrow D, Li Y-Y and Fukuhara H (1994). Linear DNA plasmids from *Pichia etchellsii*, *Debaryomyces hansenii* and *Wingea robetsiae*. *Microbiology*, **140**, 1327–35.

Connolly V and Simchen G (1968). Linkage to the incompatibility factors and maintenance of genetic variation in selection lines of *Schizophyllum commune*. *Heredity*, **23**, 387–402.

Cooke DEL, Drenth A, Duncan JM, Wagels G and Brasier CM (2000). A molecular phylogeny of *Phytophthora* and related Oomycetes. *Fungal Genetics and Biology*, **30**, 17–32.

Cooke RC and Rayner ADM (1984). *Ecology of saprophytic fungi*. Longmans, London.

Correll JC (1991). The relationship between formae specialis, races, and vegetative compatibility groups in *Fusarium oxysporum*. *Phytopathology*, **81**, 1061–4.

Correll JC, Puhalla JE and Schneider RW (1986). Identification of *Fusarium oxysporum* f.sp. *apii* on the basis of colony size, virulence and vegetative compatibility. *Phytopathology*, **76**, 396–400.

Correll JC, Klittich CJR and Leslie JF (1989). Heterokaryon self-incompatibility in *Gibberella fujikuroi* (*Fusarium moniliforme*). *Mycological Research*, **93**, 21–7.

Couch BC and Kohn LM (2000). Clonal spread of *Sclerotium cepivorum* in onion production with evidence of past recombination events. *Phytopathology*, **90**, 514–21.

Cove DJ (1976a). Chlorate toxicity in *Aspergillus nidulans*: studies of mutants altered in nitrate assimiliation. *Molecular and General Genetics*, **146**, 147–59.

Cove DJ (1976b). Chlorate toxicity in *Aspergillus nidulans*: the selection and characterization of chlorate resistant mutants. *Heredity*, **36**, 191–203.

Cox BS and Parry EM (1967). Natural selection in an artificial population of *Sordaria fimicola*. *New Phytologist*, **66**, 643–52.

Crandall KA and Templeton AR (1996). Applications of intraspecific phylogenies. In *New uses for new phylogenies* (ed. PH Harvey and S Nee), pp. 81–99. Oxford University Press.

Crawford MS, Chumley EG, Weaver CG and Valent B (1986). Characterization of the heterokaryotic and vegetative diploid phases of *Magnaporthe grisea*. *Genetics*, **114**, 1111–29.

Croft JH and Jinks JL (1977). Aspects of the population genetics of *Aspergillus nidulans*. In *Genetics and physiology* of Aspergillus (ed. JE Smith and JA Pateman), pp. 339–60. Academic Press, London

Croft JH and Dales RBG (1984). Mycelial interactions and mitochondrial interactions in *Aspergillus nidulans*. In *The ecology and physiology of the fungal mycelium* (ed. DH Jennings and ADM Rayner), pp. 433–50. Cambridge University Press.

Croft JH and Simchen GW (1965). Natural variation among monokaryons of *Collybia velutipes*. *American Naturalist*, **49**, 451–62.

Crow JF and Kimura M (1966). Evolution in sexual and asexual populations. *American Naturalist*, **99**, 439–50.

Crowe LK (1960). The exchange of genes between nuclei of a dikaryon. *Heredity*, **15**, 397–405.

Culberson CF and Elix JA (1989). Lichen substances. In *Methods in plant biochemistry* (ed. PM Dey and JB Harborne), pp. 509–35. Academic Press, London.

Cummins GB and Hiatsuka Y (1992). *Illustrated genera of rust fungi*. APS Press, St Paul, MN.

Cutter VM (1942a). Nuclear behaviour in the Mucorales. I. The *Mucor* pattern. *Bulletin of the Torrey Botanical Club*, **69**, 480–508.

Cutter VM (1942b). Nuclear behaviour in the Mucorales. II. The *Rhizopus*, *Phycomyces* and *Sporodinia* patterns. *Bulletin of the Torrey Botanical Club*, **69**, 592–616.

Daboussi MJ (1996). Fungal transposable elements: generators of diversity and genetic tools. *Journal of Genetics*, **75**, 325–39.

Daboussi MJ and Langin T (1995). Transposable elements in the fungal plant pathogen *Fusarium oxysporum*. *Genetica*, **93**, 49–59.

Dahlberg A, Jonsson L and Nylund J-E (1997). Species diversity and distribution of biomass above and below ground among ectomycorrhizal fungi in an old-growth Norway spruce forest in south Sweden. *Canadian Journal of Botany*, **75**, 1323–35.

Dale MRT (2000). *Spatial pattern analyisis in plant ecology*. Cambridge University Press.

Dalmacio SC, McKenzie DR and Nelson RR (1979). Heritability of differences in virulence between races 2 and 3 of *Cochliobolus carbonum*. *Phillippine Phytopathology*, **15**, 47–50.

Darnall DW and Klotz IM (1972). Protein subunits: a table (revised edition). *Archives of Biochemistry and Biophysics*, **149**, 1–14.

Darwin C (1859). *On the origin of species by means of natural selection*. John Murray, London.

Davet P and Rouxel F (2000). *Detection and isolation of soil fungi*. Science Publishers, Enfield, NH.

Davis JI (1999). Monophyly, populations and species. In *Molecular sytematics and plant evolution* (ed. PM Hollingsworth, RM Bateman and RJ Gornall), pp. 139–70. Taylor and Francis, London.

Davis RH (1959). Asexual selection in *Neurospora crassa*. *Genetics*, **44**, 1291–1308.

Davis TM, Yu H, Haigis KM and McGowan PJ (1995). Template mixing—a method of enhancing detection and interpretation of codominant RAPD markers. *Theoretical and Applied Genetics*, **91**, 582–8.

Day PR (1974a). *Genetics of host–parasite interaction*. WH Freeman, San Francisco, CA.

Day PR (1974b). *Botanical Gazette*, **141**, 313–20.

Deacon JW (1997). *Modern mycology*, Blackwell Science, Oxford.

Deacon JW and Donaldson SP (1993). Molecular recognition in the homing responses of zoosporic fungi, with special reference to *Pythium* and *Phytophthora*. *Mycological Research*, **97**, 1153–71.

Deahl KL, Goth RW, Young R, Sinden SL and Gallegly ME (1990). Occurrence of the A2 mating type of *Phytophthora infestans* in the United States and Canada. *Phytopathology*, **80**, 67 (abstract).

de Serres FJ and Kolmark HG (1973). A direct method for forward mutation rates in *Neurospora crassa. Nature*, **182**, 1249–50.

Detloff P, Sieber J and Petes TD (1991). Repair of specific base pair mismatches formed during meiotic recombination in the yeast *Saccharomyces cerevisiae. Molecular and Cellular Biology*, **11**, 737–45.

De Vallavieille C and Erselius LJ (1984). Variation in protein profiles of *Phytophthora*: survey of a composite population of three species on citrus. *Transactions of the British Mycological Society*, **78**, 227–38.

Dickinson CH (1976). Fungi on the aerial surfaces of higher plants. In *Microbiology of aerial plant surfaces* (ed. CH Dickinson and TF Preece), pp. 293–324. Academic Press, London.

Dickman A and Cook S (1989). Fire and forest in a mountain hemlock forest. *Canadian Journal of Botany*, **67**, 2008–16.

Dodd JL and Hooker AL (1990). Previously undescribed pathotype of *Bipolaris zeicola* on corn. *Plant Disease*, **74**, 530 (abstract).

Dodge BO (1942). Heterocaryotic vigour in *Neurospora. Bulletin of the Torrey Botanical Club*, **69**, 75–91.

Drake JW (1991). A rate of spontaneous mutation common to DNA-based microbes. *Proceedings of the National Academy of Sciences of the United States of America*, **88**, 7160–4.

Drenth A, Tas ICQ and Gowers F (1994). DNA fingerprinting uncovers new sexually reproducing populations of *Phytophthora infestans. European Journal of Phytopathology*, **100**, 97–107.

Duncan EC and Macdonald JA (1967). Micro-evolution in *Auricularia auricula. Mycologia*, **59**, 808–13.

Dutta SK (1976). DNA homologies among heterothallic species of *Neurospora. Mycologia*, **68**, 388–401.

Earl AJ, Turner G, Croft JH *et al.*(1981). High frequency transfer of species specific mitochondrial DNA sequences between members of the Aspergillaceae. *Current Genetics*, **3**, 221–8.

Egger KN (1995). Molecular analysis of ectomycorrhizal fungal communities. *Canadian Journal of Botany*, **73** (Supplement 1), S1415–22.

Eigel A and Feldmann H (1982). Ty1 and delta elements occur adjacent to several tRNA genes in yeast. *EMBO Journal*, **1**, 1245–50.

Ekman S (1999). PCR optimization and troubleshooting, with special reference to the amplification of ribosomal DNA in lichenized fungi. *Lichenologist*, **31**, 517–31.

Elliott CG (1994). *Reproduction in Fungi: genetical and physiological aspects*. Chapman & Hall, London.

Elias KS and Schneider RW (1991). Vegetative compatibility groups in *Fusarium oxysporum* f.sp. *lycopersici Phytopathology*, **81**, 159–62.

Emara YA and Sidhu G (1974). Polygenic inheritance of aggressiveness in *Ustilago hordei. Heredity*, **32**, 219–24.

Emerson BC, Paradis E and Thébaud C (2001). Revealing the demographic histories of species using DNA sequences. *Trends in Evolution and Ecology*, **16**, 707–18.

Emerson R (1941) An experimental study of the life cycles and taxonomy of Allomyces. *Lloydia*, **4**, 77–144.

Emerson R and Wilson CM (1949). The significance of meiosis in *Allomyces. Science*, **110**, 86–8.

Emerson R and Wilson CM (1954). Interspecific hybrids and cytogenetics and cytotaxonomy of *Euallomyces. Mycologia*, **46**, 393–434.

Emerson S (1952). Biochemical models of heterosis in *Neurospora*. In *Heterosis* (ed. JW Gowen), pp. 199–217. Iowa State University Press, Ames, IA.

Endler JA (1986). *Natural selection in the wild*. Princeton University Press.

Ennos RA and McConnell KC (1995). Using genetic markers to investigate natural selection in fungal populations. *Canadian Journal of Botany*, **73** (Supplement 1), S302–10.

Ennos RA and Swales KW (1987). Estimation of the mating system in a fungal pathogen *Crumenulopsis soraria* (Karst.) Groves using isozyme markers. *Heredity*, **59**, 423–30.

Ennos RA and Swales KW (1991). Genetic variability and population structure in the canker pathogen *Crumenulopsis soraria. Mycological Research*, **95**, 521–5.

Eriksson J (1894). Über die Specialisirung des Parasitismus bei den Getreide-rostpilzen. *Bericht der Deutschen Botanischen Gesellschaft*, **12**, 292.

Erselius LJ and de Vallavieille C (1984). Variation in protein profiles of *Phytophthora*: comparison of six species. *Transactions of the British Mycological Society*, **83**, 463–72.

Eshed N and Wahl I (1970). Host ranges and interrelationships of *Erysiphe graminis hordei*, *E. graminis tritici* and *E. graminis avenae. Phytopathology*, **60**, 628–34.

Eshed N and Wahl I (1975). Role of wild grasses in epidemics of powdery mildews on small grains in Israel. *Phytopathology*, **65**, 57–63.

Eslava AP, Alvarez MA and Delbrück M (1975). Meiosis in *Phycomyces. Proceedings of the National Academy of Sciences of the United States of America*, **72**, 4076–80.

Evans HJ (1959). Nuclear behaviour in the cultivated mushroom. *Chromosoma*, **10**, 115–35.

Excoffier L, Smouse PE and Quattro JM (1992). Analysis of molecular variance inferred from metric distances among DNA haplotypes: application to human mitochondrial DNA restriction data. *Genetics*, **131**, 479–91.

Falahati-Rastegar M, Manners JG and Smart J (1981). Effects of temperature and inoculum density on competition between races of *Puccinia hordei*. *Transactions of the British Mycological Society*, **77**, 359–68.

Falconer DS and Mackay TFC (1996). *Introduction to quantitative genetics*. Addison Wesley Longman, Edinburgh.

Federici BA (1985). Experimental systematics. In *The genus Coelomomyces* (ed. JN Couch and CE Bland), pp. 299–320. Academic Press, New York.

Feldman H (1999). The yeast *Saccharomyces cerevisiae*: insights from the first complete eukaryotic genome. In *Molecular fungal biology* (ed. RP Oliver and M Schweizer), pp. 78–134. Cambridge University Press.

Felsenstein J (1978). Cases in which parsimony or compatibility methods will be positively misleading. *Systematic Zoology*, **27**, 401–10.

Felsenstein J (1981). Evolutionary trees from DNA sequences: a maximum likelihood approach. *Journal of Molecular Evolution*, **17**, 368–76.

Felsenstein J (1985). Confidence limits on phylogenies: an approach utilizing the bootstrap. *Evolution*, **39**, 783–91.

Felstenstein J (1993). *PHYLIP (Phylogeny Inference Package)*. Department of Genetics, University of Washington, Seattle, WA.

Ferrandino FJ (1998). Past nonrandomness and aggregation to spatial correlation: 2DCORR, a new approach for discrete data. *Phytopathology*, **88**, 84–91.

Fischer GW and Holton CS (1957). *The biology and control of the smut fungi*. Ronald Press, New York.

Fisher MC, Koenig GL, White TJ and Taylor JW (2002). Molecular and phenotypic description of *Coccidioides posadasii* sp.nov., previously recognized as the non-Californian population of *Coccidioides immitis*. *Mycologia*, **94**, 85–98.

Fisher RA (1930). *The genetical theory of natural selection*. Clarendon Press, Oxford.

Fitch WM and Margoliash E (1967). Construction of phylogenetic trees. *Science*, **155**, 279–84.

Flor HH (1942). Inheritance of pathogenicity in *Melampsora lini*. *Phytopathology*, **32**, 653–69.

Flor HH (1953). Epidemiology of flax rust in the north central states. *Phytopathology*, **43**, 624–8.

Flor HH (1956). The complementary genic systems in flax and flax rust. *Advances in Genetics*, **8**, 267–382.

Flossdorf J (1983). A rapid method for the determination of the base composition of bacterial DNA. *Journal of Microbiological Methods*, **1**, 305–11.

Ford EB (1965). *Genetic polymorphism*. Faber & Faber, London.

Fordyce C and Green RJ (1964). Mechanism of variation in *Verticillium albo-atrum*. *Phytopathology*, **54**, 795–8.

Förster H, Kinscherf TG, Leong SA and Maxwell DP (1989). Restriction fragment length polymorphisms of the mitochondrial DNA of *Phytophthora megasperma* isolated from soybean, alfalfa, and fruit trees. *Canadian Journal of Botany*, **67**, 529–37.

Förster H, Oudemans P and Coffey MD (1990). Mitochondrial and nuclear DNA diversity within six species of *Phytophthora*. *Experimental Mycology*, **14**, 18–31.

Francis DM and St Clair DA (1993). Outcrossing in the homothallic oomycete *Pythium ultimum* detected with molecular markers. *Current Genetics*, **24**, 100–6.

Frank SA (1991). Ecological and genetic models of host–pathogen coevolution. *Heredity*, **67**, 73–83.

Frank SA (1993). Coevolutionary genetics of plants and pathogens. *Evolutionary Ecology*, **7**, 45–75.

Franke RG (1973). Electrophoresis and the taxonomy of saprophytic fungi. *Bulletin of the Torrey Botanical Club*, **100**, 287–96.

Franken P and Gianninazzi-Pearson V (1996). Construction of genomic phase libraries of the arbuscular mycorrhizal fungi *Glomus mosseae* and *Scutellospora castanea* and isolation of ribosomal genes. *Mycorrhiza*, **6**, 167–73.

Frankland JC, Poskitt JM and Howard DM (1995). Spatial development of populations of a decomposer fungus, *Mycena galopus*. *Canadian Journal of Botany*, **73** (Supplement 2), S1339–1406.

Franklin AL, Filion WG and Anderson JB (1983). Determination of DNA content in fungi using mithramycin; vegetative diploidy in *Armillaria mellea* confirmed. *Canadian Journal of Microbiology*, **29**, 1179–83.

Fries N (1948). Viability and resistance of spontaneous mutations in *Ophiostoma* representing different degrees of heterotrophy. *Physiologia Plantarum*, **1**, 330–41.

Fries N and Aschan K (1952). The physiological heterogeneity of the dikaryotic mycelium of *Polyporus abietinus* investigated with the aid of microsurgical technique. *Svensk Botaniske Tidskrift*, **46**, 429–45.

Frisvad JC, Bridge PD and Arora DK (1998). *Chemical fungal taxonomy*. Marcel Dekker, New York.

Furnier GR, Anderson NA and Skilling D (1984). Genetic variation in *Gremmeniella abietina*. In *Scleroderris canker of conifers* (ed. PD Manion), pp. 59–67. Martinus Nijhoff–Dr W Junk Publishers, The Hague.

Fry WE, Goodwin SB, Matuszak JM, Spielman LJ, Milgroom MG and Drenth A (1992). Population genetics and intercontinental migrations of *Phytophthora infestans*. *Annual Review of Phytopatholgy*, **30**, 107–29.

Fullerton RA and Nielsen J (1974). *Canadian Journal of Plant Science*, **54**, 253–7.

Gams W (1992). In *Fungi in vegetation science* (ed. W Winterhoff), p. 183. Kluwer Academic, Dordrecht.

Garbelotto M, Ratcliff A, Bruns TD, Cobb FW and Otrosina WJ (1996). Use of taxon-specific competitive priming PCR to study host specificity, hybridization, and intergroup gene flow in intersterility groups of *Heterobasidion annosum*. *Phytopathology,* **86**, 543–51.

Garbelotto M, Lee HK, Slaughter G, Popenuck T, Cobb FW and Bruns TD (1997). Heterokaryosis is not required for virulence of *Heterobasidion annosum*. *Mycologia,* **89**, 92–102.

Garber RC and Yoder OC (1983). Isolation of DNA from filamentous fungi and separation into nuclear, mitochondrial, ribosomal, and plasmid components. *Analytical Biochemistry,* **135**, 416–22.

Garber RC and Yoder OC (1984). Mitochondrial DNA of the filamentous ascomycete *Cochliobolus heterostrophus*. Characterization of the mitochondrial chromosome and population genetics of a restriction enzyme polymorphism. *Current Genetics,* **8**, 621–8.

Gardes M and Bruns TD (1992). ITS primers with enhanced specificity for basidiomycetes—application to identification of mycorrhizae and rusts. *Molecular Ecology,* **2**, 113–18.

Gardes M and Bruns TD (1996). Community structure of ectomycorrhizal fungi in a *Pinus muricata* forest: above and below ground views. *Canadian Journal of Botany,* **74**, 1572–83.

Garnjobst L (1953). Genetic control of heterokaryosis in *Neurospora crassa*. *American Journal of Botany,* **40**, 607–14.

Garnjobst L (1955). Further analysis of heterokaryosis in *Neurospora crassa*. *American Journal of Botany,* **42**, 444–8.

Garnjobst L and Wilson JF (1956). Heterocaryosis and protoplasmic incompatibility in *Neurospora crassa*. *Proceedings of the National Academy of Sciences of the United States of America,* **42**, 613–18.

Gaudet DA, Gusse J and Larsche A (1998). Origins and inheritance of chromosome length polymorphisms in the barley covered smut fungus *Ustilago hordei*. *Current Genetics,* **33**, 216–24.

Gauger W (1965). The germination of zygospores of *Mucor hiemalis*. *Mycologia,* **57**, 634–41.

Gauger W (1966). Further studies on sexuality in azygosporic strains of *Mucor hiemalis*. *Transactions of the British Mycological Society,* **64**, 113–18.

Gauger W (1977). Meiotic gene segregation in *Rhizopus stolonifer*. *Journal of General Microbiology,* **101**, 211–17.

Geiser DM, Arnold ML and Timberlake WE (1994). Sexual origins of British *Aspergillus nidulans* isolates. *Proceedings of the National Academy of Sciences of the United States of America,* **91**, 2349–52.

Geiser DM, Arnold ML and Timberlake NE (1996). Wild chromosomal variants in *Aspergillus nidulans*. *Current Genetics,* **29**, 293–300.

Geiser DM, Frisvad JC and Taylor JW (1998). Evolutionary relationships in *Aspergillus* section *Fumigati* inferred from partial $\beta$-tubulin and hydrophobin DNA sequences. *Mycologia,* **90**, 831–45.

Genovesi AD and Magill DW (1976). Heterokaryosis and parasexuality in *Pyricularia oryzae*. *Canadian Journal of Microbiology,* **22**, 531–6.

Gillespie JH (1991). *The causes of molecular evolution.* Oxford University Press.

Giovannetti M, Azzolini D and Citernesi AS (1999). Anastomosis formation and nuclear and protoplasmic exchange in arbuscular mycorrhizal fungi. *Applied and Evironmental Microbiology,* **65**, 5571–5.

Giovannetti M, Fortuna P, Citernesi AS, Morini S and Nuti MP (2001). The occurrence of anastomosis formation and nuclear exchange in intact arbuscular mycorrhizal networks. *New Phytologist,* **151**, 717–24.

Giraud T, Fortini T, Lewis C *et al.* (1997). RFLP markers show genetic recombination in *Botryotinia fuckeliana* (*Botrytis cinerea*) and transposable elements reveal two sympatric species. *Molecular Biology and Evolution,* **14**, 1977–99.

Giraud T, Fournier E, Vautin D *et al.* (2002). Isolation of 44 polymorphic microsatellite loci in three host races of the phytopathogenis fungus, *Microbotryum violaceuam*. *Molecular Ecology Notes,* **2**, 142–6.

Glass NI and Kuldau GA (1992). Mating type and vegetative incompatibility in filamentous Ascomycetes. *Annual Review of Phytopathology,* **30**, 201–4.

Glass NI, Metzenberg RL and Raju NB (1990). Homothallic Sordariaceae from nature: the absence of strains containing only the *a* mating type sequence. *Experimental Mycology,* **14**, 274–89.

Glass NI, Vollmer SJ, Staben C, Grotelueschen J, Metzenberg RL and Yanofsky C (1988). DNAs of the two mating-type alleles of *Neurospora crassa* are highly dissimilar. *Science,* **241**, 570–3.

Goddard MV (1976). Production of a new race, 105 E 137, of *Puccinia striiformis* in glass-house experiments. *Transaction of the British Mycological Society,* **67**, 395–8.

Goggioli V, Capretti P, Hamelin RC and Vendramin GG (1998). Isozyme and RAPD polymorphisms in *Heterobasidion annosum* in Italy. *European Journal of Forest Pathology,* **28**, 63–74.

Golin JE and Esposito MS (1977). Evidence for joint genic control of spontaneous mutation and genetic recombination during mitosis in Saccharomyces. *Molecular and General Genetics*, **150**, 127–35.

Gonzalez M, Rodriguez R, Zavala ME *et al.* (1998). Characterization of Mexican isolates of *Colletotrichum lindemuthianum* by using differential cultivars and molecular markers. *Phytopathology*, **88**, 292–9.

Goodwin PH and Annis SL (1991). Rapid identification of genetic variation and pathotype of *Leptosphaeria maculans* by random amplified polymorphic DNA assay. *Applied Environmental Microbiology*, **57**, 2482–6.

Goodwin SB (1997). The population genetics of *Phytophthora*. *Phytopathology*, **87**, 462–73.

Goodwin SB, Cohen PA and Fry WE (1994). Panglobal distribution of a single clonal lineage of the Irish Potato Famine fungus. *Proceedings of the National Academy of Sciences of the United States of America*, **91**, 11591–5.

Goodwin SB, Saghai-Maroof MA, Allard RW and Webster RK (1993). Isozyme variation within and among populations of *Rhynchosporium secalis* in Europe, Australia and the United States. *Mycological Research*, **97**, 49–58.

Gordon SA and Petersen RH (1991). Mating systems in *Marasmius*. *Mycotaxon*, **41**, 371–86.

Gordon SA, Desjardin DE and Peterson RH (1994). Mating sytems in *Marasmius*: additional evidence to support sectional consistency. *Mycological Research*, **98**, 200–4.

Gordon TR and Martyn RD (1997). The evolutionary biology of *Fusarium oxysporum*. *Annual Review of Phytopathology*, **35**, 111–28.

Gordon TR, Okamoto D and Milgroom MG (1992). The structure and interrelationship of fungal populations in native and cultivated soils. *Molecular Ecology*, **1**, 241–9.

Gordon WL (1954). Geographical distribution of mating types in *Gibberella cyanogena* (Desm.) Sacc. *Nature*, **173**, 505.

Gottlieb DJC and von Borstel RC (1976). Mutations in *Saccharomyces cerevisiae, mut 1–1, mut 1–2* and *mut 1–3*. *Genetics*, **83**, 653–66.

Gräser Y, Volovsek M, Arrington J *et al.* (1996). Molecular markers reveal that population structure of the human pathogen *Candida albicans* exhibits both clonality and recombination. *Proceedings of the National Academy of the Sciences of the United States of America*, **93**, 12473–7.

Gray J (1985). The microfossil record of early land plants: advances in understanding of early terrestrialization, 1970–1984. *Philosophical Transactions of the Royal Society of London, Series B*, **309**, 167–95.

Green GJ (1971). Hybridization between *Puccinia graminis tritici* and *Puccinia graminis secalis* and its evolutionary implications. *Canadian Journal of Botany*, **49**, 2089–95.

Greene DR, Koenig G, Fisher MC and Taylor JW (2000). Soil isolation and molecular identification of *Coccidioides immitis*. *Mycologia*, **92**, 406–10.

Gregory PH (1944). The multiple infection transformation. *Annals of Applied Biology*, **35**, 412–17.

Gregory PH (1966). The fungus spore: what it is and what it does. In *The fungus spore* (ed. MF Madelin), pp. 1–14. Butterworths, London.

Gregory PH (1973). *Microbiology of the atmosphere*, Leonard Hill, Aylesbury.

Greig-Smith P (1952). The use of random and contiguous quadrats in the study of the structure of plant communities. *Annals of Botany*, **16**, 293–316.

Griffiths AJF (1992). Fungal senescence. *Annual Reviews of Genetics*, **26**, 351–72.

Griffiths AJF and Yang X (1993). Senescence in natural populations of *Neurospora intermedia*. *Mycological Research*, **97**, 1379–87.

Griffiths AJF, Kraus SR, Barton R, Court DA, Myers CJ and Bertrand H (1990). Heterokaryotic transmission of senescence plasmid DNA in *Neurospora*. *Current Genetics*, **17**, 139–45.

Griffiths AJF, Yang X, Debets FJ and Wei Y (1995). Plasmids in natural populations of *Neurospora*. *Canadian Journal of Botany*, **73** (Supplement 1), S186–92.

Griffiths RC and Marjoram P (1996). Ancestral inference from samples of DNA sequences with recombination. *Journal of Computational Biology*, **3**, 479–502.

Grigorova R and Norris JR (1990). *Methods in microbiology*. Academic Press, New York.

Grindle M (1963a). Heterokaryon compatibility of unrelated strains in the *Aspergillus nidulans* group. *Heredity*, **18**, 191–204.

Grindle M (1963b). Heterokaryon compatibility of closely related strains in the *Aspergillus nidulans* group. *Heredity*, **18**, 397–405.

Groppe K, Sanders I, Wiemken A and Boller T (1995). A microsatellite marker for studying the ecology and diversity of fungal endophytes (*Epichlöe* spp.) in grasses. *Applied and Environmental Microbiology*, **61**, 3943–9.

Groth JV and Barrett JA (1980). Estimating parasite fitness. A reply. *Phytopathology*, **70**, 840–2.

Groth JV and Roelfs AP (1982). Effect of sexual and asexual reproduction on race abundance in cereal rust fungus populations. *Phytopathology*, **72**, 1503–7.

Groth JV and Roelfs AP (1986). Analysis of virulence diversity in populations of plant pathogens. In *Populations of plant pathogens: their dynamics and genetics* (ed. MS Wolfe and CE Caten), pp. 63–74. Blackwell Scientific, Oxford.

Groth JV and Roelfs AP (1987). The concept and measurement of phenotypic diversity in *Puccinia graminis* f.sp. *tritici. Phytopathology,* **77**, 1395–9.

Gryta H, Debaud J-C, Efforse A, Gay G and Marmeisse R (1995). Fine scale structure of populations of the ectomycorrhizal fungus *Hebeloma cylindrosporum* in coastal sand dune forest ecosystems. *Molecular Ecology,* **4**, 353–64.

Guidot A, Debaud J-C and Marmeise R (2001). Correspondence between genetic diversity and spatial distribution of above- and below-ground populations of the ectomycorrhizal fungus *Hebeloma cylindrosporum. Molecular Ecology,* **10**, 1121–31.

Guillaumin JJ (1998). Genetic control of somatic incompatibility in *Armillaria ostoyae*—a preliminary analysis. In *Ninth conference on root and butt-rot of forest trees,* vol. 89 (ed. C Delatour, JJ Guillaumin, Lung-Escarmet and B Marçais), pp. 31–8. INRA France.

Guillaumin JJ, Anderson JB and Korhonen K (1991). Life cycle, interfertility and biological species. In *Armillaria root disease,* Vol. 691 (ed. GC Shaw and GA Kile), pp. 10–20. US Department of Agriculture, Washington, DC.

Guillaumin JJ, Anderson JB, Legrand P, Gharari S and Berthelay S (1996). A comparison of different methods for the identification of genets of *Armillaria* spp. *New Phytologist,* **133**, 333–43.

Guillaumin JJ, Lung B, Romagnesi H *et al.* (1985). Systématique des Armillaires du group Mellea. Conséquences phytopathologiques. *European Journal of Forest Pathology,* **15**, 268–77.

Guttman DS and Dykhuizen DE (1994). Clonal divergence in *Escherichia coli* as a result of recombination, not mutation. *Science,* **266**, 1380–3.

Gwynne-Vaughan HCI and Williamson HS (1930) Contributions to the study of *Humaria granulata. Annals of Botany,* **49**, 127–45.

Haigh J (1978). The accumulation of deleterious genes in a population—Muller's ratchet. *Theoretical Population Biology,* **14**, 251–67.

Hale ME (1974). *The biology of lichens.* Edward Arnold, London.

Hall BG (2001). Phylogenetic trees made easy: a how-to manual for molecular biologists. Sinauer Associates, Sunderland, MA.

Hallenberg N (1984). A taxonomic analysis of the *Sistotrema brinkmannii* complex (Corticiaceae, Basidiomycetes). *Mycotaxon,* **21**, 389–411.

Hallenberg N (1995). Dispersal abilities and distributional patterns in Aphyllophorales, with emphasis on corticoid fungi. *Symbolae Botanicae Upsalienses,* **30**, 95–100.

Hallenberg N and Larsson E (1992). Mating biology in *Peniophora cinerea* (Basidiomycetes). *Canadian Journal of Botany,* **70**, 1758–64.

Hamada N (1989). The effect of various culture conditions on depside production by an isolated lichen mycobiont. *Bryologist,* **92**, 310–13.

Hamelin RC, Lecours N, Hannson P, Hellgren M and Laflamme G (1996). Genetic differentiation within the European race of *Gremmeniella abietina. Mycological Research,* **100**, 49–56.

Hamelin RC, Lecours N and Laflamme G (1998). Molecular evidence of distinct introductions of the European race of *Gremmeniella abietina* into North America. *Phytopathology,* **88**, 582–8.

Hamelin RC and Rail J (1997). Phylogeny of *Gremmeniella* species based on sequences of the 5.8S rDNA and internal spacer region. *Canadian Journal of Botany,* **75**, 693–8.

Hamer JE (1991). Molecular probes for rice blast disease. *Science,* **252**, 632–3.

Hames BD and Rickwood D (1990). *Gel electrophoresis of proteins: a practical approach.* IRL Press, Oxford.

Hanlin RT (1990). *Illustrated genera of Ascomycetes: I.* APS Press, St Paul, MN.

Hanlin RT (1998a). *Illustrated genera of Ascomycetes: II.* APS Press, St Paul, MN.

Hanlin RT (1998b). *Combined keys to illustrated genera of Ascomycetes I and II.* APS Press, St Paul, MN.

Hansen EM (1987). Speciation in plant pathogenic fungi. The influence of agricultural practice. *Canadian Journal of Plant Pathology,* **9**, 403–10.

Hansen EM and Hamelin RC (1999). Population structure of basidiomycetes. In *Structure and dynamics of fungal populations* (ed. J. T. Worrall), pp. 251–81. Kluwer Academic, Dordrecht.

Hansen EM and Maxwell DP (1991). Species of the *Phytophthora megasperma* complex. *Mycologia,* **83**, 376–81.

Hansen EM, Brasier CD, Shaw DS and Hamm PB (1986). The taxonomic stucture of *Phytophthora megasperma*: evidence for emerging biological species groups. *Transactions of the British Mycological Society,* **87**, 557–73.

Hansen EM, Stenlid J and Johansson M (1993). Genetic control of somatic incompatibility in the root rotting basidiomycete *Heterobasidion annosum. Mycological Research,* **97**, 1229–33.

Hansen EM, Stenlid J and Johansson M (1994). Somatic incompatibility in *Heterobasidion annosum* and *Phellinus weirii.* In *Proceedings of 8th IUFRO Root and Butt Rot Conference* (ed. M Johansson and J Stenlid), pp. 323–33. Swedish University of Agricultural Sciences, Uppsala.

Hansen HN (1938). Dual phenomenon in imperfect fungi. *Mycologia,* **30**, 442–55.

Hantula J and Müller MM (1997). Variation within *Gremmeniella abietina* in Finland and other countries as determined by random amplified microsatellites (RAMS). *Mycological Research*, **101**, 169–75.

Hantula J, Dusabenyagasani M and Hamelin RC (1996). Random amplified microsatellites (RAMS)—a novel method for characterizing genetic variation within fungi. *European Journal of Forest Pathology*, **26**, 159–66.

Harberd DJ (1961). Observations on population structure and longevity of *Festuca rubra*. *New Phytologist*, **60**, 184–206.

Harberd DJ (1962). Some observations on natural clones in *Festuca ovina*. *New Phytologist*, **61**, 85–100.

Harder R (1927). Über mikrochirurgische Operationen an Hymenomyceten. *Zeitschrift für Wissenschaftliche und Mikroskopische Technik*, **44**, 173–82.

Harrington TC and McNew DL (1997). Self-fertility and uni-directional mating-type switching in *Ceratocystis coerulescens*, a filamentous ascomycete. *Current Genetics*, **32**, 52–9.

Harrington TC and McNew DL (1998). Partial infertility among the *Ceratocystis* species on conifers. *Fungal Genetics and Biology*, **25**, 44–53.

Harrington TC and Rizzo DM (1999). Defining species in the Fungi. In *Structure and dynamics of fungal populations* (ed. J. J. Worrall), pp. 43–71. Kluwer Academic, Dordrecht.

Harrington TC and Wingfield BD (1998). The *Ceratocystis* species on conifers. *Canadian Journal of Botany*, **76**, 1446–7.

Harrington TC, Worrall JJ and Rizzo DM (1989). Compatibility among host-specialized isolates of *Heterobasidion annosum* from Western North America. *Phytopathology*, **79**, 290–6.

Harrington TC, Steimel JP, Wingfield MJ and Kile GA (1996). Isozyme variation and species delimitation in the *Ceratocystis coerulescens* complex. *Mycologia*, **88**, 104–13.

Harrington TC, Stenlid J and Korhonen K (1997). Evolution in the genus *Heterobasidion*. In *Proceedings of the 9th IUFRO Root and Butt Rot Conference*, pp. 64–74. INRA, Paris.

Harry IB and Clarke DD (1986). Race-specific resistance in groundsel (*Senecio vulgaris*) to the powdery mildew *Erysiphe fischeri*. *New Phytologist*, **103**, 167–75.

Harry IB and Clarke DD (1987). The genetics of race-specific resistance in groundsel (*Senecio vulgaris*) to the powdery mildew fungus *Erysphe fischeri*. *New Phytologist*, **107**, 715–23.

Harvey PR, Langridge P and Marshall DR (2001). Genetic drift and host-mediated selection cause genetic differentiation among *Gaeumannomyces graminis* populations infecting cereals in southern Australia. *Mycological Research*, **105**, 927–35.

Hasegawa M, Kishino H and Yano T-O (1985). Dating the human–ape splitting by a molecular clock of mitochondrial DNA. *Journal of Molecular Evolution*, **22**, 160–74.

Hastie AC (1967). Mitotic recombination in conidiophores of *Verticillium albo-atrum*. *Nature*, **214**, 249–52.

Hastie AC (1973). Hybridization of *Verticillium albo-atrum* and *Verticillium dahliae*. *Transactions of the British Mycological Society*, **60**, 511–23.

Hastings PJ, Quah S-K and von Borstel RC (1976). Spontaneous mutation by mutagenic repair of spontaneous lesions in DNA. *Nature*, **264**, 719–722.

Hayashi K (1991). PCR-SSCP: a simple and sensitive method for the detection of mutations in the genomic DNA. *PCR Methods and Applications*, **1**, 34–8.

Haymaker HH (1928). Pathogenicity of two strains of the tomato-wilt fungus *Fusarium lycopersici*. *Journal of Agricultural Research*, **36**, 675–95.

Heald FD (1921). The relation of spore load to the percent of stinking smut in the crop. *Phytopathology*, **11**, 269–78.

Hellgren M and Högeberg N (1995). Ecotypic variation of *Gremmeniella abietina* in northern Europe: disease patterns reflected by DNA variation. *Canadian Journal of Botany*, **73**, 1531–9.

Helmi S and Lamb BC (1983). The interactions of three widely separated loci controlling conversion properties of *w* locus 1 in *Ascobolus immersus*. *Genetics*, **104**, 23–40.

Hennig W (1950). *Grundzüge einer Theorie der phylogenetischen Systematik*. Deutscher Zentralverlag, Berlin.

Henrion B, Le Tacon F and Martin F (1992). Rapid identification of genetic variation of ectomycorrhizal fungi by amplification of ribosomal RNA genes. *New Phytologist*, **122**, 289–98.

Henson JM and Caesar-Ton That TC (1995). Mitochondrial plasmids of the *Gauemannomyces–Phialophora* complex and their detection by primed, *in situ*, fluorescence labelling. *Experimental Mycology*, **19**, 263–74.

Hepper CM, Sen R and Maskall CS (1986). Identification of vesicular–arbuscular mycorrhizal fungi in leek (*Allium porrum*) and maize (*Zea mais*) on the basis of enzyme mobility during polyacrylamide gel electrophoresis. *New Phytologist*, **102**, 529–39.

Hermanns J and Osiewacz HD (1996). Induction of longevity by cytoplasmic transfer of a linear plasmid in *Podospora anserina*. *Current Genetics*, **29**, 250–6.

Hibbett DS, Murakami S and Tsuneda A (1994). Post-meiotic behaviour in *Lentinus, Panus* and *Neolentinus*. *Mycologia*, **86**, 725–32.

Hibbett DS, Fukumasa-Nakai Y, Tsunedo A and Donoghue MJ (1995). Phylogenetic diversity in shiitake inferred from nuclear ribosomal DNA sequences. *Mycologia*, **87**, 618–38.

Hicks JB, Strathern JN and Herskowitz I (1977). Interconversion of yeast mating types. III: Action of the homothallism (*HO*) gene in cells homozygous for the mating type locus. *Genetics*, **85**, 395–405.

Higgins DG, Bleasby AJ and Fuchs R (1991). CLUSTAL V: improved software for multiple sequence alignment. *Computer Applications in Bioscience*, **8**, 189–91.

Hijri M, Hosney M, van Tuinen D and Dulieu H (1999). Intraspecific ITS polymorphism in *Scutellospora castanea* (Glomales, Zygomycetes) is structured within multinucleate spores. *Fungal Genetics and Biology*, **26**, 141–51.

Hillis DM, Moritz C and Mable BK (1996). *Molecular systematics*. Sinauer Associates, Sunderland, MA.

Hintz WEA, Mohan M, Anderson JB and Horgen PA (1985). The mitochondrial DNAs of *Agaricus*: heterogeneity in *A. bitorquis* and *A. brunnescens*. *Current Genetics*, **9**, 127–32.

Hirst JM, Stedman OJ and Hurst GW (1967). Long-distance spore transport: vertical sections of spore clouds over the sea. *Journal of General Microbiology*, **48**, 357–77.

Hiura U (1978). Genetic basis of *formae speciales* in *Erysiphe graminis*. In *The powdery mildews* (ed. DM Spencer), pp. 101–28. Academic Press, London.

Hodnett B and Anderson JB (2000). Genomic stability of two individuals of *Armillaria gallica*. *Mycologia*, **92**, 894–9.

Hoeben P and Clark-Walker GD (1986). An approach to yeast classification by mapping yeast mitochondrial DNA from *Dekkara/Brettanomyces* and *Eeniella* genera. *Current Genetics*, **9**, 127–32.

Hoegger PJ, Rigling D, Holdenrieder O and Heiniger U (2000). Genetic structure of newly established populations of *Cryphonectria parasitica*. *Mycological Research*, **104**, 1108–16.

Hoffman P and Esser K (1978). Genetics of speciation in the basidiomycetous genus *Polyporus*. *Theoretical and Applied Genetics*, **53**, 273–82.

Högberg N and Stenlid J (1999). Population genetics of *Fomitopsis rosea*—a wood-decay fungus of the old growth Europaen taiga. *Molecular Ecology*, **8**, 703–10.

Hohl HR and Iselin K (1984). Strains of *Phytophthora infestans* from Switzerland with A2 mating behaviour. *Transactions of the British Mycological Society*, **83**, 529–31.

Holton CS (1944). Inheritance of chlamydospore and sorus characters in species and races of *Tilletia caries* and *T. foetida*. *Phytopathology*, **34**, 586–92.

Holton CS (1951). Methods and results of studies on heterothallism and hybridization in *Tilletia caries* and *T. foetida*. *Phytopathology*, **41**, 511–21.

Holton CS (1954). Natural hybridization between common and darf bunt as related to the problem of delimitation of species of *Tilletia* occurring on wheat. *Phytopathology*, **44**, 493 (abstract).

Holton CS (1967). *Plant Disease Reporter*, **50**, 62–3.

Hood ME and Antonovics J (2000). Intratetrad mating, heterozygosity and the maintenance of deleterious alleles in *Microbotryum violaceum* (*Ustilago violacea*). *Heredity*, **85**, 231–41.

Hooker AL, Smith DR, Lim SM and Musson MD (1970). Physiological races of *Helminthosporium maydis* and disease resistance. *Plant Disease Reporter*, **54**, 1109–10.

Hora FB (1959). Quantitative experiments on toadstool production in woods. *Transactions of the British Mycological Society*, **42**, 1–14.

Hosney M, Hijri M, Passerieux E and Dulieu H (1999). rDNA units are highly polymorphic in *Scutellospora castanea* (Glomales, Zygomycota). *Gene*, **226**, 61–71.

Hott TW, Durgan ME and Michelmore RW (1987). Genetics of virulence in Californian populations of *Bremia lactucae* (lettuce downy mildew). *Phytopathology*, **77**, 1381–6.

Howlett BJ (1990). Pulsed field gel electrophoresis as a method for examining phylogenetic relationships between organisms: its application to the genus Phytophthora. *Australian Journal of Systematic Botany*, **3**, 75–80.

Hsiang T and Wu C (2000). Genetic relationships of pathogenic *Typhula* species assessed by RAPD, ITS-RFLP and ITS sequencing. *Mycological Research*, **104**, 16–22.

Hsieh WH, Smith SN and Snyder WC (1977). Mating groups in *Fusarium moniliforme*. *Phytopathology*, **67**, 1041–3.

Hua-Van A, Davière J-M, Kaper F, Langin T and Daboussi M-J (2000). Genome organization in *Fusarium oxysporum*: clusters of Class II transposons. *Curent Genetics*, **37**, 329–47.

Huber DH and Fulbright DW (1994). Preliminary investigations on the effect of *vic* genes upon the transmission of *ds*RNA in *Cryphonectria parasitica*. In *Proceedings of the International Chestnut Conference* (ed. ML Double and WL MacDonald), pp. 15–19. West Virginia University Press, Morgantown, WV.

Hudson RR (1990). Gene genealogies and the coalescent process. In *Oxford surveys in evolutionary biology*, Vol. 7 (ed. D Futuyama and J Antonovics), pp.1-44. Oxford University Press.

Hudspeth MES, Shumard DS, Braford CJR and Grossman LI (1980). Rapid purification of yeast mitochondrial DNA in high yield. *Biochimica et Biophysica Acta*, **610**, 221–8.

Huelsenbeck JB, Bull JJ and Cunningham CW (1996). Combining data in phylogenetic analysis. *Trends in Ecology and Evolution*, **11**, 152–8.

Huff DR, Bunting TE and Plumley KA (1994). Use of random amplified polymorphic DNA markers for the detection of genetic variation in *Magnaporthe poae*. *Phytopathology*, **88**, 1312–16.

Hughes KW, McGhee LL, Methven AS, Johnson JE and Petersen RH (1999). Patterns of geographic speciation in the genus *Flammulina* based on sequences of the ribosomal ITS1–5.8S–ITS2 area. *Mycologia*, **91**, 978–86.

Hunter I, Cotter RR and Royle DJ (1999). The teleomorph stage, *Mycosphaerella graminicola*, in epidemics of *Septoria tritici* blotch on winter wheat in the UK. *Plant Pathology*, **48**, 51–7.

Huss MJ (1996). Isozyme analysis of population structure and diversity in the puffball species *Lycoperdon pyriforme*. *Mycologia*, **88**, 977–85.

Hyakumachi M and Ui T (1987). Non-self-anastomosing isolates of *Rhizoctonia solani* obtained from fields of sugar beet monoculture. *Transactions of the British Mycological Society*, **89**, 155–9.

Ingold CT (1971). *Fungal spores: their liberation and dispersal*. Clarendon Press, Oxford.

Ingram R (1968). *Verticillium dahliae* var *longisporum*, a stable diploid. *Transactions of the British Mycological Society*, **51**, 339–41.

Innis MA, Gelfand DH, Sninsky JJ and White TJI (1990). *PCR protocols: a guide to methods and applications*. Academic Press, San Diego, CA.

Jaccard P (1908). Nouvelles recherches sur la distribution florale. *Bulletin de la Société de Vaud. Sciences Naturelles*, **44**, 223–70.

Jacobson DJ (1992). Control of mating type heterokaryon incompatibility by the *tol* gene in *Neurospora crassa* and *N. tetrasperma*. *Genome*, **35**, 347–53.

Jacobson DJ (1995). Sexual dysfunction associated with outcrossing in *Neurospora tetrasperma*, a pseudo-homothallic ascomycete. *Mycologia*, **87**, 604–17.

Jacobson KM and Miller OK Jr (1994). Postmeiotic mitosis in the basidia of *Suillus granulatus*: implications for population structure and dispersal. *Mycologia*, **86**, 511–16.

Jacobson KM, Miller OK Jr and Turner BJ (1993). Randomly amplified polymorphic DNA markers are superior to somatic incompatibility tests for discriminating genotypes in natural populations of the ectomycorrhizal fungus *Suillus granulatus*. *Proceedings of the National Academy of Sciences of the United States of America*, **90**, 9159–63.

Jahnke K-D (1984). Artbegrenzung durch DNA Analyse bei einigen Vertreten der Strophariaceae (Basidiomycetes). *Bibliotheca Mycologia*, **96**, 1–183.

Jahnke K-D and Bahnweg G (1986). Assessing natural relationships in the Basidiomycetes by DNA analysis. *Transactions of the British Mycological Society*, **87**, 175–91.

James TY and Vilgalys R (2001). Abundance and diversity of *Schizophyllum commune* spore clouds in the Caribbean detected by selective sampling. *Molecular Ecology*, **10**, 471–9.

James TY, Porter D, Hamrick JL and Vilgalys R (1999). Evidence for limited intercontinental gene flow in the cosmopolitan mushroom, *Schizophyllum commune*. *Evolution*, **53**, 1665–77.

Jarosz AM and Burdon JJ (1991). Host pathogen interactions in natural populations of *Linum marginale* and *Melampsora lini*. II. Local and regional variation in patterns of resistance and racial structure. *Evolution*, **45**, 1618–27.

Jarosz AM and Burdon JJ (1992). Host–pathogen interactions in natural populations of *Linum marginale* and *Melampsora lini*. III. Influence of pathogen epidemics on host survivorship and flower production. *Oecologia*, **89**, 53–61.

Ji H, Moore DP, Blomberg MA *et al.* (1993). Hot spots for unselected Ty1 transposition effects on yeast chromosome III are near tRNA genes and LTR sequences. *Cell*, **73**, 1007–13.

Jin J, Hughes KW and Petersen RH (2001). Biogeographical patterns in *Panellus stypticus*. *Mycologia*, **93**, 309–16.

Jinks JL (1952). Heterokaryosis: a system of adaptation in wild fungi. *Proceedings of the Royal Society of London, Series B*, **140**, 83–99.

Jinks JL (1957). Selection for cytoplasmic differences. *Proceedings of the Royal Society of London, Series B*, **146**, 517–40.

Jinks JL (1966). Mechanisms of inheritance. 4. Extranuclear inheritance. In *The Fungi*, Vol. II (ed. GC Ainsworth and AS Sussman), pp. 619–60. Academic Press, London.

Jinks JL, Caten CE, Simchen G and Croft JH (1966). Heterokaryon incompatibility and variation in wild populations of *Aspergillus nidulans*. *Heredity*, **21**, 227–39.

Johannesson H, Gustafsson M and Stenlid J (2001). Local population structure of the wood decay ascomycete *Daldinia loculata*. *Mycologia*, **93**, 440–6.

Johnson LF, Curl EA, Bond JH and Fribourg HA (1959). *Methods for studying soil microflora-plant disease relationships*. Burgess, Minneapolis, MN.

Johnson T (1949). Intervarietal crosses in *Puccinia graminis*. *Canadian Journal of Research*, **C27**, 45–65.

Jones RN (1995). B chromosomes in plants. *New Phytologist*, **131**, 411–36.

Joosten MHAJ, Cozijsnsen TJ and de Wit PJGM (1994). Host resistance to a fungal tomato pathogen lost by a single base-pair change in an avirulence gene. *Nature*, **167**, 384–6.

Jukes TH and Cantor CR (1969). *Evolution of protein molecules*. Academic Press, New York.

Jurand MK and Kemp RFO (1973). An incompatibility sytem determined by three factors in a species of *Psathyrella*. *Genetical Research*, **22**, 125–34.

Kachroo P, Leong SA and Chattoo BB (1995). MG-SINE— a short interspersed nuclear element from the rice blast fungus, *Magnaporthe grisea*. *Proceedings of the National Academy of Sciences of the United States of America*, **92**, 11125–9.

Kadish D and Cohen Y (1988). Fitness of *Phytophthora infestans* isolates from metalaxyl-sensitive and resistant populations. *Phytopathology*, **78**, 912–15.

Käfer, E (1977). Meiotic and mitotic recombination in *Aspergillus* and its chromosomal aberrations. *Advances in Genetics*, **19**, 33–131.

Kaltz O and Shykoff JA (1997). Sporidial mating type ratios of teliospores from natural populations of anther smut fungus *Microbotryum violaceum* (*Ustilago violacea*). *International Journal of Plant Science*, **158**, 575–84.

Kaltz O and Shykoff JA (1999). Selfing propensity of the fungal pathogen *Microbotryum violaceum* varies across *Silene latifolia* host plants. *Journal of Evolutionary Biology*, **12**, 340–9.

Karapapa VK, Bainbridge BW and Heale JB (1997). Morphological and molecular characterization of *Verticillium longisporum* comb.nov., pathogenic to oil-seed rape. *Mycological Research*, **101**, 1281–94.

Karlsson, J.-O. and Stenlid, J. (1991). Pectic isozyme profiles of intersterility groups in *Heterobasidion annosum*. *Mycological Research*, **95**, 531–536.

Kasuga T, Taylor JW and White TJ (1999). Phylogenetic relationships of varieties and geographical groups of the human pathogenic fungus, *Histoplasma capsulatum* Darling. *Journal of Clinical Microbiology*, **37**, 653–63.

Katan T (2000). Vegetative compatibility in populations of *Verticillium*—an overview. In *Advances in Verticillium research and diseases* (ed. E Tjamos, RC Rowe, JB Heale and D Fravel), pp. 77–94. American Phytopathological Society, St Paul, MN.

Katsuya K and Green GJ (1967). Reproductive potentials of races 15B and 56 of wheat stem rust. *Canadian Journal of Botany*, **45**, 1077–91.

Kay E and Vilgalys R (1992). Spatial distribution and genetic relationships among individuals in a natural population of the oyster mushroom *Pleurotus ostreatus*. *Mycologia*, **84**, 173–82.

Kayser T and Wöstenmeyer J (1991). Electrophoretic karyotype of the zygomycete *Absidia glauca*: evidence for differences between mating types. *Current Genetics*, **19**, 279–84.

Keller SM, McDermott JM, Pettway RE, Wolfe MS and McDonald BA (1997). Gene flow and sexual reproduction in the wheat glume blotch pathogen *Phaeosphaeria nodorum* (anamorph *Stagnospora nodorum*). *Phytopathology*, **87**, 353–8.

Kelly A, Alcalá-Jiménez AR, Bainbridge BW, Heale JB, Pérez-Artés E and Jiménez-Díaz RM (1994). Use of genetic fingerprinting and random amplified polymorphic DNA to characterize pathotypes of *Fusarium oxysporum* f.sp. *ciceris* infecting chickpea. *Phytopathology*, **84**, 1293–8.

Kemp RFO (1975). Oidia, plasmogamy and speciation in basidiomycetes. *Biological Journal of the Linnean Society*, **7**, 57–69.

Kemp RFO (1977). Oidial homing and the taxonomy and speciation of Basidiomycetes with special reference to the genus *Coprinus*. In *Bibliotheca Mycologica*. Vol. 61: *The species concept in Hymenomycetes* (ed. H. Clémençon), pp. 259–73. J. Cramer (A. R. Gartner Verlag), Vaduz.

Kemp RFO (1980). The genetics of the A–B–C type heterokaryon incompatibility in *Coprinus bisporus*. *Transactions of the British Mycological Society*, **75**, 29–35.

Kemp RFO (1983). Incompatibility and speciation in the genus*Coprinus*. *Revista de Biologia*, **12**, 179–86.

Kemp RFO (1985). Do fungal species really exist? A study of basidiomycete species with special reference to those in *Coprinus* section *Lanatuli*. *Bulletin of the British Mycological Society*, **19**, 34–9.

Kemp RFO (1995). Incompatibility in basidiomycetes: the heterogenic pentax. *Edinburgh Journal of Botany*, **52**, 71–89.

Kempken F (1995). Plasmid DNA in mycelial fungi. In *The Mycota*, Vol. 2 (ed. U Kück), pp. 169–87. Springer-Verlag, Berlin.

Kendrick B (ed.) (1979). *The whole fungus: the sexual–asexual synthesis*. National Museums of Canada, Ottawa.

Kennedy ME and Burnett JH (1956). Amphithallism in fungi. *Nature*, **177**, 882–3.

Kidston R and Lang WH (1921). On the Old Redstone plants of the Rhynie Chert. V: The thallophyta occurring in the peat bed: the succession of accumulation and preservation of the deposit. *Transactions of the Royal Society of Edinburgh*, **52**, 855–902.

Kile GA (1983). Identification of genotypes and the clonal development of *Armillaria luteobubalina* Watling and Kile in eucalypt forests. *Australian Journal of Botany*, **31**, 657–71.

Kile GA and Brasier CM (1990). Inheritance and inter-relationship of fitness characters in progeny of an aggressive × non-aggressive cross of *Ophiostoma ulmi*. *Mycological Research*, **94**, 514–22.

Kim DH, Martyn RD and Magill CW (1993). Chromosomal polymorphism in *Fusarium oxysporum* f.sp. *niveum*. *Phytopathology*, **83**, 1209–16.

Kim M-S, Klopfenstein NB, McDonald GI, Arumuganathan K and Vidavar AK (2000). Characterization of North American *Armillaria* species by nuclear DNA content and RFLP analysis. *Mycologia*, **92**, 874–83.

Kimura M (1968). Evolutionary rate at the molecular level. *Nature*, **217**, 624–6.

Kimura M (1980). A simple method for estimating evolutionary rates of base substitutions through comparative studies of nucleotide sequences. *Journal of Molecular Evolution*, **16**, 111–20.

Kimura M (1983). *The neutral theory of molecular evolution*. Cambridge University Press.

Kimura M and Ohta T (1971). *Theoretical aspects of population genetics*. Princeton University Press.

Kingman JFC (1982). On the genealogy of large populations. *Journal of Applied Probability*, **19A**, 27–43.

Kinscherf TC and Leong SA (1988). Molecular analysis of the karyotype of *Ustilago maydis*. *Chromosoma (Berlin)*, **96**, 427–33.

Kinsey JA (1990). *Tad*, a LINE-like transposable element of *Neurospora* can transpose between nuclei in a heterokaryon. *Genetics*, **126**, 317–23.

Kinsey JA and Helber J (1989). Isolation of a transposable element from *Neurospora crassa*. *Proceedings of the National Academy of Sciences of the United States of America*, **86**, 1929–33

Kirby GC (1984). Breeding systems and heterozygosity in populations of tetrad forming fungi. *Heredity*, **52**, 35–41.

Kirk PM, Cannon PF, David JC and Stalpers JA (2001). *Ainsworth and Bisby's Dictionary of the Fungi* (9th edn). CAB International, Wallingford.

Kishino H and Hasegawa M (1989). Evaluation of the maximum likelihood estimate of the evolutionary tree topologies from DNA sequence data, and the branching order in Hominoidea. *Journal of Molecular Evolution*, **16**, 170–9.

Kistler HC (1997). Genetic diversity in the plant-pathogenic fungus *Fusarium oxysporum*. *Phytopathology*, **87**, 474–9.

Kistler HC and Benny U (1992). Autonomously replicating plasmids and chromosome rearrangement during transformation of *Nectria haematococca*. *Gene*, **117**, 81–91.

Kistler HC and Miao VPW (1992). New modes of genetic change in filamentous fungi. *Annual Review of Phytopatholoy*, **30**, 131–52.

Kistler HC, Alabouvette C, Baayen RP *et al*. (1998). Systematic numbering of vegetative compatibility groups in the plant pathogenic fungus *Fusarium oxysporum*. *Phytopathology*, **88**, 30–2.

Kitani Y and Olive LS (1976a). Genetics of *Sordaria fimicola*.VI. Gene conversion at the g locus in mutant x wild type crosses. *Genetics*, **57**, 767–82.

Kitani Y and Olive LS (1976b). Genetics of *Sordaria fimicola*. VII. Gene conversion at the g locus in inter allelic crosses. *Genetics*, **62**, 23–66.

Koch AL (1971). The adaptive response of *Escherichia coli* to feast and famine existence. *Advances in Microbial Physiology*, **6**, 147–217.

Köhler F (1935). Genetische Studien an *Mucor mucedo* Brefeld.*Zeitschrift für Induktiv Abstammungs- und Vererbungslehre*, **70**, 1–54.

Kohli Y, Morrall RAA, Anderson JB and Kohn LM (1992). Local and trans-Canadian clonal distribution of *Sclerotium sclerotiorum* on canola. *Phytopathology*, **82**, 875–80.

Kohli Y, Brunner LJ, Yoell H *et al*. (1995). Clonal dispersal and spatial mixing in populations of the plant pathogenic fungus, *Sclerotium sclerotiorum*. *Molecular Ecology*, **4**, 69–77.

Kohlmeyer J and Kohlmeyer E (1974). Distribution of *Epichloë typhina* (Ascomycetes) and its parasitic fly. *Mycologia*, **66**, 77–86.

Koltin Y and Leibowitz MJ (1988). *Viruses of fungi and simple eukaryotes*. Marcel Dekker, New York.

Korhonen K (1978a). Intersterility groups of *Heterobasidion annosum*. *Communications of the Institute of Forestry Fennica*, **94**, 1–25.

Korhonen K (1978b). Interfertility and clonal size in the *Armillaria mellea* complex. *Karstenia*, **18**, 31–42.

Korhonen K and Hintikka V (1974). Cytological evidence for somatic diploidization in dikaryotic cells of *Armillariella mellea*. *Archiv für Mikrobiologie*, **59**, 187–92.

Kozlowski M, Bartnik E and Stepien PP (1982). Restriction enzyme analysis of mitochondrial DNA of members of the genus *Aspergillus* as an aid in taxonomy. *Journal of General Microbiology*, **128**, 471–6.

Krumlauf R and Marzluf GA (1980). Genome organization and characterization of the repetitive and inverted DNA sequences in *Neurospora crassa*. *Journal of Biological Chemistry*, **255**, 1138–45.

Kück U (1989). Mitochondrial DNA rearrangements in *Podospora anserina*. *Experimental Mycology*, **13**, 111–20.

Kües U and Casselton LA (1992). Fungal mating type genes—regulators of sexual development. *Mycological Research*, **96**, 993–1006.

Kuhlman EG (1982). Varieties of *Gibberella fujikuroi* with anamorphs in *Fusarium* section *Liseola*. *Mycologia*, **74**, 759–68.

Kühner R (1945). Le problème de la filiation des Agaricales à la lumière de nouvelles observations d'ordre cytologique sur les Agaricales leucosporées. *Bulletin de la Société Linneanne de Lyon*, **14**, 160–8.

Kuldau GA, Tsai H-F and Schardl LC (1999). Genome sizes of *Epichloë* species and anamorphic hybridism. *Mycologia*, **91**, 776–82.

Kumar J, Nelson RJ and Zeigler RS (1999). Population structure and dynamics of *Magnaporthe grisea* in the Indian Himalayas. *Genetics*, **152**, 971–84.

Kumar S, Tamura K and Nei M (1993). *MEGA (molecular evolutionary genetics analysis)*. Pennsylvania State University, University Park, PA.

Kurkela T, Hanso M and Hantula J (1999). Differentiating characters between *Melampsoridium* rusts infecting birch and alder leaves. *Mycologia*, **91**, 987–92.

Kurtzman CP (1985). Molecular taxonomy of the fungi. In *Gene manipulations in fungi* (ed. JW Bennett and LL Lasure), pp. 35–63. Academic Press, London.

Kwan H-S, Chiu S-W, Pang K-M and Cheng S-C (1992). Strain typing in *Lentinula edodes* by polymerase chain reaction. *Experimental Mycology*, **16**, 163–6.

Laane MM (1974). Nuclear behaviour during vegetative stage and zygospore formation in *Absidia glauca*. *Norwegian Journal of Botany*, **21**, 125–35.

Lamb BC (1984). The properties of meiotic gene conversion are important in its effects on evolution. *Heredity*, **53**, 113–38.

Lamb BC (1986). Gene conversion disparity: factors controlling its direction and extent, with test of assumptions and predictions in its evolutionary effects. *Genetics*, **114**, 611–32.

Lamb BC (1996). Ascomycete genetics: the part played by ascus segregation phenomena in our understanding of the mechanisms of recombination. *Mycological Research*, **100**, 1025–59.

Lamb BC and Helmi S (1982). The extent to which gene conversion can change allele frequencies in populations. *Genetical Research*, **39**, 199–217.

Lamb BC and Helmi S (1989). The effects of gene conversion control factors on conversion-induced changes in allele frequencies in populations and on linkage disequilibrium. *Genetica*, **78**, 167–81.

Lamb BC, Saleem M, Scott W, Thapa N and Nevo E (1998). Inherited and environmentally induced differences in mutation frequencies between wild strains of *Sordaria fimicola* from 'Evolution Canyon'. *Genetics*, **149**, 87–99.

Lamboy WF (1994a). Computing similarity coefficients from RAPD data: the effects of PCR artefacts. *PCR Methods and Application*, **4**, 31–7.

Lamboy WF (1994b). Computing similarity coefficients from RAPD data: correcting for the effects of PCR artefacts caused by variation in experimental conditions. *PCR Methods and Applications*, **4**, 38–43.

Lanfranco L, Delpeio N and Bonfante P (1999). Intrasporal variability of ribosomal sequences in the endomycorrhizal fungus *Gigaspora margarita*. *Molecular Ecology*, **8**, 37–45.

Lange M (1952). Species concept in the genus *Coprinus*. *Dansk Banisk Arkiv*, **14**, 1–164.

Langton FA and Elliott TJ (1980). Genetics of secondarily homothallic basidiomycetes. *Heredity*, **45**, 99–106.

Largo E and Vezinhet F (1993). Chromosomal rearrangements during vegetative growth of a wild strain of *Saccharomyces cerevisiae*. *Applied and Environmental Microbiology*, **59**, 323–6.

Larraya LM, Pêrez G, Iribarren I, *et al.* (2001). Relationship between monokaryotic growth rate and mating type in the edible basidiomycete *Pleurotus ostreatus*. *Applied and Environmental Microbiology*, **67**, 3385–90.

Larraya LM, Idareta E, Arana D, Ritter E, Pisabarro AG and Ramírez L (2002). Quantitative trait loci controlling vegetative growth rate in the edible basidiomycete *Pleurotus ostreatus*. *Applied and Environmental Microbiology*, **68**, 1109–14.

Latgé JP, Debeaupuis JP, Sarfati J and Paris S (1998). Variability in the human opportunistic fungal pathogen *Aspergillus fumigatus*. In *Molecular variability of fungal pathogens* (ed. P Bridge, Y Couteaudier and J Clarkson), pp. 121–31. CAB International, Wallingford.

Lawrence GJ (1989). Flax rust from *Linum marginale*: pathogenicity reaction on the *Linum usitatissimum* set of differential varieties. *Canadian Journal of Botany*, **67**, 3187–91.

Lawrence GJ and Burdon JJ (1989). Flax rust from *Linum marginale*: variation in a natural host–pathogen interaction. *Canadian Journal of Botany*, **67**, 3192–8.

Layton AC and Kuhn DN (1990). *In planta* formation of heterokaryons of *Phytophthora megasperma* f.sp. *glycinea*. *Phytopathology*, **80**, 602–6.

Leach J, Tegtmeier KJ, Daly JM and Yoder CO (1982). Dominance at the *Tox1* locus controlling T-toxin production by *Cochliobolus heterostrophus*. *Physiological Plant Pathology*, **21**, 327–33.

Leal-Lara H and Eger-Hummel G (1982). A monokaryotisation method and its use for genetic studies in wood-rotting basidiomycetes. *Theoretical and Applied Genetics*, **61**, 65–8.

Lee BTO and Pateman JA (1959). Linkage of polygenes controlling size of ascospore in *Neurospora crassa*. *Nature*, **183**, 698–9.

Lee BTO and Pateman JA (1961). Studies on the inheritance of ascospore length in *Neurospora crassa*. I. Studies on large-spored strains. *Australian Journal of Biological Sciences*, **14**, 223–30.

Lee BTO and Pateman JA (1962). Studies concerning the inheritance of ascospore length in *Neurospora crassa*. II. Selection experiments with wild-type strains. *Australian Journal of Biological Sciences*, **15**, 160–5.

Lee SB, Milgroom MG and Taylor JW (1988). A rapid, high yield mini-prep method for isolation of total genomic DNA from fungi. *Fungal Genetics Newsletter*, **35**, 24.

Lee SB and Taylor JW (1990). Isolation of DNA from fungal mycelia and single spores. In *PCR protocols: a guide to methods and applications* (ed. MA Innis, DH Gelfand, JJ Sninsky and TJ White), pp. 282–7. Academic Press, San Diego, CA.

Lemke PA and Nash CH (1970). Isolation and characterization of a double-stranded ribonucleic acid from *Penicillium chrysogenum*. *Journal of Virology*, **6**, 813–19.

Leonard KJ (1977). Selection pressures and plant pathogens. *Annals of the New York Academy of Sciences*, **287**, 207–22.

Leonard KJ (1987). The host factor as a selective factor. In *Populations of plant pathogens, their dynamics and genetics* (ed. MS Wolfe and CE Caten), pp. 163–79. Blackwell Scientific, Oxford.

Leonard KJ and Czochor RJ (1980). Theory of genetic interactions among populations of plants and their pathogens. *Annual Review of Phytopathology*, **18**, 237–58.

Leonard KJ and Leath S (1990). Genetic diversity in field populations of *Cochliobolus carbonum* on corn in North Carolina. *Phytopathology*, **80**, 1154–9.

Leslie JF (1991). Mating populations of *Gibberella fujikuroi* (*Fusarium* section *Liseola*). *Phytopathology*, **81**, 1058–60.

Leslie JF (1993). Fungal vegetative incompatibility. *Annual Review of Phytopathology*, **31**, 127–50.

Leslie JF (1995). *Gibberella fujikuroi*: available populations and variable traits. *Canadian Journal of Botany*, **73** (Supplement 1), S282–91.

Leslie JF and Klein KK (1996). Female fertility and mating type effects on effective population size and evolution in filamentous fungi. *Genetics*, **144**, 557–67.

Leuchtmann A and Clay K (1989). Morphological, cultural and mating studies in *Atkinsonella*, including *A. texensis*. *Mycologia*, **81**, 692–701.

Leuchtmann A and Clay K (1996). Isozyme evidence for host races of the fungus *Atkinsonella hypoxylon* (Clavicipitaceae) infecting the Danthonia (Poaceae) complex in the Southern Appalachians. *American Journal of Botany*, **83**, 1144–52.

Leung H, Borromeo ES, Bernardo MA and Notteghm JL (1988). Genetic analysis of virulence in the rice blast fungus *Magnaporthe grisea*. *Phytopathology*, **78**, 1227–33.

Leupold U (1956). Tetraploid inheritance in *Saccharomyces*. *Journal of Genetics*, **54**, 411–26.

Levin DA (2000). *The origin, expansion and demise of plant species*. Oxford University Press.

Levine MN and Cotter RU (1931). A synthetic production of *Puccinia graminis hordei* F and J. *Phytopathology*, **21**, 107.

Levis C, Fortini D and Brygoo Y (1997). Flipper, a mobile Fot1-like transposable element in *Botrytis cinerea*. *Molecular and General Genetics*, **254**, 674–80.

Levy M, Correa-Victoria FJ, Zeigler RS, Xu S and Hamer JE (1993). Genetic diversity of the rice blast fungus in a disease nursery in Colombia. *Phytopathology*, **83**, 1427–33.

Levy M, Romao J, Marchetti MA and Hansen JE (1991). DNA fingerprinting with a dispersed repetitive sequence resolves pathotype diversity in the rice blast fungus. *Plant Cell*, **3**, 95–102.

Lewontin RC (1988). On measures of gametic disequilibrium. *Genetics*, **120**, 849–52.

Li Z-Q and Shang H-S (1989). *Wheat rusts and their control*. Shanghai Science and Technology Press, Shanghai.

Liebman SW and Picologlou S (1988). Recombination associated with yeast retrotransposons. In *Viruses of Fungi and simple Eukaryotes*, Vol. 7 (ed. Y Koltin and MJ Leibowitz), pp. 63–90. Marcel Dekker, New York.

Liebman SW, Singh A and Sherman F (1979). A mutator affecting the region of the iso-l-cytochrome *c* gene in yeast. *Genetics*, **92**, 783–802.

Limpert E (1987). Spread of barley mildew by wind and its significance for phytopathology, aerobiology and barley cultivation in Europe. In *Advances in aerobiology: Proceedings of the 3rd International Conference on Aerobiology* (ed. G Boehm and RM Leuschner), pp. 331–6. Experientia, Birkhäuser, Basel.

Limpert E, Andrivon D. and Fischbeck G. (1990). Virulence patterns in populations of *Erysiphe graminis* f.sp. *hordei* in Europe in 1986. *Plant Pathology*, **39**, 402–15.

Little R and Manners JG (1969a). Somatic recombination in yellow rust of wheat (*Puccinia striiformis*). I. The production and possible origin of two new physiological races. *Transactions of the British Mycological Society*, **53**, 251–8.

Little R and Manners JG (1969b). Somatic recombination in yellow rust of wheat. II. Germ tube fusions, nuclear number and nuclear size. *Transactions of the British Mycological Society*, **53**, 259–67.

Liu Y-C, Cortesi P, Double ML, MacDonald WL and Milgroom MG (1996). Diversity and multilocus genetic structure in populations of *Cryphonectria parasitica*. *Phytopathology*, **86**, 1344–51.

Lloyd-Macgilp SA, Chambers SM, Dodd JC, Fitter AH, Walker C and Young JPW (1996). Diversity of the ribosomal internal transcribed spacers within and among isolates of *Glomus mossae* and related mycorrhizal fungi. *New Phytologist*, **133**, 103–11.

LoBuglio KF, Pitt JI and Taylor JW (1993). Phylogenetic analysis of two ribosomal DNA regions indicate multiple independent losses of a sexual *Talaromyces* state among asexual *Penicillium* species in subgenus *Biverticillium*. *Mycologia*, **85**, 592–604.

Lockwood JL (1977). Fungistasis in soils. *Biological Reviews*, **52**, 1–43.

Lockwood JL (1992). Exploitation competition. In *The fungal community: its organization and role in the ecosystem* (2nd edn) (ed. GC Carroll and TD Wicklow), pp. 243–63. Marcel Dekker, New York.

Lockwood JL and Filonow AB (1981). Responses of fungi to nutrient-limiting conditions and to inhibitory substances in natural habitats. *Advances in Microbiology*, **5**, 1–61.

Lodge DJ and Leonard KJ (1984). A cline and other patterns of genetic variation in *Cochliobolus carbonum* pathogenic to corn in North Carolina. *Canadian Journal of Botany*, **62**, 995–1005.

Loegering WQ (1951). Survival of races of wheat stem rusts in mixtures. *Phytopathology*, **41**, 55–65.

Lu BC (1964). Polyploidy in the basidiomycete *Cyathus stercoreus*. *American Journal of Botany*, **51**, 343–7.

Luig NH and Watson IA (1977). The role of barley, rye and grasses in the 1973–74 wheatstem rust epiphytotic in southern and eastern Australia. *Proceedings of the Linnean Society of New South Wales*, **101**, 65–76.

Lynch M (1990). The similarity index and DNA fingerprinting. *Molecular Biology and Evolution*, **7**, 401–10.

MacArthur RH and Wilson EO (1967). *The theory of island life*. Princeton University Press.

McCain JW and Groth JV (1989). Patterns of intrapopulation isozyme variation in the bean rust fungus. *Phytopathology*, **75t**, 579 (abstract).

McCain JW, Groth JV and Roelfs AP (1992). Inter- and intrapopulation isozyme variation in collections from sexually reproducing populations of the bean rust fungus, *Uromyces appendiculatus*. *Mycologia*, **84**, 329–40.

McClintock B. (1945). *Neurospora*. I. Preliminary observations of the chromosomes of *Neurospora crassa*. *American Journal of Botany*, **32**, 671–8.

McClure MA (1999). The retroid agents: disease, function and evolution. In *Origin and evolution of viruses* (ed. E Domingo *et al.*), pp. 163–95. Academic Press, London.

McCluskey K and Mills D (1990). Identification and characterization of chromosome length polymorphisms amiong strains representing fourteen races of *Ustilago hordei*. *Molecular Plant–Microbe Interactions*, **3**, 366–73.

McCluskey K, Russell BW and Mills D (1990). Electrophoretic karyotyping without the need for generating protoplasts. *Current Genetics*, **18**, 385–6.

McCluskey K, Agnon J and Mills D (1994). Characterization of genome plasticity in *Ustilago hordei*. *Current Genetics*, **26**, 486–93.

McDermott JM and McDonald BA (1993). Gene flow in plant pathosystems. *Annual Review of Phytopathology*, **31**, 353–73.

McDermott JM, McDonald BA, Allard RW and Webster RK (1989). Genetic variability for pathogenicity, isozyme, ribosomal DNA and colony colour variants in a population of *Rhynchosporium secalis*. *Genetics*, **122**, 561–5.

McDermott JM, Brändle U, Dutly F, *et al.* (1994). Genetic variation in powdery mildew of barley: development of RAPD, SCAR and VTNR markers. *Phytopathology*, **84**, 1316–21.

McDonald BA and Martinez JP (1990a). DNA restriction fragment length polymorphisms among *Mycosphaerella graminicola* (anamorph *Septoria tritici*) isolates collected from a single wheat field. *Phytopathology*, **80**, 1368–73.

McDonald BA and Martinez JP (1990b). Restriction fragment length polymorphisms in *Septoria tritici* occur at high frequency. *Current Genetics*, **17**, 133–8.

McDonald BA and Martinez JP (1991). Chromosome length polymorphisms in a *Septoria tritici* population. *Current Genetics*, **19**, 265–71.

McDonald BA, Miles J, Nelson LR and Pettway RE (1994). Genetic variability in nuclear DNA in field populations of *Stagnospora nodorum*. *Phytopathology*, **84**, 250–5.

McDonald BA, Pettway RE, Chen RS, Boeger JM and Martinez JP (1995). The population genetics of *Septoria tritici* (teleomorph *Mycosphaerella graminicola*). *Canadian Journal of Botany*, **73** (Supplement 1), S292–301.

McGinnis RC (1956). Cytological studies of chromosomes of rust fungi. *Journal of Heredity*, **47**, 257–9.

McKay MCR and MacNeill BH (1979). Spectrum of sensitivity to dodine in field populations of *Venturia inaequalis*. *Canadian Journal of Plant Pathology*, **1**, 76–8.

McKeen C (1952). A cultural and taxonomic study of three species of *Peniophora*. *Canadian Journal of Botany*, **30**, 764–87.

MacKenzie DR (1978). Estimating parasitic fitness. *Phytopathology*, **68**, 9–13, 1108.

McLaughlin DJ, McLaughlin EG and Lemke PA (2001a). *Systematics and evolution. Part A: The Mycota* (ed. K Esser and PA Lemke), Vol. 7a. Springer-Verlag, Berlin.

McLaughlin DJ, McLaughlin EG and Lemke PA (2001b). *Systematics and evolution. Part B: The Mycota* (ed. K Esser and PA Lemke), Vol. 7b. Springer-Verlag, Berlin.

McNabb SA and Klassen GR (1988). Uniformity of mitochondrial DNA complexity in Oomycetes and the evolution of the inverted repeat. *Experimental Mycology*, **12**, 233–42.

MacNish GC and Sweetingham MW (1993). Evidence that each Rhizoctonia bare patch is dominated by an individual zymogram group (ZG) of *Rhizoctonia solani*. *Australian Journal of Agricultural Research*, **44**, 1175–94.

MacNish GC, Carling DE and Brainard KA (1993a). Characterization of *Rhizoctonia solani* AG-8 from bare patches by pectic isozyme (zymogram) and anastomosis techniques. *Phytopathology*, **83**, 922–7.

MacNish GC, McLernon CK and Wood DA (1993b). The use of zymogram and anastomosis techniques to follow the expansion and demise of two coalescing bare patches caused by *Rhizoctonia solani*. *Australian Journal of Agricultural Research*, **44**, 1161–73.

Macrae R (1937). Interfertility phenomena of the American and European forms of *Panus stipticus* (Bull.) Fries. *Nature*, **139**, 674.

Macrae R (1967). Pairing incompatibility and other distinctions among *Hirschioporus* (*Polyporus*) *abietinus*, *H. fusco-violaceus* and *H. laricinus*. *Canadian Journal of Botany*, **45**, 1371–98.

Machlis L (1958a). Evidence for a sexual hormone in *Allomyces*. *Physiologia Plantarum*, **11**, 181–92.

Machlis L (1958b). A study of sirenin, the chemotactic sexual hormone from the water mold *Allomyces*. *Physiologia Plantarum*, **11**, 845–54.

Magee BB and Magee PT (1987). Electrophoretic karyotypes and chromosome numbers in *Candida* species. *Journal of General Microbiology*, **133**, 425–30.

Mainwaring HR and Wilson IM (1968). The life cycle and cytology of an apomictic *Podospora*. *Transactions of the British Mycological Society*, **51**, 663–77.

Maire R (1902). Recherches cytologiques et taxonomiques sur les Basidiomycètes. Thesis, Université de Paris.

Majer D, Mithen R, Lewis BG, Vos P and Oliver RP (1996). The use of AFLP fingerprinting for the detection of genetic variation in fungi. *Mycological Research*, **100**, 1107–11.

Majer D, Lewis BG and Mithen R (1998). Genetic variation among field isolates of *Pyrenopeziza brassicae*. *Plant Pathology*, **47**, 22–8.

Mallett KI and Harrison LM (1988). The mating system of the fairy ring fungus *Marasmius oreades* and the genetic relationship of fairy rings. *Canadian Journal of Botany*, **66**, 1111–16.

Maltby AD and Mihail JD (1997). Competition among *Sclerotinia sclerotiorum* genotypes within canola stems. *Canadian Journal of Botany*, **75**, 462–8.

Mandel M and Marmur J (1968). The use of ultraviolet absorbance-profile for determining the guanine plus cytosine content of DNA. In *Methods in enzymology*, **12B**, 195–206.

Man in t'Veld WA, Veenbas-Rijks WJ, Ilieva AW, de Cock AWA, Bonants PJM and Pieters R (1998). Natural hybrids of *Phytophthora nicotianae* and *Phytophthora cactorum* demonstrated by isozyme analysis and random amplified polymorphic DNA. *Phytopathology*, **88**, 922–9.

Manners JG (1988). *Puccinia striiformis*, yellow rust (stripe rust) of cereals and grasses. *Advances in Plant Pathology*, **6**, 373–87.

Marçais B, Caël O and Delatour C (2000). Genetics of somatic incompatibility in *Collybia fusipes*. *Mycological Research*, **104**, 304–10.

Marmur J and Doty P (1962). Thermal denaturation of DNAs. *Journal of Molecular Biology*, **3**, 585–94.

Martin F, Selosse M-A, Di Battista C *et al.* (1998). Molecular markers in ecology of ectomycorrhizal fungi. **30** (Supplement 1), S333–5.

Martin F, Diez J, Dell B and Delaruelle C (2002). Phylogeography of the ectomycorrhizal *Pisolithus* species as inferred from nuclear ribosomal DNA ITS sequences. *New Phytologist*, **153**, 345–57.

Maruyama T and Kimura M (1980). Genetic variability and effective population size when local extinction and repopulation of subpopulations are frequent. *Proceedings of the National Academy of Sciences of the United States of America*, **77**, 6710–14.

Mather K (1943). Polygenic inheritance and natural selection. *Biological Reviews*, **18**, 32–64.

Mather K (1973). *Genetical structure of populations*. Chapman & Hall, London.

Mathieson MJ (1952). Ascospore dimorphism and mating type in *Chromocrea spinulosa*. *Annals of Botany*, **16**, 449–66.

May G (1988). Somatic compatibility and individualism in the coprophilous basidiomycete, *Coprinus cinereus*. *Transactions of the British Mycological Society*, **91**, 443–51.

May G, Chevanto LL and Pukkila PJ (1991). Molecular analysis of the *Coprinus cinereus* mating-type A factor demonstrates an unexpectedly complex structure. *Genetics*, **128**, 529–38.

May G, Shaw F, Badrane H and Vekemans X (1999). The signature of balancing selection: fungal mating compatibility gene evolution. *Proceedings of the National Academy of Sciences of the United States of America*, **96**, 9172–7.

Mayden RL (1997). A hierarchy of species concepts: the denouement in the saga of the species problem. In *Species: the units of biodiversity* (ed. MF Claridge, HA Dawah and MR Wilson), pp. 381–424. Chapman & Hall, London.

Maynard-Smith J (1966). Sympatric speciation. *American Naturalist*, **100**, 637–50.

Maynard-Smith J and Haigh J (1974). Hitchhiking effect of a favourable gene. *Genetical Research*, **23**, 23–35.

Maynard-Smith J, Smith NH, O'Rourke M and Spratt BG (1993). How clonal are bacteria? *Proceedings of the National Academy of Sciences of the United States of America*, **90**, 4384–8.

Mayr E (1940). Speciation phenomena in birds. *American Naturalist*, **74**, 249–78.

Mayr E (1982). Speciation and macroevolution. *Evolution*, **36**, 1119–32.

Meiners J (1958). Studies on the biology of *Tilletia bromitectorum*. *Phytopathology*, **48**, 211–16.

Meinhardt F, Klempken F, Kämper J and Esser K (1990). Linear plasmids among eukaryotes. *Current Genetics*, **17**, 89–95.

Messner R, Schweigkofler W, Ibl M, Berg G and Prillinger H (1996). Molecular characterization of the plant pathogen *Verticillium dahliae* Kleb. using RAPD-PCR and sequencing of the 18S rRNA-gene. *Journal of Phytopathology*, **144**, 347–54.

Methven AS, Hughes KW and Petersen RH (2000). *Flammulina* RFLP patterns identify species and show biogeographical patterns within species. *Mycologia*, **92**, 1064–70.

Micales JA, Bonde MR and Petersen GL (1986). The use of isozyme analysis in fungal taxonomy and genetics. *Mycotaxon*, **27**, 405–49.

Michelmore RW and Sansome ER (1982). Cytological studies of heterothallism and secondary homothallism in *Bremia lactuca*. *Transactions of the British Mycological Society*, **79**, 291–7.

Mihail JD, Obert M, Taylor SJ and Bruhn JN (1994). The fractal dimension of young colonies of *Macrophomina phaseolina* produced from microsclerotia. *Mycologia*, **86**, 350–6.

Milgroom MG (1995). Analysis of population structure in fungal plant pathogen. In *Disease analysis through genetics and molecular biology: interdisciplinary bridges to improve sorghum and millet crops* (ed. JF Leslie and RA Frederiksen), pp. 213–29. Iowa University Press, Ames, IA.

Milgroom MG (1996). Recombination and the multilocus structure of fungal populations. *Annual Review of Phytopathology*, **34**, 457–77.

Milgroom MG and Cortesi P (1999). Analysis of population structure of the chestnut blight fungus based on vegetative incompatibility genotypes. *Proceedings of the National Academy of Sciences of the United States of America*, **96**, 10518–23.

Milgroom MG and Lipari SE (1995a). Spatial analysis of nuclear and mitochondrial RFLP genotypes in populations of the chestnut blight fungus, *Cryphonectria parasitica*. *Molecular Ecology*, **4**, 633–42.

Milgroom MG and Lipari SE (1995b). Population differentiation in the chestnut blight fungus, *Cryphonectria parasitica*, in Eastern North America. *Phytopathology*, **85**, 155–60.

Milgroom MG, Levin SA and Fry WE (1986). Population genetics theory and fungicide resistance. In *Plant disease epidemiology: genetics, resistance and management*, Vol. 2 (ed. KJ Leonard and WE Fry), pp. 340–67. McGraw-Hill, New York.

Milgroom MG, Lipari SE, Ennos RA and Liu Y-C (1993). Estimation of the outcrossing rate in the chestnut blight fungus, *Cryphonectria parasitica*. *Heredity*, **70**, 385–92.

Milgroom MG, Lipari SE and Powell WA (1992). DNA fingerprinting and analysis of population structure in the chestnut blight fungus, *Cryphonectria parasitica*. *Genetics*, **131**, 297–306.

Milgroom MG, MacDonald WL and Double ML (1991). Spatial pattern of analysis of vegetative compatibility groups in the chestnut blight fungus, *Cryphonectria parasitica*. *Canadian Journal of Botany*, **69**, 1407–13.

Miller JJ (1946). Cultural and taxonomic studies on certain Fusaria. I: Mutation in culture. *Canadian Journal of Research*, **C24**, 188–212.

Miller OK, Johnson JL, Burdsall HH and Flynn T (1994). Species delimitation in North American species of *Armillaria* as measured by DNA reassociation. *Mycological Research*, **98**, 1005–11.

Mills D and McCluskey K (1990). Electrophoretic karyotypes of fungi: the new cytology. *Molecular Plant–Microbe Interactions*, **3**, 351–7.

Milton JM, Rogers WG and Isaac I (1971). Application of acrylamide gel electrophoresis of soluble fungal proteins to the taxonomy of *Verticillium* species. *Transactions of the British Mycological Society*, **56**, 61–5.

Ming YN, Lin PC and Yu TF (1966). Heterokaryosis in *Fusarium fujikuroi* (Saw.) Wr. *Scientia Sinica*, **15**, 371–8.

Minter DW (1981). *Lophodermium* on pines. *CMI Mycological Paper*, **14**, 1–54.

Minter DW and Miller CS (1980a). Ecology and biology of three *Lophodermium* species on secondary needles of *Pinus sylvestris*. *European Journal of Forest Pathology*, **10**, 169–81.

Minter DW and Miller CS (1980b). A study of three pine inhabiting *Lophodermium* species in culture. *Nova Hedwigia*, **32**, 361–8.

Minter DW, Staley JM and Miller CS (1978). Four species of *Lophodermium* on *Pinus sylvestris*. *Transactions of the British Mycological Society*, **71**, 295–301.

Moore WF (1970). Origin and spread of southern corn leaf blight in 1970. *Plant Disease Reporter*, **54**, 1104–8.

Mortimer AM, Shaw DS and Sansome ER (1977). Genetical studies of secondary heterothallism in *Phytophthora dreschleri*. *Archives of Microbiology*, **111**, 255–9.

Morton A, Carder JH and Barbara DJ (1995a). Sequences of the internal transcribed spacers of the ribosomal RNA gnes and relationships between isolates of *Verticillium albo-atrum* and *Verticillium dahliae*. *Plant Pathology*, **44**, 183–90.

Morton A, Tabrett AM, Carder JH and Barbara DJ (1995b). Sub-repeat sequences in the ribosomal-RNA intergenic regions of *Verticillium albo-atrum* and *Verticillium dahliae*. *Mycological Research*, **99**, 257–66.

Mounce I and Macrae R (1938). Interfertility phenomena in *Fomes pinicola*. *Canadian Journal of Research*, **16**, 354–76.

Muller HJ (1932). Some genetic aspects of sex. *American Naturalist*, **66**, 118–38.

Muller HJ (1964). The relation of recombination to recombinational advance. *Mutation Research*, **1**, 2–9.

Müller K, McDermott JM, Wolfe MS and Limpert E (1996). Analysis of diversity in populations of plant pathogens: the barley powdery mildew pathogen across Europe. *European Journal of Plant Pathology*, **102**, 385–95.

Mueller-Dombois D and Ellenberg H (1974). *Aims and methods of vegetation ecology*. Wiley, New York.

Murakami J, Tosa Y, Kataoka T, *et al.* (2000). Analysis of host species specificity of *Magnaporthe grisea* toward wheat using a genetic cross between isolates from wheat and foxtail millet. *Phytopathology*, **90**, 1060–7.

Murphy RW, Sites Jr JW, Buth DG and Haufler H (1996). Proteins: isozyme electrophoresis. In *Molecular Systematics* (2nd edn) (ed. DM Hillis, C Moritz and BK Mable), pp. 51–120. Sinauer, Sunderland, MA.

Myers RM, Maniatis T and Lerman LS (1986). Detection and localization of single base changes by denaturing gradient gel electrophoresis. *Methods in Enzymology*, **155**, 501–27.

Mylyk OM (1975). Heterokaryon incompatibility genes in *Neurospora crassa* detected using duplication producing chromosome rearrangements. *Genetics*, **80**, 107–24.

Mylyk OM (1976). Heteromorphism for heterokaryon incompatibity genes in natural populations of *Neurospora crassa*. *Genetics*, **83**, 275–84.

Nagarajan S (1983). Race dynamics of *Puccinia graminis* f.sp. *tritici* in the USA from 1973 to 1980. *Phytopathology*, **73**, 512–14.

Nagarajan S and Singh DV (1990). Long-distance dispersion of rust pathogens. *Annual Review of Phytopathology*, **28**, 139–53.

Nakamura K and Egashira T (1961). Genetically mixed perithecia in *Neurospora*. *Nature*, **190**, 1129–30.

Nash BL and Stambaugh WJ (1989). Greater diversity of vegetative compatibility groups of *Cryphonectria parasitica* on carlet oak than on post oak in North Carolina. *Plant Disease Reporter*, **73**, 416–18.

Nei M (1972). Genetic distance between populations. *American Naturalst*, **106**, 283–92.

Nei M (1973). Analysis of gene diversity in subdivided populations. *Proceedings of the National Academy of Sciences of the United States of America*, **70**, 3321–3.

Nei M (1977). F-statistics and analysis of gene diversity in subdivided populations. *Annals of Human Genetics*, **41**, 225–33.

Nei M (1987). *Molecular evolutionary genetics*. Columbia University Press, New York.

Nei M and Chesser RK (1983). Estimation of fixation indices and gene diversities. *Annals of Human Genetics*, **47**, 253–9.

Nei M and Li W-H (1979). Mathematical model for studying genetic variation in terms of restriction endonucleases. *Proceedings of the National Academy of Sciences of the United States of America*, **76**, 5269–73.

Nelson RR (1964) Variation in mating capacities among isolates of *Cochliobolus carbonum*. *Phytopathology*, **54**, 867–77.

Nelson RR (1970a). Variation in mating capacities among isolates of *Cochliobolus carbonum*. *Canadian Journal of Botany*, **48**, 261–3.

Nelson RR (1970b). Genes for pathogenicity in *Cochliobolus carbonum*. *Phytopathology*, **60**, 1335–7.

Nelson RR and Kline DM (1963). Gene systems for pathogenicity and pathogenic potentials. I: Interspecific hybrids of *Helminthosporium carbonum* and *Helminthosporium victoriae*. *Phytopathology*, **53**, 101–5.

Nelson RR and Kline DM (1969). The identification of genes for pathogenicity in *Cochliobolus carbonum*. *Phytopathology*, **59**, 164–7.

Nelson RR and Kline DM (1971). The pathogenicity of 52 isolates of *Cochliobolus carbonum* to 22 gramineous species. *Plant Disease Reporter*, **55**, 325–7.

Nelson RR, Ayres JE, Cole H and Paterson DH (1970). Studies and observations on the past occurrence and geographical distribution of isolates of race T of *Helminthosporium maydis*. *Plant Disease Reporter*, **54**, 1123–6.

Newcombe G, Stirling B, McDonald S and Bradshaw HD (2000). *Melampsora × columbina*, a natural hybrid of *M. medusae* and *M. occidentalis*. *Mycological Research*, **104**, 261–74.

Newton CR and Graham GA (1997). *PCR*. Bios Scientific, Oxford.

Newton MR, Kinkel LL and Leonard KJ (1996). Density- and frequency-dependent fitness in coexisting foliar pathogens. *Phytopathology*, **86** (Supplement), S85.

Newton MR, Kinkel LL and Leonard KJ (1997). Competition and density-dependent fitness in a plant parasitic fungus. *Ecology*, **78**, 1774–84.

Newton MR, Kinkel LL and Leonard KJ (1998). Determinants of density- and frequency-dependent fitness in competing plant pathogens. *Phytopathology*, **88**, 45–51.

Nicholson P, Rezanoor HN and Hollins TW (1993). Classification of a world-wide collection of isolates of *Pseudocercosporella herpotrichoides* by RFLP analysis of mitochondrial and ribosomal DNA and host range. *Plant Pathology*, **42**, 58–66.

Niederhauser JS (1956). The blight, the blighter, and the blighted. *Transactions of the New York Academy of Sciences, Series II*, **19**, 55–63.

Nielsen K and Yohalem DS (2001). Origin of a polyploid *Botrytis* pathogen through interspecific hybridization between *Botrytis aclada* and *B. byssoidea*. *Mycologia*, **93**, 1064–71.

Nielsen K, Justesen AF, Jensen DF and Yohalem DS (2001). Universally primed polymerase chain reaction alleles and internal transcribed spacer restriction fragment length polymorphism distinguish two subgroups in *Botrytis aclada* distinct from *B. byssoidea*. *Phytopathology*, **91**, 527–33.

Nitta N, Farman ML and Leong SA (1997). Genome organization of *Magnaporthe grisea*: integration of genetic maps, clustering of transposable elements and identification of genome duplications and rearrangements. *Theoretical and Applied Genetics*, **95**, 20–32.

Notteghem JL and Silué D (1992). Distribution of the mating type alleles in *Magnaporthe grisea* populations pathogenic on rice. *Phytopathology*, **82**, 412–24.

Nuss DL (1992). Biological control of chestnut blight: an example of virus mediated attenuation of fungal pathogenesis. *Microbiological Reviews*, **56**, 561–76.

Nuss DL and Koltin Y (1990). Significance of dsRNA genetic elements in plant pathogenic fungi. *Annual Review of Phytopathology*, **28**, 37–58.

O'Dell M, Flavell RB and Hollins TW (1992). The classification of isolates of *Gaeumannomyces graminis* from wheat, rye and oats using restriction fragment length polymorphisms in families of repeated DNA sequences. *Plant Pathology*, **41**, 554–62.

O'Donnell K (2000). Molecular phylogeny of the *Nectria haematococca–Fusarium solani* species complex. *Mycologia*, **92**, 919–38.

O'Donnell K and Cigelnik E (1997). Two divergent intragenomic rDNA ITS2 types within a monophyletic lineage of the fungus *Fusarium* are nonorthologous. *Molecular Phylogenetics and Evolution*, **7**, 103–16.

O'Donnell K, Cigelnik E and Nirenberg HI (1998). Molecular systematics and phylogeography of the *Gibberella fujikuroi* species complex. *Mycologia*, **90**, 465–93.

O'Hara RB and Brown JKM (1996). Frequency- and density-dependent selection in wheat powdery mildew. *Heredity*, **77**, 439–47.

Ohta T (1992). The nearly neutral theory of molecular evolution. *Annual Review of Ecology and Systematics*, **23**, 263–86.

Olive LS (1956). Genetics of *Sordaria fimicola*. I: Ascospore color mutants. *American Journal of Botany*, **43**, 97–107.

Ono I and Ishino-Arao Y (1988). Inheritance of chromosome length polymorphisms in *Saccharomyces cerevisiae*. *Current Genetics*, **14**, 413–18.

Orbach MJ, Vollrath D, Davis RW and Yanofsky C (1988). An electrophoretic karyotype of *Neurospora crassa*. *Molecular and Cellular Biology*, **8**, 1469–73.

Orbach MJ, Chumley FG and Valent B (1996). Electrophoretic karyotypes of *Magnaporthe grisea* pathogens of diverse grasses. *Molecular Plant–Microbe Interactions*, **9**, 261–71.

Ota Y, Fukuda K and Suzuki K (1998a). The non-heterothallic life cycle of Japanese *Armillaria mellea*. *Mycologia*, **90**, 396–405.

Ota Y, Intini M and Hattori T (2000). Genetic characterization of heterothallic and non-heterothallic *Armillaria mellea sensu stricto*. *Mycological Research*, **104**, 1046–54.

Ota Y, Matsushita N, Nagasawa E, Terashita T, Fukuda K and Suzuki K (1998b). Biological species of *Armillaria* in Japan. *Plant Disease*, **82**, 537–43.

Otrosina WJ, Chase TE and Cobb FWJ (1992). Allozyme differentiation of intersterility groups of *Heterobasidion annosum* isolted from conifers in the western United States. *Phytopathology*, **82**, 540–5.

Oudemans PV, Alexander HM, Antonovics J, Alitzer S, Thrall PH and Rose L (1998). The distribution of mating type bias in natural populations of the anther-smut fungus *Ustilago violacea* on *Silene alba* in Virginia. *Mycologia*, **90**, 372–81.

Overholts LO (1953). *Polyporaceae of the United States, Alaska and Canada*. University of Michigan Press, Ann Arbor, MI.

Page DM and Holmes EC (1998). *Molecular evolution: a phylogenetic approach*. Blackwell Science, Oxford.

Panaccione DG, Sheets NL, Miller SP and Cumming JR (2001). Diversity of *Cenococum geophilum* isolates for serpentine and non-serpentine soils. *Mycologia*, **93**, 645–52.

Papa KE, Srb AM and Federer WT (1966). Selection for improved growth rate in inter- and intra-strain crosses of *Neurospora*. *Heredity*, **21**, 595–613.

Papa KE, Srb AM and Federer WT (1967). Inheritance of growth rate in *Neurospora crassa*: reverse selection in an improved strain. *Heredity*, **22**, 285–96.

Paquin C and Adams J (1983a). Frequency of fixation of adaptive mutations is higher in evolving diploid than haploid yeast populations. *Nature*, **302**, 495–500.

Paquin C and Adams J (1983b). Relative fitness can decrease in evolving asexual populations of *Saccharomyces cerevisiae*. *Nature*, **306**, 368–71.

Paran I and Michelmore RW (1991). Development of reliable PCR-based markers linked to downy mildew resistance genes in lettuce. *Theoretical and Applied Genetics*, **85**, 985–93.

Parlevliet JE (1977). Evidence of differential interaction in the polygenic *Hordeum vulgare-Puccinia hordei* relation during epidemic development. *Phytopathology*, **67**, 776–8.

Parlevliet JE and Zadoks JC (1977). The integrated concept of disease resistance: a new view including horizontal and vertical resistance in plants. *Euphytica*, **26**, 5–21.

Pateman JA (1955). Polygenic inheritance in *Neurospora*. *Nature*, **176**, 1274–5.

Pateman JA (1959). The effect of selection on ascospore size in *Neurospora crassa*. *Heredity*, **13**, 1–21.

Pateman JA and Lee BTO (1960). Segregation of polygenes in ordered tetrads. *Heredity*, **15**, 351–61.

Paterson RRM, King JG and Bridge PD (1990). High resolution thermal denaturation studies DNA from 14 *Penicillium* strains. *Mycological Research*, **94**, 152–6.

Peberdy JF, Eyssen H and Anné J (1977). Interspecific hybridization between *Penicillium chrysogenum* and *Penicillium cyano-fulvum* following protoplast fusion. *Molecular and General Genetics*, **157**, 281–4.

Peever TL and Milgroom MG (1994). Genetic structure of *Pyrenophora teres* populations determined with random amplified polymorphic DNA markers. *Canadian Journal of Botany*, **72**, 915–23.

Pegler DN (1983). The genus *Lentinula* (Tricholomataceae tribe Collybieae). *Sydowia*, **36**, 227–39.

Penner GA, Bush A, Wise R *et al.* (1993). Reproducibility of random amplified polymorphic DNA. *PCR Methods and Applications*, **2**, 341–5.

Perkins DD (1985). Aspects of the *Neurospora* genome. In *Molecular genetics of filamentous fungi* (ed. WE Timberlake), pp. 277–94. Liss, New York.

Perkins DD (1987). Mating-type switching in filamentous ascomycetes. *Genetics*, **115**, 384–92.

Perkins DD (1994). How should the infertility of interspecific crosses be designated? *Mycologia*, **86**, 758–61.

Perkins DD (1997). Chromosomal rearrangements in *Neurospora* and other filamentous fungi. *Advances in Genetics*, **36**, 239–398.

Perkins DD and Barry EG (1977). The cytogenetics of *Neurospora*. *Advances in Genetics*, **19**, 133–285.

Perkins DD and Turner BC (1988). *Neurospora* from natural populations: toward the population biology of a haploid eukaryote. *Experimental Mycology*, **12**, 91–131.

Person C (1966). Genetic polymorphism in parasitic systems. *Nature*, **212**, 266–7.

Petersen RH (1995a). There's more to a mushroom than meets the eye: mating studies in the Agaricales. *Mycologia*, **87**, 1–17.

Petersen RH (1995b). Contributions of mating studies to mushroom systematics. *Canadian Journal of Botany*, **73**, S831–42.

Petersen RH and Bermudes D (1992). *Panellus stypticus*: geographically separated interbreeding populations. *Mycologia*, **84**, 209–13.

Petrini O, Toti L, Petrini LE and Henniger W (1990). *Gremmeniella abietina* and *G. laricina* in Europe: characterization and identification of isolates and laboratory strains by soluble protein electrophoresis. *Canadian Journal of Botany*, **68**, 2629–35.

Pimental G, Carris LM and Peever TL (2000). Characterization of interspecific hybrids between *Tilletia controversa* and *T.bromi*. *Mycologia*, **92**, 411–20.

Pipe ND, Azcoitia V and Shaw DS (2000). Self-fertility in *Phytophthora infestans*: heterokaryons segregate several phenotypes. *Mycological Research*, **104**, 676–80.

Pipe ND, Buck KW and Brasier CD (1997). Comparison of the cerato-ulmin (cu) gene sequences of the Himalayan Dutch elm disease fungus *Ophiostoma himal-ulmi* with those of *O. ulmi* and *O. novo-ulmi* suggests that the cu gene of *O. novo-ulmi* is unlikely to have been acquired recently from *O. himal-ulmi*. *Mycological Research*, **101**, 415–21.

Pipe ND, Brasier CM and Buck KW (2000). Evolutionary relationships of the Dutch elm disease fungus *Ophiostoma novo-ulmi* to the *Ophiostoma* species investigated by RFLP analysis of the rDNA region. *Journal of Phytopathology*, **148**, 533–9.

Pirozynski KA and Dalpé Y (1989). Geological history of the Glomaceae with particular reference to mycorrhizal symbiosis. *Symbiosis*, **7**, 1–36.

Pittenger TH and Atwood KC (1956). Stability of nuclear proportions during growth of *Neurospora* heterokaryons. *Genetics*, **41**, 227–41.

Pittenger TH, Kimball AW and Atwood KC (1955). Control of nuclear ratios in *Neurospora* heterokaryons. *American Journal of Botany*, **42**, 954–8.

Plummer KM and Howlett BJ (1993). Major chromosomal length polymorphisms are evident after meiosis in the phytopathogenic fungus *Leptosphaeria maculans*. *Current Genetics*, **24**, 107–13.

Plummer KM and Howlett BJ (1995). Inheritance of chromosomal length polymorphisms in the ascomycete *Leptosphaeria maculans*. *Molecular and General Genetics*, **247**, 416–27.

Pontecorvo G (1946). Genetic systems based on heterocaryosis. *Cold Spring Harbor Symposia on Quantitative Biology*, **11**, 193–201.

Pontecorvo G (1959). *Trends in genetic analysis*. Columbia University Press, New York.

Pontecorvo G and Gemmell AR (1944). Genetic proof of heterokaryosis in *Penicillium notatum*. *Nature*, **154**, 532–34.

Pontecorvo G and Käfer E (1958). Genetic analysis by means of mitotic recombination. *Advances in Genetics*, **9**, 71–104.

Poon A and Otto SP (2000). Compensating for our load of mutations: freezing the meltdown of small populations. *Evolution*, **54**, 1467–79.

Posada D, Crandall KA and Templeton AR (2000). GEODIS: a program for the cladistic nested analysis of the georgraphical distribution of genetic haplotypes. *Molecular Ecology*, **9**, 487–8.

Pratt RG and Green RJ (1973). The sexuality and population structure of *Pythium sylvaticum*. *Canadian Journal of Botany*, **51**, 429–36.

Pringle A, Moncalvo J-M and Vilgalys R (2000). High levels of variation in ribosomal DNA sequences within and among spores of a natural population of the arbuscular mycorrhizal fungus *Acaulospora colossica*. *Mycologia*, **92**, 259–68.

Pugh GJF (1980). Strategies in fungal ecology. *Transactions of the British Mycological Society*, **75**, 1–14.

Pugh GJF and Buckley NG (1971). The leaf surface as a substrate for colonization by fungi. In *Ecology of leaf surface microorganisms* (ed. TF Preece and CH Dickinson), pp. 431–46. Academic Press, London.

Puhalla JE (1979). Classification of isolates of *Verticillium dahliae* based on heterokaryon incompatibility. *Phytopathology*, **69**, 1186–9.

Puhalla JE (1985). Classification of strains of *Fusarium oxysporum* on the basis of vegetative compatibility. *Canadian Journal of Botany*, **63**, 179–83.

Puhalla JP and Hummel M (1983). Vegetative compatibility groups within *Verticillium dahliae*. *Phytopathology*, **73**, 1305–8.

Pukkila PJ and Casselton LJ (1991). Molecular genetics of the agaric, *Coprinus cinereus*. In *More gene manipulations in fungi* (ed. JW Bennett and LL Lasure), pp. 126–50. Academic Press, San Diego, CA.

Quintanilha A (1937). Contribution à l'étude génétique du phénomène de Buller. *Comptes Rendus Hebdomadaires des Séances de l'Académie des Sciences*, **205**, 745.

Raju NB (1978). Meiotic nuclear behaviour and ascospore formation in five homothallic species of *Neurospora*. *Canadian Journal of Botany*, **56**, 754–63.

Raju NB (1992). Functional heterothallism resulting from homokaryotic conidia and ascospores in *Neurospora tetrasperma*. *Mycological Research*, **96**, 103–16.

Raju NB (1994). Ascomycete spore killers: chromosomal elements that distort genetic ratios among the products of meiosis. *Mycologia*, **86**, 461–73.

Raju NB and Newmeyer D (1977). Giant croziers and abnormal croziers in a mutant of *Neurospora crassa*. *Experimental Mycology*, **1**, 152–65.

Ramsdale M (1999). Genomic conflict in fungal mycelia. In *Structure and dynamics of fungal populations* (ed. JJ Worrall), pp. 139–74. Kluwer Academic, Dordrecht.

Ramsdale M and Rayner ADM (1994). Distribution patterns of nuclei in conidia from heterokaryons of *Heterobasidion annosum* (Fr.)Bref. and their interpretation in terms of genomic conflict. *New Phytologist*, **128**, 123–34.

Ramsdale M and Rayner ADM (1996). Imbalanced nuclear ratios, post germination mortality and phenotype–genotype relationships in allopatrically-derived heterokaryons of *Heterobasidion annosum*. *New Phytologist*, **133**, 303–19.

Randerson PE (1993). Ordination. In *Biological data analysis: a practical approach* (ed. JC Fry), pp. 173–217. IRL Press, Oxford.

Raper JR (1939). Sexual hormones in *Achlya*. I. Indicative evidence for a hormonal coordinating mechanism. *American Journal of Botany*, **26**, 639–50.

Raper JR (1966). *Genetics of sexuality in higher fungi*. Ronald Press, New York.

Raper JR and Ellingboe AH (1962). The Buller phenomenon in *Schizophyllum commune*: nuclear selection in fully compatible dikaryotic–homokaryotic matings. *American Journal of Botany*, **49**, 538–46.

Raper JR and Flexer AS (1970). The road to diploidy with emphasis on a detour. In *Organization and control in prokaryotic and eukaryotic cells* (ed. HP Charles and BCJ Knight), pp. 401–32. Cambridge University Press.

Raper JR, Baxter MG and Middleton RB (1958). The genetic structure of the incompatibility factors in *Schizophyllum commune*. *Proceedings of the National Academy of Sciences of the United States of America*, **46**, 833–42.

Rayner ADM and Todd NK (1978). Polymorphism in *Coriolus versicolor* and its relation to interfertility and intraspecific antagonism. *Transactions of the British Mycological Society,* **71**, 99–106.

Rayner ADM and Todd NK (1979). Population and community structure and dynamics of fungi in decaying wood. *Advances in Botanical Research,* **7**, 333–420.

Reynolds DR and Taylor JW (ed.) (1993). The fungal holomorphic species: mitotic, meiotic and pleomorphic speciation in fungal systematics. CAB International, Wallingford.

Rishbeth J (1968). The growth rate of *Armillaria mellea. Transactions of the British Mycological Society,* **51**, 575–86.

Rizet G and Esser K (1953). Sur des phénomènes d'incompatibilité entre souches d'origines différentes chez *Podospora anserina. Comptes Rendus Hebdomadaires des Séances de l'Académie des Sciences,* **237**, 760–1.

Rizzo DM, Rentmeester RM and Burdsall HH (1995). Sexuality and somatic compatibility in *Phellinus gilvus. Mycologia,* **87**, 805–20.

Robertson SJ, Bond DJ and Read ND (1998). Homothallism and heterothallism in *Sordaria brevicollis. Mycological Research,* **102**, 1215–23.

Roeder DS and Fink GR (1983). Transposable elements in yeast. In *Mobile genetic elements* (ed. JA Shapiro), pp. 299–328. Academic Press, New York.

Roelfs AP (1982). Effects of barberry eradication on stem rust in the United States. *Plant Disease Reporter,* **66**, 177–81.

Roelfs AP and Groth JV (1980). A comparison of virulence phenotypes in wheat stem rust populations reproducing sexually and asexually. *Phytopathology,* **70**, 855–62.

Roelfs AP, McCallum B, McVey DV and Groth JV (1997). Comparison of virulence and isozyme phenotypes of Pgt-QCCJ and Great Plains races of *Puccinia graminis* f.sp. *tritici. Phytopathology,* **87**, 910–14.

Rogers HJ, Buck KW and Brasier CE (1988). dsRNA and disease factors of the aggressive subgroup of *Ophiostoma ulmi.* In *Viruses of fungi and simple eukaryotes* (ed. Y Koltin and MJ Leibowitz), pp. 327–51. Marcl Dekker, New York.

Rogers JD (1968). *Xylaria curta*: cytology of the ascus. *Canadian Journal of Botany,* **46**, 1337–40.

Rogers JD (1973). Polyploidy in fungi. *Evolution,* **27**, 153–60.

Rogers SO and Rogers MM (1999). Gene flow in fungi. In *Structure and dynamics of fungal populations* (ed. JJ Worrall), pp. 97–121. Kluwer Academic, Dordrecht.

Roll-Hansen F and Roll-Hansen H (1973). *Cleroderris lagerbergii* in Norway. Hosts, distribution, perfect and imperfect state and mode of attack. *Meddelelser Fra Det Norske Skogforsoksvesen,* **124**, 441–59.

Rosendahl S and Sen R (1992). Isozyme analysis of mycorrhizal fungi and their ectomycorrhiza. In *Techniques for the study of mycorrhiza,* vol. 24 (ed. JR Norris, DJ Read and AK Varma), pp.167–94. Academic Press, London.

Rosendahl S and Taylor JW (1997). Development of multiple genetic markers for studies of genetic variation in arbuscular mycorrhizal fungi using AFLP. *Molecular Ecology,* **6**, 821–9.

Roumen E, Levy M and Notteghm JL (1997). Characterization of the European population of *Magnaporthe grisea* by DNA fingerprinting and pathotype analysis. *European Journal of Plant Pathology,* **103**, 363–71.

Roux KH (1995). Optimization and trouble shooting in PCR. *PCR Methods and Applications,* **4**, 184–95.

Rowe RC (1995). Recent progress in understanding relationships between *Verticillium* and subspecific groups. *Phytoparasitica,* **23**, 31–8.

Rowell JB and Roelfs AP (1971). Evidence for an unrecognized source of overwintering stem rust in the United States. *Plant Disease Reporter,* **55**, 990–2.

Roy BA (1993). Floral mimicry by plant pathogens. *Nature,* **362**, 56–8.

Russell PJ, Wagner S, Rodland KD *et al.* (1984). Organization of the ribosomal nucleic acid genes in various wild-type strains and wild collections of *Neurospora. Molecular and General Genetics,* **196**, 275–82.

Rustchenko-Bulgac EP (1991). Variation of *Candida albicans* electrophoretic karyotypes. *Journal of Bacteriology,* **173**, 6586–96.

Rustchenko-Bulgac EP, Sherman F and Hicks JB (1990). Chromosomal rearrangements associated with morphological mutants provide a means for genetic variation of *Candida albicans. Journal of Bacteriology,* **172**, 1276–83.

Rychlik W, Spencer WJ and Rhoades RE (1990). Optimization of the annealing temperature for DNA amplification *in vitro. Nucleic Acids Research,* **18**, 6409–12.

Rychlik W, Spencer WJ and Rhoades RE (1991). Corrigendum. Optimization of the annealing temperature for DNA amplification *in vitro. Nucleic Acids Research,* **19**, 698.

Sackville-Hamilton NR, Schmid B and Harper JL (1987). Life-history concepts and the population biology of modular organisms. *Proceedings of the Royal Society of London, Series B,* **232**, 35–57.

Saenz GS, Stam JG, Jacobson DJ and Natvig DO (2001). Hetroallelism at the *het-c* locus contributes to sexual dysfunction in outcrossed strains of *Neurospora tetrasperma. Fungal Genetics and Biology,* **34**, 123–9.

Saiki RK, Gelfand DH, Stoffel S, *et al.* (1988). Primer-directed enzymatic amplification of DNA with a thermostable DNA polymerase. *Nature*, **239**, 487–97.

Saitou N and Nei M (1987). The neighbour-joining method: a new method for reconstructing phylogenetic trees. *Molecular Biology and Evolution*, **4**, 406–25.

Sanders IR (1999). No sex please, we're fungi. *Nature*, **399**, 737–9.

Sanders JPM, Heyting C, DiFranco A *et al.* (1976). The organization of genes in yeast mitochondrial DNA. In *The genetic function of mitochondrial DNA* (ed. C Saccone and AM Kroon), pp. 259–72. North Holland Publishing, Amsterdam.

Sanders IR, Alt M, Groppe K, Boller T and Wiemken A (1995). Identification of ribosomal DNA polymorphism among and within spores of the Glomales: application to studies on the genetic diversity of arbuscular mycorrhizal fungal communities. *New Phytologist*, **130**, 419–27.

Sanderson FR (1972). A *Mycosphaerella* species as the ascogenous stage of *Septoria tritici* Rob. and Desm. *New Zealand Journal of Botany*, **10**, 707–9.

Sanderson FR (1976). *Mycosphaerella graminicola* (Fuckel) Sanderson comb. nov., the ascogenous state of *Septoria tritici* Rob. and Desm. *New Zealand Journal of Botany*, **14**, 359–60.

Sanger F, Nicklen S and Coulson AR (1977). DNA sequencing with chain terminating inhibitors. *Proceedings of the National Academy of Sciences of the United States of America*, **74**, 5463–7.

Sansome ER (1959). Pachytene in *Puccinia kraussiana* Cooke, on *Smilax kraussiana*. *Nature*, **184**, 1820–1.

Sansome ER (1961). Meiosis in the oogonium and antheridium of *Pythium debaryanum*. *Nature*, **191**, 827–8.

Sansome ER (1965). Meiosis in diploid and polyploid sex organs of *Phytophthora* and *Achlya*. *Cytologia*, **30**, 103–17.

Sansome ER (1977). Polyploidy and induced gametangial formation in British isolates of *Phytophthora infestans*. *Journal of General Microbiology*, **99**, 311–16.

Sansome ER (1987). Fungal chromosomes as observed with the light microscope. In *Evolutionary biology of the Fungi* (ed. ADM Rayner, CM Brasier and D Moore), pp. 97–113. Cambridge University Press.

Sansome ER and Brasier CM (1974). Polyploidy associated with varietal differentiation in the *megasperma* complex of *Phytophthora*. *Transactions of the British Mycological Society*, **63**, 461–7.

Sansome ER, Brasier CM and Hamm PB (1991). *Phytophthora media* may be a species hybrid. *Mycological Research*, **95**, 273–7.

Satina S and Blakeslee AF (1930). Imperfect sexual reactions in homothallic and heterothallic forms. *Botanical Gazette*, **90**, 299–311.

Saunders GA and Hancock JH (1994). Self-sterile isolates of *Pythium* mate with self-fertile isolates of *Pythium ultimum*. *Mycologia*, **86**, 660–6.

Saupe SJ, Kuldan GA, Smith ML and Glass NL (1996). The product of the *het-c* heterokaryon incompatibility gene of *Neurospora crassa* has characteristics of a glycine-rich cell wall protein. *Genetics*, **143**, 1589–1616.

Saville BJ, Yoell H and Anderson JB (1996). Genetic exchange and recombination in populations of the root infecting fungus *Armillaria gallica*. *Molecular Ecology*, **5**, 485–98.

Schäfer W and Yoder OC (1994). Organ specificity of fungal pathogens on host and non-host plants. *Physiological and Molecular Plant Pathology*, **45**, 211–18.

Schäfer W, Stanley D, Ciuffetti L, Van Etten HD and Yoder OC (1989). One enzyme makes a fungal pathogen, but not a saprophyte, virulent on a new host. *Science*, **246**, 247–9.

Schardl CL, Leuchtman H-F, Tsai MA, Collett MA, Watt DM and Scott DB (1994). Origin of a fungal symbiont of perennial ryegrass by interspecific hybridization of a mutualist with the ryegrass choke pathogen, *Epichloë typhina*. *Genetics*, **136**, 1307–17.

Schipper MAA (1973). A study of variabiity in *Mucor hiemalis* and related species. *Studies in Mycology*, **4**, 1–40.

Schipper MAA (1975). *On Mucor mucedo, Mucor flavus and related species*. *Studies on Mycology*, **10**, 1–33.

Schipper MAA, Gauger W and van den Ende H (1985). Hybridization of *Rhizopus* species. *Journal of General Microbiology*, **131**, 2359–65.

Schneider S, Keuffer J-M, Roessli D and Excoffier L (1997). *Arlequin: software for population genetic data analysis*. Genetics and Biometry Laboratory, University of Geneva.

Schüepp H and Küng M (1981). Stability of tolerance to MBC in populations of *Botrytis cinerea* in vineyards of northern and eastern Switzerland. *Canadian Journal of Plant Pathology*, **3**, 180–1.

Schwarzbach E (1978). A high throughput jet trap for collecting mildew spores on living leaves. *Phytopathologische Zeitung*, **94**, 165–71.

Scott PR, Sanderson FR and Benedikz PW (1988). Occurrence of *Mycosphaerella graminicola*, a teleomorph of *Septoria tritici*, on wheat debris in the UK. *Plant Pathology*, **37**, 285–90.

Selosse M-A *et al.* (1996). *Current Genetics*, **30**, 332–7.

Selosse M-A, Jacquot D, Bouchard D, Martin F and Le Tacon F (1998a). Temporal persistence and spatial distribution of an American strain of the ectomycorrhizal basidiomycete *Laccaria bicolor* in European forest plantations. *Molecular Ecology*, **7**, 561–76.

Selosse M-A, Martin F and Le Tacon F (1998b). Survival of an introduced ectomycorrhizal *Laccaria bicolor* strain in a European forest plantation monitored by mitochondrial ribosomal DNA analysis. *New Phytologist*, **140**, 753–61.

Selosse M-A, Martin F, Bouchard D and Le Tacon F (1999). Structure and dynamics of experimentally introduced and naturally occurring *Laccaria* spp. discrete genotypes in a Douglas fir plantation. *Applied and Environmental Microbiology*, **65**, 2006–14.

Sen R (1990). Intraspecific variation in two species of *Suillus* from Scots pine (*Pinus sylvestris* L.) forests based on somatic incompatibility and isozyme analyses. *New Phytologist*, **114**, 607–16.

Shabi E, Katan T and Marton K (1983). Inheritance of resistance to benomyl in isolates of *Venturia inaequalis*. *Plant Pathology*, **32**, 207–11.

Shan W-X, Chen S-Y, Kang Z-S, Wu L-R and Li Z-Q (1998). Genetic diversity in *Puccinia striiformis* Westend. f.sp. *tritici* revealed by pathogen genome-specific repetitive sequence. *Canadian Journal of Botany*, **76**, 587–95.

Shannon CE and Weaver W (1949). *The mathematical theory of communication*. University of Illinois Press, Urbana, IL.

Shattock RC, Tooley PW and Fry WE (1986). Genetics of *Phytophthora infestans*: characterisation of single-oospore cultures form A1 isolates induced to self by intraspecific stimulation. *Phytopathology*, **76**, 407–10.

Shaw DV and Brown AHD (1982). Optimum number of marker loci for estimating outcrossing in plant populations. *Theoretical and Applied Genetics*, **61**, 321–5.

Shaw DV, Kahler AL and Allard RW (1981). A multilocus estimator of mating system parameters in plant populations. *Proceedings of the National Academy of Sciences of the United States of America*, **78**, 1298–1302.

Shaw MW and Royle DJ (1989). Airborne inoculum as a major source of *Septoria tritici* (*Mycosphaerella graminicola*) infections in winter wheat crops in the UK. *Plant Pathology*, **38**, 35–43.

Shear CL and Dodge BO (1927). Life histories and heterothallism of the red bread-mould fungus of the *Monilia sitophila* group. *Journal of Agricultural Research*, **34**, 1014–42.

Sheldon AL (1969). Equitability indices: dependence on the species count. *Ecology*, **50**, 466–7.

Shepherd HS (1992). Linear, non-mitochondrial plasmids of *Alternaria alternata*. *Current Genetics*, **21**, 169–72.

Shi YL, Loomis P, Christian D, Carris LM and Leung H (1996). Analysis of genetic relationships among the wheat bunt fungi using RAPD and ribosomal DNA markers. *Phytopathology*, **86**, 311–18.

Shimomura N, Hasebe K, Nakai-Fukumasa Y and Komatsu M (1992). Intercompatibility between geographically distant strains of shiitake. *Report of the Tottori Mycological Institute*, **30**, 26–9.

Shirane N, Masuko M and Hayasahi Y (1989). Light microscopic observations on nuclei and mitotic chromosomes of *Botrytis* species. *Phytopathology*, **79**, 728–30.

Shykoff JA, Meyhöfer A and Bücheli E (1999). Genetic isolation among host races of the anther smut fungus *Microbotryum violaceum* on three plant host species. *International Journal of Plant Science*, **160**, 907–16.

Sidenko LV (1965). Breeding maize against maize smut (in Russian). In *IVth All-Union conference on immunity in agricultural plants* (ed. Kisiner I), pp. 207–8. Moscow.

Sidhu G and Person C (1971). Genetic control of virulence in *Ustilago hordei*. II: Segregations for higher levels of virulence. *Canadian Journal of Genetics and Cytology*, **13**, 173–8.

Simchen G (1965). Variation in a dikaryotic population of *Collybia velutipes*. *Genetics*, **51**, 709–21.

Simchen G (1966). Fruiting and growth rate among dikaryotic progeny of single wild isolates of *Schizophyllum commune*. *Genetics*, **53**, 1151–65.

Simchen GW (1967). Independent evolution of a polygenic system in isolated populations of the fungus *Schizophyllum commune*. *Evolution*, **21**, 310–15.

Simchen GW and Jinks JL (1964). The determination of dikaryotic growth rate in the Basidiomycete *Schizophyllum commune*. *Heredity*, **19**, 629–49.

Simpson EH (1949). Measurement of diversity. *Nature*, **163**, 688.

Skinner DZ, Budde AD and Leong SA (1991). Molecular karyotype analysis of fungi. In *More gene manipulations in fungi* (ed. JW Bennett and LL Lasure), pp. 86–103. Academic Press, San Diego, CA.

Sjöwall M (1945). *Studien über Sexualität, Vererbung und Zytologie bei einigen diözischen Mucoreen*. Thesis, pp. 97. Botanisches Instituts, University of Lund.

Skinner DZ, Budde AD, Farman ML, Smith JR, Leung H and Leong SA (1993). Genome organization of *Magnaporthe grisea*—genetic map, electrophoretic karyotype and occurrence of repeated DNAs. *Theoretical and Applied Genetics*, **87**, 545–57.

Skupski MP, Jackson DA and Natvig DO (1997). Phylogenetic analysis of heterothallic *Neurospora* species. *Fungal Genetics and Biology*, **21**, 153–62.

Slatkin M (1985). Gene flow in natural populations. *Annual Review of Ecology and Systematics*, **16**, 393–430.

Slatkin M (1993). Isolation by distance in equilibrium and non-equilibrium populations. Evolution, 47, 264–278.

Slatkin M and Barton NH (1989). A comparison of three indirect methods for estimating the average levels of gene flow. Evolution, 43, 1349–67.

Smart CD, Willmann MR, Mayton H et al. (1998). Self-fertility in two clonal lineages of Phytophthora infestans. Fungal Genetics and Biology, 25, 134–42.

Smart CD, Mayton H, Mizubuti ESG, Willmann MR and Fry WE (2000). Environmental and genetic factors influencing self-fertility in Phytophthora infestans. Phytopathology, 90, 987–94.

Smith BR (1966). Genetic controls of recombination. Heredity, 21, 481–98.

Smith ML, Bruhn JN and Anderson JB (1992). The fungus Armillaria bulbosa is amongst the largest and oldest living organisms. Nature, 356, 428–31.

Smith ML, Bruhn JN and Anderson JB (1994). Relatedness and spatial distribution of Armillaria genets infecting red pine seedlings. Phytopathology, 84, 822–9.

Smith SE and Read DJ (1997). Mycorrhizal symbiosis. Academic Press, London.

Sneath PHA and Sokal RR (1973). Numerical taxonomy. WH Freeman, San Francisco, CA.

Snyder WC, Georgopoulos SG, Webster RK and Smith SN (1975). Sexuality and genetic behaviour in the fungus Hypomyces (Fusarium) solani f.sp. cucurbitae. Hilgardia, 43, 161–85.

Sokal RR and Oden NL (1978). Spatial autocorrelation in biology. 1: Methodology. Biological Journal of the Linnean Society, 10, 199–228.

Sokal RR and Sneath PHA (1963). Principles of numerical taxonomy. WH Freeman, San Francisco, CA.

Southern EM (1975). Detection of specific sequences among DNA fragments separated by gel electrophoresis. Journal of Molecular Biology, 98, 503–17.

Specht CA, Novotny, CP and Ullrich RC (1984). Strain specific differences in ribosomal DNA from the fungus Schizophyllum commune. Current Genetics, 8, 42–9.

Spielman LJ, Drenth A, Davidse LC et al. (1991). A second world-wide migration and population displacement of Phytophthora infestans? Plant Pathology, 40, 422–30.

Spiers AG and Hopcroft DH (1994). Comparative studies of the poplar rusts Melampsora medusae, M. larici-populina and their interspecific hybrid M. medusae-populina. Mycological Research, 98, 889–903.

Spieth PT (1975). Population genetics of allozyme variation in Neurospora intermedia. Genetics, 80, 785–805.

Sreenivasaprasad S, Mills PR, Meehan BM and Brown AE (1996). Phylogeny and systematics of 18 Colletotrichum species based on ribosomal DNA spacer sequences. Genome, 39, 499–512.

Stadler D, Macleod H and Dillon D (1991). Spontaneous mutation at the mtr locus of Neurospora: the spectrum of mutant types. Genetics, 129, 39–47.

Stakman EC (1914). A study in cereal rusts. Physiological races. Minnesota Agricultural Experiment Station Bulletin, 138, 27 pp.

Stakman EC (1933). Epidemiology of cereal rusts. In Proceedings of the 5th Pacific Science Congress, Victoria and Vancouver, 1933, Vol. 4, pp. 3177–84, National Research Council of Canada.

Stakman EC and Harrar JG (1957). Principles of plant pathology. Ronald Press, New York.

Stakman EC and Levine MN (1922). The determination of biologic forms of Puccinia graminis on Triticum spp. Technical Bulletin 8, University of Minnesota Agricultural Experimental Station.

Stasz TE and Martin SP (1988). Insensitivity of thick walled-oospores of Pythium ultimum to fungicides, methyl bromide, and heat. Phytopathology, 78, 1409–12.

Stasz TE, Weeden NF and Harman GE (1988). Methods of isozyme electrophoresis for Trichoderma and Gliocladium species. Mycologia, 80, 870–74.

Staub T (1991). Fungicide resistance: practical experience with antiresistance strategies and the role of integrated use. Annual Review of Phytopathology, 29, 421–42.

Stearns SC (1982). The role of development in the evolution of life histories. In Evolution and development (ed. JT Bonner), pp. 237–58. Springer-Verlag, New York.

Stenlid J (1985). Population structure in Heterobasidion annosum as determined by somatic incompatibility, sexual compartibility, and isoenzyme patterns. Canadian Journal of Botany, 63, 2268–73.

Stenlid J (1987). Controlling and predicting the spread of Heterobasidion annosum from infected stumps and trees of Picea abies. Scandinavian Journal of Forest Research, 2, 187–98.

Stenlid J and Karlsson J-O (1991). Partial intersterility in Heterobasidion annosum. Mycological Research, 95, 1153–9.

Stenlid J and Swedjemark G (1988). Differential growth of S- and P-isolates of Heterobasidion annosum on Pices abies and Pinus sylvestris. Transactions of the British Mycological Society, 90, 209–13.

Stipes RJ, Emert GH and Brown RD Jr (1982). Differentiation of Endothia gyrosa and Endothis parasitica by disc electrophoresis of intramycelial enzymes and proteins. Mycologia, 74, 138–41.

Stoddart JA and Taylor JF (1988). Genotypic diversity: estimation and prediction in samples. Genetics, 118, 705–11.

Sussman AS (1968). Longevity and survival of fungi. In *The Fungi: an advanced treatise*. Vol. II: *The fungal population* (ed. GC Ainsworth and AS Sussman), pp. 447–85. Academic Press, New York.

Swift MJ (1958). Inhibition of rhizomorph development by *Armillaria mellea* in Rhodesian forest soils. *Transactions of the British Mycological Society*, 51, 241–7.

Swofford DL (1998). PAUP*: *Phylogenetic Analysis Using Parsimony (and other methods)*. Sinauer Associates, Sunderland, MA.

Swofford DL (2000). PAUP*: *Phylogenetic Analyses Using Parsimony and other methods*. Sinauer Associates, Sunderland, MA.

Swofford DL, Olsen GJ, Waddell PJ and Hillis DM (1996). Phylogenetic inference. In *Molecular systematics* (ed. DM Hillis, C Moritz and BK Mable), pp. 407–514. Sinauer Associates, Sunderland, MA.

Taga M and Murata M (1994). Visualization of mitotic chromosomes in filamentous fungi by fluorescence staining and fluorescence *in situ* hybridization. *Chromosoma*, 103, 408–13.

Taga M, Murata M and Saito H (1998). Comparison of different karyotyping methods in filamentous ascomycetes—a case study of *Nectria haematococca*. *Mycological Research*, 102, 1355–64.

Talbot NJ (1998). Molecular variation of fungal pathogens: using the rice blast fungus as a case study. In *Molecular variability of fungal pathogens* (ed. PD Bridge, Y Couteaudier and JM Clarkson), pp. 1–18. CAB International, Wallingford.

Talbot NJ, Salch YP, Ma M and Hamer JE (1993). Karyotype variation within clonal lineages of the rice blast fungus, *Magnaporthe grisea*. *Applied and Environmental Microbiology*, 59, 585–93.

Tanksley SD and Orton TJ (1983). *Isozymes in plant genetics and breeding*. Elsevier, Amsterdam.

Tariq V-N, Gutteridge CS and Jefferies P (1985). Comparative studies of cultural and biochemical characteristics used for distinguishing species within *Sclerotinia*. *Transactions of the British Mycological Society*, 84, 381–97.

Tatum EL, Barratt RW, Fries N and Bonner DM (1950). Biochemical mutant strains of *Neurospora* produced by physical and chemical treatment. *American Journal of Botany*, 37, 38–46.

Taylor JW (1986). Fungal evolutionary biology and mitochondrial DNA. *Experimental Mycology*, 10, 259–69.

Taylor JW, Jacobson DJ and Fisher MC (1999). The evolution of asexual fungi: reproduction, speciation and classification. *Annual Review of Phytopathology*, 37, 187–246.

Taylor JW, Jacobson DJ, Kroken S *et al.* (2000). Phylogenetic species recognition and species concepts in fungi. *Fungal Genetics and Biology*, 31, 21–3.

Tehler A, Farris J, Lipscomb DL and Källersjò M (2000). Phylogenetic analyses of the fungi based on large rDNA data sets. *Mycologia*, 92, 459–74.

Templeton AR (1998). Nested clade analysis of phylogeographic data: testing hypotheses about gene flow and population history. *Molecular Ecology*, 7, 381–97.

Templeton AR and Sing CF (1993). A cladistic analysis of phenotypic associates with haplotypes inferred from restriction endonuclease mapping and DNA sequence data. IV: Nested analysis with cladogram uncertainty and recombination. *Genetics*, 134, 659–69.

Templeton AR, Boerwinkle E and Sing CF (1987). A cladistic analysis of phenotype associations with haplotypes inferred from restriction endonuclease mapping. I: Basic theory and analysis of alcohol dehydrogenase activity in Drosophila. *Genetics*, 117, 343–51.

Templeton AR, Sing CF, Kessling A and Humphries S (1988). A cladistic analysis of phenotype associations with haplotypes inferred from restriction endonuclease mapping. II: The analysis of natural populations. *Genetics*, 120, 1145–54.

Templeton AR, Crandall KA and Sing CF (1992). A cladistic analysis of phenotypic associates with haplotypes inferred from restriction endonuclease mapping and DNA sequence data. III: Cladogram estimation. *Genetics*, 132, 619–33.

Templeton AR, Routeman E and Phillips CA (1995). Separating population structure from population history: a cladistic analysis of the geographical distribution of mitochondrial haplotypes in the Tiger Salamander, *Ambystoma tigrinum*. *Genetics*, 140, 767–82.

Teow SC and Upshall A (1983). A spontaneous mutant approach to genetic study of filamentous fungi. *Transactions of the British Mycological Society*, 81, 513–21.

Thoday JM (1972). Disruptive selection. *Proceedings of the Royal Society of London, Series B*, 182, 109–43.

Thompson Jr. JR. (1975). Quantitative variation and gene number. *Nature*, 258, 665–8.

Thompson JD, Gibson JT, Plewniak F, Jeanmougin F and Higgins DG (1997). The Clustal X–Windows interface: flexible strategies for multiple sequence alignment aided by quality analysis tools. *Nucleic Acids Research*, 25, 57–86.

Thon MR and Royse DJ (1999). Partial ß-tubulin gene sequences for evolutionary studies in the Basidiomycotina. *Mycologia*, 91, 468–74.

Thurston HD (1961). The relative survival abilities of races of *Phytophthora infestans* in mixtures. *Phytopathology*, 51, 748–55.

Tinline RD (1988). *Cochliobolus sativus*, a pathogen of wide host range. In *Genetics of plant pathogenic fungi*, Vol. 6 (ed. GS Sidhu), pp. 113–21. Academic Press, London.

Todd NK and Aylmore RC (1985). Cytology of hyphal interactions and reactions in *Schizophyllum commune*. In *Developmental biology of higher fungi* (ed. D Moore, LA Casselton, AD Wood and JC Frankland), pp. 231–48. Cambridge University Press.

Tommerup IC (1988). The vesicular–arbuscular mycorrhizas. In *Genetics of plant pathogenic fungi*, Vol. 6 (ed. GS Sidhu), pp. 81–91. Academic Press, London.

Tommerup IC (1990). *Laccaria* species ectomycorrhizal with eucalypts: why does the ecology of bisporic species differ from quadrisporic species? In *Proceedings of the eighth North American conference on mycorrhizae* (ed. M Allen and SM Williams), p. 284. Agricultural Experimental Station, Laramie, University of Wyoming.

Tommerup IC and Sivasithamparam K (1990). Zygospores and asexual spores of *Gigaspora decipiens*, an arbuscular–mycorrhizal fungus. *Mycological Research*, **94**, 897–900.

Tooley PW and Therrien CD (1991). Variation in ploidy in *Phytophthora infestans*. In *Phytophthora* (ed. JA Lucas, RC Shattock, DS Shaw and LR Cooke), pp. 205–17. Cambridge University Press.

Tooley PW, Fry WE and Villareal Gonzalez MJ (1985). Isozyme characterization of sexual and asexual *Phytophthora infestans* populations. *Journal of Heredity*, **76**, 431–5.

Trinci APJ (1994). Evolution of the Quorn® myco-protein fungus, *Fusarium graminearum*. *Microbiology*, **140**, 2181–8.

Trinci APJ, Robson GD, Wiebe MG, Cunliffe B and Naylor TW (1990). Growth and morphology of *Fusarium graminearum* and other fungi in batch and continuous culture. In *Microbial growth dynamics* (ed. RK Pole, MJ Bazin and CW Keevil), pp. 17–38. IRL Press, Oxford.

Tudzynski P and Esser K (1985). Mitochondrial DNA for gene cloning in Eukaryotes. In *Gene manipulations in Fungi* (ed. JW Bennet and LL Lasure), pp. 403–16. Academic Press, Orlando, FL.

Turgeon BG (1998). Application of mating type gene technology to problems in fungal biology. *Annual Review of Phytopathology*, **36**, 115–37.

Turner BC (1987). Two ecotypes of *Neurospora intermedia*, *Mycologia*, **79**, 425–32.

Turner BC, Perkins DD and Fairfield A (2001). *Neurospora* from natural populations: a global study. *Fungal Genetics and Biology*, **32**, 1–92.

Tzeng T, Chang H and Bronson CR (1989). Electrophoretic confirmation of chromosome rearrangements in *Cochliobolus heterostrophus*—translocation at *Tox1*. *Phytopathology*, **79**, 1204.

Tzeng T, Lynglolm LK, Ford CF and Bronson CR (1992). A restriction fragment length polymorphism map and electrophoretic karyotype of the fungal maize pathogen, *Cochliobolus heterostrophus*. *Genetics*, **139**, 81–96.

Uhm JY and Fujii H (1983a). Ascospore dimorphism in *Sclerotinia trifoliorum* and cultural characters of strains from different sized spores. *Phytopathology*, **73**, 565–9.

Uhm JY and Fujii H (1983b). Heterothallism and mating type mutation in *Sclerotinia trifoliorum*. *Phytopathology*, **73**, 569–72.

Ullrich RC (1973). Sexuality, incompatibility, and intersterility in the biology of the *Sistotrema brinkmannii* aggregate. *Mycologia*, **65**, 1234–9.

Ullrich RC and Anderson JB (1978). Sex and diploidy in *Armillaria mellea*. *Experimental Mycology*, **2**, 119–29.

Uotila A (1992). Mating system and apothecia production in *Gremmeniella abietina*. *European Journal of Forest Pathology*, **22**, 410–17.

Uotila A, Hantula J, Väänänen AK and Hamelin RC (2000). Hybridization between two biotypes of *Gremmeniella abietina* in artificial pairings. *Forest Pathology*, **30**, 211–19.

Vaillancourt LJ and Hanau RM (1992). Genetic and morphological comparisons of *Glomerella* (*Colletotrichum*) isolates from maize and from sorghum. *Experimental Mycology*, **16**, 219–29.

Vainio EJ and Hantula J (2000). Genetic differentiation between European and North American populations of *Phlebiopsis gigantea*. *Mycologia*, **92**, 436–46.

van der Auwera G, de Baere R, van der Peer Y, de Rijk P, van den Broeck I and de Wachter R (1995). The phylogeny of the Hyphochytriomycota as deduced from ribosomal RNA sequences of *Hyphochytrium catenoides*. *Molecular and Biological Evolution*, **12**, 671–8.

Van der Gaag M, Debets AJM, Oosterhof J, Slakhorst M, Thijssen JAGM and Hoekstra RF (2000). Spore-killing meiotic drive factors in a natural population of the fungus *Podospora anserina*. *Genetics*, **156**, 592–605.

Van der Plank JE (1968). *Disease resistance in plants*. Academic Press, New York.

Van der Plank JE (1978). *Genetic and molecular basis of plant pathogenesis*. Springer-Verlag, Berlin.

Van der Plank JE (1982). *Host–pathogen interactions in plant disease*. Academic Press, New York.

van der Walt JP and Johannsen E (1979). A comparison of infertility and *in vitro* DNA–DNA reassociation as criteria for speciation in the genus *Kluveromyces*. *Antonie van Leeuwenhoek: Journal of Microbiology and Serology*, **45**, 281–91.

Van Etten HD, Funnell-Baerg D, Wasmann C and McClusky K (1994). Location of pathogenicity genes on dispensible chromosomes in *Nectria haematococca* MP VI. *Antonie van Leeuwenhoek: Journal of Microbiology and Serology*, **65**, 263–7.

Van Horn R and Clay K (1995). Mitochondrial DNA variation in the fungus *Atkinsonella hypoxylon* infecting sympatric *Danthonia* grasses. *Evolution*, **49**, 360–71.

Vos P, Hogers R, Bleeker M *et al.* (1995). AFLP: a new technique for DNA fingerprinting. *Nucleic Acids Research*, **23**, 4407–14.

Vaillancourt LJ and Hanau RM (1991). A method for genetic analysis of *Glomerella graminicola* from maize. *Phytopathology*, **81**, 530–4.

Vaillancourt LJ and Hanau RM (1992). Genetic and morphological comparisons of *Glomerella* (*Colletotrichum*) isolates from maize and from sorghum. *Experimental Mycology*, **16**, 219–29.

Valent B and Chumley FG (1991). Molecular genetic analysis of the rice blast fungus, *Magnaporthe grisea*. *Annual Review of Phytopathology*, **29**, 443–67.

Vandendries R (1927). Le comportment sexuel du Corpin Micacé dans ses rapports avec la dispersion de l'espèce. *Bulletin de la Société Royale de Belge*, **60**, 62–5.

Vasiliauskas R and Stenlid J (1997). Population structure and genetic variation in *Nectria fuckeliana*. *Canadian Journal of Botany*, **75**, 1707–13.

Vilgalys R (1991). Speciation and species concepts in the *Collybia dryophila* complex. *Mycologia*, **83**, 758–73.

Wahl I, Eshed N, Segal A and Sobel Z (1978). Significance of wild relatives of small grains and other wild grasses in cereal powdery mildews. In *The powdery mildews* (ed. DM Spencer), pp. 83–100. Academic Press, London.

Wang SS and Anderson NA (1972). A genetic analysis of sporocarp production in *Pleurotus sapidus*. *Mycologia*, **64**, 521–8.

Wang X-R (1997). Genetic variability in the canker pathogen fungus, *Gremmeniella abietina*. Contribution of sexual compared with asexual reproduction. *Mycological Research*, **101**, 1195–1201.

Wang X-R, Ennos RA, Szmidt AE and Hanson P (1997). Genetic variability in the canker pathogen fungus, *Gremmeniella abietina*. 2: Fine-scale investigation of the population genetic structure. *Canadian Journal of Botany*, **75**, 1460–9.

Watling R (1977). An analysis of the taxonomic characters used in defining species of the Bolbitaceae. In *The species concept in Hymenomycetes*, Vol. 61 (ed. H. Clémençon), pp. 11–43. J. Cramer (A. R. Gantner Verlag KG), Vaduz.

Watling R (1988). Larger fungi and some of Earth's major catastrophes. *Proceedings of the Royal Society of Edinburgh, Series B*, **94**, 49–59.

Watson IG and Luig NH (1963). The classification of *Puccinia graminis* var. *tritici* in relation to breeding resistant varieties. *Proceedings of the Linnean Society of New South Wales*, **88**, 235–58.

Webber JF and Gibbs JN (1984). Colonization of elm bark by *Phomopsis oblonga*. *Transactions of the British Mycological Society*, **82**, 348–52.

Webster RK and Nelson RR (1968). The genetics of *Cochliobolus spicifer*. I: Genetic inhibition of perithecium and ascus formation. *Canadian Journal of Botany*, **46**, 197–202.

Weir BS (1996a). Intraspecific differentiation. In *Molecular systematics* (ed. DM Hillis, C Moritz and Marble BK), pp. 385–405. Sinauer Associates, Sunderland, MA.

Weir BS (1996b). *Genetic data analysis II*. Sinauer Associates, Sunderland, MA.

Weir B and Cockerham CC (1984). Estimating *F*-statistics for the analysis of population structure. *Evolution*, **38**, 1358–70.

Weising K, Nybom H, Wolff K and Meyer W (1994). *DNA fingerprinting in plants and fungi*. CRC Press, Boca Raton, FL.

Wellings CR and McIntosh RA (1990). *Puccinia striiformis* f.sp.*tritici* in Australia: pathogenic changes in the first 10 years. *Plant Pathology*, **39**, 316-25.

Welz HG and Leonard KJ (1993). Phenotypic variation and parasitic fitness of races of *Cochliobolus carbonum* on corn in North Carolina. *Phytopathology*, **83**, 593–601.

Welz HG, Köhler W and Leonard KJ (1994). Isozyme variation within and among pathogenic races of *Cochliobolus carbonum* on corn in North Carolina. *Phytopathology*, **84**, 31–8.

Welz HG and Leonard KJ (1995). Gametic phase disequilibrium in populations of races 2 and 3 of *Cochliobolus carbonum*. *European Journal of Plant Pathology*, **10**, 301–10.

Whalley AJS and Edwards RL (1995). Secondary metabolites and systematic arrangement within the Xylariaceae. *Canadian Journal of Botany*, **73** (Supplement 1), S802–10.

Wheeler H (1956). Sexual versus asexual reproduction in *Glomerella*. *Mycologia*, **48**, 349–53.

Wheeler H and McGahen JW (1952). Genetics of *Glomerella*. X: Genes affecting sexual reproduction. *American Journal of Botany*, **39**, 110–19.

White TJ, Bruns T, Lee S and Taylor JW (1990). Amplification and direct sequencing of fungal ribosomal genes for phylogenetics. In *PCR Protocols: a guide to methods and applications* (ed. MA Innis, DH Gelf, JJ Sninsky and TJ White), pp. 315–22. Academic Press, San Diego.

Wicklow DT and Donahue JE (1984). Sporogenic germination of sclerotia in *Aspergillus flavus* and

*A. parasiticus. Transactions of the British Mycological Society*, **82**, 621–4.

Wicklow DT and Wilson DM (1986). Germination of *Aspergillus flavus* sclerotia in a Georgia maize field. *Transactions of the British Mycological Society*, **87**, 651–3.

Wicklow TD (1992). Interference competition. In *The fungal community: its organization and role in the ecosystem* (2nd edn) (ed. GC Carroll and TD Wicklow), pp. 265–74. Marcel Dekker, New York.

Wiebe MG, Trinci APJ, Cunliffe B, Robson GD and Oliver SG (1991). Appearance of morphological (colonial) mutants in glucose-limited continuous flow cultures of *Fusarium graminearum. Mycological Research*, **95**, 1284–8.

Wiebe MG, Robson GD, Cunliffe B, Oliver SG and Trinci APJ (1993). Periodic selection in long term continuous-flow cultures of the filamentous fungus *Fusarium graminearum. Journal of General Microbiology*, **139**, 2811–17.

Wiebe MG, Blakeborough ML, Craig SH, Robson GD and Trinci APJ (1996). How do highly branched (colonial) mutants of *Fusarium graminearum* A3/5 arise during Quorn® mycoprotein fermentation. *Microbiology*, **142**, 525–32.

Wiley EO and Mayden RL (1997). The evolutionary species concept. In *Species concepts and phylogenetic theory* (ed. QD Wheeler and R Meier). Columbia University Press, New York.

Willets HJ and Wong JA-L (1980). The biology of *Sclerotinia sclerotiorum, S. trifoliorum* and *S. minor* with emphasis on specific nomenclature. *Botanical Review*, **46**, 101–65.

Williams END, Todd NK and Rayner ADM (1984). Characterization of the spore rain of *Coriolus versicolor* and its ecological significance. *Transactions of the British Mycological Society*, **82**, 323–6.

Williams JGK, Kubelik AR, Livak KJ, Rafalski JA and Tingey SV (1990). DNA polymorphisms amplified by arbitrary primers are useful as genetic markers. *Nucleic Acids Research*, **18**, 6531–5.

Wilson CM (1952). Meiosis in *Allomyces. Bulletin of the Torrey Botanical Club*, **79**, 139–59.

Winge Ö (1935). On haplophase and diplophase in some Saccharomycetes. *Comptes Rendus des travaux du Laboratoire de Carlsberg, Série Physiologique*, **21**, 77–111.

Win-Tin and Dick MW (1975). Cytology of Oomycetes. Evidence for meiosis and multiple chromosomal associations in Saprolegniaceae and Pythiaceae, with an introduction to the cytology of *Achlya* and *Pythium. Archives of Microbiology*, **105**, 283–93.

Witthuhn RC, Wingfield BD, Wingfield BJ, Wolfaard M and Harrington TC (1998). Monophylly of the conifer species in the *Ceratocystis coerulea* complex using DNA sequence data. *Mycologia*, **90**, 96–101.

Wolfe MS and McDermott JM (1994). Population genetics of plant pathogen interactions: the example of *Erysiphe graminis–Hordeum vulgare* pathosystem. *Annual Review of Phytopathology*, **32**, 89–113.

Wolfe MS and Schwarzbach E (1975). The use of virulence analysis in cereal mildews. *Phytopathologische Zeitung*, **82**, 297–307.

Wolfe KH and Shiels DC (1997). Molecular evidence for an ancient duplication of the entire yeast genome. *Nature*, **387**, 708–13.

Wood HA and Bozarth RF (1973). Heterokaryon transfer of virus-like particles associated with cytoplasmically inherited determinant in *Ustilago maydis. Phytopathology*, **63**, 1019–21.

Worrall JJ (1994). Population structure of *Armillaria* species in several forest types. *Mycologia*, **86**, 401–7.

Worrall JJ (1997). Somatic incompatibility in Basidiomycetes. *Mycologia*, **89**, 24–36.

Wright S. (1931). Evolution in Mendelian populations. *Genetics*, **16**, 97–159.

Wright S (1938). Size of population and breeding structure in relation to evolution. *Science*, **87**, 430–1.

Wright S (1943). Isolation by distance. *Genetics*, **28**, 114–38.

Wright S (1951). The genetical structure of populations. *Annals of Eugenics*, **15**, 323–54.

Wright S (1960). The distribution of self-incompatibility alleles in populations. *Evolution*, **18**, 609–18.

Wright S (1965). The interpretation of population structure by *F*-statistics with special regard to systems of mating. *Evolution*, **19**, 395–420.

Wu J, Saupe SW and Glass NL (1998). Evidence for balancing selection operating at the *het-C* heterokaryon incompatibility locus in a group of filamentous fungi. *Proceedings of the National Academy of Sciences of the United States of America*, **95**, 12398–403.

Xu J, Kerrigan RW, Callac, P, Horgen PA and Anderson JB (1997). Genetic structure of natural populations of *Agaricus bisporus*, the commercial button mushroom. *Journal of Heredity*, 88, 482–8.

Xu J, Mitchell TG and Vigalys R (1999). PCR restriction fragments length polymorphism (RFLP) analyses reveal both extensive clonality and local genetic differences in *Candida albicans. Molecular Ecology*, **8**, 59–73.

Xu J, Vigalys R and Mitchell TG (2000). Multiple gene genealogies reveal recent disturbance and hybridization in the human pathogenic fungus *Cryptococcus neoformans. Molecular Ecology*, **9**, 1471–82.

Yeh FC, Yang R-C, Boyle TBJ and Mao JX (1997). *Shareware for population genetic analysis*. Molecular Biology and Biotechnology Centre, University of Alberta, Edmonton.

Yoon C-S, Glawe DA and Shaw PD (1991). A method for rapid small-scale preparation of fungal DNA. *Mycologia*, **82**, 835–9.

Young TW (1987). Killer yeasts. In *The yeasts*. Vol. 2: *Yeasts and the environment* (ed. AH Rose and JS Harrison), pp. 131–64. Academic Press, London.

Yu M-Q and Ko W-H (1999). The inheritance of mating type and growth rate in *Choanephora cucurbitarum*. *Canadian Journal of Botany*, **77**, 1178–84.

Zambino P, Groth JV, Lukens L, Garton JR and May G (1997). Variation at the *b* mating type locus of *Ustilago maydis*. *Phytopathology*, **87**, 1233–9.

Zeigler RS, Cuoc LX, Scott RP, *et al.* (1995). The relationship between lineage and virulence in *Pyricularia grisea* in the Philippines. *Phytopathology*, **85**, 443–51.

Zeigler RS, Scott RP, Leung H, Bordeos AA, Kumar J and Nelson RJ (1997). Evidence of parasexual exchange of DNA in the rice blast fungus challenges its exclusive clonality. *Phytopathology*, **87**, 284–94.

Zézé A, Sulistyowáti E, Ophel-Keller K, Barker S and Smith S (1997). Intrasporal genetic variation of *Gigaspora margarita*, a vesicular arbuscular mycorrhizal fungus revealed by M13 minisatellite-primed PCR. *Applied and Environmental Microbiology*, **63**, 676–8.

Zhan J, Mundt CC and McDonald BA (1999). Measuring immigration and sexual reproduction in field populations of *Mycosphaerella, graminicola*. *Phytopathology*, **88**, 1330–7.

Zhong S and Steffenson BJ (2001). Genetic and molecular characterization of mating type genes in *Cochliobolus sativus*. *Mycologia*, **93**, 852–63.

Zhou Z, Miwa M and Hogetsu T (2001). Polymorphism of simple sequence repeats reveals gene flow within and between ectomycorrhizic *Suillus grevillei* populations. *New Phytologist*, 149, 339–48.

Zhu HK, Higginbotham O and Dancik BP (1998). Intraspecific variability of isozymes in the ectomycorrhizal fungus *Suillus tomentosus*. *Canadian Journal of Botany*, **66**, 588–94.

Zietkiewicz E, Rafalski A and Labuda D (1994). Genome fingerprinting by simple sequence repeat (SSR)-anchored polymerase chain reaction amplification. *Genomics*, **20**, 176–83.

Zillig H. (1921). Über spezialisierte Formen bei Antherenbrand, *Ustilago violacea* (Pers.) Fuckel. *Zentralblatt für Bakteriologie, Parasitenkunde und Infektionskrankheiten*, **53**, 83–95.

Zimmer EA, Martin SL, Beverly SM, Kan YK and Wilson AC (1980). Rapid duplication and loss of genes coding for alpha chains of haemoglobin. *Proceedings of the National Academy of Sciences of the United States of America*, **77**, 2158–62.

Zimmer M, Lückermann G, Lang BF and Wolf K (1984). The mitochondrial genome of the fission yeast *Schizosaccharomyces pombe*. 3: Gene mapping in strain *EF1(CBS 356)* and analysis of hybrids between the strains *EF1* and *ade7–50h⁻*. *Molecular and General Genetics*, **196**, 473–81.

Zolan M (1995) Chromosome length polymorphism in fungi. *Microbial Reviews*, **59**, 686–98.

# Abbreviations

| | |
|---|---|
| AFLP | amplified fragment-length polymorphism |
| agg. | aggregate species |
| AMOVA | analysis of molecular variance |
| ANOVA | analysis of variance |
| CHEF | contour-clamped homogenous electric field |
| DAPI | 4′-6-diamidino-2-phenyindole |
| dsRNA | double-stranded RNA |
| EF-1α | elongation factor-1α |
| f.sp. | *forma specialis* |
| G | genotypic diversity |
| GD | gametic disequilibrium |
| GCPSR | genealogical concordance phylogenetic species recognition |
| HET | heterokaryon incompatibility |
| HS | hyphal swelling |
| HSI | heterokaryon self-incompatibility |
| ICG | intercompatibility group |
| IF | infection frequency |
| IGS | intergenic spacer |
| ITS | internal transcribed spacer |
| LA | lesion area |
| LTT | large tree type |
| LTT plots | lineage through time plots |
| MP | mating population |
| MRCA | most recent common ancestor |
| mtDNA | mitochondrial DNA |
| MTF | mating-type factor |
| NANOVA | nested ANOVA |
| nDNA | nuclear DNA |
| OTU | operational taxonomic unit |
| PAGE | polyacrylamide gel electrophoresis |
| PCR | polymerase chain reaction |
| PGFE | pulse-field gel electrophoresis |
| PTLPT | parsimony tree length permutation test |
| QTL | quantitative trait locus |
| RAPD | random amplified polymorphic DNA |
| rDNA | ribosomal DNA |
| RDZ | resource depletion zone |
| RFLP | restriction fragment-length polymorphism |
| s.str. | *sensu stricto* |
| SC | sporulation capacity |
| SCAR | sequence-characterized amplified regions |
| SI | somatic incompatibility |
| SSCP | single-strand conformational polymorphism |
| SSR | short-sequence repeat |
| STT | small tree type |
| TAFE | transverse alternating field electrophoresis |
| tRNA | transfer RNA |
| TSCP-PCR | taxon-specific competitive PCR primers |
| UPGMA | unweighted pair-group method with arithmetic means |
| VCG | vegetative incompatibility group |

# Mycological glossary

**Agaric** – common name for a toadstool or member of Agaricales.

**Amixis** – the apomictic condition in fungi.

**Anamorph** – the asexual stage in the life of a pleomorphic fungus (*see* holomorph, teleomorph, pleomorphic).

**Anastomosis** – a term used to describe the fusion between two hyphae, usually initiated at a short distance through the development of one or two very short lateral hyphae which fuse at their tips.

**Antheridium** – a male gametangium as in Straminipila.

**Antherozoid** – the motile male gamete produced within an antheridium.

**Apothecium** – the cup-shaped sexual reproductive structure in some Ascomycotina (*see* proto-apothecium).

**Ascocarp** – *see* Ascoma.

**Ascogenous hyphae** – dikaryotic cells arising from plasmogamy in the Ascomycotina and giving rise to the asci (*see* Plasmogamy, ascus, Figure 1.6(c)).

**Ascoma** – the reproductive body of Ascomycotina in which fertilization and meiosis occurs. Apart from the asci and ascogenous hypha the structure is derived from the 'maternal' parental mycelium.

**Ascospore** – spore derived from meiosis in an ascus, normally in some multiple of four; commonly eight per ascus (*see* ascus).

**Ascus** – an elongated cylindrical cell born penultimately on ascogenous hyphae of the Ascomycotina in which both karyogamy and meiosis occur (*see* Figure 1.6(c)).

**Azygospore** – a zygospore-like body which usually develops from the zygophore of one partner in an incomplete or non-fertile interspecific mating in the Zygomycotina (*see* zygophore, zygote).

**Barrage reaction** – an intermycelial reaction between interacting vegetatively incompatible strains (*see* vegetative incompatibility).

**Basidiocarp** – *see* Basidioma.

**Basidioma** – the reproductive structure in the Basidiomycotina in which the basidia develop (*see* Figure 1.6(d)).

**Basidiospore** – the exogenously borne spore, usually containing a single product of meiosis in the Basidiomycotina.

**Basidium** – the club-shaped or cylindrical cell in which fertilization and meiosis occurs in the Basidiomycotina (*see* Figure 1.6(d)).

**Bifactorial heterothallism** – *see* heterothallism.

**Bipolar sexuality** – *see* heterothallism, diaphoromixis.

**Bolete** – the common name for the agaric-like members of the Boletaceae where gills are replaced by pores.

**Chlamydospore** – a thick-walled perennating asexual spore developed from one or a few hyphal compartments.

**Chytrid** – a common name for a member of the Chytridiomycotina.

**Clamp connection** – a unique form of lateral anastomosis between adjacent dikaryotic cells in some Basidiomycotina (*see* Figure 1.2).

**Coenocyte** – in fungi, a mycelium lacking septa (*see* septum).

**Compartment** – a term used to describe the 'cells' of many fungi, *i.e.* those lengths of the hyphae bounded by septa.

**Conidiophore** – a structure which can assume many forms that bears conidia (*see* Figure 1.5(b)).

**Conidium** – an asexual exogenously borne, non-motile spore, characteristic of Ascomycotina and most Fungi Anamorphici.

**Diaphoromixis (diaphoromictic)** – a fungus in which many allelic MTFs occur at either one or two loci (*see* heterothallism).

**Dikaryon** – a mycelium in which two sexually compatible nuclei, which divide synchronously, occur regularly in each compartment.

**Dimixis (dimictic)** – a fungus having only two kinds of sexually compatible mating types (*see* heterothallism).

**Dolipore septum** – complex septum found in some Basidiomycetes in which the lip of the central pore is flanged and a septal pore cap is also present (*see* Figure 1.2(c)).

**Ectomycorrhiza** – a mycorrhiza which develops an enveloping sheath of hyphae over the root tips of its host plant (*see* mycorrhiza).

**Endomycorrhiza** – a mycorrhiza which infects a root and penetrates the root cells (*see* mycorrhiza).

**Endophyte** – a fungus which lives within the cellular spaces within its host plant usually without penetrating its cells.

**Flagellum** – a hair-like structure which propels a motile cell (*see* Figures 1.5(d) and 1.5(e)).

*Forma specialis* – a pathogen specialized to a specific host plant, often differing very slightly from other *formae speciales* which occur on similar plants.

**Germ tube** – one or more short hyphae which emerge when a spore germinates and develops into the mycelium.

**Gill** – the radiating lamella on the underside of the pileus on which the basidia develop in an Agaric.

**Heterogenic incompatibility** – *see* somatic incompatibility.

**Heterokaryosis (heterokaryotic)** – a mycelium which includes genetically different nuclei (*see* Figure 8.1).

**Heterothallism** – sexual differentiation which is not expressed morphologically. In the simplest case two kinds of individual occur, each carrying a different idiomorphic MTF (dimixis). In others there are one, two or even three mating-type loci, each of which can be occupied by a different MTF. They can be referred to generally as diaphoromictic or more specifically as uni-factorial (bipolar), bifactorial (tetrapolar) or tri-factorial forms. In all cases the MTFs must be *different* in the two partners for a compatible mating.

**Holomorph** – a fungus possessing alternating anamorphic and teleomorphic phases (*see* anamorph, teleomorph).

**Homothallism** – the condition of self-fertility in fungi.

**Hymenium** – a parallel layer of basidia or asci in a reproductive body.

**Hypha** – the fine ($10\,\mu\text{m}+$) cylindrical cells which grow at their apex and are the basic structural element of the fungus.

**Idiomorph** – a term used to describe the MTFs which behave as if allelomorphic although of substantially different nucleotide sequences.

**Imperfect fungi** – those lacking a sexual phase completely.

**Karyogamy** – nuclear fusion at fertilization.

**Lichen** – an apparent single and distinct organism involving the close symbiotic relationship of a fungus and either a green alga or a cyanobacterium.

**Mating-type factors (MTFs)** – the genetic sequences that determine the mating behaviour of the fungi that carry them.

**Meiospore** – a spore produced after meiosis, usually carrying a single product of meiosis.

**Microconidium (microspore)** – a small uninucleate asexually produced spore or conidium which usually can only act as a fertilizing agent.

**Mitospore** – an asexual spore or conidium produced as a result of mitotic division.

**Monokaryon** – a mycelium which carries a single, usually genetically identical, nucleus in each compartment; rarely, the apical cell may carry more than one nucleus.

**Mycorrhiza** – a symbiotic structure adopting various morphological forms which develops between the absorbent organs, usually roots, of a green plant and a fungus.

**Oogamous** – sexual reproduction, as in Straminipila, involving an oogonium.

**Oogonium** – the female gametangium containing one or more non-motile eggs.

**Oospore** – the fertilized product of an oosphere.

**Parasexuality** – the process of intra-hyphal nuclear fusion followed by mitotic crossing over and de-diploidization of the nucleus so formed.

**Parthenosome** – the septal pore cap of endoplasmic reticulum on a complex septal pore (*see* Figure 1.2(b)).

**Perfect fungi** – fungi possessing a sexual stage.

**Perithecium** – a closed ascoma, commonly flask-shaped, which opens by an apical pore (*see* ostiole).

**Pheromone** – a hormone which promotes the growth of sexually compatible cells or hyphae towards each other.

**Pileus** – the cap-shaped part of the basidioma of Agarics and Boletes.

**Plasmogamy** – the process(es) whereby compatible cells become attached.

**Pleomorphism** – the condition where a fungus exists in more than one state, e.g. the anamorphic and teleomorphic, or produces more than one kind of asexual spore or conidium.

**Polypore** – common name for bracket fungi or 'conks' with pores rather than gills.

**Protoapothecium** or **protoperithecium** – the small structures that develop on a mycelium and determine the site where an apothecium or perithecium will develop, if fertilized.

**Pycnidiospore** – conidia derived from a pycnidium.

**Pycnidium** – a hollow structure whose internal walls are lined with conidia.

**Rhizomorph** – a differentiate strand of more or less parallel hyphae forming a structure which usually grows more rapidly than a single hypha and is often capable of penetrating a host or other organic material.

**Rust fungi** – common name for the Urediniomycetes because of the colour of their uredospores.

**Saprophobe** – a fungus capable of exploiting dead organic material.

**Sclerotium** – a perennating structure derived through repeated hyphal divisions and compaction, usually with a hard exterior and the interior cells filled with reserve carbohydrates and oil globules (*see* Figure 1.4(b)).

**Secondary homothallism** – a fungus derived from a heterothallic form but carrying compatible MTFs so that it is capable of self-fertilization.

**Septum** – the, usually incomplete, septum which grows inwards from the cylindrical walls of a hypha; intact septa often delimit sporangiophores or conidiophores and sexual structures.

**Smut fungi** – common name for the Ustilaginomycetes because of the dark sooty colour of their spores.

**Somatic incompatibility** – a genetic system within a species based on compatible partners carrying identical vegetative compatibility factors which enables them to form anastomoses. In Ascomycotina it operates between the monokaryotic hyphae, but in Basidiomycotina it operates most commonly between dikaryotic mycelia. (*See* vegetative compatibility, heterogenic incompatibility.)

**Sorus** – a mass of spores or conidia formed in an open structure, especially in the Urediniomycetes.

**Sporangiospore** – a non-motile spore that develops in a sporangium.

**Sporangium** – an aerial asexual reproductive structure within which spores are delimited by cleavage. The spore may be non-motile as in Zygomycotina or motile as in Chytridiomycotina.

**Sporophore** – a structure that bears sporangia.

**Sterigma** – the short extension from the basidium or basidial cells which bears the basidiospores exogenously.

**Stipe** – the stalk of a basidioma.

**Stroma** – a compact mass of hyphae on or in which reproductive structures often develop.

**Teleomorph** – the sexual phase of a fungus.

**Teleutospore** – the thick-walled resting spore of rust and smut fungi which eventually give rise to basidia and basidiospores.

**Tetrapolar sexuality** – *see* heterothallism, diaphoromixis.

**Thallus** – a simple structure devoid of stems, roots or leaves—the mycelium in a fungus.

**Thigmotropism** – touch sensitivity which promotes cell fusion.

**Tinsel flagellum** – a flagellum bearing numerous very fine hairs along its length (*see* Figure 1(e)).

**Trichogyne** – the highly receptive hypha forming part of a protoapothecium or protoperithecium which either attracts the growth of fertilizing microconidia to it or is attracted to such cells.

**Unifactorial heterothallism** – *see* heterothallism, diaphoromixis.

**Uredospore** – the dikaryotic spore of a rust fungus.

**Vegetative incompatibility** – *see* somatic incompatibility.

**Whiplash flagellum** – a flagellum which beats like a whiplash.

**Zoosporangium** – a sporangium which gives rise to motile spores.

**Zoospore** – a motile spore found only in the Chytridiomycotina or Straminipila amongst the Fungi.

**Zygophore** – special hyphae that are mutually attractive between sexually compatible Zygomycotina.

**Zygospore** – the name given to the zygote of the Zygomycotina.

**Zygotropism** – the process of attraction between mutually compatible fungal sexual partners, usually mediated by one or more pheromones.

# Classification of fungi in the text

The species are arranged in accordance with the classification adopted in *Ainsworth and Bisby's Dictionary of the Fungi* (Kirk *et al.* 2001). Abbreviations used: FA – Fungi Anamorphici; T – Teleomorph

## EUMYCOTA

## Chytridiomycotina

5 orders/32 families including
Blastocladiales   5 families including
Blastocladiaceae   *Allomyces*—*A. anomala,*
*A. arbuscula, A. macrogyna, A.* × *javanicus*
Coelomomycetaceae   *Coelomomyces*—*C. dodgei,*
*C. punctatus*
Chytridiales   4 families
Monoblepharidales   4 families
Neocallimastigales   1 family
Spizellomycetes   3 families

## Zygomycotina

Zygomycetes   10 orders/32 families including
Mucorales   12 families including
Choanephoraceae   *Choamephora cucurbitarum*
Mucoraceae   *Absidia*—*Absidia glauca*
*Mucor*—*M. mucedo, M. hiemalis, M. genevensis,*
*M. piriformis, M. rammanianus*
*Rhizopus*—*R. microsporus, R. rhizopodiformis,*
*R. sexualis, R. stolonifer (R. microsporus* ×
*R. rhizopodiformis)*
*Sporodina ( = Szygites)*—*S. grandis (Szygites*
*megalocarpus)*
*Zygorrhynchus*
Phycomycetaceae   *Phycomyces*—*P. nitens*
*Glomales   3 families
Acaulosporaceae   *Acaulospora*—*A. colossica*
Gigasporaceae   *Gigaspora*—*G. decipiens, G. margarita*
*Scutellospora*—*S. castanea*
Glomaceae   *Glomus*—*G. mosseae*
(* Recently the Glomales have been recognized as a
distinct phylum Glomamycotina)
Trichomycetes   3 orders/6 families

## Ascomycotina

Ascomycetes   50 orders/275 families including
Mycosphaerellales   1 family

Mycosphaerellacae   *Mycosphaerella*
*(Micronectriella)*—*M. graminicola*
(see FA—*Septoria*), *M. fijiensis*
Pleosporales   19 families including
Leptosphaeriaceae   *Leptosphaeria*—*L. maculans,*
*L. nodorum*
Pleosporaceae   *Cochliobolus*—*C. carbonum,*
*C. heterostrophus, C. homomorphus, C. luttrelli,*
*C. sativus, C. spicifer*
*Pyrenophora*—*P. teres*
*Setosphaeria*—*S. turcica*
Venturiaceae   *Venturia*—*V. inaequalis*
Erysiphales   1 family
Erysiphaceae   *Blumeria* (see FA = *Erysiphe*)—
*B. graminis (E. graminis)* & f.sp. *avenae, hordei,*
*tritici*
*Erysiphe*—*E. fischeri*
Elaphomycetales
Elaphomycetaceae   *Elaphomyces*
Eurotiales   1 family
Trichomacaceae   *Emericella* (see FA *Aspergillus*)
*E. nidulans*
*Talaromyces*—*T. wortmannii*
Onygenales   3 families including
Onygenaceae   *Ajellomyces*—*A. capsulatum* & vars.
*capsulatum, farciminosum, A. duboisii*
Lecanorales   42 families (lichens) including
Parmeliaceae—*Parmelia* — *P. plitii*
Helotiales   15 families including
Dermatateaceae   *Pyrenopeziza*—*P. brassicae*
Helotiaceae   *Crumenulopsis*—*C. soraria*
*Gremmeniella*—*G. abietina* &
var. *cembrae*
Sclerotiniaceae   *Botryotinia* (see FA *Botrytis*)   *B. allii,*
*B. byssoidea*
*Sclerotinia*—*S. minor, S. sclerotiorum, S. trifoliorum,*
*S. sp.'I'*
Rhytismatales   2 families including
Rhytismataceae   *Hypoderma*—*H. mutata, H. populnea*
*Lophodermium*—*L. conigenum, L. pinastri,*
*L. seditiosum*
Pezizales   15 families including

Morchellaceae  *Morchella*
Pyronemataceae  *Humaria*—*H. granulata*
Tuberaceae  *Tuber*
Diaporthales  2 families including
Valsaceae  *Cryphonectria*—*C. parasitica*
*Endothia*—*E. gyrosa*
Hypocreales  6 families including
Clavicipitaceae  *Atkinsonella*—*A. hypoxylon*,
*A. texensis*
*Cordyceps*
*Epichloë* (see FA *Neotyphodium*)—*E. amrillans*,
*E. baconii*, *E. bromicola*, *E. clarkii*, *E. festucae*,
*E. typhina*, *E. sylvatica*
*Torrubiella*
Hypocreaceae  *Hypocrea* ( = *Chromocrea*)
*C. spinulosa*
Nectriaceae  *Calonectria*
*Gibberella*—*G. cyanogenea*, *G. fujikuroi*
*Nectria*—*N. fuckeliana*, *N. haematococca*
Microascales  3 families including
Ceratocystidaceae  *Ceratocystis*—*C. coerulescens*,
*C. douglassii*, *C. laricicola*, *C. pinastri*, *C. polonica*,
*C. resinifera*, *C. virescens*
(*C. coerulescens* × *C. resinifera*)
Ophiostomatales  2 families including
Ophiostomaceae  *Ophiostoma*—*O. himal-ulmi*,
*O. multiannulatum*, *O. novo-ulmi* & sspp.
*americana*, *novo-ulmi*, *O. ulmi*, (*O. novo-ulmi*
*americana* × *O. ulmi*)
Sordariales  7 families including
Coniochaetaceae  *Ephemeroascus*
Lasiosphaeraceae  *Podospora*—*P. anserina*,
*P. arizoensis*
Sordariaceae  *Gelasinospora* ( = *Anixiella*)
*Neurospora*—*N. africana*, *N. crassa*, *N. intermedia*,
*N. sitophila*, *N. terricola*, *N. tetrasperma*,
(*N. crassa* × *N. tetrasperma*, *N. intermedia* ×
*N. intermedia*)
*Sordaria*—*S. brevicollis*, *S. fimicola*
Xylariales  8 families including
Xylariaceae  *Daldinia*—*D. loculata*
*Xylaria*—*X. curta*
Families of uncertain affinities include
Amorphothecaceae  *Amorphotheca* (see FA
*Hormoconis*)—*A. resinae*,
Glomerellaceae  *Glomerella* (See FA
*Colletotrichum*)—*G. cingulata*, *G. graminicola*
Magnaporthaceae  *Magnaporthe* (see FA
*Pyricularia*)—*M. oryzae* Couch ( = *M. grisea* auct)
Neolectomycetes  1 order/1 family
Pneumocystidiomycetes  1 order/1 family including
Pneumocystidiales
Pneumocystidaceae  *Pneumocystis*—*P. carinii*
Saccharomycetes  1 order/11 families including
Saccharomycetales
Saccharomycetaceae  *Candida*—*C. albicans*,
*C. stellatoides*, *Kluveromyces*, *Pichia*

*Saccharomyces*—*S. cereviseae*
Schizosaccharomycetes  1 order/1 family including
Schizosaccharomycetales
Schizosaccharomycetaceae  *Schizosaccharomyces*—
*S. pombe*
Taphrinomycetes  1 order/2 families

## Basidiomycotina

Ustilaginomycetes  10 orders/31 families including
Tilletiales  1 family
Tilletiaceae  *Tilletia*—*T. bromi*, *T. caries*, *T. controversa*,
*T. foetida*, *T. laevis*, *T. tritici*, (*T. controversa* ×
*T. bromi*)
Urocystales  2 families including
Urocystaceae  *Urocystis*—*U. appendiculatus*
Ustilaginales  2 families including
Ustilaginaceae  *Ustilago*—*U. avenae*, *U. hordei*,
*U. kolleri*, *U. maydis*, *U. segetum* & vars. *avenae*,
*segetum*
*Microbotryum*—*M. violaceum* ( = *U. violacea*)
*Uromyces*—*U. appendiculatus*
Urediniomycetes  5 orders/25 families including
Uredinales  14 families including
Cronartiaceae *Cronartium*
Melampsoraceae  *Melampsora*—*M. occidentalis*,
*M. larici-populnea*, *M. lini*, *M. medusae*,
*M.* × *columbiana* ( = *M. medusae* × *M. occidentalis*),
*M.* × *medusae-populina* ( = *M. meduse* × *M. larici-
populina*)
Pucciniaceae  *Puccinia*—*P. coronata* f.sp. *avenae*,
*P. graminis* & f.sp. *avenae, secalis, tritici, P. hordei,*
*P. kraussiana*, *P. polysora*, *P. recondita*, *P. striiformis*
& f.sp.*tritici*
Puccinistraceae  *Melampsoridium*
Sphaerophragmiaceae *Triphragmium*—*T. ulmariae*
Basidiomycetes
Agaricomycetideae  8 orders/84 families including
Agaricales  26 families including
Agaricaceae  *Agaricus*—*A. bisporus*
Bolbitaceae  *Conocybe*—*C. farinacea*
*Hebeloma*—*H. crustuliniforme*, *H. cylindrosporum*,
*H. incarnulatum*, *H. velutipes*
Coprinaceae  *Coprinus*—*C. bisporus*, *C. cinereus*,
*C. micaceus*
*Psathyrella*—*P. candolleana*, *P. coprobia*
Cortinariaceae  *Galerina*—*G. mutabilis*
Hydnangiaceae  *Laccaria*—*L. bicolor*, *L. laccata*
Lycoperdaceae *Lycoperdon*—*L. pyriforme*
Marasmiaceae  *Armillaria*—*A. bulbosa*,
*A. calvescens*, *A. cepistipes*, *A. gallica*,
*A. luteobulbulina*, *A. mellea*, *A. ostoyae*, *A. sinapina*
*Flammulina*—*F. velutipes*
*Lentinula*—*L. edodes*
*Marasmius*—*M. androsaceus*, *M. rotula*

Nidulariaceae  *Cyathus—C. stercoreus*
  *Mycocalia—N. denudata*
  *Nidularia—N. denudata*
Pleurotaceae  *Pleurotus—P. ostreatus, P. sapidus*
Pluteaceae  *Amanita—A. muscaria, A. franchetii*
  *Volvariella—V. volvacea*
Strophariaceae  *Psilocybe (Hypholoma),*
  *H. fasciculare*
Schizophyllaceae  *Schizophyllum—S. commune*
Tricholomataceae  *Clitocybe—C. aurantiaca*
  *Collybia—C. dryophila, C. fusipes, C. velutipes*
  *Mycena—M. epipterygia, M. galopus, M. rosea*
  *Panellus—P. stypticus*
Typhulaceae  *Typhula—T. idahoensis,*
  *T. ishikarensis*
Boletales   18 families including
  Coniophoraceae  *Serpula—S. lacrymans*
  Suillaceae  *Suillus—S. bovinus, S. grandis,*
  *S. granulatus, S. grevillei, S. pungens*
Hymenochaetales
  Hymenochaetaceae  *Phellinus—P. gilvus,*
  *P. weirrii*
Phallales   5 families including
  Geastraceae  *Sphaerobolus*
  Phallaceae  *Phallus—P. impudicus*
Polyporales   23 families including
  Fomitopsidaceae  *Fomitopsis—F. rosea*
  *Piptoporus—P. betulinus*
  Ganodermataceae  *Ganoderma*
  Polyporaceae  *Fomes—F. cajanderi, F. pinicola*
  *Lentinus—L. tigrinus*
  *Phaeolus—P. schweinitzii*
  *Polyporus—P. arcularis, P. brumalis, P. ciliatus,*
  *P. abietinus* ( = *Trichaptum*)
  *Trametes (Coriolus), C. versicolor*
  *Trichaptum* ( = *Hirschioporus*)—*H. abietinus,*
  *H. fusco-violaceus, H. laricinus*
  Sistotremataceae  *Sistotrema—S. brinkmannii*
Russulales   11 families including
  Auriscalpiaceae  *Lentinellus—L. edodes, L. laterita,*
  *L. novaezelandiae*
  Bondarzewiaceae  *Heterobasidion—H. annosum*
  Lachnocladiaceae  *Vararia—V. granulata*
  Peniophoraceae  *Peniophora—P. cinerea. P. gigantea,*
  *P. heterocystidea, P. ludoviciana, P. mutata,*
  *P. populnea*
  Russulaceae  *Russula—R. amoenolens*
  Stereaceae  *Stereum—S. gausapatum,*
  *S. hirsutum*
Tremellomycetideae   8 orders/19 families including
  Auriculariales   1 family
  Auriculariaceae  *Auricularia—A. auricula*
Filobasidiales
  Filobasidiaceae  *Filobasidiella* (see FA
  *Cryptococcus*)—*F. neoformans* & vars. *gattii,*
  *grubbii, neoformans*

# Fungi Anamorphici

In alphabetical order
  *Acremonium* ( = *Neotyphodium*)
  *Alternaria—A. alternata*
  *Aspergillus—A. amstelodami, A. flavus, A. glaucus,*
  *A.(Emericella)nidulans*
  *Bipolaris—B. maydis, B. sacchari, B. zeicola*
  *Botrytis—B. aclada* AI & AII, *B. allii, B. byssoidea,*
  *B. squamosa, B. tulipae,* (*Botrytis aclada*AI ×
  *Botrytinia byssoidea*)
  *Cenococcum—C. geophilum*
  *Cladosporium—C. fulvum*
  *Coccidioides—C. immitis*
  *Colletotrichum—C. gloeosporides* (T = *Glomerella*),
  *C. lindemuthianum*
  *Cryptococcus* (T = *Filobasidiella*)—*C. neoformans*
  *Curvularia*
  *Cylindrocarpon*
  *Fusarium—F. graminearum, F. lactis, F. moniliforme,*
  *F. oxysporum* & f.spp. *albedinis, apii, conglutinans,*
  *ciceris, conglutinans, lycopersici, vasinfectum,*
  *F. ramigenum, F. solani* & f.sp. *cucurbitae,*
  *F. subglutinans, F. sulphureum*
  *Helminthosporium—H. maydis, H. sativum*
  *Histoplasma* (T = *Ajellomyces*)—*H. capsulatum*
  *Hormoconis* (T = *Amorphotheca*)—*H. resinae*
  *Macrophomina—M. phaseolina (Rhizoctonia*
  *baticola*)
  *Neotyphodium—N. aotearoae, N. australiense,*
  *N. coenophialum, N. loliy, N. melicicola, N.occul-*
  *tans, N. tembladeros, N. uncinatum, N.* LPTG-2
  *Penicillium*  *P. chrysogenum (P. notatum),*
  *P. cyaneo-fulvum, P. cyclopium, P. islandicum,*
  *P. variable*
  *Phoma*  *P. ligam*
  *Phomopsis*
  *Pseudocercosporella—P. herpotrichoides*
  (T = *Mycosphaerella*)
  *Rhizoctonia—R. solani, R. batticola*
  (see *Macrophomina*)
  *Rhodotorula*
  *Sclerotium—S. cepivorum*
  *Septoria—S. nodosum, S. tritici*
  *Stagnospora—S. nodorum*
  *Trichophyton—T. rubrum* (T = *Arthroderma*)
  *Verticillium*  *V. albo-atrum, V. dahliae, V. lecanii,*
  *V. longisporum (V. dahliae* var. *longisporum),*
  *V. nigrescens, V. tricorpus*

# STRAMINIPILA

Hypochytriomycota   1 order/2 families
Labyrinthulomycota   3 orders/2 families
Oomycota   8 orders/24 families including

Perenosporales
    Perenosporaceae, *Albugo*
    *Bremia*—*B. lactuca*
Pythiales
    Pythiaceae    *Phyophthora*—*P. cactorum, P. cambivora,*
    *P. capsici, P. citricola, P. citrophora, P. cryptogea,*
    *P. dreschleri, P. fragariae* var. *rubi, P. gonapodydes,*

*P. infestans, P. meadii, P. megasperma* & vars
*megasperma, sojae, P. nicotianae, P.* 'Alder'sp.,
*P.* sp.'*h*' (*P. cambivora* × *P. fragariae* var. *rubi*),
(*P. nicotianae* × *P. cactorum*)
*Pythium    P. sylvaticum, P. ultimum*
Saprolegniales
    Saprolegniaceae    *Achlya*—*A. heterosexualis*

# Species index

# General index